PARTICLE ADHESION: APPLICATIONS AND ADVANCES

PARTICLE ADHESION: APPLICATIONS AND ADVANCES

Edited by
David J. Quesnel
Donald S. Rimai
and
Louis H. Sharpe

TAYLOR & FRANCIS
Founded 1798

USA	Publishing Office:	TAYLOR & FRANCIS 29 West 35th Street New York, NY 10001 Tel: (212) 216-7800 Fax: (212) 564-7854
	Distribution Center:	TAYLOR & FRANCIS 7625 Empire Drive Florence, KY 41042 Tel: 1-800-634-7064 Fax: 1-800-248-4724
UK		TAYLOR & FRANCIS 27 Church Road Hove E. Sussex, BN3 2FA Tel.: +44 (0) 1273 207411 Fax: +44 (0) 1273 205612

PARTICLE ADHESION: APPLICATIONS AND ADVANCES

1 2 3 4 5 6 7 8 9 0

Printed by Edwards Brothers, Lillington, NC, 2001.
Cover photo: Figure 6B, page 330 of the article "Interactions Between Micron-sized Glass Particles and poly(dimethyl siloxane) in the Absence and Presence of Applied Load", by Toikka et al.
The articles appearing in this book were originally published in Volume 74, Numbers 1-4 of *The Journal of Adhesion.*

A CIP catalog record for this book is available from the British Library.
The paper in this publication meets the requirements of the ANSI Standard Z39, 48-1984 (Permanence of Paper).

Library of Congress Cataloging-in-Publication Data

Available from publisher

ISBN 90-5699-725-4 (case)

Dedication

With each new generation, science continues to provide a stepping stone to reach new heights. This book is dedicated to Alicia Quesnel (medicine), Benjamin Rimai (mechanical engineering), and the three Sharpe scientist-children: Louis (physics), Nancy (chemistry), and Amy (toxicology).

CONTENTS

Advances in Controlling the Attachment and Removal of Groups of Particles

PREFACE

The consequences of particle adhesion are far reaching. The adhesion of particles can destroy, as when dirt particles occlude the pattern during semiconductor fabrication or when cancer cells adhere strongly to new locations during metastasis. The adhesion of particles can produce wonderful images, as when electrically charged toner particles migrate in response to the electric fields within a toner-based color print engine. The adhesion of particles can even promote life when the fertilized egg adheres to the womb or the body responds to an injury.

With cleanliness next to Godliness, where is the need to understand what causes particles to adhere and release from surfaces? Virtually all fields of scientific endeavor can benefit from understanding the basic interactions between the small groups of atoms that we call particles and how these particles interact with macroscopic objects.

This book on applications and advances in particle adhesion has a very specific mission: to share the knowledge of particle adhesion and promote its use over a wide range of disciplines. One compelling example is the use of particle adhesion concepts in biological applications. The papers presented in this book reflect the desire of the editors to provide a thorough background in particle adhesion concepts spanning a very broad range of topics. With its wealth of references to fundamental concepts and its discussion of recently achieved results, this text can empower both the experienced practitioner and the novice alike to make innovative contributions to the infrastructure, bringing fresh ideas, and developing new ways of improving the quality of life.

This book brings together a selection of papers, most of which were presented at the Particle Division symposia of the 22nd and 23rd annual meetings of the Adhesion Society. The papers were chosen to reflect advances in fundamental understanding of particle adhesion and to relate particle adhesion to other disciplines, most notably what the editors refer to as the emerging field of bioparticle adhesion—that is, the use of particle adhesion concepts to address biological issues involving particulates. These papers were all reviewed and first published in *The Journal of Adhesion*, edited by Dr. Louis H. Sharpe. However, rather than strictly following the form of typical research papers, at the request of the editors, the authors of each paper included an extensive introduction and sufficient references so as to make the papers accessible to novices in the field. As such, it is the hope of the editors that this volume serves to establish and promote connections between the field of particle adhesion and a variety of other scientific disciplines where the interactions of small portions of matter do indeed matter.

The papers solicited for this book represent the wide range of activities in particle adhesion that were in progress at the turn of the century and should serve as a rich means of idea generation for you, the reader. You are encouraged to read through not only those papers specifically related to the topics with which you may already be familiar, but to explore those from other areas, so that the cross-pollination of idea generation can work its effect. To give you a sense of the contents, let us take a moment to consider some of the concepts laid out in this book.

Biological Applications of Particle Adhesion Concepts

First is the notion that plants, animals, and the majority of living things can be conceived of as collections of particles held together by adhesive forces. When you cut yourself, chemical messengers are released that suppress adhesion between undamaged cells that make up several of the surface layers adjacent to either side of the cut. These non-adhering cells float around in the wound fluids and eventually adhesion begins again as the free floating cells begin to grow, leading to the formation of living material across the gap between the two sides of the cut. This is your body using adhesion to heal itself. The details of the behavior control the degree of scarring that develops. Similar things happen during the development of cancers. Two papers in this text address the issues associated with adhesion in the metastasis, or spread, of cancers. The first, by Love and Forsten, focuses on the cellular detachment process that allows cancer cells to float free from a main tumor. The second, by Moss and Anderson, focuses on the issues of cellular adhesion that produce initial adhesion and eventually firm attachment of cancerous cells, indeed particles in their own right, as they float through the body. Understanding of these phenomena will make substantial inroads to blocking the spread of cancers.

Adhesion also occurs between blood cells and, to different degrees, in different species. Changes in this adhesion can be both a symptom and a problem in its own right that can have significant health implications. Attenborough and Kendall present the results of their experimental work on cell-cell adhesion based on measurement of the statistically stable distribution of clusters of cells that result when cells collide in a fluid. They review techniques used to assay adhesion between cells, many of which can be applied to non-biological applications as well.

When man-made materials are used in the body, a variety of particles are introduced to the living system. Carter *et al.*, discuss the biological response to wear particles from implants and implant adhesives. Many of these particles elicit unexpected responses including the mobilization of cells in the body that can digest bone. Dust particles from talc-coated surgical gloves have even been shown to pose problems when they contaminate the tissues. In the next two papers that complete the set of biologically oriented presentations, Baier *et al.*, discuss the implication of particulates associated with intentional implants and the consequences of inadvertent implants that occur when respirable particles become embedded in lung tissue. These papers make it clear that the utmost precautions should be taken in controlling the degree of particulate exposure and residual particulates in the manufacture of implants and highlight the range of biological responses that the exposure to particulates can elicit.

Elastic and Viscoelastic Contributions to Understanding Particle Adhesion

Experimental and theoretical contributions continue to enrich our understanding of the fundamental physics controlling particle adhesion well beyond the pioneering work of the early contact mechanicians. The behavior of polymeric particles is of particular technical importance so that the role of viscoelasticity in understanding particle adhesion has moved to center stage. Reitsma, Craig, and Simon report clever experiments where a polystyrene particle, covered with smaller particles, is shown never actually to contact the planar substrate to which it adheres. Plastic deformation of nanometer-scale contacts is observed on the smaller beads in proportion to the loads they experience, loads that occur over a protracted period of time requiring the concept of viscous flow. Unertl presents a analysis of the nanometer-scale contacts, showing that there is a clear delay time between the time of maximum load application and the time of maximum displacement, suggesting that internal friction and damping can occur as a result of particulate deformations. Fundamental work on the axisymmetric elastic contact problem in the presence of adhesive interactions is given in the other two papers of this section. Huguet and Barthel present a synthetic approach to the problem of adhesive contact that directly yields most of the useful adhesion models. The approach even allows for situations where the bodies are not in actual contact but where there is still interaction. Quesnel and Rimai show that finite elements can be used to solve contact mechanics problems in a way that produces results effectively identical to the equations of the JKR solution. This opens the door for finite elements to be used for all types of problems of irregular geometries and spatial gradients in the properties of the system.

Particle Surface Interactions That Influence Adhesion

The adhesion of particles is influenced by many factors, including how the particles are placed on the surface. Brach, Dunn, and Li give an extensive presentation of experiments and engineering models that detail the influence of particle impact on the subsequent adhesion. They provide clear answers to questions of when particles bounce and when particles are captured. They detail issues involving rotation of particles, the role of friction, and the behavior of particles impinging on rough and smooth surfaces. The relative contributions of electrostatic and van der Waals forces in the adhesion of irregular particles is given by Rimai, Quesnel, and Reifenberger in an experimental paper that uses an ultracentrifuge to measure the force of removal. Attention to the role of asperities is given in explaining the results where silica particles acting as spacers reduced the effective contact of the primary particles.

Electrical Conductivity Through Particles

Electrical and thermal conductivity of polymers often rely on the presence of particles of metal to provide a high conductivity pathway.

David Marshall's paper on copper-based conductive polymers outlines the issues that are important in providing such conductivity and reports results achieved in epoxy-based resins and certain thermoplastic resins.

Exploring Particle Adhesion with Single Particle Experiments

Toikka, Spinks and Brown report a technique to use scanning electron microscopy to examine the contact mechanics of particulate surface interactions where the loads can be determined directly from the SEM images of the test system. The role of shear in dislodging particles from the surface was explored by using AFM-like cantilevers of a variety of stiffnesses. Our cover photo is from the work of this group. Schaefer and Gomez present a review of the use of AFM to measure adhesion and introduce a new "Jump Mode" technique to obtain adhesion maps of surfaces. Studies of the contact between polyethylene particles and mineral surfaces such as quartz are reported by Drelich *et al.*, to gain further insight into the interaction of oil droplets and soil. They use the polyethylene as a model system to represent oil in a surfactant-rich environment, finding limitations to traditional thinking based on the Young-Dupré equation when changes in surfactant concentrations are used to change the surface energies. For more rigid particles, Dickinson, Hariadi, and Langford report the detachment of NaCl particles from glass surfaces over a range of humidities. They suggest that higher humidity provides a chemically assisted mechanism for the crack growth that leads to detachment of the particle.

Advances in Controlling the Attachment and Removal of Groups of Particles

Quite the opposite effect of humidity on the ease of particle removal is reported by Tang and Busnaina. Here the high moisture produced capillary forces on the 22 micron polystyrene particle that were sufficient to produce creep deformation, increasing the effect of the van der Waals interaction cited to be responsible for the adhesion when water was not present. In a related work, Busnaina and Elsawy show that adhesion can be high at both high and low humidity, dominated by electrostatics at low humidity (below 45%) and by capillary forces at high humidity (above 70%). Privman reviews the deposition process of particles on surfaces, discussing issues ranging from dilute systems where diffusion dominates to crowded systems where interparticle collisions are of significant interest. Some attention is given to microstructures that can form in two-dimensional particle systems adhering to substrates. Fluid forces on particles are discussed with attention to lifts, drags, and torques on particles induced by fluid motions in the final paper by Zhang and Amadi. Numerical simulations are used to evaluate the various mechanisms that control particle removal and re-entrainment.

Closure

The papers that make up this book provide a rich source of research ideas and a variety of approaches to widely different problems, all of which share a central theme of particle adhesion. The authors have made an effort to provide depth to the references that will make further exploration of topics an easy task. We look at this compendium on the many applications and advances in particle adhesion as both a milestone and an invitation for further research. We hope you share our vision.

David J. Quesnel
Donald S. Rimai
Louis H. Sharpe

Biological Applications of Particle Adhesion Concepts

A Particle Adhesion Perspective on Metastasis

B. J. LOVE

Department of Materials Science and Engineering, Virginia Polytechnic Institute and State University, Blacksburg, Virginia 24061-0237, USA

K. E. FORSTEN

Department of Chemical Engineering, Virginia Polytechnic Institute and State University, Blacksburg, Virginia 24061-0211, USA

One area of interest in bioadhesion that has not been emphasized within the adhesion science community relates to the disaggregation of cells which occurs when cancer metastases arise. Metastasis involves the distribution of cancerous tumor cells from a large localized tumor. The resulting spatial separation of the cancerous masses makes treatment more difficult. Making use of biochemical and cellular assays, detailed mechanisms for the cell detachment processes involving cadherin cell adhesion molecules have been proposed. This paper reviews proposed mechanisms for metastasis from a cellular adhesion perspective and the testing methodologies that have been utilized. Additional understanding might be gleaned from considering the loss of cell adhesion molecules and the overall disaggregation process from a particle adhesion perspective. Several pertinent theories are presented as well as a brief discussion of areas for future effort.

INTRODUCTION

Organized agglomerations of cells form the basis for mammalian tissue and organ architecture. Cellular regulatory mechanisms normally maintain a delicate balance of cell proliferation, quiescence, and death. This balance is disturbed in tumorigenesis. Tumor cells characteristically have exhibited increased growth and inhibited cell death, ultimately leading to an overall accumulation of tumor mass [1]. Malignant cells can detach from the growing mass and travel, *via* the bloodstream or lymphatic system, to distant sites where secondary tumors or metastatic lesions can be initiated (Fig. 1). Metastasis is a complex series of events involving a number of important factors; however, we will concentrate in this review on E-cadherin, one cell adhesion molecule thought to have a role in detachment from the growing tumor.

Detachment, whether active or simply the lack of cell adhesion, is a critical step in cancer progression [3]. Mammalian cells express (have present) a number of different adhesion molecules on their cell surface and the type and level are tightly regulated. The surface adherent molecules allow cells to bind specifically to particular cells and matrix components within the tissue, thereby promoting cell organization [4, 5] (see Fig. 2). Most tumors are derived from a specific type of cell known as an epithelial cell and the cadherin superfamily of adhesion molecules normally plays a role in epithelial cell connections [6]. Cadherin molecules span the cell membrane and their large extracellular domain can specifically interact with similar molecules on adjacent cells (Fig. 3). This is known as homotypic binding. E-cadherin, a member of the cadherin family involved in tight junction formation between neighboring epithelial cells, interacts with cell cytoskeletal components *via* molecules known as the catenins. This linkage

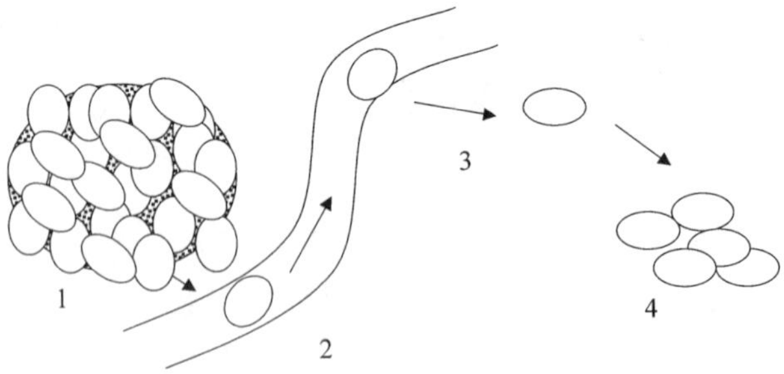

FIGURE 1 Metastasis is initiated when a cancer cell (1) detaches from the tumor mass, (2) enters the circulatory systems, (3) is transported to a distal site and invades the surrounding tissue, and (4) proliferates. Malignant tumors have this ability to leave the local environment and form metastatic lesions. It should be noted that a growing tumors mass establishes its own capillary network and that entry into the circulatory system need not occur outside the tumor itself.

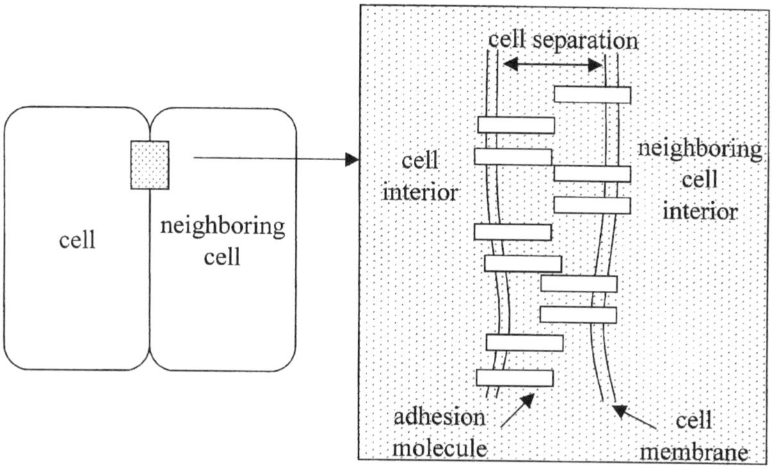

FIGURE 2 Changes in the synthesis or degradation rate of cell surface adhesion molecules can impact cell agglomeration. High levels of adhesion molecules increase the probability of cell binding and overall strength of attachment between neighboring cells. Reduced levels can result in poor adhesion due to both a reduction in overall bond strength and an increased probability of cell detachment.

to the cytoskeleton can provide rigidity and cell–cell attachment strength. Interruption of this linkage *via* changes in catenin interactions can lead to reduced cell adhesion [7, 8] and reduction in α-catenin may assist in tumor cell metastasis [9, 10]. In addition, altered expression of E-cadherin has been found in both primary and metastatic tumors from a number of different epithelial-derived cancers including pancreatic [11], prostate [12], gastric [13], and breast cancer [14, 15]. Further, E-cadherin expression and activity has been linked to growth factor stimulation, an additional potential link to tumorigenesis [16–18]. There is evidence that additional E-cadherin expression *via* cell transfection can reverse the invasiveness of malignant cells [19] although it has been suggested that the loss of E-cadherin, must be coupled with the expression of other cadherins in order to promote the proper environmental interactions needed by the invading cell [20]. There clearly exists a connection between overall metastatic activity and variations in the expression of specific adhesion and detachment factors with E-cadherin being an important example molecule.

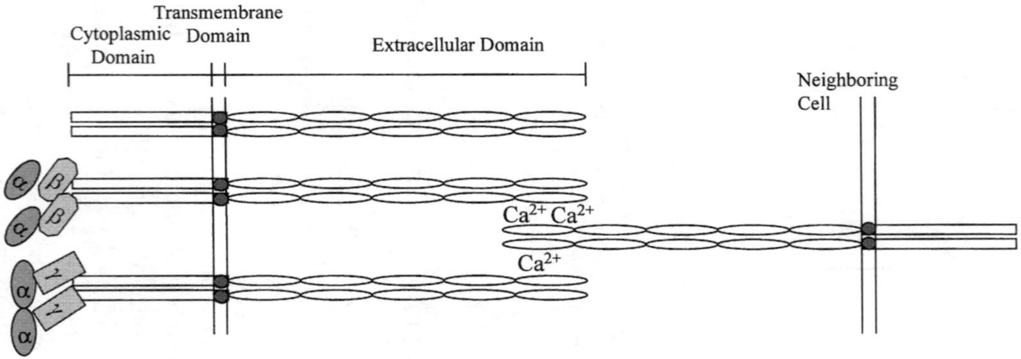

FIGURE 3 E-cadherin is a transmembrane molecule which forms homodimers in the presence of calcium. The extracellular domain contains 5 cadherin repeat units which can interact, in the presence of calcium, with E-cadherin molecules on neighboring cells. Homotypic (both E-cadherins) rather than heterotypic (E-cadherin binding with a different type of cadherin) adhesion is typical. The cytoplasmic tail of E-cadherin can interact directly with β-catenin (β) or γ-catenin (γ) which can then interact with α-catenin (α). Linkage to the actin cytoskeleton occurs through the catenin molecules and is required for cadherin-mediated cell adhesion.

Biological Adhesion Assays

Assessment of cell adhesion and detachment within the biological community is somewhat different from what one typically sees with a particle adhesion approach. Biological measurements of metastatic potential have focused on overall cell assays rather than specific bond measurements. For example, microscopy has been used to observe cell aggregation on surfaces [21, 22] and cell invasiveness into collagen gel [19]. Reaggregation assays have been frequently used in which agglomeration of dissociated cells is assayed using a Coulter Counter [21, 23, 24] (Fig. 4). Transfection of cells to express mutated E-cadherin [25] or α-catenin [7] or excess copies of E-cadherin [26] is another way in which the importance of these molecules with regard to aggregation have been studied. Alternatively, the resistance to cell removal by rinsing [7, 22, 27, 28] or centrifugation [8] is another way to assess cell adherence. These assays focus on the force required to eliminate cell association although, perhaps, a more direct analysis of binding could be made using a buoyancy-based assay which uses floatation to remove non-adherent cells [29]. In all cases, an overall rather than the individual interaction is being measured.

While some quantification can be associated with these biological assays, there is a limited ability to extract the true individual bond strength for cadherin interactions. Micromechanical methods to study individual cell–cell contact and deformation have been used extensively with blood cells to determine overall cell-adhesion energies [30–33]. Measurements of specific cell detachment based on surface interactions have been carried out using flow chambers to quantify the shear force required to cause cell detachment [34, 35]. Studies done under flow suggest that the heterogeneous nature of the cell glycocalyx, represented by antibody-coated beads, can affect detachment [36] and could play a role in cell detachment from a growing tumor. Recently, flow studies were performed to analyze the homotypic interaction between cadherin-coated surfaces and cadherin-coated spheres and, for example, an increase in binding frequency with cadherin surface density was observed [37]. Alternatively, optical tweezers are being investigated as tools for trapping both cells [38, 39] and individual molecules [40, 41] in order to evaluate the strength and specificity of the interaction. A recent paper by Sako *et al.*,

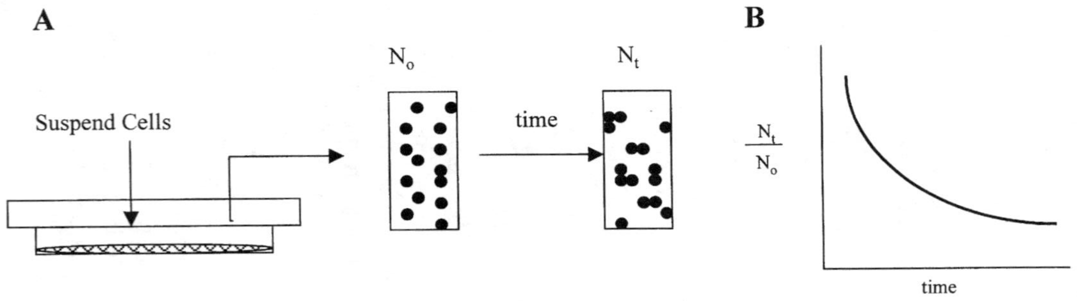

FIGURE 4 Reaggregation assay – (A) Cells are removed from the culture dish, generally *via* enzymatic dissociation, and resuspended in buffer. A sample is counted using a Coulter Counter to determine an initial number of cells (N_0). The solution is then incubated at 37°C on a shaker table to promote cell suspension and mixing. At time t, glutaraldehyde is added to fix the cells and a sample is counted in the Coulter Counter to determine the number of particles (N_t). (B) The amount of aggregation is determined from the reduction in particle number at time t compared to the original sample [24].

used laser trapping combined with particle tracking to evaluate diffusion of E-cadherin in the cell membrane and the role of cytoskeletal binding in retarding movement [42]. Finally, both atomic force microscopy (AFM) and the Surface Force Apparatus have been used to investigate molecular interactions and receptor-ligand binding and hold promise for quantifying adhesion binding exhibited by E-cadherin [43–45]. A challenge in evaluating the effect of specific cell adhesion molecules as they regulate cell adhesion is the interplay of other binding reactions, both extracellular and membrane-associated. The difficulty is in evaluating the relative contributions of each and how effective interventional strategies might be designed based on the pivotal associations.

CELL–CELL ASSOCIATION: A COLLOIDAL PERSPECTIVE

The agglomeration characteristics of cancer cells and the process of metastatic redistribution are of both clinical and fundamental interest and importance. Clinically, preventing deagglomeration of primary tumors could significantly ease surgical treatment and patient aftercare. Fundamentally, one would like to understand how attachment and detachment are regulated as these processes are important for tissue differentiation as well as metastasis. From an adhesion science perspective, the primary question is whether surface chemistry dominates binding across cell surfaces or if there is a structural lock-and-key effect controlled by specific cell adhesion molecules. This is a simplified picture, but one helpful for establishing a basic foundation.

Greater understanding can arise by linking experimental evidence with available theories. Three main approaches/theories exist to address attachment/detachment from a fundamental colloidal perspective. The first approach is to evaluate cell–cell interactions as colloidal agglomerations using some variation of the Derjaguin–Landau–Verwey–Overbeek (DLVO) theory. The second approach evaluates particle interactions on surfaces making use of the Johnson–Kendall–Roberts (JKR) theory. The final approach is based on surface energetics.

DLVO Theory

DLVO theory relating colloidal stability of particles [46, 47] has withstood the test of time as a generalized model to describe many problems in particle dynamics [48]. The theory has, as its basis, that particles in a medium, much like those dissolved in solution, change the chemical potential of the system. Surfactants, ions, pH, temperature and electrostatic interactions all can impact the relative isolation characteristics of each particle. As particle concentration increases, the number of cell–cell interactions that can be sensed by colligative properties like osmotic pressure increases [48, 49]. Further, direct measurement of particle aggregation can be made using various scattering techniques such as X-rays [50], light [51, 52], or neutrons [53]. The most important parameter that can be extracted from these types of measurements is the pair-correlation function, which gives the probability of finding the center of an interacting particle at a distance r from the center of a reference particle. This parameter is central to DLVO theory. The simplest DLVO structural analysis is based on spherical scattering sources (cell or particle) [54]; however, alternative scattering profiles have been determined for anisotropic particles [55, 56], which may be more appropriate for certain cell types. The ability to test the theory fully for idealized particles or cells has improved, given that polymerization processes yielding large numbers of relatively spherical by-shaped particles are readily available [49]. Again, however, some heterogeneity both *in vivo* and *in vitro* with isolated mammalian cells is to be expected and will add uncertainty to theoretical predictions.

DLVO theory for particle interaction has been broken down along two main efforts primarily related to the relative stiffness of the interacting particles [48]. The classic hard-sphere model shows relatively small interactions in potential energy until contact between the particles occurs. Continued compression following contact leads to a rapid rise in the repulsive force pushing these particles apart. The deformation characteristics and membrane strength of mammalian cells confound use of this type of model, suggesting that soft sphere models would be more applicable to address cancer cell interactions. These models demonstrate a balance between the repulsive (steric and electrostatic) forces favoring separation and the attractive dispersion forces favoring agglomeration. The potential barrier for agglomeration is the sum of these different potentials (Fig. 5).

With the soft-sphere model, the attractive dispersion force, $PE_{attractive}$, is given as:

$$PE_{attractive} = -\frac{A}{6}\left\{\frac{2a^2}{H(4a+H)} + \frac{2a^2}{(2a+H)^2} + \ln\left[\frac{H(4a+H)}{(2a+H)^2}\right]\right\}$$

where A is the Hamaker constant, typically in the range of 1×10^{-20} Joules, a is the particle radius and H is the distance between particles [57].

The corresponding repulsive force, $PE_{repulsive}$, comes from Overbeek [58, 59] and makes use of a moderate electrostatic potential. It is expressed as:

$$PE_{repulsive} = 2\pi a\varepsilon\varepsilon_0\left(\frac{4kT}{ze}\gamma\right)^2 e^{-\kappa H}$$

where κ is the Debye Parameter, H remains the interparticle distance, γ is a lumped parameter term, ε is the permittivity, ε_0 is the permittivity in vacuum, a is the particle radius, z is the ion valence, k is Boltzmann's constant, and e is the charge of an electron. Soft-sphere

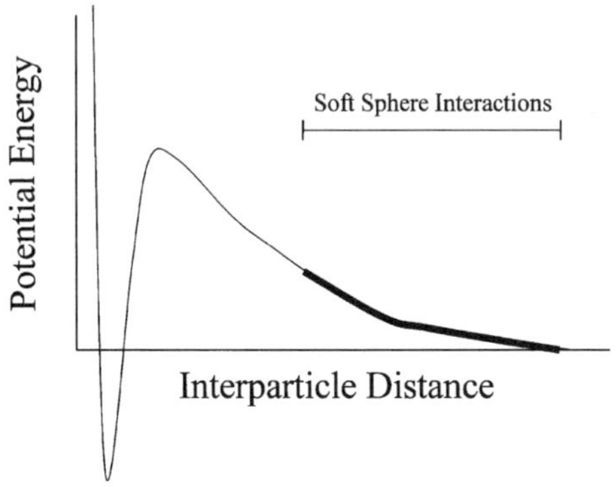

FIGURE 5 Potential energy diagram for soft-sphere interactions from a colloidal perspective. There are electrostatic interactions leading to repulsive forces between particles at relatively large interparticle spacings. There are also attractive interactions that favor interparticle wetting as the spacing decreases. Soft interactions can lead to cell–cell deformation which can increase repulsive forces if not capable of overcoming the potential energy barrier.

interaction models have been applied to latex particles and other colloidal dispersions as a function of pH and ionic strength of the suspension and appear to be appropriate for describing cellular interactions [60, 61]. Gingell and Fornes [62, 63] provided the first evidence of the validity of using DLVO in cell–cell interactions. Recent work by Molina–Bolivar *et al.*, has shown that the DLVO theory can be applied in a direct way to protein-mediated particle adhesion by interpreting the structure factors and clustering phenomena associated with protein-coated polymeric latex particles [57]. Evidence of the theory's validity is available; however, application to viable cells may prove difficult given the complexities of the normal cell glycocalyx interaction.

The independent variables associated with the theory relate primarily to the surface energy of the interaction and the charge associated with the colloidal interaction. These independent variables will define the interparticle spacing. To link theory with experiment, the goal would be to establish how changing the charge and surface energy of the cell mass leads to changes in the interparticle spacing between cells. In the event that stronger interactions beyond charge attraction are holding the cell mass together, varying charge and surface energy between cells in the mass will likely not lead to any observable change in the interparticle spacing. This would indicate that DLVO theory does not capture the essential phenomena and validates the importance of specific binding between adhesion molecules. Should interparticle spacing change appreciably with charge, it would indicate a reduced importance for molecules such as E-cadherin in the dissociation process, although this must be tempered with the knowledge that charge attractions may be involved in E-cadherin binding as well.

JKR Model

One could alternatively envision the detachment of a cell from a tumor mass as the loss of adhesion of a cell to a "tumor surface". There are a multitude of papers investigating particle-surface as well as cell-surface interactions [64–72]. Under this model, the interaction and corresponding adhesion of a particle to a surface is regulated by a series of contributions including the surface energetics of the contact and the mechanical stiffness of the contacting media. JKR theory of particle contact on a solid surface has been applied to systems having larger surface energies and lower elastic moduli. Focus has been on determining for what regimes the theory is valid based on experimental studies of particles interacting on surfaces.

The theory is based on establishment of an equilibrium contact radius for the cell with the surface resulting from contributions relating to elastically-stored energy, the energy associated with surface forces of interaction, and the mechanical energy applied from an externally-applied load (Fig. 6). From a fundamental perspective, Quesnel *et al.* indicated that this theory is based on mathematical interpretation of particle adhesion, measured on a macroscopic level, with the physics of the interaction occurring on a microscopic or molecular level [73]. This is certainly the case for cellular adhesion, where cells may be idealized as spherical particles on the macroscopic level and textured surfaces with specific receptor proteins spanning the cell membrane on the molecular level.

JKR analysis predicts a relationship for this contact radius as [74, 75]:

$$a^3 = \frac{R}{K}[P + 3w_a\pi RP + (6w_a\pi RP + (3w_a\pi R)^2)]^{0.5}$$

where a is the contact radius formed from the particle wetting the surface, R is the particle radius, P is the applied load, w_a is the work

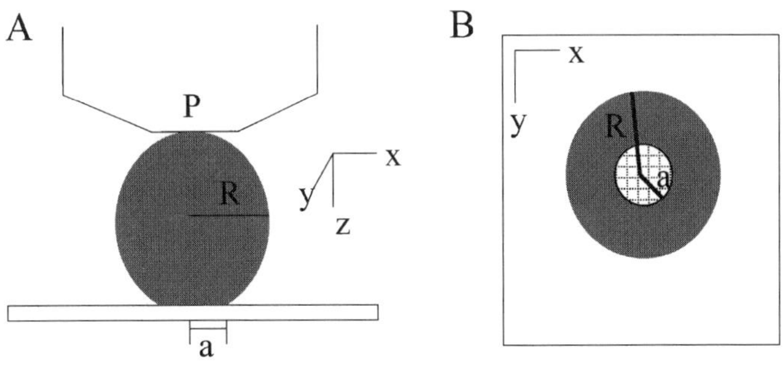

FIGURE 6 Particle adhesion configuration that considers the pressure-induced wetting of a hard sphere of radius R onto a soft substrate of known surface energy. Pressure-induced contact creates a circle of contact with radius, a, under applied load, P. The three-dimensional configuration of the apparatus is shown in part A and a two-dimensional projection looking up from the bottom of the apparatus is shown in part B.

of adhesion (a surface energy term), and K is an effective stiffness corresponding to the stiffness and Poisson's ratio of the constituents making up the particle/substrate interaction.

The work of adhesion or the energy and distance required to separate the particle from the surface can reasonably be represented by a Lennard–Jones Potential, where the exponents of the attractive and repulsive terms of the interaction are variables much like the energetics of molecular bonding [73]. This potential has been shown to represent polymeric interactions well in terms of their elastic properties and surface energy but has yet to be shown valid for a much more deformable body like a mammalian cell. For that matter, it is not clear whether all cell types would have similar surface energetics. These are gaps in knowledge which need to be explored.

One way to test the application of JKR theory to cell adhesion molecules and metastasis would be to spin-coat or deposit cell adhesion molecules onto a surface and then perform cell deformation experiments with the protein-coated surface. By controlling the force one applies, a cell or small tumor cluster could be brought into contact with the protein-coated surface and the contact radius measured as a function of the application pressure.These experiments could be done as a function of temperature or as a function of electrolyte concentration in the solution where both temperature and concentration should affect the "work of adhesion" term between cell and surface. Difficulties in performing these types of experiments are linked to the inelastic deformation of the cell during pressure application which may lead to altered force-contact curves at different locations of the cell. Further, at high pressures, cell rupture could occur. There may be other unexpected consequences for live cells due to the dynamic nature of the glycocalyx to pressure application, as well as other difficulties which may be due to variations in the orientation of the protein on the surface which may or may not be physiologically relevant. While the experiments are conceptually simple, data generated will need to be analyzed carefully to avoid inappropriate conclusions.

Surface Energetics of Cellular Interactions

Another interpretation for disaggregation occurring with metastasis could be made through the use of a surface energetics argument.

The surface energy of the agglomerated cell mass could be sufficient to allow for the detachment of smaller sections (metastases) which distribute and proliferate on their own. The application of surface energetics is well known in the biomaterials literature relating to eucaryotic cell adhesion to surfaces [76, 77]. Under this theory, cells interacting with a surface will spread if there is a thermodynamic driving force to increase cell-surface interactions. The process of cell wetting on a surface can be interpreted through contact angle measurements and has been elegantly described by the Young–Dupre equation [77–79]. Essentially, the contact angle is inversely related to the wettability of a particular surface and, as the cell spreads more onto a surface, an equilibrium develops. The amount of wetting has been related to the independent contributions of the surface energies of the cell and surface with the environment and the corresponding surface energy between cell and surface [77–79]. Spreading on the surface is a separate event following cell wetting and occurs over a much longer time scale. The difficulties in applying a surface energetics approach to metastasis are the inability to generate either measurements of the contact angle between cells *in vivo* and the lack of knowledge about what should represent the surface and its surface energy. Measurements could focus on whether tumor size impacts individual cell spreading and the ability of the cell to disengage from the mass. Another component that could be probed would be whether sectioned tumor masses exhibit the same level of cell–cell interaction or if adhesion is spatially dependent, particularly taking into account blood vessel location within the mass.

Application of Particle Aggregation Theory to Cadherins and Other Cell Adhesion Molecules in Cancer Metastasis

Biological aggregation is couched in very different language from colloidal adhesion theory; however, there is still tremendous overlap. Surface chemistry and morphology certainly play an important role in the fundamental phenomena in both sciences. However, the type and level of adhesion molecules present on the mammalian cell surface can change in response to environmental signals from neighboring cells as well as the extracellular matrix [80, 81], lending a complexity not generally seen in colloidal systems. In addition, the specific binding of cadherins and other adhesion molecules augments considerably the attractive forces and decreases the probability of dissociation between the cells. These changes in the bond characteristics or number of bonds will likely impact the overall cell agglomeration.

There are a series of questions that need to be addressed to apply particle adhesion theory to cancer cell metastasis. The first is whether metastasis is a group/collective event or whether individual pairs of metastatic cells dissociate as well. It is conceivable that active metastasis may only occur when there is a large enough group of cells to trigger a reduction in the synthesis of cell adhesion molecules. Thus, experiments need to be conducted on cells in a wide range of densities to insure that the proper regime is investigated. Further, it is unlikely that cells will be homogeneous with regard to adhesion molecule levels and the issue of whether an isolated individual cell change is sufficient or if a global tumor change is needed must be examined. Therefore, while it may be simplistic, it is important to evaluate initially the effect of cadherin on cell aggregation and dissociation on uniform cells in isolation. *In vivo*, there are likely molecular and enzymatic cellular activities which can interact with cadherins and may increase the relative chance for metastasis or, alternatively, other adhesion molecules that could interfere with cadherin homotypic binding. Initial testing with coated polymeric beads and increasing the coating complexity incrementally would allow for the fundamental studies needed to analyze the complex cellular environment. Further, it would be advantageous to initiate studies in a fibrous or gelatin-like environment more representative of the tissue architecture. The lack of suitable isolation techniques and the cost for isolation of proteins such as E-cadherin can stymie these types of experimental efforts but does not diminish the need for these types of studies.

CONCLUSION

Cancer cell metastasis is a complex biochemical process, in which cell adhesive proteins affect the progression of the disease. Assays and flow cell measurements have been used to evaluate binding mechanisms and there are available theories that have been proposed that could be applied to cadherins and the process of metastasis.

The difficulty in applying theory to experiment in dealing with metastasis is an inherent problem in viewing behavior of an ensemble of particles relative to individual cell–cell interactions. "Colloidal" type experimental measurements can lend insight into the controlling phenomena of cell detachment and may facilitate the development of novel treatment protocols on a localized level.

Acknowledgements

The authors would like to thank the Center for Biomedical Engineering at Virginia Tech for its general support. In addition, K. E. Forsten would like to thank The Whitaker Foundation for continued support through a Biomedical Engineering Research Grant.

References

[1] Benitez-Bribiesca, L., In: *When Cells Die*, Lockshin, R. A., Zakeri, Z. and Tilly, J. L., Eds. (Wiley-Liss, New York, 1998), pp. 453–482.
[2] Hart, I. R., Goode, N. T. and Wilson, R. E., *Biochim. Biophys. Acta* **989**, 65–84 (1989).
[3] Shiozaki, H., Oka, H., Inoue, M., Tamura, S. and Monden, M., *Cancer* **77**, 1605–1613 (1996).
[4] Takeichi, M., *Science* **251**, 1451–1455 (1991).
[5] Petruzzelli, L., Takami, M. and Humes, H. D., *Am. J. Med.* **106**, 467–476 (1999).
[6] Christofori, G. and Semb, H., *Trends Biochem. Sci.* **24**, 73–76 (1999).
[7] Imamura, Y., Itoh, M., Maeno, Y., Tsukita, S. and Nagafuchi, A., *J. Cell Biol.* **144**, 1311–1322 (1999).
[8] Angres, B., Barth, A. and Nelson, W. J., *J. Cell Biol.* **134**, 549–557 (1996).
[9] Kadowaki, T., Shiozaki, H., Inoue, M., Tamura, S., Oka, H., Doki, Y., Iihara, K., Matsui, S., Iwazawa, T., Nagafuchi, A., Tsukita, S. and Mori, T., *Cancer Res.* **54**, 291–296 (1994).
[10] Shimoyama, Y., Nagafuchi, A., Fujita, S., Gotoh, M., Takeichi, M., Tsukita, S. and Hirohashi, S., *Cancer Res.* **52**, 5770–5774 (1992).
[11] Karayiannakis, A. J., Syrigos, K. N., Chatzigianni, E., Papanikolaou, S., Alexiou, D., Kalahanis, N., Rosenberg, T. and Bastounis, E., *Anticancer Res.* **18**, 4177–4180 (1998).
[12] Umbas, R., Schalken, J. A., Aalders, T. W., Carter, B. S., Karthaus, H. F. M., Schaafsma, H. E., Debruyne, F. M. J. and Isaacs, W. B., *Cancer Res.* **52**, 5104–5109 (1992).
[13] Mayer, B., Johnson, J. P., Leitl, F., Jauch, K. W., Heiss, M. M., Schildberg, F. W., Birchmeier, W. and Funke, I., *Cancer Res.* **53**, 1690–1695 (1993).
[14] Siitonen, S. M., Kononen, J. T., Helin, H. J., Rantala, I. S., Holli, K. A. and Isola, J. J., *Am. J. Clin. Pathol.* **105**, 394–402 (1996).
[15] Lipponen, P., Saarelainen, E., Ji, H., Aaltomaa, S. and Syrjanen, K., *J. Pathol.* **174**, 101–109 (1994).
[16] Al Moustafa, A. E., Yansouni, C., Alaoui-Jamali, M. A. and O'Connor-McCourt, M., *Clin. Cancer Res.* **5**, 681–686 (1999).

[17] Guvakova, M. A. and Surmacz, E., *Exp. Cell Res.* **231**, 149–162 (1997).
[18] Takahashi, K. and Suzuki, K., *Exp. Cell Res.* **226**, 214–222 (1996).
[19] Frixen, U. H., Behrens, J., Schs, M., Eberle, G., Voss, B., Warda, A., Lochner, D. and Birchmeier, W., *J. Cell Biol.* **113**, 173–185 (1991).
[20] Pishvaian, M. J., Feltes, C. M., Thompson, P., Bussemakers, M. J., Schalken, J. A. and Byers, S. W., *Cancer Res.* **59**, 947–952 (1999).
[21] Deman, J. J., Van Larebeke, N. A., Bruyneel, E. A., Bracke, M. E., Vermeulen, S. J., Vennekens, K. M. and Mareel, M. M., *In Vitro Cell. Dev. Biol.* **31**, 633–639 (1995).
[22] Komatsu, M., Carraway, C. A. C., Fregien, N. L. and Carraway, K. L., *J. Biol. Chem.* **272**, 33245–33254 (1997).
[23] Ozawa, M., Ringwald, M. and Kemler, R., *Proc. Natl. Acad. Sci. USA* **87**, 4246–4250 (1990).
[24] Takeichi, M., *J. Cell Biol.* **75**, 464–474 (1977).
[25] Nagafuchi, A. and Takeichi, M., *EMBO* **7**, 3679–3684 (1988).
[26] Nagafuchi, A., Shirayoshi, Y., Okazaki, K., Yasuda and Takeichi, M., *Nature* **329**, 341–343 (1987).
[27] Bauer, J. S., Schreiner, C. L., Giancotti, F. G., Ruoslahti, E. and Juliano, R. L., *J. Cell Biol.* **116**, 477–487 (1992).
[28] St. John, J. J., Schroen, D. J. and Cheung, H. T., *J. Immunol. Methods* **170**, 159–166 (1994).
[29] Goodwin, A. E. and Pauli, B. U., *J. Immunol. Methods* **187**, 213–219 (1995).
[30] Berk, D. and Evans, E., *Biophys. J.* **59**, 861–872 (1991).
[31] Evans, E., In: *Physical Basis of Cell–Cell Adhesion*, Bongrand, P., Ed. (CRC Press, Inc., Boca Raton, 1988), pp. 174–189.
[32] Tozeren, A., Sung, K. L., Sung, L. A., Dustin, M. L., Chan, P. Y., Springer, T. A. and Chien, S., *J. Cell Biol.* **116**, 997–1000 (1992).
[33] Shao, J. Y. and Hochmuth, R. M., *Biophys. J.* **71**, 2892–2901 (1996).
[34] Kuo, S. C. and Lauffenburger, D. A., *Biophys. J.* **65**, 2191–2200 (1993).
[35] Cozens-Roberts, C., Quinn, J. A. and Lauffenburger, D. A., *Biophys. J.* **58**, 857–872 (1990).
[36] Saterbak, A., Kuo, S. C. and Lauffenburger, D. A., *Biophys. J.* **65**, 243–252 (1993).
[37] Pierres, A., Benoleil, A. and Bongrand, P., *J. Phys. III* **6**, 807–824 (1996).
[38] Mammen, M., Helmerson, K., Kishore, R., Choi, S.-K., Phillips, W. D. and Whitesides, G. M., *Chem. Biol.* **3**, 757–763 (1996).
[39] Bronkhorst, P. J., Grimbergen, J., Brakenhoff, G. J., Heethaar, R. M. and Sixma, J. J., *Br. J. Haematol.* **96**, 256–258 (1997).
[40] Mehta, A. D., Rief, M., Spudich, A. D., Smith, D. A. and Simmons, R. M., *Science* **283**, 1689–1695 (1999).
[41] Finer, J. T., Mehta, A. D. and Spudih, J. A., *Biophys. J.* **68**, 291S–296S (1995).
[42] Sako, Y., Nagafuchi, A., Tsukita, S., Takeichi, M. and Kusumi, A., *J. Cell Biol.* **9**, 1227–1240.
[43] Hinterdorfer, P., Baumgartner, W., Gruber, H. J., Schilcher, K. and Schindler, H., *Proc. Natl. Acad. Sci. USA* **93**, 3477–3481 (1996).
[44] Wong, J. Y., Kuhl, T. L., Israelachvili, J. N., Mullah, N. and Zalipsky, S., *Science* **275**, 820–822 (1997).
[45] Willemsen, O. H., Snel, M. M., Kuipers, L., Figdor, C. G., Greve, J. and DeGrooth, B. G., *Biophys. J.* **76**, 716–724 (1999).
[46] Derjaguin, B. and Landau, L., *Acta Physicochim URSS* **14**, 633–662 (1941).
[47] Verwey, E. J. W. and Overbeek, J. T. G., *Theory of Stability of Lyophobic Colloids* (Elsevier, Amsterdam, 1948).
[48] Otterwill, R. H., *Faraday Discuss. Chem. Soc.* **90**, 1–15 (1990).
[49] Hunt, W. and Zukoswki, C., *J. Coll. Interf. Sci.* **210**, 332–342 (1999).
[50] Delfort, B., Daoudal, B. and Barre, L., *Tribol. T.* **42**, 296–302 (1999).
[51] Burns, J., Yan, Y., Jameson, G. and Biggs, S., *Langmuir* **13**, 6413–6420 (1997).

[52] Lehner, D., Worning, P., Fritz, G., Ogendal, L., Bauer, R. and Glatter, O., *J. Colloid Interf. Sci.* **213**, 445–456 (1999).
[53] Brunner-Popela, J., Mittelbach, R., Strey, R., Schubert, K., Kaler, E. and Glatter, O., *J. Chem. Phys.* **110**, 10623–10632 (1999).
[54] Debye, J. P., *J. Appl. Phys.* **15**, 338–342 (1944).
[55] Trinkhaus, J. P., *Cells into Organs, The Forces that Shape the Embryo* (Prentice Hall, Englewood Cliffs, NJ 1984).
[56] Cabannes, J. and Rochard, Y., *La Diffusion Moleculaire de la Lumiere* (Les Presses Universataires de France, Paris 1929).
[57] Molina-Bolivar, J., Galisto-Gonzalez, F. and Hidalgo-Alvarez, R., *J. Colloid Interf. Sci.* **208**, 445–454 (1998).
[58] Overbeek, J. T. G., *Adv. Colloid Interface Sci.* **16**, 17 (1982).
[59] Overbeek, J. T. G., *Pure Appl. Chem.* **52**, 1151 (1980).
[60] Kendall, K., Liang, W. and Stainton, C., *J. Adhesion* **67**, 97–109 (1998).
[61] Kendall, K., Liang, W. and Stainton, C., *Proc. Royal Soc. London A* **454** (1977), 2529–2533 (1998).
[62] Gingell, D. and Fornes, J. A., *Biophys. J.* **16**, 1131 (1976).
[63] Gingell, D. and Fornes, J. A., *Nature* **256**, 210 (1975).
[64] Bowers, V. M., Fisher, L. R. and Francis, G. W., *J. Biomed. Mat. Res.* **23**, 1453–1473 (1989).
[65] Bruil, A., Terlingen, J., Beugeling, T., van Aken, W. and Feijen, J., *Biomaterials* **13**, 915–923 (1992).
[66] Bruil, A., Brenneisen, L., Terlingen, J., Beugeling, T., van Aken, W. and Feijen, J., *J. Colloid lnterf. Sci.* **165**, 72–81 (1994).
[67] Tirrell, M., Falsafi, A., Bates, F. and Pocius, A., In: *Proceedings of the Adhesion Society* **20**, 121–122, Hilton Head Island (1997).
[68] Scheuerman, T. R., Camper, A. K. and Hamilton, M. A., *J. Colloid Interf. Sci.* **208**, 23–33 (1998).
[69] Seyfert, S., Voigt, A. and Kabbeck-Kupijai, D., *Biomaterials* **16**, 201–207 (1995).
[70] Schakenraad, J. M. In: *Biomaterials Science – An Introduction to Materials in Medicine*, Ratner, B. D., Hoffman, A. S., Schoen, F. J. and Lemons, J. E., Eds. (Academic Press, San Diego, 1996), pp. 141–147.
[71] Tamada, Y. and Ikada, Y., *J. Colloid Interf. Sci.* **155**, 334–339 (1993).
[72] Brach, R., Li, X. and Dunn, P., *J. Adhesion* **69**, 181–200 (1999).
[73] Quesnel, D. J., Rimai, D. S. and DeMejo, L. P., *J. Adhesion* **67**, 235–257 (1998).
[74] Quesnel, D. J., Rimai, D. S., Gady, B. and DeMejo, L. P., In: *Proceedings of the Adhesion Society* **21**, 290–292, Savannah (1998).
[75] Johnson, K. L., Kendall, K. and Roberts, A. D., *Proc. R. Soc. London, Ser. A* **324**, 301–313 (1971).
[76] Ratner, B. D., In: *Biomaterials Science – An Introduction to Materials in Medicine*, Ratner, B. D., Hoffman, A. S., Schoen, F. J. and Lemons, J. E., Eds. (Academic Press, San Diego, 1996), pp. 21–35.
[77] Schakenraad, J. M., Busscher, H. J., Wildevuur, C. R. H. and Arends, J., *Cell Biophyics* **13**, 75–91 (1988).
[78] van Wachem, P. B., Beugeling, T., Feijen, J., Bantjes, A., Detmers, J. P. and van Aken, W. G., *Biomaterials* **6**, 403–408 (1985).
[79] Jansen, J. A., van der Waerden, J. P. C. M. and de Groot, K., *Biomaterials* **10**, 604–608 (1989).
[80] Dejana, E., *J. Clin. Invest.* **98**, 1949–1953 (1996).
[81] Bongrand, P., Capo, C., Mege, J.-L. and Benoliel, A.-M., In: *Physical Basis of Cell–Cell Adhesion*, Bongrand, P., Ed. (CRC Press, Boca Raton, 1988), pp. 61–90.

Adhesion of Cancer Cells to Endothelial Monolayers: A Study of Initial Attachment *Versus* Firm Adhesion

MELISSA A. MOSS

Department of Research, Mayo Clinic Jacksonville, Jacksonville, FL 32224, USA

KIMBERLY W. ANDERSON

Department of Chemical and Materials Engineering, 177 Anderson Hall, University of Kentucky, Lexington, KY40506-0046, USA

For most cancer patients, the ultimate cause of death is not the primary tumor itself, but metastasis, or the spread of cancer from the primary tumor throughout the body. The formation of tumor foci at sites distant from the primary tumor is a multistep process which includes dissemination of the cancer cells through the blood stream and hence, interactions with the endothelium lining the blood vessels walls. At least two theories have been proposed for explaining the interaction between cancer cells and endothelium. According to one theory, the tumor cells roll along the endothelium and the rolling velocity decreases until the cells become firmly attached to the vessel wall. In another theory, the circulating cancer cells must first lodge inside small vessels before they attach to the endothelium. In the latter case, the cells would only metastasize in the smaller vessels where lodging can occur. To gain further insight into the process of metastasis, the adhesion of human breast cancer cells to human umbilical vein endothelial monolayers was investigated in terms of both initial attachment followed by firm adhesion and firm adhesion following incubation in a static environment. The parallel plate flow chamber was employed to perform two different adhesion assays that would simulate these two adhesion mechanisms. Adhesion assays were carried out at a variety of physiological shear stresses found in the microvasculature and both highly metastatic and nonmetastatic cells were investigated. Results showed that initial attachment was only observed at very low shear stresses whereas firm adhesion occurred at a number of physiological shear stresses. These results suggest that the adhesion of the human breast cancer cells used in this study to endothelium most likely takes place *via* a lodging-firm adhesion mechanism in the capillaries and venules. However, it is important to note that other factors such as pulsatility and vessel compliance may contribute to the attachment. It was also shown that, for these specific breast cancer cells, adhesion did not correlate with metastatic potential. This suggests that while blocking the adhesion of highly metastatic cells may inhibit their ability to metastasize, adhesion properties alone do not provide an indication as to whether a cell is metastatic or nonmetastatic under the conditions studied here.

INTRODUCTION

Breast cancer is the most common cancer among women in the USA with one in eight women falling victim to the disease (Statistics from the National Cancer Institute). For most cancer patients, the ultimate cause of death is not the primary tumor itself, but metastasis, or the spread of cancer from the primary tumor throughout the body. Consequently, an important step in cancer research is understanding the metastatic cascade. Correlations between tumor microvessel density and degree of metastasis indicate that breast cancer metastasizes primarily *via* the blood vessels, rather than through the lymphatic system [1]. Metastasis, or the formation of tumor foci at sites distant from the primary tumor, is a multistep process. This process is depicted in Figure 1 for metastasis *via* the blood stream. To metastasize, cancer cells must detach from the primary tumor, intravasate through the vessel wall to enter the blood stream, disseminate through the blood stream, and extravasate back through the vessel wall to reestablish in the interstitial tissue. Many cancers will metastasize preferentially to certain organs. For example, prostate cancers metastasize most frequently to bone and small-cell lung

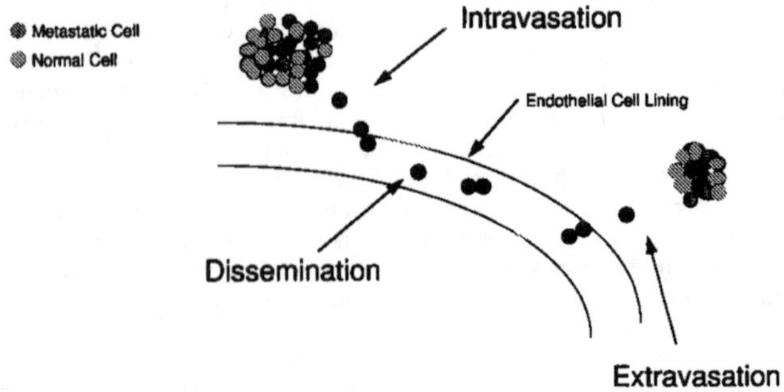

FIGURE 1 Schematic representing the various steps of the metastatic cascade in the microvasculature.

carcinomas most often to the brain. This organ-specific nature of metastasis has been attributed to interactions during the dissemination step between endothelial adhesion molecule ligands and their specific adhesion molecule receptors expressed on cancer cells [2]. Such interactions are involved in the attachment of circulating cancer cells to the endothelium that lines the vessel walls. Changes in adhesion molecule expression or binding affinity will enhance or decrease the adhesion potential of a receptor-ligand pair. Identification of adhesion molecules involved in the interaction of a cell pair is important to understanding adhesion. The adhesion between different cancer cell-endothelial cell pairs is likely to be mediated by different receptor-ligand combinations. In addition, more than one receptor-ligand pair may be responsible for a single cell-cell adhesion event. A review of the surface molecules responsible for adhesion can be found in [3].

At least two theories by which cancer cells adhere to the endothelium have been suggested in the literature. One hypothesis, set forth by Honn and Tang [4], is known as the ‘docking and locking’ hypothesis. According to this hypothesis, cancer cells first ‘dock’ and roll along the endothelium by rapidly forming and breaking adhesion molecule interactions with endothelial cells [5]. The rolling of tumor cells along the endothelium may also serve to activate additional adhesion molecules involved in the second step of cell attachment. As adhesion interactions form and break, the rolling velocity of the cancer cell decreases until the cell velocity is slow enough that more firm adhesion forms in the second ‘locking’ step of adhesion. These interactions are able to resist rheological forces and, thus, render the circulating cell stationary. In support of this hypothesis, cancer cell rolling followed by attachment has been observed *in vivo* for activated endothelium [6]. In addition, several researchers have employed *in vitro* dynamic adhesion assays to observe tumor cell rolling on endothelial monolayers [7–12].

In another theory, circulating cancer cells must first lodge inside small vessels before they adhere to the endothelium. When a circulating cancer cell encounters a vessel diameter smaller than that of the cancer cell itself, the cell will deform similar to noncancerous cells in the microvasculature in an attempt to pass through the vessel. If deformation of the cell does not exceed the elasticity limits of the cell membrane, the cancer cell will successfully pass through the vessel. If the elasticity limits of the cell membrane are exceeded, however, the cell membrane will rupture and the cancer cell will be killed [13–15]. Those cells that are not killed can be trapped in the vasculature. Entrapment of deformed cancer cells has been observed in the microcirculation of model animals [13]. After the cancer cell has stopped as a result of lodging, firm adhesion is able to take place between the stationary cancer cell and the endothelium. These interactions serve to stabilize the arrested cell so that rheological forces fail to dislodge the cell and the cell remains adherent. *In vivo* observations of cancer cell arrest without stimulation of adhesion molecules on endothelial cells reveal an absence of cell rolling along the endothelium and attribute cancer cell arrest to lodging of the cells within the microvasculature [6, 16, 17]. It remains to be shown which of these hypotheses most accurately reflects the mechanism by which cancer cells adhere to the endothelium, an essential step in the metastatic cascade.

Many studies which have investigated the adhesion of cancer cells to endothelial monolayers or extracellular matrix proteins have employed a static adhesion assay to quantify adhesive strength [18–30]. In a static adhesion assay, cancer cells are allowed to settle and adhere to the endothelial monolayer for a given period of time and nonadherent cells are removed through a series of manual washes. Difficulties in quantification arise from the fact that forces imposed during manual washing usually are not measured and may not be uniform. As a result, static adhesion assays provide only qualitative comparisons of adhesive properties. Furthermore, this system permits only the measurement of firm adhesion under static conditions. The event of initial attachment of cancer cells in fluid flow to a stationary endothelial monolayer or extracellular matrix protein cannot be observed.

Studies of the adhesion of cancer cells to endothelial monolayers or extracellular matrix proteins have also employed flow chambers [8–12, 31–33]. The hydrodynamic forces acting upon the cells within these flow chamber systems can be both measured and controlled. Consequently, accurate quantitative measurements of cell adhesion are possible. Flow chambers also provide an adhesion environment that better mimics the conditions under which cancer cells attach to the endothelium *in vivo* by allowing observation of the attachment of cells under flow conditions. Because adhesion can differ under static and flow conditions, it is important to study adhesion under conditions that best mimic the *in vivo* situation. For example, Felding-Habermann *et al.* [33] observed that while M21 melanoma cells adhere to a collagen matrix under static conditions, adhesion fails to occur under dynamic flow conditions, even at a low shear stress of 2 dynes cm^{-2}. Similarly, Kojima *et al.* [8] demonstrated that differences in the adhesion of B16 melanoma variant cells to endothelial monolayers were more readily observed in a dynamic assay than in a static assay. Furthermore, it was deduced that the mechanism responsible for adhesion differed in the two systems.

In any flow chamber system, both attachment and detachment assays may be used. In our specific detachment assay, cells are allowed to settle under zero flow for a given period of time. The cells are then removed with a measurable shear stress and the detached cells are quantified. In contrast, during an attachment assay, cell adhesion takes place as the cells are moving with the shearing fluid and the adherent cells are quantified.

To study the mechanism for adhesion in metastasis, we used the parallel plate flow chamber to investigate the adhesion of human breast cancer cells to human umbilical vein endothelial cell (HUVEC) monolayers using the attachment and detachment assay types to simulate initial attachment and post-lodging firm adhesion, respectively. These two mechanisms for adhesion were studied at a variety of shear stresses corresponding to the shear stresses that the cancer cells experience in different vessel types within the body. The adhesive properties of two different breast cancer cell lines with different metastatic potentials, MCF-7 (nonmetastatic) and MDA-MB-435 (highly metastatic), were studied to elucidate whether variations in adhesive strength with assay type and shear stress are dependent on metastatic potential. The rest of this manuscript will focus on the procedure for performing these studies and the results obtained.

MATERIALS AND METHODS

Cell Lines and Cell Culture Techniques

Endothelial monolayers were composed of human umbilical vein endothelial cells (HUVECs) obtained from Cell Systems (Kirkland, WA). Cells were purchased at passage 1 and used in adhesion studies up to passage 8. Preliminary studies showed that adhesion of the breast cancer cells used in this study to HUVECs was consistent up to passage 10. HUVECs were grown in 75 cm^2 Costar tissue culture flasks coated with CSC attachment factor (Cell Systems). HUVECs were sustained in CSC Complete Medium (Cell Systems) and maintained in a humidified incubator which provided an atmosphere of 5% CO_2 and 95% air at a constant temperature of 37 C.

The MCF-7 human breast cancer cell line was obtained from American Type Cell Culture (ATCC, Rockville, MD) at passages ranging between 120 and 180. This cell line has been shown to be nonmetastatic in a nude mouse model [34]. The MDA-MB-435 human breast cancer cell line was obtained from the laboratory of Janet Price (M. D. Anderson Cancer Center, Houston, TX) at passages ranging between 50 and 100. This cell line has been shown to be highly metastatic in both spontaneous and experimental metastasis models [35, 36]. Because the cells were received at high passage, they were checked periodically to confirm their metastatic potential. Both human breast cancer cell lines were grown as monolayers in either 75 cm^2 or 150 cm^2 Corning tissue culture flasks. Cells were maintained in Minimum Essential Media (MEM) (Gibco, Grand Island, NY) supplemented with 10% fetal bovine serum (FBS) (Sigma, St. Louis, MO), 10,000 units ml^{-1} penicillin (Gibco), 10 $\mu g\ ml^{-1}$ streptomycin (Gibco), 1 mM glutamine (Gibco), 0.3% sodium bicarbonate (Gibco), and 2.5 mM nonessential amino acids (Gibco). Human breast cancer cells were maintained in a humidified incubator which provided an atmosphere of 5% CO_2 and 95% air at a constant temperature of 37 C.

Preparation of the HUVEC Monolayer

HUVECs were seeded onto Permanox slides (Fisher) coated with 2% gelatin (Sigma) and endothelial cell attachment factor (Cell Systems). Seeded monolayers were sustained in CSC Complete Medium (Cell Systems) and maintained in a humidified incubator which provided an atmosphere of 5% CO_2 and 95% air at a constant temperature of 37 C. HUVEC monolayers were allowed to reach confluency (3 – 4 days) under static conditions on the Permanox slides prior to assembly of the flow chamber. HUVEC monolayers were stimulated by replacing cell culture media with 240 U of tumor necrosis factor-alpha

(TNF-α) (Promega, Madison, WI) suspended in 2 ml of CSC Complete Medium 4 hours prior to flow chamber assembly. TNF-α is a 52 kDa protein and is named for its ability to induce necrosis in primary tumors [37]. It has been shown that exposure of endothelial cells to TNF-α leads to an increase in expression of adhesion molecules such as ICAM-1, VCAM-1, *E*-selectin, *P*-selectin, and hyaluronate [38–43]. We chose to stimulate our endothelial cells with TNF-α because several *in vivo* sources of TNF-α exist and previous results have shown that the cancer cells used in these studies do not attach to unstimulated endothelial cells under flow conditions [10, 44]. The monolayers were returned to the cell culture incubator for the 4-hour incubation period. Prior to experimentation, the TNF-α was removed and the endothelial cells were washed with media.

Preparation of Human Breast Cancer Cells

Prior to experiments, confluent monolayers of human breast cancer cells were trypsinized and resuspended at a concentration of either 5.0×10^5 cells ml^{-1} for attachment experiments or 1.25×10^6 cells ml^{-1} for detachment experiments. Resuspension took place in a solution of dextran (Sigma) dissolved in CSC Complete Medium. A calibrated amount of dextran was added to this medium so that the viscosity of the cell suspension would match that of blood (3.9 cp). The cell suspension was maintained at 37 C.

Flow Chamber and Related Equipment

To assess the adhesive strength between human breast cancer cells and endothelial monolayers, two different adhesion assay types, an initial attachment assay and a detachment assay, were carried out inside a parallel plate flow chamber. The parallel plate flow chamber, depicted in Figure 2, consisted of a polycarbonate plate and a Permanox slide with a confluent layer of endothelial cells. The two plates were separated by a Silastic gasket. The system was held together by a vacuum to ensure a constant height and, therefore, a constant wall shear stress, along the length of the flow chamber. The Permanox/cell surface composed the bottom of the flow chamber. Two pressure ports located in the polycarbonate base enabled measurement of the pressure drop across

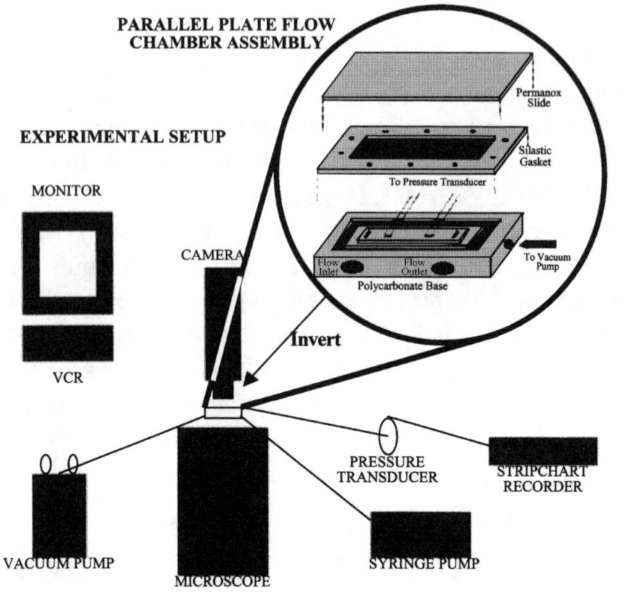

FIGURE 2 Schematic of experiment setup employed in the adhesion studies. The parallel-plate flow chamber is inverted on the microscope stage such that the endothelial monolayer comprises the bottom of the chamber.

the chamber. These pressure ports were connected *via* fluid filled tubing to a variable reluctance differential pressure transducer (Validyne Engineering, Northridge, CA). The transducer transmitted a signal to a digital indicator (Validyne Engineering) and stripchart recorder (Linseis, Princeton, NJ) so that the pressure drop across the monolayer could be continuously monitored and recorded. This pressure drop was related to the shear stress imposed upon the endothelial monolayer using the theory of plane Poiseuille flow, or flow between two infinite parallel plates.

Inlet and outlet ports within the flow chamber permitted the entrance and exit of media and cell suspensions. Inlet tubing consisted of both a primary and a secondary line. The primary line connected directly to the flow chamber itself, while the secondary line formed a T-connection with the primary tubing. During experimentation, the primary line was employed to introduce cell-free media into the flow chamber, and the secondary line was used to introduce the human breast cancer cell suspensions. All flow solutions contained endothelial

CSC Complete Medium, a native environment for the HUVEC monolayer. A calibrated amount of dextran (Sigma) was dissolved in this medium to increase the fluid viscosity to match that of blood (3.9 cp). All flow solutions were introduced using a Harvard syringe pump. Media and cell suspensions were maintained at 37 C.

The shear stress imposed upon the endothelial monolayer, or the shear stress at the wall, was calculated using:

$$\tau = \frac{\Delta P d}{2L}. \tag{1}$$

Here, τ is the shear stress in dynes cm^{-2}, ΔP is the measured pressure drop in mm Hg, and L is the distance between the pressure ports (2.33 cm). The height of the flow chamber, d, in cm, was calculated by:

$$d^3 = \frac{12\mu Q L}{b\Delta P}. \tag{2}$$

Here, μ is the fluid viscosity (3.9 cp), Q is the flow rate in ml min^{-1} controlled by the syringe pump, b is the chamber width (1.45 cm), and ΔP is the measured pressure drop in mm Hg. The height varied in each experiment depending on the quality of vacuum and the compressibility of the gasket. The range was 0.018 – 0.025 cm.

Once assembled, the flow chamber was inverted so that the endothelial monolayer served as the bottom of the chamber. The flow chamber was then mounted on the stage of an inverted, phase-contrast, light microscope (Zeiss, Batavia, IL). This microscope was equipped with a video camera (Phonic Microscopy, Oak Brooke, IL), black and white monitor (Sony, Teaneck, NJ), VCR (Panasonic, Secaucus, NJ), and time-date generator (Panasonic) so that experiments could be documented for analysis at a later time. A schematic of the complete system is shown in Figure 2.

Adhesion Assays

Initial Attachment

To quantify the initial attachment of human breast cancer cells to endothelial monolayers, an attachment assay was carried out inside the parallel plate flow chamber. Following assembly of the flow chamber, the primary line was used to perfuse the monolayer at a flow rate of 1.0 ml min^{-1} (shear stress of approximately 8 dynes cm^{-2}) for 2 minutes to remove loose endothelial cells or cell fragments resting on the monolayer. The monolayer was then perfused at the flow rate required to impose the desired shear stress of 0.25, 1, 5, 10 or 15 dynes cm^{-2}. The monolayer was perfused for 3 minutes to ensure that the desired shear stress was obtained. The secondary line was then employed to pass human breast cancer cells at a concentration of 5.0×10^5 cells ml^{-1} across the monolayer. Attachment of human breast cancer cells was allowed to proceed at this desired shear stress for a period of 30 minutes and the experiment was terminated. This experiment simulated the initial attachment of cancer cells to the endothelium *in vivo*. Rolling cancer cells were observed to slow and eventually stop and adhere to the endothelial monolayer. The attachment period was recorded and the video was analyzed at a later time for total number of stationary cells at 0, 5, 10, 15, 20, 25 and 30 minutes. Details of the data analysis are given in the section below.

This attachment protocol was employed to evaluate the conditions of initial attachment between the HUVEC monolayer and both MCF-7 and MDA-MB-435 human breast cancer cells using 5 different shear stresses (0.25, 1, 5, 10 and 15 dynes cm^{-2}). All experiments in which attachment was observed were completed with 6 repetitions. Experiments where rolling but no attachment was observed were completed with 3 repetitions.

Detachment

A detachment assay carried out inside the parallel plate flow chamber was employed to assess the firm adhesion of human breast cancer cells to HUVEC monolayers. A suspension of 1.25×10^6 human breast cancer cells ml^{-1} was introduced at a low flow rate corresponding to a shear stress of 0.025 dynes cm^{-2} using the secondary line. Running at this low shear stress also permitted the removal of loose endothelial cells and cell fragments from the monolayer. The flow was then stopped and the cancer cells were allowed to settle and adhere to the endothelial monolayer for a period of 30 minutes. Because a cell that becomes trapped in a vessel probably occludes or partially occludes the vessel, this stationary settling simulated the lodging of cancer cells within the vasculature. At the end of the settling time, the desired shear stress of 1, 5, 10 or 15 dynes cm^{-2} was imposed upon the monolayer to assess the strength of adhesion. Shear exposure continued for a period of 9 minutes, a duration that allowed detachment to reach an extinction value. The 9-minute detachment period was recorded and the video was analyzed at a later time.

This detachment assay was employed to evaluate the relative strength of adhesion between the HUVEC monolayer and both MCF-7 and MDA-MB-435 human breast cancer cells after a static 30-minute exposure period. Studies were performed at 4 different detachment shear stresses (1, 5, 10 and 15 dynes cm^{-2}). All experiments were completed with 6 repetitions.

Data Analysis

Initial Attachment

The attachment experiments were analyzed by counting the total number of stationary, adherent cells at 0, 5, 10, 15, 20, 25 and 30 minutes. The results showed a linear trend of attachment *vs* time and experimental results were reported as the total number of adherent cells at the end of the 30-minute attachment period. This value was used for statistical analysis of the effects of shear stress and metastatic potential on attachment.

Detachment

The detachment experiments were analyzed by counting the total number of cells present on the endothelial monolayer at the end of the 30-minute settling time and then counting the total number of cells remaining following the onset of the desired shear stress at 5, 30, 60, 180, 300, 420, 480 and 540 seconds. Results were plotted as fraction of cells retained *versus* time. The initial number of cells prior to shear stress ranged from 80–100. All results exhibited a leveling off of the value of fraction retained by the end of the 9-minute detachment time, indicating that an extinction value had been reached. Results were, thus, recorded as the extinction value at the end of the 9-minute detachment period. These extinction values were used for statistical analysis of the effects of shear stress and metastatic potential on detachment.

For both attachment and detachment experiments, the final magnification on the video monitor was 200X. One field per plate was analyzed with an area of 0.012 cm^2. The area was chosen in the center of the plate to ensure that the cells were exposed to fully developed flow (entrance length = 0.003 cm).

Statistics

Statistical significance was assessed using a two-way ANOVA followed by Student-Newman-Keuls *post-hoc* tests to compare the effect and interaction of cell type and shear stress for both initial attachment and detachment experiments.

RESULTS

Initial Attachment

Results for the initial attachment of human breast cancer cells to stimulated endothelial monolayers under shear stresses of 0.25, 1, 5, 10, and 15 dynes cm^{-2} are shown in Figure 3 for both nonmetastatic MCF-7 cells and highly metastatic MDA-MB-435 cells. These results indicated that the initial attachment of highly-metastatic human breast cancer cells to endothelial monolayers was significant only at a shear stress of 0.25 dynes cm^{-2}. In addition, the number of highly metastatic cells adhering at this shear stress was significantly greater than the number of attached lowly-metastatic cells. Approximately 100 MDA-MB-435 cells had attached after 30 minutes of flow compared with 60 MCF-7 cells. At higher shear stresses, zero attachment of the highly-metastatic cells was observed. Initial attachment of nonmetastatic cells was significant at both 0.25 and 1.0 dynes cm^{-2} although the number of cells observed at 1 dyne cm^{-2} was 69% less than at 0.25 dynes cm^{-2}.

Attachment under higher shear stresses of 5, 10 and 15 dynes cm^{-2} yielded zero attachment to endothelial cell monolayers although

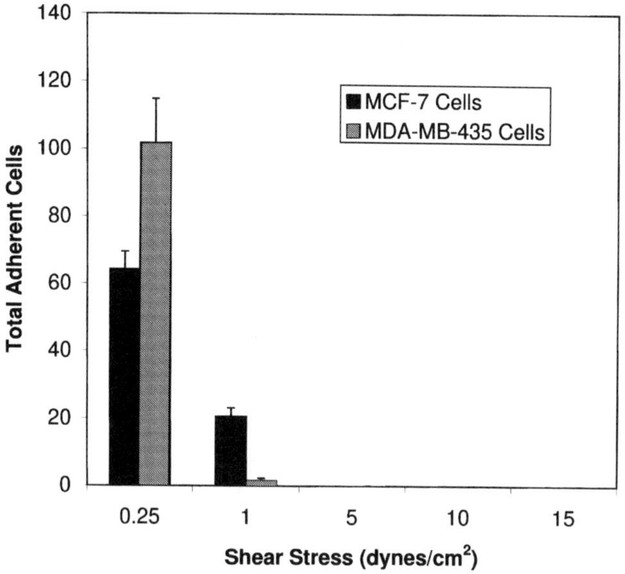

FIGURE 3 Total number of adherent cells *vs* shear stress applied in the attachment assay for both nonmetastatic (MCF-7) and highly-metastatic (MDA-MB-435) cells. Error bars indicate standard error. $N=6$ for 0.25 and 1 dyne cm^{-2}. $N=3$ for 5, 10 and 15 dyne cm^{-2}.

rolling along the monolayer was observed. This was true for both nonmetastatic MCF-7 cells and highly-metastatic MDA-MB-435 cells.

Detachment

Results for detachment of human breast cancer cells to endothelial monolayers under detachment shear stresses of 1, 5, 10 and 15 dynes cm^{-2} are shown in Figure 4 for nonmetastatic MCF-7 cells and highly metastatic MDA-435 cells. These results showed that at a shear stress of 15 dynes cm^{-2}, less than 20% of the cells remained attached to the endothelial monolayer. At detachment shear stresses of 10 dynes cm^{-2} or lower, however, more than 30% of cells remained attached to the endothelial cells. There was no significant difference between the nonmetastatic cells and highly-metastatic cells except at 10 dynes cm^{-2} where the nonmetastatic cells were slightly more adherent and at 5 dynes cm^{-2} where the highly-metastatic cells were slightly more adherent.

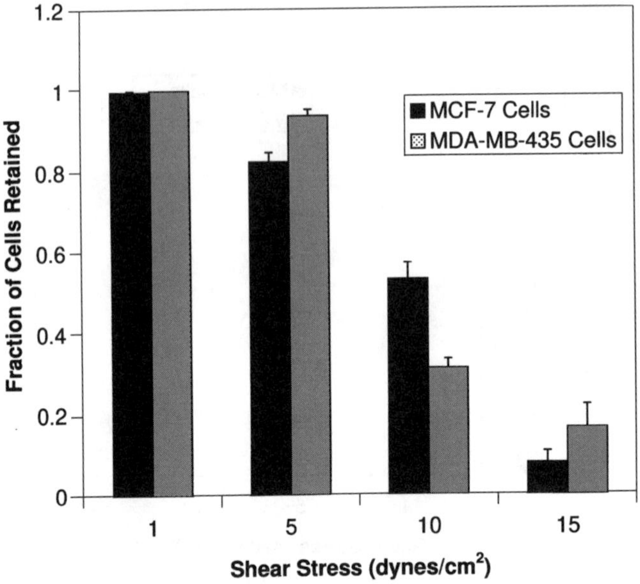

FIGURE 4 Fraction of cells retained *vs* shear stress applied in the detachment experiments for both nonmetastatic (MCF-7) and highly-metastatic (MDA-MB-435) cells. Error bars indicate standard error. $N = 6$ for all shear stresses.

DISCUSSION

Adhesion of cancer cells to vascular endothelium and its role in metastasis formation has been a focus of interest in cancer research for a long time. This research has led to at least two different theories concerning the means by which cancer cells adhere to the endothelium. Many of the studies that have investigated the adhesion of cancer cells to the endothelium have employed a static adhesion assay to evaluate adhesive strength [18, 20–22, 45]. In a static adhesion assay, cancer cells are allowed to settle and adhere to the endothelial monolayer for a given period of time and nonadherent cells are removed through a series of manual washes. Usually, the detachment forces imposed upon adherent cells are neither controlled nor measured. Furthermore, a static system permits only the evaluation of relative adhesion. The initial attachment of cancer cells in fluid flow to a stationary endothelial monolayer cannot be observed. The parallel-plate flow chamber design overcomes these obstacles. Inside the parallel-plate flow chamber, fluid flow between two parallel plates can be observed. This flow is described by plane, Poiseuille flow, and fluid flow can be modeled, measured, and controlled. Consequently, cells suspended in the fluid can attach or adherent cells can be detached under a constant, designated shear stress.

A number of researchers have employed this flow chamber geometry to study the adhesion of cancer cells to endothelial monolayers, extracellular matrix proteins, and ligand coated surfaces [7–10, 12, 31, 33]. Aigner *et al.* [12] employed the parallel-plate flow chamber to study the rolling of breast carcinoma cells on *P*-selectin coated surfaces. Tozeren *et al.* [10] used the parallel-plate flow chamber to carry out both detachment and attachment assays at several shear stresses to study the adhesion of a variety of breast cancer cells to TNF-α stimulated endothelial monolayers. In a separate study, Tozeren and coworkers employed the parallel-plate flow chamber at a variety of shear stresses to demonstrate that the integrin $\alpha_6\beta_4$ was capable of supporting both stable and dynamic attachment of several tumor cell lines [32]. Giavazzi *et al.* [9] studied the dynamic interaction of colon carcinoma, ovarian carcinoma, and breast carcinoma cells with IL-1 activated endothelial monolayers under incrementally decreasing shear stresses using the parallel-plate flow chamber. Patton *et al.* [31] employed the parallel-plate flow chamber, along with computerized analysis, to observe the attachment and subsequent stabilization of tumor cells on a laminin matrix. Kojima *et al.* [7] used the parallel-plate flow chamber over a range of shear stresses to demonstrate that adhesion of HL60 leukemia cells to *E*-selectin coated surfaces is mediated by different ligands at low and high shear stresses. In a separate study, Kojima and coworkers [8] employed the parallel-plate flow chamber and a dynamic attachment assay to correlate the initial attachment of B16 melanoma cell variants to endothelial monolayers with their metastatic capability. Felding-Habermann and coworkers employed the parallel plate flow chamber to investigate the involvement of platelets in the adhesion of melanoma cells to collagen [33].

In this study, the parallel-plate flow chamber was employed to observe both the initial attachment of cells in fluid flow to an endothelial monolayer and the detachment of adherent cells from an endothelial monolayer, such that both the initial attachment under flow conditions and detachment following exposure to a 30-minute settling time of human breast cancer cells to the endothelium could be evaluated.

Experiments carried out inside the parallel-plate flow chamber indicated that the initial attachment of highly metastatic human breast cancer cells to endothelial monolayers was significant only at a shear stress of 0.25 dynes cm^{-2}. At higher shear stresses, zero attachment was observed even though the cells did roll along the monolayer. Nonmetastatic cells were able to attach at 0.25 and 1.0 dynes cm^{-2}; however, the number of cells remaining attached at 1.0 dynes cm^{-2} was less than the number at 0.25 dynes cm^{-2}. Parallel detachment experiments showed that at a shear stress of 15 dynes cm^{-2}, less than 20% of the cells remained attached to the endothelial monolayer. At detachment shear stresses of 10 dynes cm^{-2} or lower, however, more than 30% of cells remained attached with 100% of the cells retained at 1 dyne cm^{-2}. These experimental results were compared with physiological parameters for several different human vessel types. Findings are summarized in Table I. Here, the average diameter and wall shear stress for each vessel type are given. The wall shear stress was calculated from the average blood velocity, the vessel diameter, and the viscosity of plasma, or the fluid viscosity at the wall [45]. The wall shear stress was compared with results from the initial attachment and detachment experiments at various shear stresses to determine whether initial attachment and/or detachment could occur in each vessel type. Finally, using the vessel diameter, the possibility of cell lodging was determined by comparing vessel diameter with cell size. An average tumor cell diameter of 20 μm was used [46] which corresponds to the average diameter of the cancer cells used in this study.

As indicated in Table I, our results support the hypothesis that initial attachment of human breast cancer cells to endothelium is not

TABLE I Comparison of physiological parameters in various human vessel types with results from initial attachment and firm adhesion studies. Average values for diameters and shear stresses in the vessels were obtained from Ref. [45]

Vessel type	*D* (*μm*)	τ_w (*d*/*cm*^{2})	*Initial attachment*	*Firm adhesion*	*Lodging*
Arteries	4000	9.9	–	+	–
Arterioles	50	88	–	–	–
Capillaries	8	11	–	+	+
Venules	20	8.8	–	+	+
Veins	5000	1.76	–	+	–

likely to occur at the average physiological shear stress reported in the vessel types shown. In contrast, firm adhesion can occur at the shear stresses observed in the arteries, capillaries, venules, and veins. While 88 dynes cm^{-2} is beyond the limit of our experimental system, a single firm adhesion experiment performed at a detachment shear stress of 30 dynes cm^{-2} for each cell line revealed that less than 5% of cells remained firmly adherent to the endothelial monolayer (data not shown). Thus, it is unlikely that significant cellular retention would occur at an even higher shear stress of 88 dynes cm^{-2}. Before firm adhesion can occur, cancer cells must first lodge inside the vessel. The size comparison revealed that lodging is likely to occur only in capillaries and venules. The other vessel types have diameters much larger than the diameters of the cells and, hence, lodging would be unlikely except in cases of bifurcations where stagnation points could occur. Thus, results from initial attachment and detachment experiments together with physiological parameters suggest that during the metastatic cascade, the adhesion of human breast cancer cells to the endothelium is most likely to occur *via* a lodging-firm adhesion mechanism, and this adhesion could take place in capillaries or venules. These results agree with previous experimental observations of extravasation occurring most often in capillaries [47, 48]. In addition, Shioda *et al.* [49] reported entrapment and extravasation of melanoma cells in the capillary bed of a chick embryo chorio-allantoic membrane.

It should be noted that other factors may also contribute to the attachment of human breast cancer cells to the endothelium. The flow of blood through the circulatory system is pulsatile, not constant. Consequently, shear stresses at the low end of this pulse may approach 0.25 dynes cm^{-2}, where initial attachment was seen to take place in the experiments reported here. Thus, initial attachment of human breast cancer cells to the endothelium cannot be entirely excluded. Pulsatile flow may also influence both the initial attachment and firm adhesion processes as cancer cells attempt to adhere under shear stresses which are continually changing. In addition, vessels are elastic and compliant. Vessel elasticity could allow breast cancer cells to pass though vessels where they were predicted to lodge. Alternatively, vessel contraction during circulation may enable cancer cells to lodge in vessel sizes where lodging was not predicted. In addition, the cancer cell itself may deform and, thus, pass through vessels where it was predicted to lodge or deformation may lead to activation of the cell which could modify adhesion. A recent study by Shioda *et al.* [49] has shown entrapment and extravasation of melanoma cells in the chick embryo chorioallantoic membrane leads to alterations of gene expression. Many blood factors also have the potential to influence adhesion. For example, platelets present on either the cancer cell or the endothelium could enhance adhesion by presenting additional adhesion molecule interactions. Platelets could also cause cancer cell clumping, thus allowing clumps of cells to lodge in vessels where single cells would easily pass through [23].

At this point, it is unclear as to how important adhesive interactions are in lodging. Because the cell will probably initially occlude or partially occlude the vessel, the shear stress acting on the cell would be either zero or close to zero. In this case, the ability of the cell to extravasate becomes important and this process may require adhesive interactions that are distinct from interactions required to withstand shear stress. However, it is also possible that following the initial lodging, the cell adheres and spreads out on the endothelial layer. At this point, the vessel may no longer be occluded and the ability to withstand shear stress is important. Clearly, while these results from this paper provide a basis for understanding the role of initial attachment and firm adhesion on cancer cell interactions with the endothelium, much research must be done before the mechanism of adhesion between cancer cells and the endothelium can be conclusively determined.

In this study, the endothelial cell monolayer was activated with TNF-α prior to experimentation to increase expression of adhesion molecules on the surface. We chose to stimulate the cells with TNF-α because several *in vivo* sources of TNF-α exist. In addition, previous studies have shown that cancer cells do not initially attach to unstimulated endothelial cells [10, 44] at shear stresses as low as $0.25\,\text{dynes}\,\text{cm}^{-2}$ and detachment results are similar regardless of activation (unpublished results). Hence, these results indicate that even if the endothelium are not activated, lodging followed by adhesion still seems to be the governing mechanism.

Another objective of this study was to elucidate whether variations in adhesive strength with assay type and shear stress were dependent on metastatic potential. The attachment results showed that at a very low shear stress of 0.25 dynes cm^{-2}, approximately 100 highly-metastatic cells adhered to the monolayer compared with approximately 60 nonmetastatic cells. While this difference is significant, it can be concluded that a significant number of cells attached to the monolayer at this shear stress regardless of metastatic potential. In addition, it was shown that there was no significant difference between the detachment results when comparing the highly-metastatic cells with the nonmetastatic cells except at 10 dynes cm^{-2} where the nonmetastatic cells were approximately 50% more adherent and at 5 dynes cm^{-2} where the highly-metastatic cells approximately 15% more adherent. Even with these differences, the results suggest that with these specific breast cancer cells, a significant number of cells remained attached to the monolayer regardless of metastatic potential.

SUMMARY

In summary, use of the parallel-plate flow chamber as a controlled flow system allowed for the quantification of both initial attachment and firm adhesion of human breast cancer cells to endothelial monolayers at several relevant physiological shear stresses. A comparison of results from both initial attachment and detachment experiments with physiological parameters suggests that the adhesion of human breast cancer cells to endothelium most likely takes place *via* a lodging-firm adhesion mechanism in the capillaries or venules. However, it is important to note that other factors such as pulsatility may contribute to the attachment and, hence, initial attachment at low shear stresses in the absence of lodging cannot be eliminated as a possible mechanism. Further research is needed before the mechanism of adhesion between cancer cells and endothelium can be determined. Finally, it was shown that for these specific breast cancer cells, adhesion does not seem to correlate with metastatic potential. Both the initial attachment and firm adhesion results were similar for the highly-metastatic and nonmetastatic cell lines. Hence, while blocking the adhesion of highly metastatic cells may inhibit their ability to metastasize, adhesion properties alone do not provide an indication as to whether a cell is metastatic or nonmetastatic.

Acknowledgements

Melissa Moss was supported by a National Science Foundation Graduate Fellowship and a Dissertation Year Fellowship from the University of Kentucky Graduate School. We would also like to acknowledge Dr. Stephen Zimmer, Dept. of Microbiology and Immunology, University of Kentucky for helping us obtain cancer cells from Dr. Janet Price, M. D. Anderson Cancer Center, Houston, TX.

References

[1] Weidner, N. J., Semple, P., Welch, W. R. and Folkman, J., *The New England Journal of Medicine* **324**, 1–8 (1991).
[2] Pauli, B. U., Augustin-Voss, H. G., El-Sabban, M. E., Johnson, R. C. and Hammer, D. A., *Cancer and Metastasis Reviews* **9**, 175–189 (1990).
[3] Moss, M. A., Effect of TNF-alpha and shear stress stimuli on the adhesion of human breast cancer cells to endothelial monolayers. *Ph.D. Dissertation*, University of Kentucky, 2000.
[4] Honn, K. V. and Tang, D. G., *Cancer and Metastasis Reviews* **11**, 353–375 (1992).
[5] Tedder, T. F., Steeber, D. A., Chen, A. and Engel, P., *The FASEB Journal* **9**, 866–873 (1995).
[6] Scherbarth, S. and Orr, F. W., *Cancer Research* **57**, 4105–4110 (1997).
[7] Kojima, N., Handa, K., Newman, W. and Hakomori, S., *Biochemical and Biophysical Research Communications* **189**, 1686–1694 (1992).
[8] Kojima, N., Shiota, M., Sadahira, Y., Handa, K. and Hakomori, S., *J. Biological Chemistry* **267**, 17264–17270 (1992).
[9] Giavazzi, R., Foppolo, M., Dossi, R. and Remuzzi, R., *J. Clinical Investigation* **92**, 3038–3044 (1993).
[10] Tozeren, A., Kleinman, H. K., Grant, D. S., Morales, D., Mercurio, A. M. and Byers, S. W., *Internat. J. of Cancer* **60**, 426–431 (1995).
[11] Goetz, D. J., Brandley, B. K. and Hammer, D. A., *Annals of Biomedical Engineering* **24**, 87–98 (1996).
[12] Aigner, S., Ramos, C. L., Hafezi-Moghadam, A., Lawrence, M. B., Friederichs, J., Altevogt, P. and Ley, K., *FASEB Journal* **12**, 1241–1251 (1998).
[13] Weiss, L., *Cancer and Metastasis Reviews* **11**, 227–235 (1992).
[14] Weiss, L., *Invasion Metastasis* **14**, 192–197 (1994–95).
[15] Weiss, L., *J. Forensic Sciences* **35**, 614–627 (1990).
[16] Chamber, A. F., MacDonald, I. C., Schmidt, E. E., Koop, S. and Morris, V. L., *Cancer and Metastasis Reviews* **14**, 279–301 (1995).
[17] Thorlacius, H., Prieto, J., Raud, J., Gautam, N., Patarroyo, M., Hedqvist, P. and Lindbom, L., *Clinical Immunology and Immunopathology* **83**, 68–76 (1997).
[18] Lafrenie, R. M., Gallo, S., Podor, T. J., Buchanan, M. R. and Orr, F. W., *European J. of Cancer* **30A**, 2151–2158 (1994).
[19] Yamada, N., Chung, Y.-S., Sawada, T., Okuno, M. and Sowa, M., *Digestive Diseases and Sciences* **40**, 1005–1012 (1995).
[20] Mattila, P., Majuri, M.-L. and Renkonen, R., *Internat. J. of Cancer* **52**, 918–923 (1992).
[21] Kawaguchi, S., Kikuchi, K., Ishii, S., Takada, Y., Kobayashi, S. and Uede, T., *Japanese J. of Cancer Research* **83**, 1304–1316 (1992).

[22] Bliss, R. S., Kirby, J. A., Browell, D. A. and Lennard, T. W. J., *Clinical and Experimental Metastasis* **13**, 173–183 (1995).
[23] Abecassis, J., Millon-Collard, P., Klein-Soyer, C., Nicora, F., Fricker, J.-P., Beretz, A., Eber, M., Muller, D. and Cazenave, J.-P., *Internat. J. of Cancer* **40**, 525–531 (1987).
[24] Lauri, D., Giovanni, C. D., Biondelli, T., Lalli, E., Landuzzi, L., Facchini, A., Nicoletti, G., Nanni, P., Dejana, E. and Lollini, P.-L., *Brit. J. of Cancer* **68**, 862–867 (1993).
[25] Hirasawa, M., Shijubo, N., Uede, T. and Abe, S., *Brit. J. of Cancer* **70**, 466–473 (1994).
[26] Vennegoor, C. J. G. M., VanDeWiel-VanKemenade, E., Huijbens, R. J. F., Sanchez-Madrid, F., Melief, C. J. M. and Figdor, C. G., *J. of Immunology* **148**, 1093–1101 (1992).
[27] Tang, D. G., Diglio, C. A. and Honn, K. V., *Cancer Research* **54**, 1119–1129 (1994).
[28] Tomita, Y., Saito, T., Saito, K., Oite, T., Shimizu, F. and Sato, S., *Internat. J. of Cancer* **60**, 753–758 (1995).
[29] Martin-Padura, I., Mortarini, R., Lauri, D., Bernasconi, S., Sanchez-Madrid, F., Parmiani, G., Mantovani, A., Anichini, A. and Dejana, E., *Cancer Research* **51**, 2239–2241 (1991).
[30] Lafrenie, R., Shaughnessy, S. G. and Orr, F. W., *Cancer and Metastasis Reviews* **11**, 377–388 (1992).
[31] Patton, J. T., Menter, D. G., Benson, D. M., Nicolson, G. L. and McIntire, L. V., *Cell Motility and the Cytoskeleton* **26**, 88–98 (1993).
[32] Tozeren, A. H., Kleinman, K., Wu, S., Mercurio, A. M. and Byers, S. W., *J. Cell Science* **107**, 3153–3163 (1994).
[33] Felding-Habermann, B., Habermann, R., Saldivar, E. and Ruggeri, Z. M., *J. Biological Chemistry* **271**, 5893–5900 (1996).
[34] McLeskey, S. W., Kurebayashi, J., Honig, S. F., Zwiebel, J., Lippman, M. E., Dickson, R. B. and Kern, F. G., *Cancer Research* **53**, 2168–2177 (1993).
[35] Price, J. E., *Breast Cancer Research and Treatment* **39**, 93–102 (1996).
[36] Zhang, R. D., Fidler, I. J. and Price, J. E., *Invasion Metastasis* **11**, 204–215 (1991).
[37] Carswell, E. A., Old, L. J., Kassel, R. L., Green, S., Fiore, N. and Williamson, B., *Proc. National Academy of Science, USA* **72**, 3666–3670 (1975).
[38] Pober, J. S. and Cotran, R. S., *Physiological Reviews* **70**, 427–451 (1990).
[39] Mantovani, A., Bussolino, F. and Dejana, E., *FASEB Journal* **6**, 2591–2599 (1992).
[40] Swerlick, R. A., Lee, K. H., Li, L.-J., Sepp, N. T., Caughman, S. W. and Lawley, T. J., *J. Immunology* **149**, 698–705 (1992).
[41] Weller, A., Isenmann, S. and Vestweber, D., *J. Biological Chemistry* **267**, 15176–15183 (1992).
[42] Bischoff, J. and Brasel, C., *Biochemical and Biophysical Research Communications* **210**, 174–180 (1995).
[43] Haraldsen, G., Kvale, D., Lien, B., Farstad, I. N. and Brandtzaeg, P., *The Journal of Immunology* **156**, 2558–2565 (1996).
[44] Moss, M. A., Zimmer, S. and Anderson, K. W., *Anticancer Research* **20**, 1425–1434 (2000).
[45] Bergel, D. H., *Cardiovascular Fluid Dynamics* (Academic Press, New York, 1972), Vol. 2.
[46] Crissman, J. D., Cerra, R. F. and Sarkar, F., *Microcirculation in Cancer Metastasis* (CRC Press, Boston, 1991), pp. 205–215.
[47] Chew, E. C., Josepheson, R. L. and Wallace, A. C., *Fundamental Aspect of Metastasis* (American Elsevier Publishing Company Inc, New York, 1976), pp. 121–150.

[48] Koop, S., MacDonald, I. C., Luzzi, K., Schmidt, E. E., Morris, V. L., Grattan, M., Khokha, R., Chambers, A. R. and Groom, A. C., *Cancer Research* **55**, 2520–2523 (1995).
[49] Shioda, T., Munn, L. L., Fenner, M. H., Jain, R. K. and Isselbacher, K. J., *American J. Pathology* **156**, 2099–2112 (1997).

Cell–Cell Adhesion of Erythrocytes

F. R. ATTENBOROUGH and K. KENDALL

Birchall Centre, Keele University, Keele, Staffs UK ST5 5BG

Three different species of red blood cells have been tested by a new image analysis method to determine cell–cell adhesion with a high level of precision. Various literature sources have suggested that horse and rat erythrocytes adhere to form aggregates more readily than human red cells but this paper provides the first accurate quantification of this phenomenon. Addition of surface active agents such as glutaraldehyde, papain and fibronectin was also tested in order to measure the effects on cell-to-cell adhesion. Glutaraldehyde reduced adhesion whereas both papain and fibronectin increased it.

1. INTRODUCTION

Several measurement methods for determining cell adhesion have been reported over the years. Micro-pipette techniques [1, 2] gave the most direct observations, the cells being sucked onto glass probes and then pushed into contact to obtain deformation and adhesion results. The atomic force microscope has been employed more recently [3–5] to give a direct contact and measurement on the cell membrane. This has the distinct advantage that the probe can be located with nanometer precision in order to find patchy adhesion across the membrane surface. An innovation over the past three years has been the use of laser tweezers [6, 7]. In this test, a cell was held in a laser beam and brought into contact with another cell, also held by a laser. Very fine nano-manipulation of the ces was possible in this method, although the adhesion measurements were not simple to interpret. It was claimed that single macromolecular bonds could be detected, in principle, by this method. A similar claim was made by Bongrand [8, 9] for his cell flow device. He observed single cells moving over a surface in a controlled shear flow. The periodic stopping of the cell at the wall was said to be a measure of single molecule adhesion.

In 1998, a more interesting method of obtaining cell adhesion was reported, requiring no apparatus to produce the adhesion events [10]. The basis of this new idea was statistical mechanics. According to this theory, cells should move randomly with Brownian motion and should attach to neighbouring cells stochastically as random collisions occur. Doublets of cells are then produced, surviving for a certain time before Brownian collisions of sufficient force disrupt the adhesive bond between the cells. The number of cell doublets at equilibrium should, therefore, be a direct measure of cell adhesion. A high number of doublets indicates strong adhesion, whereas no doublets signify zero adhesion. This theory was verified by measuring human red cell dispersions by both optical and Coulter Counter methods [10–12]. An important feature emerging from these results was the dependence of the doublet numbers on the concentration. More doublets form at higher concentrations even though the adhesion remains constant. Therefore, it is essential in comparing cell adhesion to work at constant, known volume fraction of cells.

The purpose of this paper is to extend the studies above to compare the cell–cell adhesion of human erythrocytes with that of horse red cells, and that of other species. Various reports have suggested that horse red cells are more adherent to each other than human cells, but such observations have not been quantified [13, 14]. This paper provides the first precise measurements of horse red-cell-to-red-cell adhesion by the new Brownian method and also includes rat erythrocytes for comparison. The measurements have been interpreted using the theory outlined below.

2. THEORY

Consider a dilute dispersion of uniform spherical particles as shown in Figure 1.

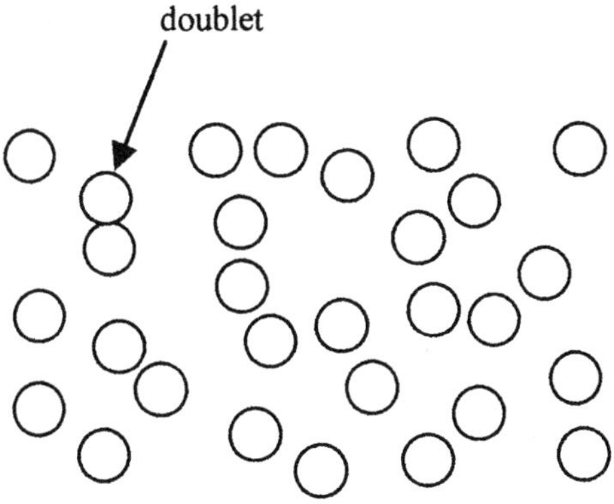

FIGURE 1 Dispersion of spherical particles.

These spheres experience Brownian motion and, therefore, diffuse in all directions, causing collisions between the particles. If there is adhesion between the particles, then a collision has a chance of creating a doublet; that is, two particles adhering together at the point of contact. If the adhesive bond is weaker than kT, then thermal collisions can break this bond in a period of time. The spheres will then separate and move apart. Thus, there is a dynamic equilibrium between joining and separation, giving a certain number of doublets in the suspension at equilibrium, after a suitable time has elapsed for diffusion to take place. High adhesion should give a larger number of doublets and lower adhesion a smaller number. Hence, there is a definite connection between sphere adhesion and the equilibrium number of doublets observed in a dilute suspension.

Of course, there are several assumptions in this argument. The main premise is that the spheres are all identical. This is not true of red cells which are known to have distributions of various molecular species on their surfaces. However, it is possible, in principle, to filter out any rogue doublets formed by unusually tacky cells. Equilibrium can then be re-established. Repeating this filtering and equilibration procedure several times should lead to a point where the remaining cells are more nearly equal.

A second assumption is that the cells are spherical and equal in diameter. In fact, human red cells are dimpled and range in size between 6 and 8 μm. The errors in the argument caused by such problems are not yet known but are being investigated by computer modeling [15].

The most interesting consequence of the above idea, that red cell adhesion may be measured by observing the number of doublets at equilibrium in a dilute suspension, is that an exact mathematical solution can be found under certain circumstances, depending on the interaction between spheres when they collide. The simplest situation is that shown in Figure 2, where a particle approaches its neighbour at constant speed until, at a certain separation, the particles are attracted to each other with an energy, ε. If this energy remains constant until the spheres touch rigidly at the point of contact, then the square well potential is revealed. The approaching sphere travels at constant speed, is accelerated into the potential well, reflects rigidly on contact, and then is decelerated as the particles move apart. This "hard sphere square well", which was first used by Alder and Wainwright [16] in 1961, can be solved exactly to predict the number of doublets in a suspension.

The mathematical result is that the ratio of doublets to singlets, N_2/N_1, is proportional to the volume fraction, ϕ, of the cells and depends on the range, λ, and the energy, ε, of the well according to the

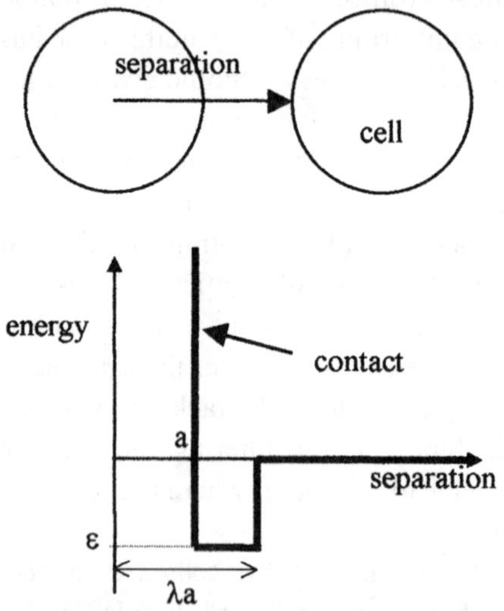

FIGURE 2 Interaction energy between approaching spheres.

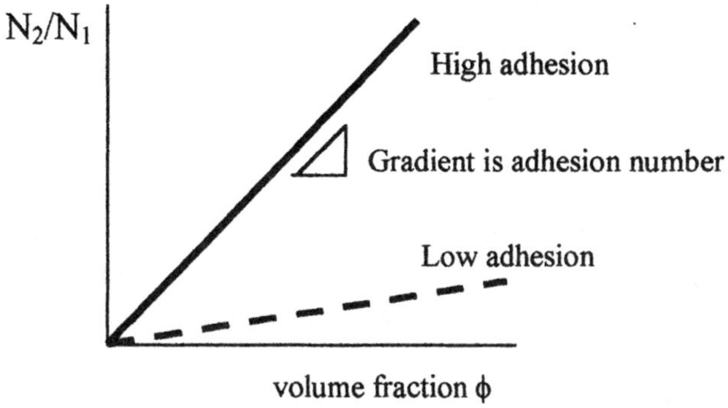

FIGURE 3 Defining the adhesion number for cell – cell interaction.

equation below.

$$NN_2/N_1^2 = 4\phi(\lambda^3 - 1)\exp(\varepsilon/\mathrm{kT}) \approx N_2/N_1 \tag{1}$$

The conclusion of this argument is that a plot of doublet-to-singlet ratio *versus* particle volume fraction should yield a straight line passing through the origin. The gradient of the line is a measure of the adhesion which depends on range and energy of the interactions. Thus, a high gradient signifies high adhesion and a low gradient low adhesion as shown below in Figure 3. Thus, an adhesion number can be defined as the gradient of this plot, to give a measure of the bonding of the cells. The experimental objective of this paper is to define this non-dimensional adhesion number for three different species of red cells, horse, rat and human.

3. EXPERIMENTAL MEASUREMENT AND RESULTS

Blood cells were prepared from three species, human blood from North Staffordshire Hospital, fresh horse blood in EDTA and fresh rat blood from Central Animal Pathology Ltd. Each blood sample was washed six to seven times in phosphate-buffered saline to remove the non-red-cell components, before suspending in physiological saline solution, then examined by both optical and Coulter tests. Tests were carried out immediately after washing since the cells degraded with time. Each species of cell was treated in three ways to judge the effect of surface adhesion molecules; by adding glutaraldehyde, fibronectin and papain.

The optical apparatus is shown in Figure 4. The cells were placed in an accurately defined 10 μm space within a glass chamber which was imaged using a video microscope at 40x magnification. Each cell could then be clearly seen moving around with Brownian movement, while not overheating as occurred at 100x magnification. Pictures of the cells were taken at random locations in the chamber and the numbers of doublets and singlets were counted by the image analysis software Sigmascan Pro [17]. By taking the ratio of doublets to singlets, the adhesion number was obtained. Cells were observed over a period of 30 minutes. They did not stick significantly to the walls of the glass cell. The assumption was made that the cells were moving randomly in three dimensions.

The collision and adhesion events could be observed in experiments as shown in Figure 5, which shows one field of view seen 30 seconds apart. Three doublets remained unchanged in this period, two broke up, while four new doublets and one new triplet were formed. The dynamic equilibrium could, thus, be evaluated.

The second set of experiments to measure the doublet numbers used the Coulter Counter, which was set up in standard mode to count the individual red cells, as shown by the results of Figure 6a. The strong peak showed a symmetrical distribution of single cells at a volume fraction near 10^{-5}.

At higher concentration, a shoulder appeared at a 13% higher diameter, and this was interpreted as a doublet peak. At still higher

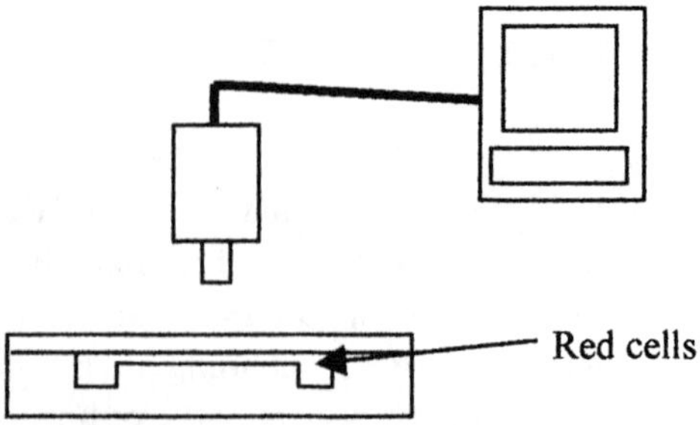

FIGURE 4 Video camera apparatus for observing red cells.

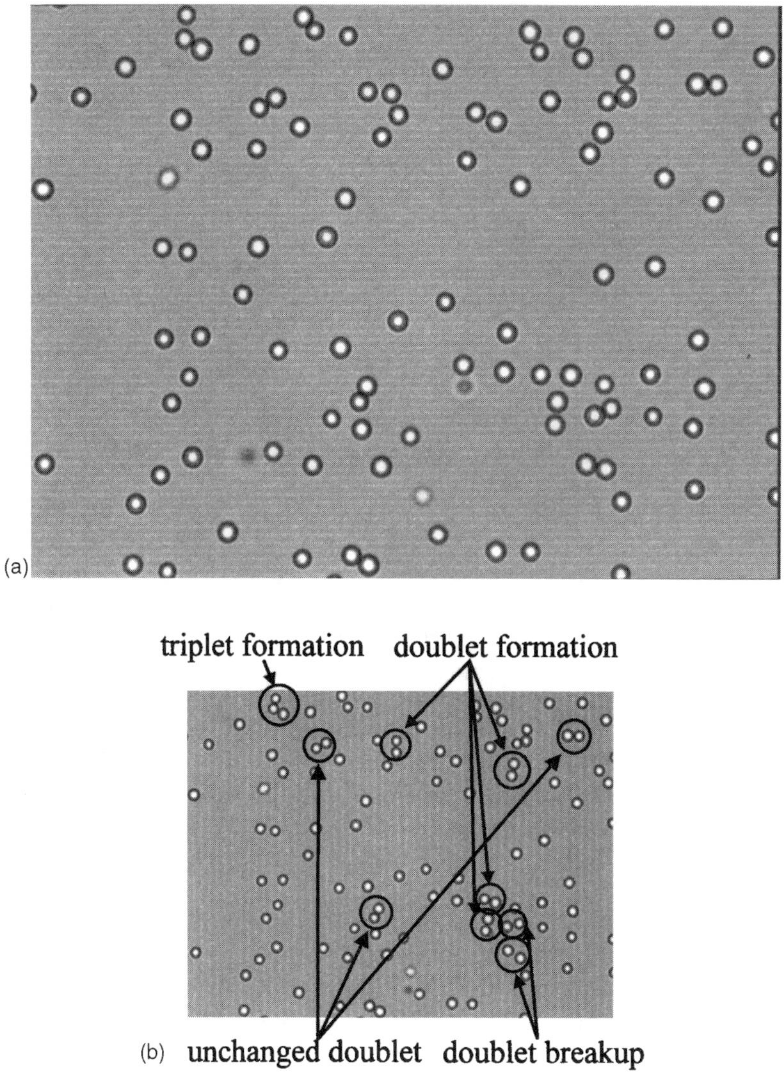

FIGURE 5 Images of horse red cells in Isoton solution; (a) time zero; (b) 30 seconds later.

concentration of the red cells, the shoulder increased in size, indicating that more doublets formed as the blood cells became more numerous. The number of doublets was measured and divided by the singlet peak

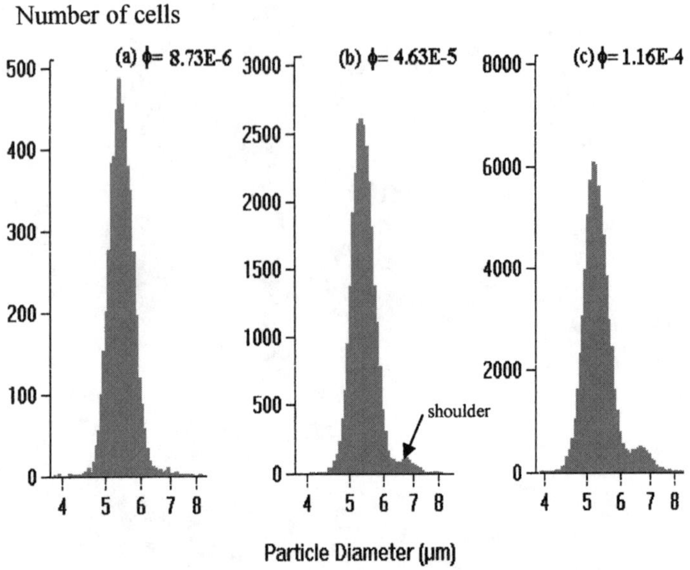

FIGURE 6 (a) Coulter counter results for human red cells; (b) Result at higher concentration showing shoulder; (c) Larger shoulder at higher concentration.

to obtain the ratio N_2/N_1. This was then plotted as a function of cell volume fraction to give the curve shown in Figure 7. The results showed the doublets increasing in proportion to concentration and allowing the adhesion number to be found by determining the gradient. For human cells this was 420.

Horse and rat erythrocytes were then tested in the same way and shown to give significantly higher adhesion. Baskurt *et al.* [13] have shown that the aggregation of such cells is increased over human cells, but volume fraction effects were not taken into account. Popel *et al.* [14] recognised that horse cells stick better and this was attributed to the athletic nature of the animal. Table I quantifies the difference of adhesion in terms of the adhesion number $N_2/N_1\phi$.

These results show conclusively that rat cells are almost twice as sticky as human red cells, while horse erythrocytes are almost twice as adhesive as rat cells. Whether this can be explained in terms of the higher energy of the bonds, as defined by Eq. (1), or in larger range of bonds remains to be determined.

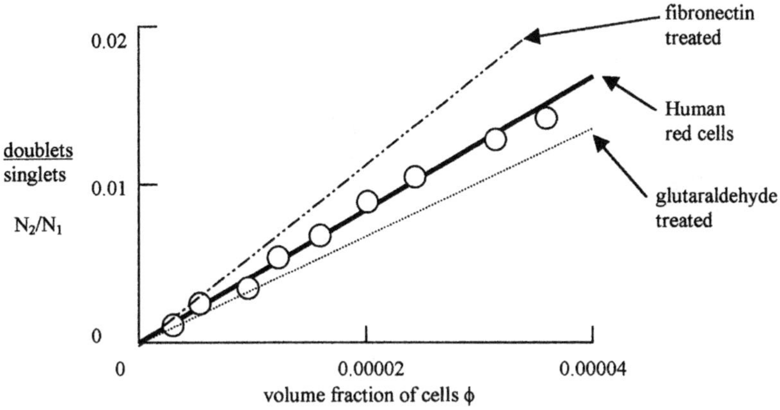

FIGURE 7 Increase in doublets with higher concentration of human red cells.

TABLE I Comparison between adhesion of various red cells

Animal	*Adhesion number* $N_2/N_1\phi$
Horse	1488 ± 200
Rat	750 ± 4
Human	420 ± 5

TABLE II Effect of surfactants on horse red cell adhesion

Horse cell treatment	*Adhesion number* $N_2/N_1\phi$
Isoton	1279 ± 203
Isoton+glutaraldehyde	1020 ± 162
Isoton+fibronectin	1399 ± 184
Isoton+papain	1513 ± 295

Addition of surface active molecules to the cell suspension was also studied. The results for human cells are illustrated in Figure 7 which shows that fibronectin increased the adhesion whereas glutaraldehyde reduced it. The effect of surfactants on horse erythrocytes is shown in Table II.

The control sample of horse cells in Isoton (Coulter Electronics, Luton, UK) showed somewhat weaker adhesion than the sample shown in Table I. Such variation was found to be common in different samples of horse blood. Differences between animals in type, age, *etc.*, and also in bod cell conditioning, had a distinct influence which will be

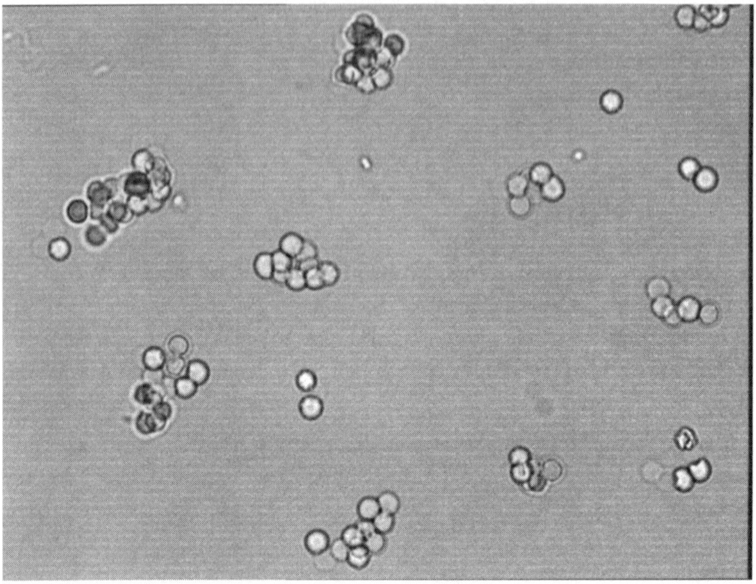

FIGURE 8 Large aggregates of horse red cells after soaking in papain.

described in separate papers. It is evident from the results that glutaraldehyde reduced the adhesion by about 25%, whereas fibronectin increased the adhesion by 10% and papain by 20%, changes which were comparable with the effects seen on human red cells but disappointingly small compared with the effects anticipated.

When the red cells were left in papain at high concentration, much larger aggregates of cells were observed, as shown in Figure 8. It is clear that adhesion has been changed substantially in this experiment. The large aggregates were also not dispersible on dilution with saline solution in this case, suggesting that the reversibility of the adhesion process had been compromised. Under such irreversible conditions, the equilibrium theory of Eq. (1) is no longer valid.

4. CONCLUSIONS

The adhesion of red blood cells has been tested by both optical analysis and by Coulter techniques. In particular, the differences between horse, rat and human cell – cell adhesion have been quantified with improved precision. Horse red cells were almost twice as adhesive as rat erythrocytes and almost four times as adhesive as human cells, under the conditions used in the experiments. Addition of surface active molecules has also been tested. Glutaraldehyde treatment decreased the adhesion significantly, whereas fibronectin and papain increased adhesion substantially, causing strong aggregation and irreversibility at high concentration.

References

[1] Evans, E. and Rawicz, W., *Phys. Rev. Lett.* **79**, 2379 – 82 (1997).
[2] Eichenbaum, G. M., Kiser, P. F., Simon, S. A. and Needham, D., *Macromol.* **31**, 5084 – 93 (1998).
[3] Evans, E. and Ritchie, K., *Biophys. J.* **72**, 1541 – 55 (1997).
[4] Israelachvili, J. N., Leckband, D., Schmitt, F.-J., Zasadzinski, J., Walker, S. and Chiruvolu, S., *Studying Cell Adhesion* (Springer Verlag, Berlin, 1994), chap. 3.
[5] Rotsch, C., Braet, F., Wiser, E. and Radmacher, M., *Cell. Biol. Int.* **21**, 685 – 96 (1997).
[6] Luo, Z. P. and An, K. N., *J. Biomech.* **31**, 1075 – 79 (1998).
[7] Bronkhorst, P. J. H., Grimbergen, J., Brakenhoff, G. J., Heethaar, R. M. and Sixma, J. J., *Brit. J. Haematology* **96**, 256 – 258 (1997).
[8] Pierres, A., Tissot, O. and Bongrand, P., *Studying Cell Adhesion* (Springer Verlag, Berlin, 1994), chap. 13.
[9] Pierres, A., Benoliel, A. and Bongrand, P., *Cell Adhesion and Communication* **5**, 375 – 95 (1998).
[10] Kendall, K., Liang, W. and Stainton, C., *Proc. R. Soc. Lond.* **A454**, 2529 – 33 (1998).
[11] Kendall, K. and Liang, W., *Brit. Cer. Trans.* **96**, 92 – 95 (1997).
[12] Kendall, K. and Liang, W., *Colloids and Surfaces* **131**, 193 – 201 (1998).
[13] Baskurt, O. K., Farley, R. A. and Meiselman, M. J., *Am. J. Physiol.: Heart and Circulatory Physiology* **42**, H2604-12 (1997).
[14] Popel, A. S., Johnson, P. C., Kavenesa, M. V. and Wild, M. A., *J. Appl. Physiol.* **77**, 1790 – 94 (1994).
[15] Stainton, C., *Ph.D. Thesis*, University of Keele, Submitted 1999.
[16] Alder, B. J. and Wainwright, T. E., *J. Chem. Phys.* **31**, 459 – 66 (1959).
[17] Jandel Scientific, Jandel GmbH, Schimmelbuschstrasse 25, 40699 Erkrath, Germany.

Particle-induced Phagocytic Cell Responses are Material Dependent: Foreign Body Giant Cells *Vs.* Osteoclasts from a Chick Chorioallantoic Membrane Particle-implantation Model

L. C. CARTER, J. M. CARTER, P. A. NICKERSON,
J. R. WRIGHT and R. E. BAIER

School of Dental Medicine and School of Medicine and Biomedical Sciences, State University of New York at Buffalo, Buffalo, NY 14214-3007, USA

There is increasing concern about particles generated from wear-prone implants that are placed in body tissues, including artificial hip, knee, and jaw joints. Although phagocytes and foreign body giant cells are associated with inhaled or embedded particulate debris, some particles also induce bone digestion by eliciting the differentiation and proliferation of highly specialized osteoclastic cells. This report describes the differential phagocytic cellular responses to four implant-related types of ground, model wear particles in a live-egg cell-response model, as implants to the chick chorioallantoic membrane (CAM): polymethylmethacrylate (PMMA), a main constituent of some temporomandibular joint (TMJ) implants and orthopedic cements used to retain artificial hips and knees; Proplast-HA, an implantable composite of polytetrafluoroethylene (PTFE) and degradable mineral (hydroxyapatite) that has been associated with bone erosion around failed TMJ implants; talc, a nondegradable mineral sometimes found in tissues as a contaminant from talc-coated surgical gloves; and authentic bone, known to induce the formation of osteoclastic cells. Light and electron microscopy of CAM tissues harvested, sectioned and stained with special reagents for the enzymes tartrate-resistant acid phosphatase (TRAP) and tartrate-resistant adenosine triphosphatase (TrATPase), and for the osteoclast-specific antigen 121F, showed that only authentic bone and the degradable HA-rich particles induced osteoclast formation. From these results, and supporting data with polypropylene particles, it is concluded that nonbiodegradable polymer particles, alone, do not induce bone dissolution. Inert polymers do induce foreign body giant cells without the external mineral digestion qualities unique to osteoclasts, however. The chick embryo model system allows quick and affordable examination of material-dependent differences in phagocytic cellular responses to implant wear debris and particles from various occupational environments.

INTRODUCTION AND BACKGROUND

Particles enter the human body by many routes and from many sources. One increasing source of concern is mechanical wear debris from various alloplastic joint replacement devices not originally expected to break down under use. In certain circumstances, foreign body giant cells appear in tissues that respond to these particles. The cells then engulf and attempt to digest the particles, internally, by merging them with hydrolytic enzymes in protected membrane-bound compartments called lysosomes. Sometimes, specific bone-digesting cells – osteoclasts – also appear with unique capabilities for external digestion of large particles resistant to engulfment. It is very important to know the differences among these cell types and their inducing agents, since bone loss can lessen a patient's ability to maintain an implant securely in place.

Skeletal defects may arise from congenital anomalies, from trauma, as the sequelae of infections/inflammatory diseases or from the treatment of neoplastic processes. Regardless of their etiology, such defects frequently lead people to attempt to improve their functional and/or aesthetic status by use of a variety of autogenous and alloplastic implants. Autogenous implants, tissues taken from other places within a person's own body, are variable in terms of their availability and success rates; certainly the additional surgery needed to harvest these tissues increases morbidity. Allografts, tissues donated from other persons, raise concerns regarding infectious disease and graft/host immunocompatibility. Thus, beginning in the 1960s, biomedical engineers and surgeons increasingly turned to a host of alloplastic materials both metallic (*e.g.*, cobalt-chromium-molybdenum alloy, Vitallium [Howmedica, West Berlin, NJ]) and polymeric (*e.g.*, medical grade silicone rubber, Silastic [Dow Corning Corporation, Midland, MI]) for the fabrication of substitute human joints.

In cases of hip, knee, and temporomandibular joint (TMJ) replacements, cobalt-chromium alloy is sometimes cemented in place with polymethylmethacrylate (PMMA) "bone cement", or PMMA is used for the actual articulating surfaces, that could spall fugitive particles and produce tissue trauma. We showed, earlier, how physical breakdown and particle production from Silastic used in TMJ interpositional implants triggered local foreign body giant cell responses accompanied by migration of particles to regional lymph nodes [1].

Other contemporary alloplastic implants incorporated a series of porous composite materials called Proplast (Vitek, Inc., Houston, TX) [2]. Based on mixtures of polytetrafluoroethylene (PTFE) with carbon or alumina (Proplast II), and later with hydroxyapatite (Proplast-HA), these composites were used for facial augmentation procedures as well as TMJ implants. Many investigators have since reported that particles shed from Proplast composites could elicit significant foreign body inflammatory giant cell responses which, in the case of Proplast HA, also were associated with severe osseous resorption [2–7].

The unfortunate clinical results were significant posterior migration of chin implants through bone, erosion of condyles and articular eminences and, in some cases, perforation of the middle cranial fossa and dura mater [3–5], exposing the brain. Marked foreign body giant cell reactions were reported to occur even in sites such as the middle ear, where tissue motion adjacent to the implant was not a plausible explanation for effects seen [6]. These particle-induced responses were so universal and severe that Proplast interpositional TMJ implants were withdrawn from the market in 1988, with the Food and Drug Administration (FDA) warning that total TMJ devices could present similar risks [7].

Today, thousands of patients retain devices containing bone cement in orthopedic prostheses and PMMA articulations in TMJ implants, and Proplast-HA in facial augmentations or TMJ reconstructions. It is important that the basic biologic response to particles of these materials be ascertained. This report focuses on the different abilities of PMMA particles and Proplast-HA particles to induce foreign body giant cells with osteoclast-like (osteolytic, bone dissolving) properties, using the embryonated chicken egg chorioallantoic membrane model developed by Krukowski and Kahn [8]. Similarly-sized particles of authentic bone (from chick tibiae) were used as "digestable" particle controls, and talc particles (as used to lubricate surgical gloves, and often inadvertently entering into wound sites) were used as non-degradable mineral controls. These results are extracted from a comprehensive series of particle studies in the same model, in which confirming data were obtained with particles of non-degradable polypropylene (suture material) and differentially degradable compositions of polyglycolic acid and polylactic acid [9].

MATERIALS AND METHODS

NIH guidelines for the care and use of laboratory animals (NIH publication #85-23 Rev. 1985) were observed during the conduct of this study.

Chorioallantoic Membrane Implantation

Chick tibiae, Proplast-HA, polymethylmethacrylate (PMMA) and talc were comminuted by cryogenic milling and sieved to the 75 – 150 μm particle size range. Talc was received as a nominally 40 μm particle size powder with many multi-particle agglomerates of larger size and was treated similarly to ensure uniformity of handling of all specimens. Two-milligram samples were sterilized by exposure to gamma irradiation at a delivered minimum dose of 3.51 Mrad and a delivered maximum dose of 4.07 Mrad for 275 minutes (Isomedix, Whippany, NJ). On day 5 of their incubation, 48 embryonated white Leghorn chicken eggs were windowed according to the methods described by Krukowski and Kahn [8]. After an additional 4 days, the chorioallantoic membranes (CAMs) were implanted with one of the particulate samples. After 9 days of implantation (day 18 of incubation), the induced cell plaques were harvested and analyzed by light microscopy, histochemistry, immunocytochemistry and electron microscopy, and were fragment-cultured using a bone slice resorption assay [10 – 12]. In addition, tibial medullary bone from 3 chick hatchlings, maintained on a low calcium diet for 4 weeks, served as an osteoclast-positive control for these assays.

Light Microscopy

Cell plaques were fixed in 10% neutral buffered formalin, dehydrated in a graded series of alcohols and embedded in paraffin. Sections were cut at 5 μm and stained with hematoxylin and eosin.

Histochemistry

Cell plaques destined for histochemistry were fixed in 2% glutaraldehyde in 0.1 M sodium cacodylate buffer with 7% sucrose at pH 7.4, embedded in JB-4 resin (Polysciences, Warrington, PA), sectioned at 3 μm and stained for either tartrate-resistant acid phosphatase (TRAP) or tartrate-resistant adenosine triphosphatasae (TrATPase) [13, 14]. Sections incubated without substrate or in the presence of sodium-ortho-vanadate served as negative controls. Sections of chicken spleen served as a positive control for TRAP and as a negative control for TrATPase reactivity. Reagents for histochemistry were purchased from Sigma Chemical Co. (St. Louis, MO).

Immunocytochemistry

Cell plaques for study by immunocytochemistry were fixed in periodate-lysine-paraformaldehyde and cryosectioned at 5 μm [15]. Using a biotinylated streptavidin peroxidase-antiperoxidase staining protocol, cryosections were incubated with primary monoclonal antibody 121F raised against chicken osteoclasts at a dilution of 1 : 500 and secondary antibody goat-antimouse lgG (1 : 250) [15]. Sections incubated in the absence of primary antibody served as negative controls. Reagents for immunocytochemistry and cell culture were obtained from GIBCO/BRL Life Technologies Inc. (Bethesda, MD).

Transmission Electron Microscopy

CAM plaques were fixed in 2% glutaraldehyde in 0.1 M sodium cacodylate buffer at pH 7.4, post-fixed in 1% osmium tetroxide and embedded in Epon-Araldite (Fullam, Inc., Schenectady, NY). Thin sections (0.05 μm) were cut with a diamond knife, mounted on copper grids, and stained with aqueous uranyl acetate and lead citrate.

Bone Slice Resorption Assay

For the bone slice resorption assay, additional CAM cell plaques or medullary bone from hypocalcemic chicks were fragment-cultured on devitalized bovine cortical bone wafers in α-MEM in HEPES buffer supplemented with penicillin and streptomycin, 10% NCS (calf serum)

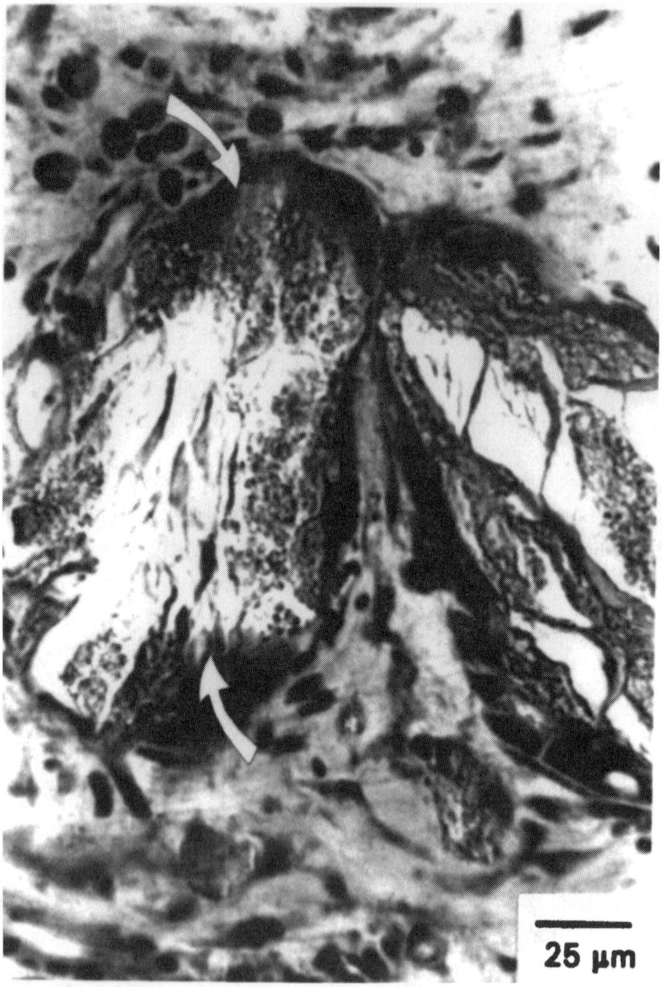

FIGURE 1a Amphophilic transition zones between giant cells and Proplast-HA are viewed (arrows) along with Pseudopodial extension of cytoplasm into the composite. Hematoxylin and eosin staining used.

at 37°C in a humidified atmosphere of 95% air and 5% CO_2 [12]. After incubation for 24 or 48 hours, wafers were fixed in 2% glutaraldehyde in 0.1 M sodium cacodylate buffer at pH 7.3, stained with 1% toluidine blue in 0.5% disodium tetraborate and examined by brightfield transmitted light microscopy for osteoclast assessment. On an additional set of wafers, cells were stripped by immersion in 5.25% NaOCl for 25 minutes followed by brief ultrasonication. These wafers then were sputter-coated with gold under an argon atmosphere for 1.5 minutes at 20 mA in a vacuum evaporator. Wafers were examined for the presence of resorption pits using darkfield reflected

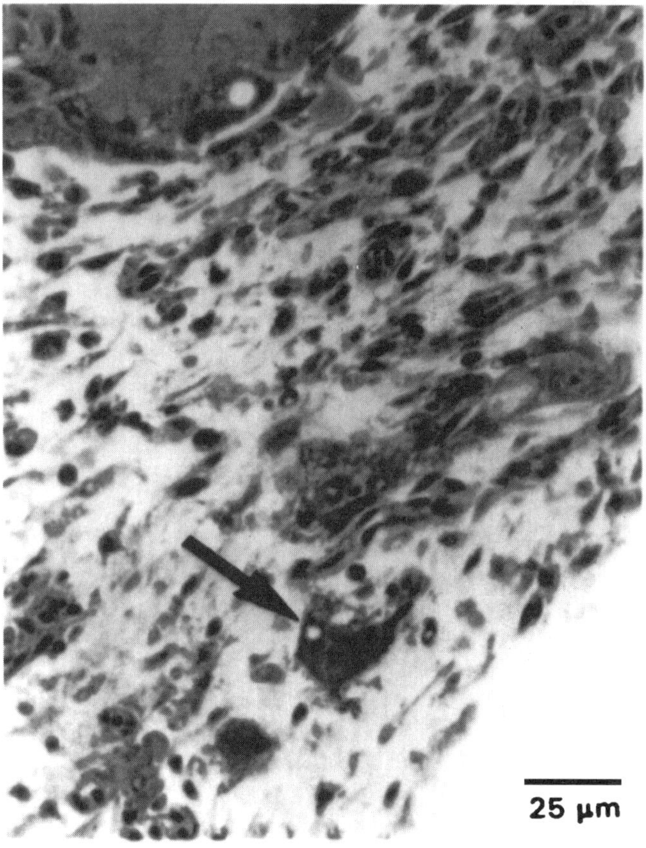

FIGURE 1b Photomicrograph reveals phagocytosis of talc particles by several polykarya including one that already harbors internalized talc (arrow). Hematoxylin and eosin staining used.

light microscopy [16]. Wafers similarly cultured in the absence of cells served as negative controls.

RESULTS

Chick Embryo Plaque Qualities

CAM foreign body giant cells induced by implantation of Proplast-HA showed an amphophilic zone of interdigitation between the cytoplasm

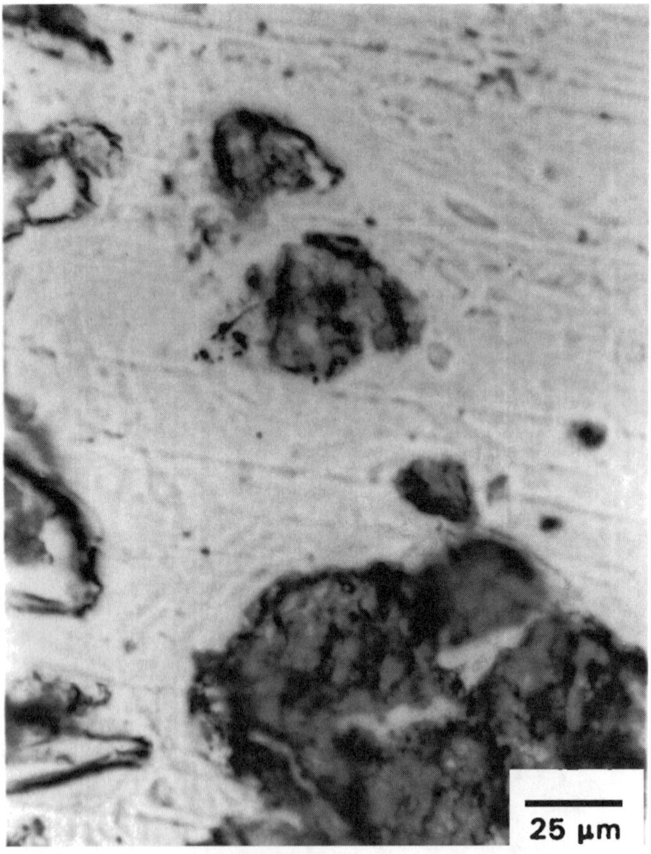

FIGURE 2 A fine, dark, granular precipitate is viewed in and vicinal to polykarya induced by implantation of particulate Proplast-HA. TrATPase, no counterstain.

and the composite, by light microscopy (Fig. 1a). In this zone, both the normal eosinophilic character of the cell cytoplasm and the normal grayish granular refractile nature of the particulate Proplast-HA material were lost. The cells' cytoplasm areas appeared foamy. Although implantation of PMMA and talc onto the CAM also elicited the formation of numerous foreign body giant cells, the appearance of the interfacial zone between particles and cells was quite different. Instead of an amphophilic transition zone, the interfaces between the PMMA or talc particles and the cells were crisp and distinct. Many of the active cells were photographed as they engaged in engulfment, phagocytosis and internalization of the PMMA or talc particles (Fig. 1b).

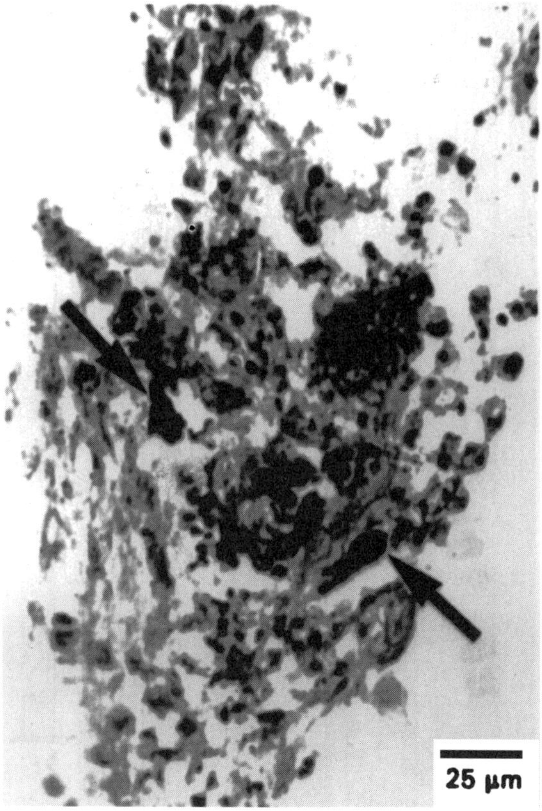

FIGURE 3a Moderate reactivity for the 121F antigen is observed in multinucleated and some mononuclear cells associated with implanted particulate Proplast-HA (arrows).

Special Stain Outcomes

Endosteal osteoclasts and many mononuclear cells from tibial medullary bone of calcium-deficient chick hatchlings displayed strong staining for TRAP, visualized as a granular maroon-colored reaction product. Staining of these cells for TrATPase, viewed as a dark brown stain with a black precipitate, also was intense, except that fewer mononuclear cells reacted, and the background osseous staining was less intense for TrATPase than for TRAP. Osteoclasts, induced after implantation of particulated tibiae on CAMs, displayed a similar

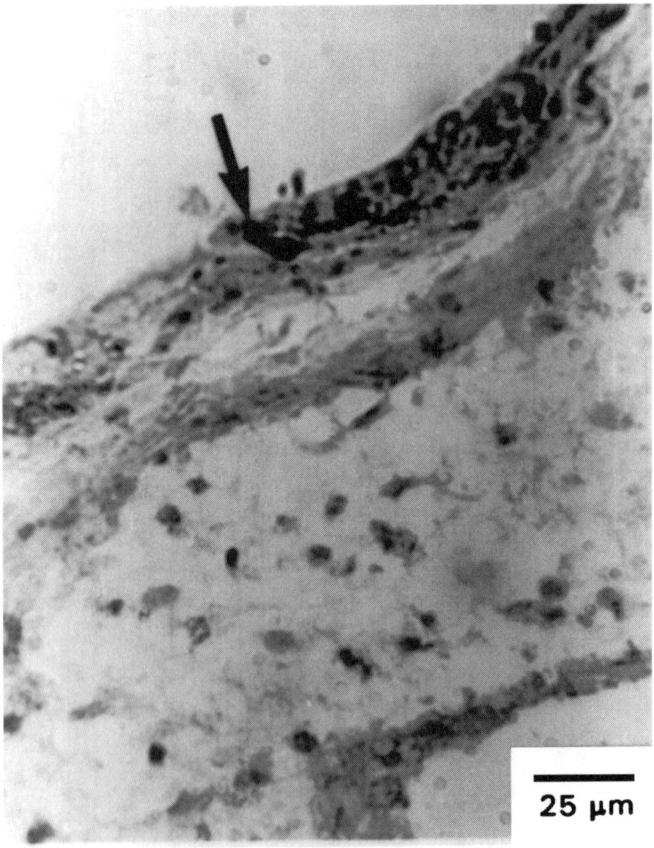

FIGURE 3b Biotinylated streptavidin immunoperoxidase stain, 121F monoclonal antibody, hematoxylin counterstain.

pattern of staining. Foreign body giant cells induced by implantation of Proplast-HA also stained intensely for both TRAP and TrATPase (Fig. 2).

On the other hand, PMMA- and talc-induced foreign body giant cells showed only mild to moderate TRAP reactivity and a total dearth of reactivity for TrATPase.

Sections of spleen displayed focal TRAP-positive cells which were distributed throughout the marginal zones of the tissue, but there was a complete absence of reactivity for TrATPase. Incubation of

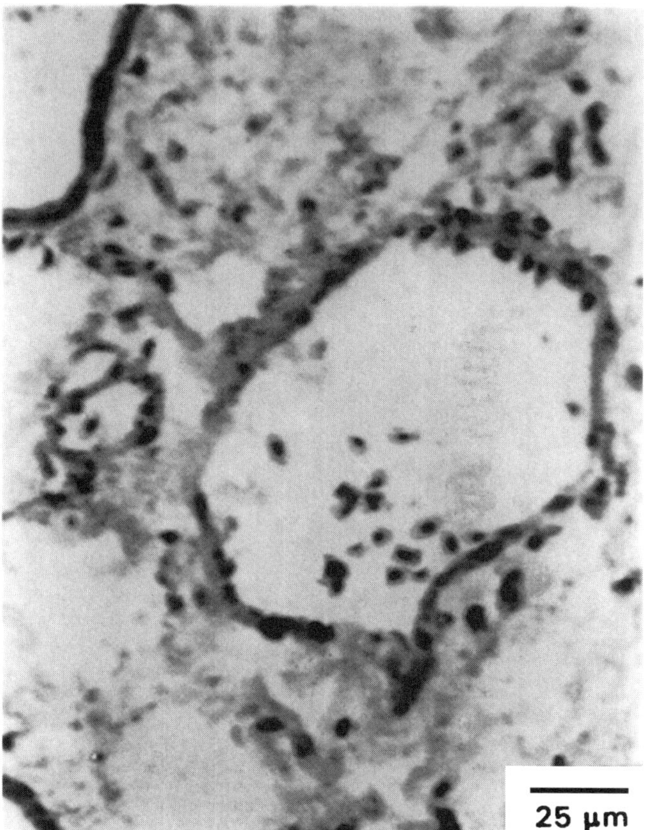

FIGURE 3c CAM polykarya induced by implantation of polymethylmethacrylate fail to react with the 121F monoclonal antibody. Biotinylated streptavidin immunoperoxidase stain, 121F monoclonal antibody, hematoxylin counterstain.

specimens with sodium-ortho-vanadate or in the absence of the particle substrates, completely extinguished the TRAP or TrATPase activities.

Antigen-antibody Reactions

Strong reactivity directed against the 121F antigen was visualized as a granular brown cytoplasmic stain in osteoclasts and mononuclear cells of hypocalcemic chick medullary bone. Multi- and

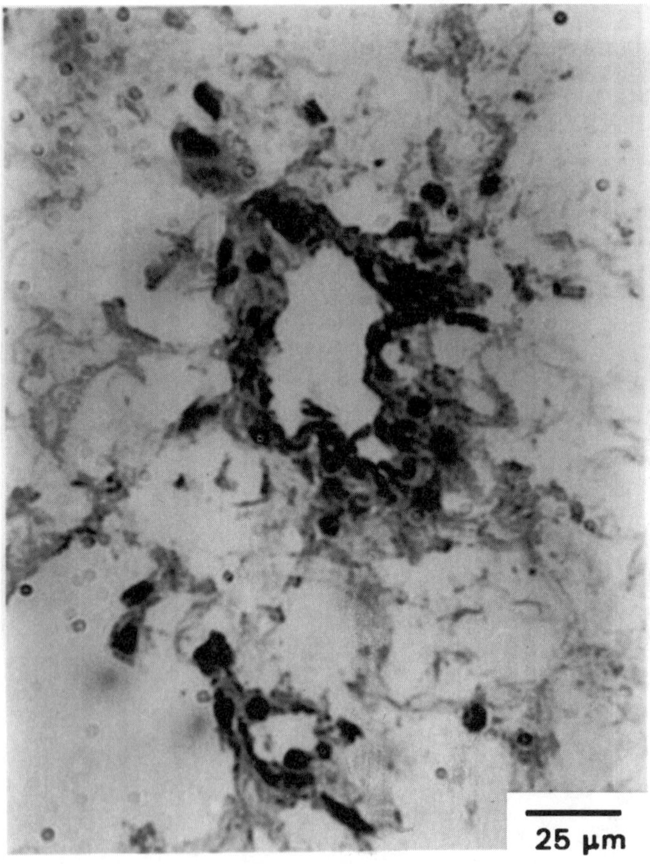

FIGURE 3d CAM polykarya induced by implantation of talc fail to react with the 121F monoclonal antibody. Biotinylated streptavidin immunoperoxidase stain, 121F monoclonal antibody, hematoxylin counterstain.

mononuclear CAM cells intimately associated with particulate chick tibiae or with Proplast-HA displayed a moderately intense positivity with the 121F monoclonal antibody (Figs. 3a, b). No reactivity was observed in CAM foreign body giant cells induced by implantation of PMMA or talc (Figs. 3c, d). Negative controls incubated in the absence of primary antibody failed to demonstrate any reactivity.

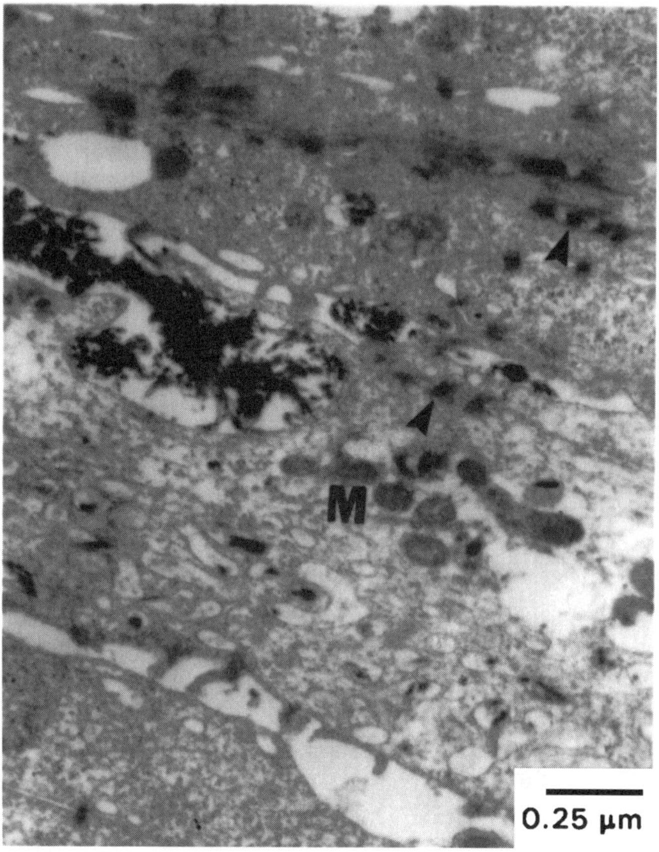

FIGURE 4a Well-developed ruffled borders are viewed in foreign body giant cells raised in response to CAM implantation of Proplast-HA particles. Many polyribosomes and mitochondria (M) as well as numerous thickened membrane specializations (arrowheads) are viewed in this electron micrograph. Extracellular channels of the ruffled border contain partially degraded substrate.

Electron Microscopic Features

Transmission electron microscopy revealed that foreign body giant cells associated with either CAM-implanted particulated chick tibiae or Proplast-HA displayed membrane ruffling against those implanted materials, this being better developed in the bone-particle specimens than in the PTFE-HA composite specimens. In both cases, partially degraded particulate matter was present within the extracellular channels of the foreign body giant cells' ruffled borders (Fig. 4a).

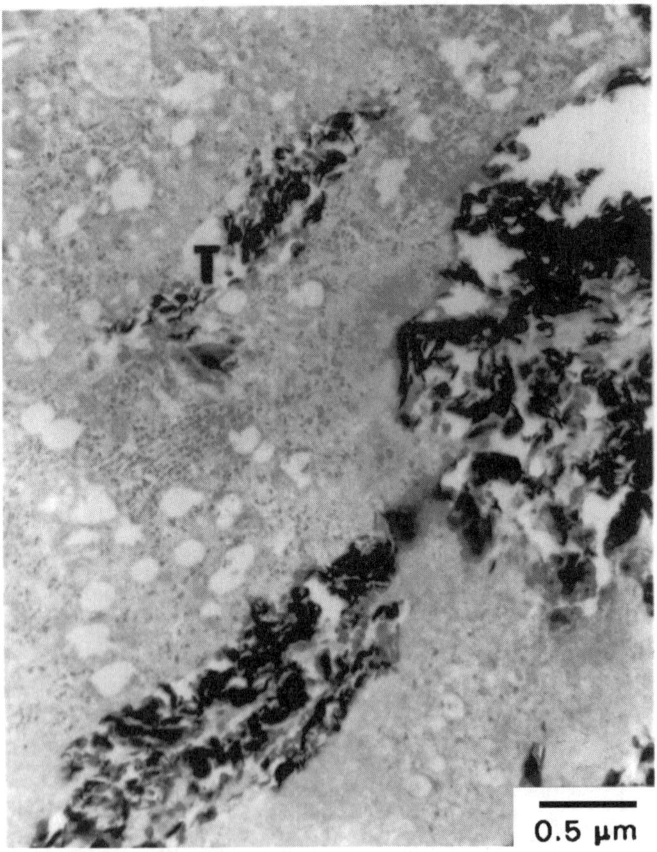

FIGURE 4b Internalized (phagocytosed) talc particles (T), and extensive rough endoplasmic reticulum are viewed within this polykaryon. Note lack of ruffled border adjacent to particulate material.

Extensive cytoplasmic complexity characterized by an abundance of mitochondria, polysomes, vacuoles and vesicles typified foreign body giant cells from both the bone particle specimens and Proplast-HA specimens. Numerous thickened membrane specializations were viewed along the degradable particles in contact with foreign body giant cells' ruffled border membranes. On the other hand, non-degradable particles induced foreign body giant cells that did not have ruffled borders. Such cells were found in close contact with talc and PMMA, and even phagocytosed some small fragments of these materials (Fig. 4b), but there was no evidence for either external

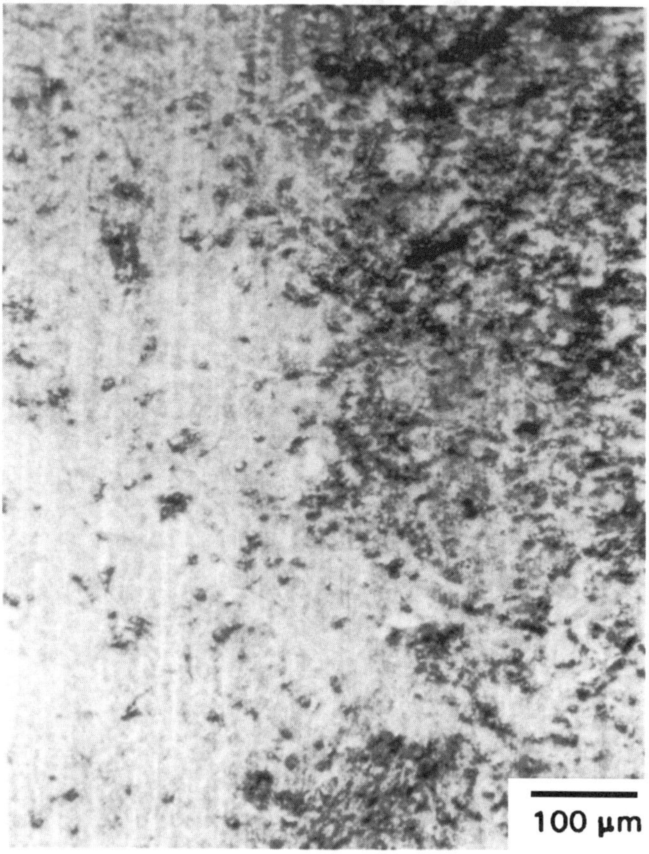

FIGURE 5a After 48 hours in culture, a bone wafer is densely populated by adherent osteoclasts from hypocalcemic chick tibiae. Toluidine blue, reflected light.

or internal material digestion. While plasmalemmal (membrane) interdigitation was viewed on the dorsolateral portions of these foreign body giant cells, raised against nondegradable particles, membrane regions in contact with the particulate PMMA and mica implants were smooth. Rough endoplasmic reticulum was quite prominent within these same cells, but overall cytoplasmic complexity was reduced in comparison with the tibiae- and Proplast-HA-induced cells. No membrane specializations of any type were found for the foreign body giant cells around the talc and PMMA particles.

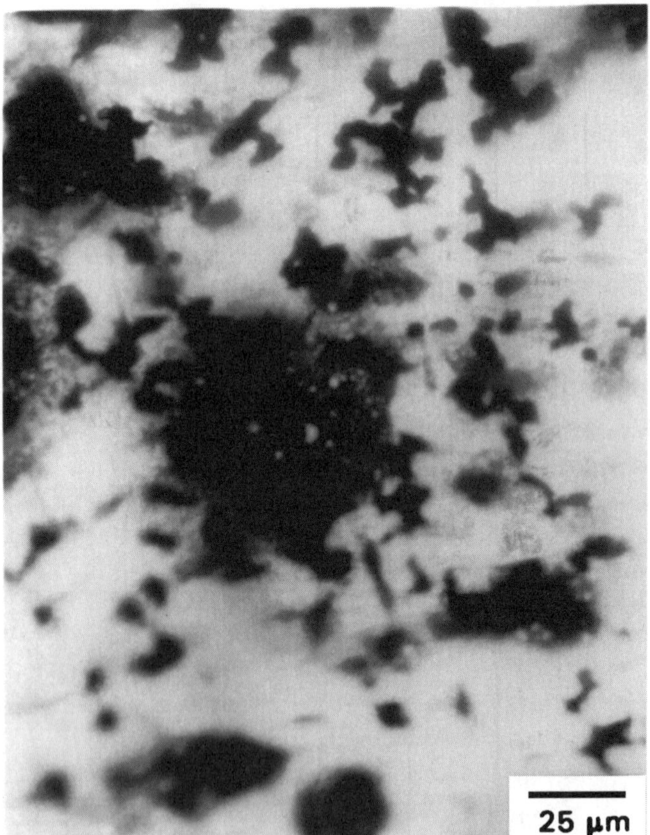

FIGURE 5b After 24 hours of culture, foreign body giant cells from CAM implanted with Proplast-HA are characterized by large, well-spread cytoplasmic skirts and profound vacuolization. Toluidine blue, transmitted light.

Bone Digestion Differences

Bone wafers, fragment-cultured with authentic osteoclasts from hypocalcemic chicks or with chick tibiae-induced CAM foreign body giant cells, displayed numerous adherent cell clusters at 24 hours which developed into a thick confluent cellular carpet by 48 hours. Adherent cells were flat, well spread and displayed prominent vacuolization. Underlying wafers became deeply excavated (dissolved or digested) and displayed numerous overlapping resorption pits and

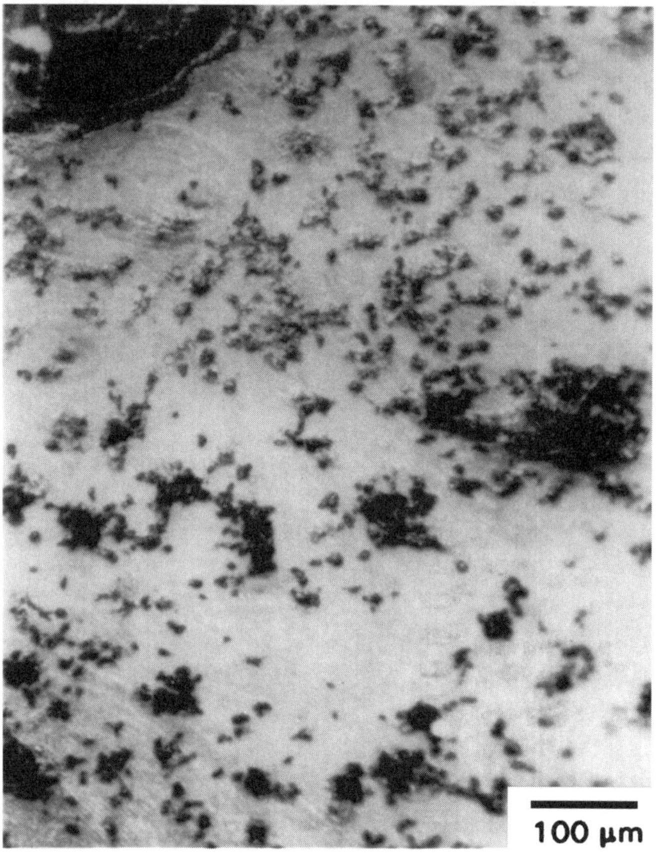

FIGURE 5c After 48 hours in culture, the population of Proplast-HA-induced CAM foreign body giant cells has greatly expanded. Confluence of resorption pits and tracks is viewed across the bone wafer. Toluidine blue, reflected light.

tracks. The excavations were most prominent in the case of wafers cultured with marrow cells from the chick hatchlings maintained on a low calcium diet (Fig. 5a). CAM multinucleated foreign body giant cells raised by implantation of Proplast-HA also displayed large, well-spread cytoplasmic skirts and profound vacuolization (Fig. 5b). After 48 hours in culture on bone wafers, extensive confluence of excavation lacunae was observed (Fig. 5c). No excavations were observed on bone wafers cultured in the absence of cells or with talc- or PMMA-induced CAM plaques (Fig. 5d).

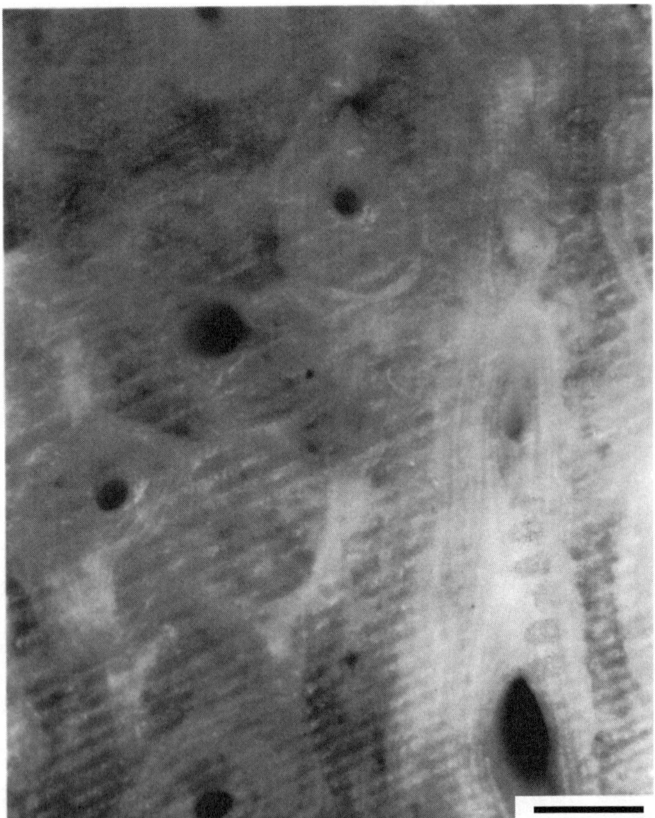

FIGURE 5d No excavations are apparent on this bone wafer cultured with CAM foreign body giant cells elicited by talc. Wafer stripped of cells after 24 hours in culture, toluidine blue, sputter-coated, reflected light.

DISCUSSION

Inequalities of Implant Materials

The question to be resolved is this: Do polymer or mineral particles have differential effects on the bone-dissolving qualities of the body's reactive phagocytes? Proplast-HA is a composite of an inert perfluorinated polymer (Teflon [DuPont, Wilmington, DE]), and acid-soluble hydroxyapatite (calcium phosphate). Perfluorocarbon polymers, as a group, display useful chemical and thermal properties as well as a low modulus of elasticity, which initially piqued the interest of the implant community [2]. Early Proplast materials, as porous composites of polytetrafluoroethylene (PTFE) with carbon or alumina, were used beginning in the mid-1960's as fillers in many non-load-bearing areas of the body [17]. Proplast interpositional implants, developed as alloplastic replacements for human TMJ discs, were inserted post-discectomy to maintain or restore vertical dimension, to avoid an anterior open bite or functional mandibular deviation, and to provide a barrier to the formation of adhesions, and ankylosis [3, 5, 18]. In the early 1980's, laminates were made to Teflon sheet, using Proplast with an admixture of either vitreous carbon fibers (Proplast I) or aluminum oxide whiskers derived from sapphire crystals (Proplast II). The porous Proplast portion was placed against the glenoid fossa to elicit bone ingrowth, while the smooth Teflon face of the laminate apposed the articular surface of the condyle [19, 20].

Particle generation occurred, but bone digestion was not reported. Later, when Proplast-HA was surface-laminated to an ultra-high-molecular-weight polyethylene articular surface, and mated against a metallic (CoCrMo) condylar head, particulate Proplast debris became associated with significant osteolysis as well as the prior-observed inflammatory responses [3, 5, 18].

Clinical Concerns

Kent noted that the great interest in use of these alloplastic devices during the years 1982–1986 slowed with an increasing number of animal and human studies reporting bone resorption, hypomobility, malocclusion and pain associated with the foreign body giant cell responses against microscopic wear debris from these newer implants [21]. Similar deleterious effects were reported, with symptoms of pain, burning sensation in the joint, crepitance, limited range of excursion, pressure around the eyes and teeth, myalgia, headache, infection, preauricular swelling, acute noninfectious lymphadenopathy and generalized weakness and malaise [5, 22–24]. Radiographic evidence confirmed the morbid sequelae: severe erosion of facial bones or destruction of condyle/fossa/base of skull complex, implant migration and fragmentation and perforation into the middle fossa with dural violation accompanied by cerebrospinal fluid leakage [3, 5, 25, 26]. How much of this problem is caused by inert wear particles of the PTFE polymer, and how much by the degradable HA mineral particles of Proplast-HA?

Reactions to Inert Materials

In 1962, Charnley abandoned the use of PTFE acetabular cups for artificial hips because particulate wear debris produced severe foreign body reactions, resulting in the production of granulomatous tissues and osseous erosion [27]. Since bone wear particles or surgical debris from bone trimming might have placed osteoclast-inducing bone particles into the same environment as the PTFE wear debris, creating a mixture not unlike that of the Proplast-HA product, no clear separation of these effects is available from the historical clinical data. Rooney and co-workers noted that, despite the general biocompatibility of PTFE, a foreign body giant cell response always is elicited and this is irrespective of the load or site of implantation, citing similar reactions in response to periurethral injections, orbital implants, vascular grafts, ossicular implants, laryngeal implants and joint replacements [28]. Spector and colleagues raised the possibility that, although porous Proplast with carbon fiber was designed to be osteoinductive, an overwhelming foreign body giant cell response played a role in inhibiting bone formation against and into the implants [29], but without notable bone dissolution.

Other polymers may be more benign than PTFE. Even in Proplast-HA systems, where 21 of 118 total joints with polyethylene articulating surfaces had been removed, there was no incidence of failure due to polyethyelene wear debris [20, 30].

Reactions to Degradable Materials

Spagnoli and Kent advised scrupulous removal of all Proplast-HA and granulomatous tissue during implant retrievals because the abundant particle debris present would continue to incite cells which elaborated inflammatory mediators IL-1 [interleukin-1] and PGE_2 [prostaglandin E_2] to stimulate bone resorption [22], in addition to the osteoclast induction noted in the present study. Carter reported that degradable poly(glycolic acid) and poly(lactic acid) particles also induced the production of foreign body giant cells with osteoclastic phenotypes [9].

Implications for Biomaterials Selection

Most reports of implant-generated particles also cite the development of an exuberant foreign body giant cell reaction, but foreign body giant cells and osteoclasts are virtually indistinguishable on light microscopic examination. The results reported here from a battery of assays, taken together, are the first to reveal distinctly different distributions of such reactive cells in relation to specific material/particle compositions. Multinuclearity and TRAP positivity proved not to be sufficiently reliable markers for the more bone-damaging osteoclast phenotype. TRAP is also expressed in splenic macrophages, pulmonary alveolar macrophages, and in multinucleated giant cells in a variety of pathologic states. Thus, although TRAP represents a marker for macrophage activation, it is not sufficiently specific – alone – to confirm cellular differentiation along the osteoclastic lineage. Mononuclear phagocytes can express the TRAP enzyme and fuse into multinuclear giant cells (polykaryons) without any apparent relationship to osteoclast differentiation.

Biomaterials choice requires that the excessive tissue damage from osteoclasts be avoided, this damage relating to their ability to degrade the biomaterials externally and to dissolve large-surface-area plates of bone. Osteoclasts degrade other objects too large to be engulfed, by exuding acids and enzymes into external pockets they seal against the material to be digested. Osteoclasts contain abundant lysosomal acid hydrolases, including TRAP, which are actively secreted into external resorption lacunae, representing the physiologic and chemical counterpart of the internal digestive lysosomes of all phagocytic cells [31]. Ascribing an osteoclastic identity to a cell type induced by any particular biomaterial requires additional morphologic evidence as well as demonstration of the appropriate antigens and ability to function in an external resorptive capacity.

TrATPase is an excellent additional indicator, as a unique member of the TRAP family of acid hydrolase isoenzymes which is secreted vectorially into the external resorption compartment and is expressed only in osteoclast ontogeny [32]. The inability of macrophagic TRAP enzymes to hydrolyze ATP as substrate renders TrATPase a more selective marker than TRAP for osteoclast identification. For example, while splenic macrophages stain for TRAP, they do not express TrATPase acitivity [33], a finding that was confirmed in our study.

Free Radical Reactions

The 121F antigen is related to the superoxide (anion, free radical) dismutase molecule and also is associated with the extracellular breakdown of resorbable or particulate material [15]. Hence, 121F antigen expression also can predict or confirm the cell's functional ability to excavate bone, and implicate further reactions of hydrogen peroxide which is generated by the superoxide dismutase reaction with the superoxide anion. Critical observations were made in our cell culture assays whereby isolated cells were allowed to excavate resorption pits (lacunae) on mineralized substrates. This test of functional activity already has served as a catalyst for major advances in osteoclast biology research [34], and will be a valuable method for future studies of material-composition-dependent differentiation of human phagocytic cells. The technique used here, whereby bone wafers were sputter-coated to reveal the topography of the resorbed surface when examined under dark field reflected light microscopy [16], provides the ability to detect recently active excavations while the cells remain *in situ*.

Production and harvesting of sufficient reactive phagocytic cells is a major requirement to further advance understanding of the body's response to small particle "implants".

Extensions to "Tissue Engineering"

In the chick CAM model, the implantation of particulate matter initiates a cascade of predictable events culminating in the induction of numerous foreign body giant cells. This implant system represents an affordable *in vivo* model for investigation of the inductive specificity of various biomaterials for polykaryon ontogeny and ectopic osteoclast differentiation. Only implant materials capable of being degraded and resorbed extracellularly (Proplast-HA, poly(glycolic acid), and poly (lactic acid), as well as bone chips) induce cellular fusion and differentiation events indicative of osteoclast production in the CAM model [9]. Webber and colleagues reported a similar pathway for resorbable carbonate- and barium sulfate-induction of giant cells [15]. "Tissue Engineering" anticipates the exclusive use of resorbable, degradable scaffolds for cellular regeneration around high-surface-to-volume implants [35, 36] so osteoclast-like polykaryon reactions should be anticipated.

On the other hand, (1) development of TrATPase positivity, (2) expression of the 121F antigen and (3) ability to produce resorption pits on bone wafers did *not* occur with CAM foreign body giant cells elicited by nonresorbable substrata such as PMMA, polypropylene, or talc [9]. Webber and co-workers earlier reported a lack of 121F antigen expression on giant cells elicited by crosslinked Sepharose beads, PMMA and mica [15]. Osteoclast-specific features might provide morphologic and functional confirmation of an "engineered" tissue's ability to degrade its extracellular scaffold.

CONCLUSIONS

On the chick CAM, Proplast-HA particles induced multinucleated giant cells that were osteoclast-like by both morphologic and functional criteria. Osteoclast induction by degradable implant substances is the likely mechanism of severe bone erosion reported in patients with some types of implant materials, and not others, in spite of apparently equally severe inflammatory responses to particulate wear debris. Implant failures for inert materials in the TMJ, including implants fabricated from Silastic, did not lead to the severe osseous destruction reported with Proplast-HA [30, 37]. This conclusion is supported by noting that Proplast-HA uniquely contained a degradable mineral component, calcium phosphate, rather than the inert silica filler of Silastic, or the inert carbon or alumina components of Proplast I and II. Furthermore, bone erosion also was seen adjacent to facial augmentation prostheses fabricated of Proplast-HA where the implants were not under obvious loads. Since bone erosion was not reported for the earlier Proplast versions, and PTFE polymer (Teflon) was the common material in all three Proplast products, the degradable HA mineral particles were the most probable osteoclast-inducing factor.

Acknowledgments

We extend special thanks to Dr. Philip Osdoby of Washington University (St. Louis, MO) for providing the 121F monoclonal antibody and, along with Dr. C. Z. Liu of the University of Washington (Seattle, WA), for kind advice and continuing research guidance. We gratefully acknowledge Janet Gorfien, Maria Kozak, and Sandra Mendel for expert technical assistance and Elisabeth Lawson for preparation of the micrographs. This study was supported in part by an American Fellowship from the American Association of University Women and the Mark Diamond Research Fund (GSA) of the State University of New York at Buffalo.

References

[1] Hartman (Carter), L., Bessette, R., Baier, R., Meyer, A. and Wirth, J., *J. Biomed. Mater. Res.* **22**, 475 (1988).
[2] Homsy, C., In: *Biocompatibility of clinical implant materials* (CRC Press, Inc. Boca Raton, 1982), pp. 59–77.
[3] Berarducci, J., Thompson, D. and Scheffer, R., *J. Oral Maxillofac. Surg.* **48**, 496 (1990).
[4] Stambaugh, K., *Arch. Otolaryngol. Head Neck Surg.* **118**, 682 (1992).
[5] Chuong, R. and Piper, M., *Oral Surg., Oral Med., Oral Pathol.* **74**, 422 (1992).
[6] Brennan, G., Kerr, A. and Smyth, G., *Clin. Otolaryngol.* **9**, 229 (1984).
[7] AAOMS: TMJ Implant Advisory. Rosemont, IL., AAOMS, August, 1992.
[8] Krukowski, M. and Kahn, A., *Calcif. Tissue Int.* **34**, 474 (1982).
[9] Carter, L. C., *Analysis of the Cellular Healing Response of the Chick Chorioallantoic Membrane to Implanted Poly(Glycolic Acid)*, Doctoral Dissertation, State University of New York at Buffalo (1993).
[10] Arnett, T. and Dempster, D., *Endocrinology* **119**, 119 (1986).

[11] Arnett, T. and Dempster, D., *Endocrinology* **120**, 602 (1987).
[12] Dempster, D., Murrills, R., Horbert, W. and Arnett, T., *J. Bone Miner. Res.* **2**, 443 (1987).
[13] Barka, T. and Anderson, P., *J. Histochem. Cytochem.* **10**, 741 (1962).
[14] Andersson, G. and Marks, S., *J. Histochem. Cytochem.* **37**, 115 (1989).
[15] Webber, D., Osdoby, P., Hauschka, P. and Krukowski, M., *J. Bone Miner. Res.* **5**, 401 (1990).
[16] Walsh, C., Beresford, J., Birch, M., Boothroyd, B. and Gallagher, J., *J. Bone Miner. Res.* **6**, 661 (1991).
[17] Wagner, J. and Mosby, E., *J. Oral Maxillofac. Surg.* **48**, 1140 (1990).
[18] Gallagher, D. and Wolford, L., *J. Oral Maxillofac. Surg.* **40**, 672 (1982).
[19] Florine, B., Gatto, D., Wade, M. and Waite, D., *J. Oral Maxillofac. Surg.* **46**, 183 (1988).
[20] Kent, J., Block, M., Halpern, J. and Fontenot, M., *J. Oral Maxillofac. Surg.* **51**, 408 (1993).
[21] Kent, J., *J. Oral Maxillofac. Surg.* **49**, 443 (1991).
[22] Spagnoli, D. and Kent, J., *Oral Surg., Oral Med., Oral Pathol.* **74**, 411 (1992).
[23] Lagrotteria, L., Scapino, R., Granston, R. and Felgenhauer, D., *J. Craniomandib. Pract.* **4**, 172 (1986).
[24] Feinerman, D. and Peicuch, J., *Int. J. Oral Maxillofac. Surg.* **22**, 11 (1993).
[25] Valentine, J., Reiman, B., Nat, R., Beuttenmuller, E. and Donovan, M., *J. Oral Maxillofac. Surg.* **47**, 689 (1989).
[26] Heffez, L., Mafee, M., Rosenberg, H. and Langer, B., *J. Oral Maxillofac. Surg.* **45**, 657 (1987).
[27] Charnley, J., *Fed. Proc.* **25**, 1079 (1962).
[28] Rooney, T., Haug, R., Toor, A. and Inaresano, A., *J. Oral Maxillofac. Surg.* **46**, 240 (1988).
[29] Spector, M., Harmon, S. and Kreutner, A., *J. Biomed. Mater. Res.* **13**, 677 (1979).
[30] Bronstein, S., *Oral Surg., Oral Med., Oral Pathol.* **64**, 135 (1987).
[31] Andersson, G., Ek-Rylander, B. and Minkin, C., "Acid phosphatases". In: Rifkin, B. and Gay, C. (Eds.), *Biology and Physiology of the Osteoclast* (CRC Press, Boca Raton, 1992), pp. 55–73.
[32] Flanagan, A. and Chambers, T., *J. Pathol.* **159**, 53 (1989).
[33] Linduger, A., Mackay, C., Ek-Rylander, B., Andersson, G. and Marks, S., *Bone and Mineral* **10**, 109 (1990).
[34] Boyde, A., Ali, N. and Jones, S., *Br. Dent. J.* **56**, 216 (1985).
[35] Langer, R., *Accounts Chem. Res.* **33**, 94 (2000).
[36] Langer, R. and Vacanti, J. P., *Science* **260**, 920 (1993).
[37] Timmis, D., Aragon, S., Van Sickels, J. and Aufdemorte, T., *J. Oral Maxillofac. Surg.* **44**, 541 (1986).

The Body's Response to Deliberate Implants: Phagocytic Cell Responses to Large Substrata *Vs.* Small Particles

R. BAIER, E. AXELSON, A. MEYER and L. CARTER

State University of New York at Buffalo, Industry/University Center for Biosurfaces and Department of Oral Diagnostic Sciences

and

D. KAPLAN and G. PICCIOLO

Food and Drug Administration

S. JAHAN

University of Memphis

It is important to characterize possible inflammatory responses to small particles, and to separate clearly these effects from responses to larger objects nearby. This research used a chemiluminescent assay, scanning electron micrographs, and energy dispersive X-ray spectra to monitor inflammation-related reactive oxygen intermediate (ROI) production and morphological alterations of human monocyte-derived macrophages interacting with the walls of apolar and polar polystyrene cuvettes, in the absence and presence of small particles of surface-characterized Teflon^{TM}, polyethylene, Co-Cr-Mo alloy, titanium and alumina. The two types of polystyrene substrata represent the "bacterial" (as produced) and "tissue culture" (gas-plasma-treated [GPT]) materials widely used in biological testing and tissue culture. Monocyte-derived macrophage spreading during contact with the higher-surface-energy, more polar substratum suppressed "oxidative bursts" to lower levels than expressed from rounded cells in contact with the lower-energy, apolar substratum. Particulate matter engulfed by both rounded and spread cells did not significantly enhance ROI production beyond levels observed for no-particle controls during the one-hour exposure time. Biocompatibility of some implants might be related to cell-spreading-induced suppression of ROI production, improving the tissue integration of GPT implants.

INTRODUCTION

Medical and dental implants, their associated surgical placement debris, and loosened cement and wear debris particles all can be present simultaneously at sites of human prostheses. Even in the absence of infective microorganisms, inflammatory responses involving phagocytic cells often are associated with prosthetic hip, knee, and temporomandibular joint implants that show aseptic loosening [1]. This investigation set out to identify, using a simple *in vitro* model, the relative influences of engulfable small particles and of adjacent nonengulfable larger substrata on one early phase of phagocytic cell behavior: relative generation of reactive oxygen intermediates (ROI). The experiments simulated cases when biodegradation-resistant implants and biodegradation-resistant, engulfable particles would be simultaneously presented to arriving monocyte-derived macrophages. Human monocyte-derived macrophages were exposed to contact (10,000 cells per square centimeter) with reference polystyrene or gas-plasma-treated (GPT) polystyrene, serving as surrogate implant materials, while the monocyte-derived macrophages were allowed to contact and engulf small test particles (supplied at approximately 10 particles per cell) of titanium, cobalt/chromium/molybdenum alloy, Teflon, polyethylene, or alumina.

For 60 minutes after first "foreign materials" contact, monocyte-derived macrophage production of two key reactive oxygen intermediates, superoxide anion and hydrogen peroxide, was monitored in a luminometer assay developed by Kaplan and Picciolo [2]. In parallel experiments, scanning electron microscopy and energy-dispersive-x-ray analyses were used to monitor the changes in cell shapes and to confirm cellular engulfment and identity of test particles in each combined-contact experiment.

ROI production of human monocyte-derived macrophages was most strongly correlated ($p < .001$) with their spread areas in contact with the nonengulfable (cuvette walls or separate test specimens) substrata, suggesting that (a) variable cell spreading might trigger variable cell-membrane-based configuration changes of catalytic macromolecules known to be important in the early inflammatory responses of phagocytes [3–5], or (b) ROI was not detectable by the probe due to sequestration.

MATERIALS AND METHODS

Human Cells

Monocyte-derived macrophages were prepared and used in the morphological and oxidative studies, in accord with detailed protocols previously published [6]. Monocytes were isolated from adult humans by leukophoresis. The cell population was enriched for monocytes by elutriation. The cells were cultured in polypropylene tubes for 3-8 days in the presence of M-CSF (macrophage colony stimulating factor) and human AB serum to differentiate the monocytes to monocyte-derived macrophages [6]. Half the media were changed every other day.

Test Substrata

The nonengulfable substrata utilized in this study were reference polystyrene (PS) and gas-plasma-treated polystyrene (GPTPS), both previously characterized in detail with regard to their utility and properties for cell adhesion experiments [7]. For the cell contact studies, these materials were used in a flat, sheet form. For the oxidative studies, these materials were in a cuvette form allowing easy use in the luminometer assay [2].

Test Particles

Five types of implant-relevant, engulfable particles were used, first prepared in dry form as powder-like materials. Three types were gifts from Smith Nephew Orthopedics (Memphis, TN) including ASTM F-75 (Co-Cr-Mo), high density polyethylene (HDPE) and titanium, which are referred to herein as F-75, PE and Ti, respectively. These materials were represented to be relevant to orthopedic industry interests in biological responses to wear debris near articulating segments of orthopedic implants, including the total joint replacement implants of the hip and knee [8]. Alumina was a gift from Ivoclar North America, as representative of the material widely used to surface-abrade many implant materials and found embedded in implant surfaces [9]. Impact-fractured Teflon (Tef) was used, from a supply first prepared to assess phagocytic responses in the chick chorioallantoic membrane model [10]. Teflon particulate is clinically relevant, since Teflon wear debris from failed temporomandibular joint implants has been implicated as a causative agent in stimulating and sustaining inflammatory reactions that led to implant rejection [11]. Multisizer data and scanning electron microscope images showed that all the test particles were between 1 – 10 μm, a range suitable for particle internalization by monocyte-derived macrophages (phagocytosis).

Particle/Monocyte-derived Macrophage Exposure Experiments

Approximately 50 million monocyte-derived macrophages in 1 mL of prepared complete medium (FBS, human AB serum, M-CSF, glutamine, glucose) were prepared by a 10x dilution with complete medium and stored at room temperature. Reference and GPT polystyrene specimens were individually placed in labeled dishes. Milligram quantities of particles of each type were placed on separate pieces of reference and GPT polystyrene. With a glass pipette, approximately 500 μL of cell suspension was aliquoted and carefully applied to each particle type. Pipetting up and down formed suspensions in which cell-particle contacts were maximized. These suspensions then were re-applied to the original GPT or reference polystyrene substrata, in approximately 150 μL drops. "No-particle" control specimens were included in each test series and were treated in the same manner as cells + particles samples. The samples were covered and incubated at 37°C for 1 hour to allow for cell attachment and spreading over the substratum and concurrent phagocytosis of the particles. After 1 hour, liquid was removed from each drop-covered area with a glass pipette, and the remaining substrata with attached cells and particles were processed for SEM by fixation in 2.5% glutaraldehyde (GA), followed by dehydration in a graded series of ethanol/water mixtures, and critical-point drying from hexamethyldisilazane (HMDS). The GA, ethanol/water mixtures, and HMDS were applied in sequential drop-wise fashion over the original cell- and particle-exposed areas.

Scanning Electron Microscopy/Energy Dispersive X-ray (SEM/EDX) Analysis

Samples were further processed for SEM/EDX by mounting the drop-exposed, fixed specimens on aluminum stubs with double-sided, conductive carbon tape. The specimens then were overcoated with sputtered carbon. SEM images and EDX spectra were recorded for substrata, cells, and particles. Most SEM images were collected at 20 keV acceleration voltage and EDX spectra were collected at 100,000x magnification (100 kx). Additional SEM images and EDX spectra were collected at 3 keV acceleration voltage to resolve better the cell and particle morphology. SEM images of a 10 μm diffraction grating were used to calibrate the magnification scales for quantitative analyses of the shape parameters of attached monocyte-derived macrophages, including cell diameter and surface area, on the GPT *versus* reference polystyrene substrata.

Particle Stocks

Test particle suspensions of 25 mg/mL in physiologic saline were prepared. High-speed vortexing was used to create homogeneous suspensions prior to addition to the test cuvettes.

Particle counts were obtained using a Coulter Counter (Beckman Coulter, Inc., Fullerton, CA) equipped with a 50 μm diameter aperture and software capable of generating particle size distribution curves from the recorded counts. Prior to counting, each particle stock was vortexed and 100 μL of the suspension was added to 20 mL of particle-free Isoton balanced electrolyte solution (Coulter) and counted in triplicate. The polyethylene and Teflon particle stock suspensions were unstable, as the intrinsic hydrophobicities of the particles caused clumping in the aqueous fluid. The alumina particles settled faster than did the titanium and cobalt-chromium-molybdenum particles after votexing. These difficulties were overcome by introducing pre-weighed, pre-dispersed amounts of PE, Teflon, and alumina particles to the experimental cuvettes, after determining the amounts necessary (about 10 particles per cell) to best utilize the operating range of the luminometer instrument.

Measurements also were made of particle-induced changes in contact angles and in the surface tensions of standard diagnostic liquids [12], as well as noting the liquid surface tensions associated with spontaneous particle engulfment. Contact angles of single drops of various diagnostic liquids of known surface tension first were measured as these drops rested at equilibrium upon reference-grade PTFE (polytetrafluoroethylene) film. Test particles were added in dry form to the various drops by simply sprinkling them over the drops. After addition of particles, drop contact angles were again measured while the drops were viewed through a 10x telescope for observations of particle engulfment or particle exclusion to the external surface of the drop.

IR spectra were acquired for the test particles and their extracted "residues" as dried from distilled water slurries on germanium prisms. Testing for the presence of endotoxins associated with the particles was done by a commercial Limulus Amoebocyte Lysate (LAL) *in-vitro* test.

Processing Monocyte-derived Macrophages for the Chemiluminescent (CL) Assay

Stock cells, in complete medium, were taken from a supply kept rotating in an incubator at 5% CO_2 and 37°C, within 2 days of their donation. Prior to CL assay, the cells were pelleted by gentle centrifugation (2000 rpm, approximately 250x g) for 10 minutes, to remove the supernatant and thereby minimize the contribution of medium components (*e.g.*, serum) to the measurements. Diluting medium supplemented with salts and glucose was used to resuspend the pellet. Pipetting up and down, as in the particle contact trials, and additional vortexing, fully dispersed the cells and created homogeneous suspensions. Cells used for control experiments were processed by all the steps outlined above. The numbers of viable cells present were determined by using a standard hemocytometer and 100 μL aliquots of the cell suspensions (cells stained with Trypan blue). Net viable cells were determined as the difference between total cells present and those colored blue (dead or dying).

Prior to the CL assay, test particles were placed in the luminometer cuvettes and suspended in 100 μL of diluting medium and 750 μL of luminescent "probe". Cuvette exteriors were wiped with a lint-free cloth (pre-wetted with anti-static solution) and loaded into the luminometer according to a "run list" specified in the software. Cells from stock suspensions (5×10^5 viable cells per ml) then were added to the cuvettes in 100 µl aliquots with a repetition pipettor, so that each luminometer cuvette contained 5×10^4 cells (final concentration: 5.26×10^4 cells/ml). Preliminary experiments indicated that this concentration of cells resulted in a linear CL response; no oxygen diffusion problems were noted. The door of the apparatus was closed to initiate the data collection program. The luminometer program repeatedly analyzed 25 tubes every 4 minutes, shaking each individual sample prior to its reading. Data collection was stopped after 1 hour, at the completion of the nominal "inflammatory phase" [6]. Although one hour is a very short time, in terms of host response and longer-term biocompatibility of implants, the experiments reported here were designed to indicate the *potential* for the test substrata and particles to initiate an inflammatory response. Further validation of the CL assay as a predictive tool is underway at this time.

Chemiluminescent (CL) Assay Conditions and Reagents

With reference polystyrene luminometer cuvettes, suspensions of 1, 10, and 100 particle(s) per monocyte were tested in a series of preliminary experiments. It was determined that the best use of the instrument's experimental range was for about 10 particles/cell; this proportion was used in all subsequent assays. Slight turbidity was noted only for the experiments with titanium particles. Measurements indicated that chemiluminescence could be diminished by as much as 5% by the turbidity caused by the titanium particles, if the particles stayed in suspension. The data related to titanium that are presented in this report are not corrected for turbidity, as an accurate and reproducible correction factor could not be obtained. Commercially-supplied lucigenin and luminol reagents were used to assess superoxide anion (O_2^-) and hydrogen peroxide (H_2O_2) production, respectively. Cuvettes pre-coated with a known tumorigenic phorbol ester (phorbol myristate acetate (PMA)) were used as positive controls, typically inducing generation of reactive oxygen intermediates at 3 – 5 times the amount noted for uncoated cuvettes. The reagent suspensions in reference and GPT polystyrene luminometer cuvettes without particles served as cuvette controls. Experiments were performed in triplicate.

Electron spin resonance measurements of reference PS and GPTPS, to estimate free radicals, were performed in accord with a method used to examine radiation-sterilized artificial hip components [13].

Cell Viability Using Probes

The apparent cytotoxicities of the test particles at the contact conditions used in the chemiluminescent assays were determined by using commercially-supplied calcein and ethidium probes. Cell viability was microscopically assessed after exposure to particles at concentrations and exposure times similar to those in the corresponding CL assay. Green (viable, calcein) and red (dead, ethidium bromide) cells were microscopically visualized, manually counted and videotape-recorded in five separate fields per chamber and in three separate modes including transmitted light (all cells), calcein mode, and ethidium bromide mode by adjusting filters on a bright field microscope.

Data Reduction Techniques

The CL responses for luminol and lucigenin were measured as the mean integral light emission for each sample monitored, using the instrument's automatic calculation of the area under the light emission curve. Percent variance (% variance) of 5–10% for data in triplicate was typical.

RESULTS

Cell Viability

Results from cell viability assays using calcein and ethidium bromide probes showed the Ti particles to cause between 43 and 69% cytotoxicity. F-75 particles were cytotoxic at levels between 8% and 48%. The alumina showed 0 to 25% cytotoxicity.

Teflon showed 0–31% cytotoxicity, and PE particles were 6–27% cytotoxic according to these assays.

Endotoxin Testing

The Teflon particles carried endotoxin at the level of 48 EU (standard units), compared with 0.06 EU or less for all the other test particles.

Production of Reactive Oxygen Intermediates

The differential oxidative responses of monocyte-derived macrophages are recorded in Figures 1 and 2 for cells contacting reference *versus* gas-plasma-treated polystyrene, alone, and as challenged with additions of engulfable test particles in identically-prepared cuvettes. With the exception of titanium, superoxide anion production was not enhanced by any other of the test particles. ANOVA analysis of the data demonstrated that gas-plasma-treatment (GPT) of polystyrene significantly reduced polystyrene's ability to elicit a superoxide production response from monocyte-derived macrophages, as shown in Figure 1 ($p < .001$) and Figure 2 ($p < .05$). This reduction in superoxide production response in GPTPS also was observed, at the same levels of significance, for specimens with added particles.

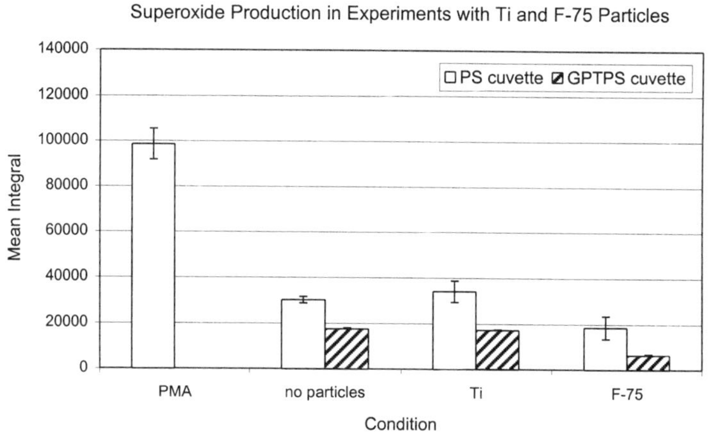

FIGURE 1 The superoxide production of human monocyte-derived macrophages exposed to the Ti and F-75 test particles in GPT surface-modified and control cuvettes using lucigenin as the CL probe (note: 10 particles/cell introduced; $\sim 1 \times 10^4$ cells/cm^2 cuvette surface; average $\pm$ standard deviation of triplicates; PMA = phorbol ester positive control). Response of cells in GPTPS cuvettes was significantly less ($p < .001$) than cells in control cuvettes.

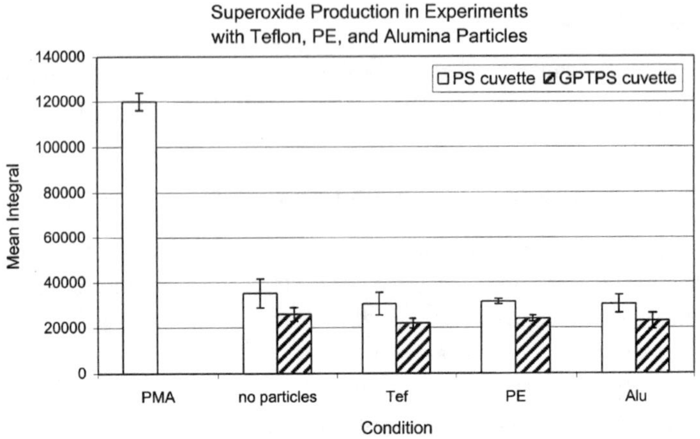

FIGURE 2 The superoxide production of human monocyte-derived macrophages exposed to the PE, Teflon, and alumina test particles in GPT surface-modified and control cuvettes using lucigenin as the CL probe (note: 2 mg introduced; $\sim 1 \times 10^4$ cells/cm^2 cuvette surface; average $\pm$ standard deviation of triplicates; PMA = phorbol ester positive control). Response of cells in GPTPS cuvettes was significantly less ($p < .05$) than cells in control cuvettes.

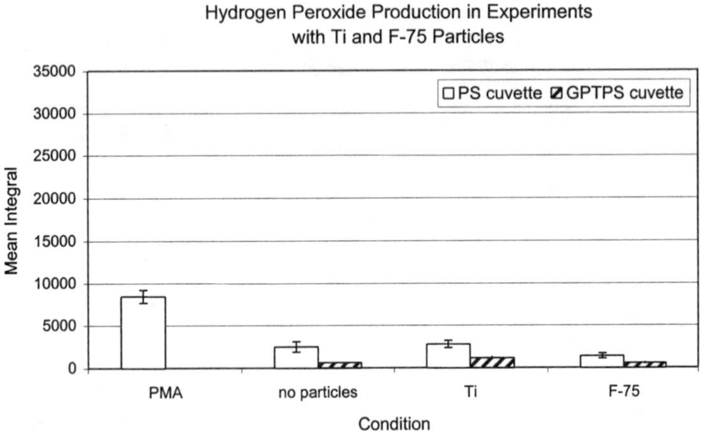

FIGURE 3 The hydrogen peroxide production of human monocyte-derived macrophages exposed to the Ti and F-75 test particles in GPT surface-modified and control cuvettes using luminol as the CL probe (note: 10 particles/cell introduced; $\sim 1 \times 10^4$ cells/cm^2 cuvette surface; average $\pm$ standard deviation of triplicates; PMA = phorbol ester positive control). Response of cells in GPTPS cuvettes was significantly less ($p < .001$) than cells in control cuvettes. Ti particles were slightly stimulatory above the reference PS control ($p < .05$).

Results for the hydrogen peroxide production response of human monocyte-derived macrophages are recorded in Figures 3 and 4. Although the Ti and alumina particles were slightly stimulatory above the reference PS control value ($p < .05$ and $p < .001$, respectively), F-75, Teflon and PE particles did not activate the cells beyond that of the reference PS. As with the superoxide cell response, contact with GPTPS reduced the ability of the cells to produce a hydrogen peroxide response; this effect also was observed for all of the particle-added cases ($p < .001$ in all cases). Table I presents data confirming that the GPT effect is a general feature of gas-plasma-induced surface energy changes for the PS, and not specific to the gas used (argon *vs.* air). Two-tailed Student t-tests demonstrated no significant differences between the two methods of gas plasma treatment evaluated in the preliminary experiments. Electron spin resonance results showed that GPT-induced superficial free radicals were present on the PS, but decayed to 75% of initial levels before the cell contact experiments were performed (Tab. II).

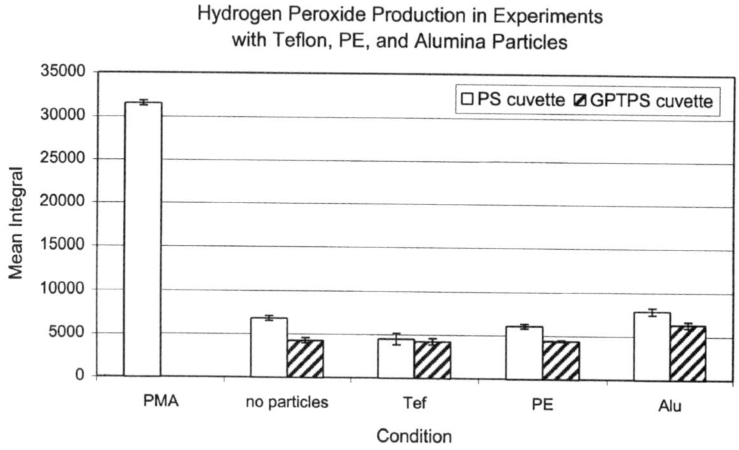

FIGURE 4 The hydrogen peroxide production of human monocyte-derived macrophages exposed to the PE, Teflon, and alumina test particles in GPT surface-modified and control cuvettes using luminol as the CL probe (note: 2 mg introduced; $\sim 1 \times 10^4$ cells/cm^2 cuvette surface; average $\pm$ standard deviation of triplicates; PMA = phorbol ester positive control). Response of cells in GPTPS cuvettes was significantly less ($p < .001$) than cells in control cuvettes. Alumina particles were slightly stimulatory above the reference PS control ($p < .001$).

TABLE I The hydrogen peroxide production of human monocyte-derived macrophages exposed to the Ti test particles in air- and argon-plasma-treated and control cuvettes using luminol as the CL probe (note: 10 particles/cell introduced; $\sim 1 \times 10^4$ cells/cm^2 cuvette surface, (+) = air-plasma-treated; (R+) = argon-plasma-treated). There is no significant difference between results from the two methods of gas plasma treatment

Luminol response for Ti in air vs. argon-plasma-treated PS cuvettes (Hydrogen Peroxide Production from Human Monocyte-derived Macrophages)		
Sample	*Mean integral values*	*% coeff. variance*
PMA	23,736	14.5
PS	3,882	12.6
PS+	2,498	10.0
PS-R+	2,454	6.5
Ti	6,407	3.9
Ti+	3,605	12.2
Ti-R+	3,559	8.2

TABLE II Relative free radical concentration in gas-plasma-treated polystyrene stored in air under ambient laboratory conditions. Data represent peak-to-peak height of signal in electron spin resonance spectra

ESR signal reduction during 72 hours after gas plasma treatment of polystyrene		
Post-GPT time (hours)	*Peak-to-peak height*	*Interval decay rate [units/hour]*
Time zero	44.15	–
1.5	44.76	–
6.5	39.18	1.12
24.5	37.65	0.09
72.0	32.87	0.10
No-GPT control	no detectable signal	–

Cell Morphology and Chemistry Changes

SEM images of human monocyte-derived macrophages on reference polystyrene substrata, after one-hour contact times, showed typical cells to be roughly spherical, but attached at multiple substratum sites with small surface membrane "retraction filaments" at their bases (Fig. 5). In contrast, SEM images of monocyte-derived macrophages from the same source, allowed to contact GPTPS substrata for one hour, showed these cells to be very flat and to have very close membrane association with the GPT substratum (Fig. 6). Cell diameters on the reference polystyrene were 8.8 μm ± 0.4 μm, and 18.0 μm ± 0.5 μm on the GPTPS. The apparent cell surface/substratum contact areas

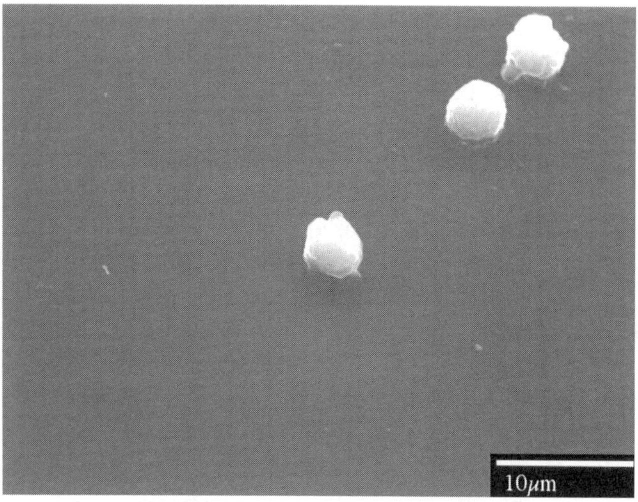

FIGURE 5 SEM photomicrograph of monocyte-derived macrophage on reference polystyrene after one hour contact. Note absence of cell spreading. Original magnification: 2000x.

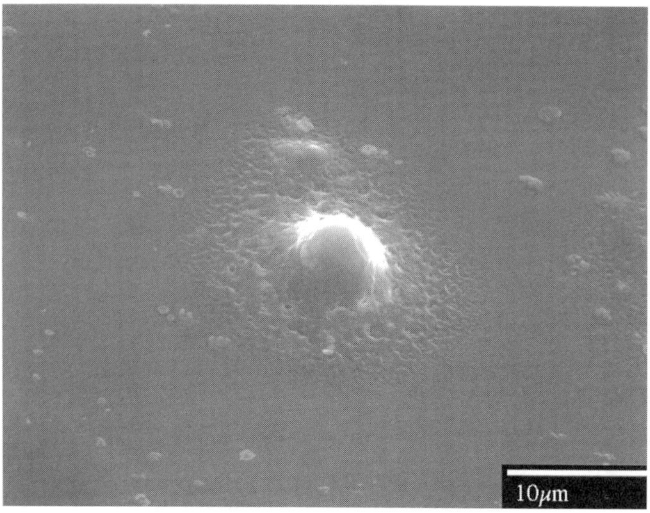

FIGURE 6 SEM photomicrograph of monocyte-derived macrophage on GPT polystyrene after one hour contact. Note spread membrane-bound cytoplasmic "skirt" which extends approximately 10 μm from the prominent nucleus of the cell. Original magnification: 2000x.

were $240\,\mu m^2 \pm 20\,\mu m^2$ and $950\,\mu m^2 \pm 50\,\mu m^2$, respectively. A two-tailed Student t-test demonstrated that the difference in cell diameters was significant ($p < .001$).

EDX spectra of cells on the reference PS contained characteristic biological salt-related peaks for Na, Cl, P, O and to a lesser extent, K and Ca (Fig. 7). These elements were absent from PS not exposed to cells and from areas of cell-exposed samples without cell attachment. A strong signal for Si also was noticed from EDX spectra of the supplied cells, suggesting some silica fine particle contamination from the elutriation and related preparative steps. Cells retained on GPT substrata similarly contained Na, Cl, P, O, K, Ca and Si, with accentuated signals for Na, Cl and Si, attributed to media residues retained around the spread cells on intrinsically polar (more wettable) specimens [14].

When test particles also were present, many particles were observed to be in close physical association with monocyte-derived macrophages (Fig. 8). Cells having attached or engulfed the various particles usually showed irregular topographical appearances but did not

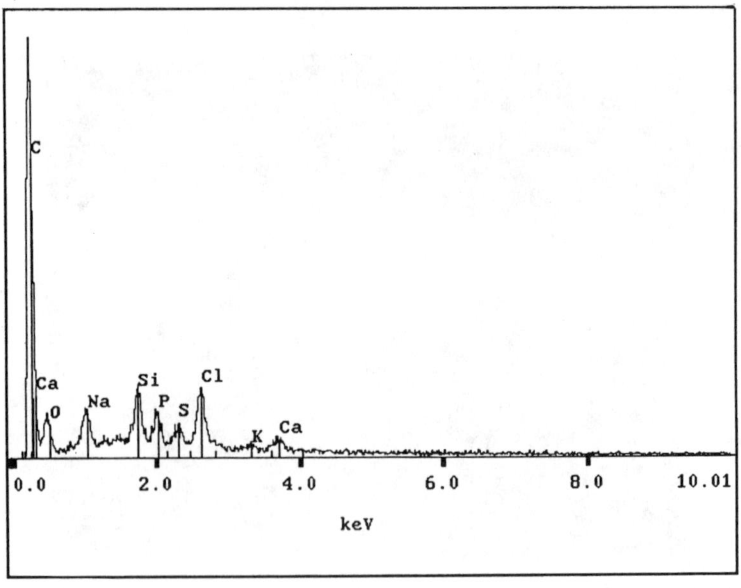

FIGURE 7 Energy-dispersive X-ray spectrum of monocyte-derived macrophages on reference polystyrene. Except for carbon, other detected elements are attributed to salts from medium.

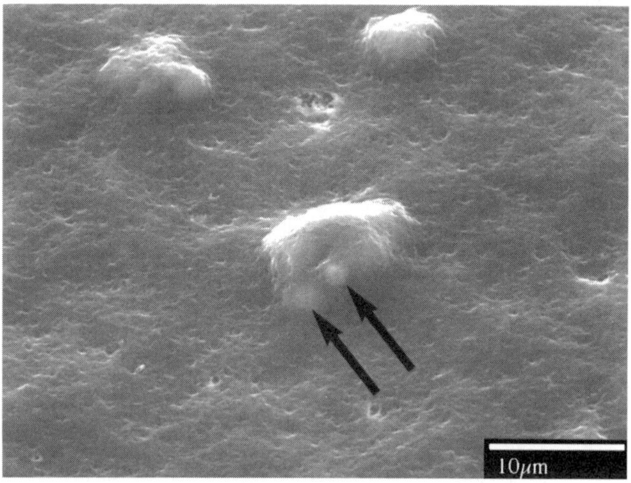

FIGURE 8 SEM photomicrograph of cells and engulfed F-75 particles (arrows) on GPTPS. Original magnification: 2000x.

change from their substratum-specific initial spread cell areas. Some irregular cell perimeters resulted from cell membrane attachments to nearby particulate matter (Fig. 9). These appeared to be mainly

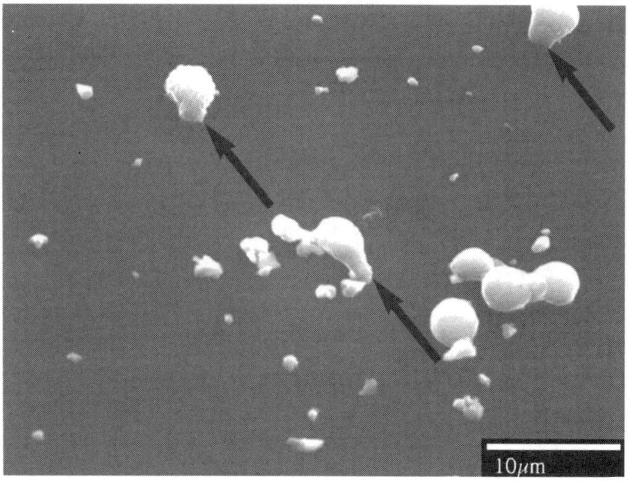

FIGURE 9 Irregular cell perimeters were observed when the cell membrane remained attached to nearby particulate matter (arrows). In the example, cells are on reference PS, and are contacting alumina particles. Original magnification: 2000x.

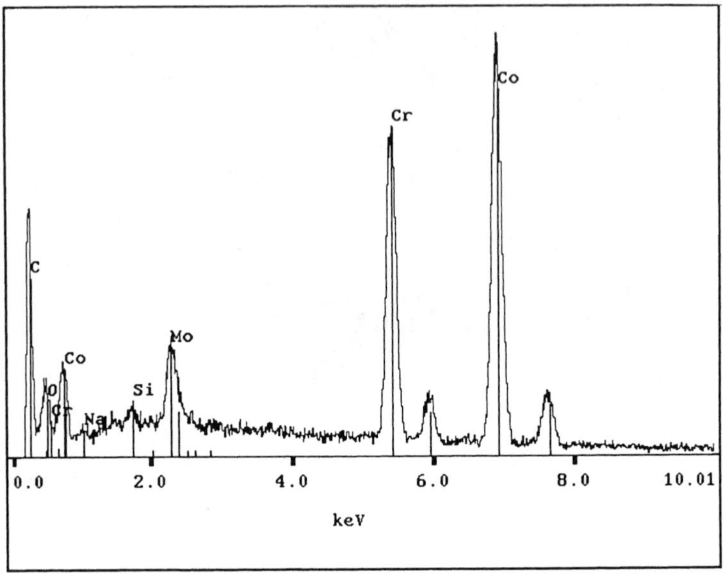

FIGURE 10 EDXray spectrum of F-75 particles engulfed by cell shown in Figure 8.

retraction fibers, and were more prominent when the higher-energy particles were on the intrinsically less adhesive (less cell spreading) reference polystyrene [15].

The various particles could be readily chemically identified, inside and outside cell bodies by using EDX spectra, with alumina, Ti, F-75, Teflon, and PE showing characteristic peaks of Al, Ti, Co-Cr-Mo, F and C, respectively. Figure 10 is an EDXray spectrum from F-75 particles within the central region of a well-spread monocyte-derived macrophage on GPTPS (see SEM photomicrograph, Fig. 8).

Surface Characteristics of Test Particles

Each dry particle preparation diminished the initial contact angles of some diagnostic liquids, first placed as drops on reference PTFE film, indicating that trace, leachable, surface-active contaminants were present in all preparations. Table III is a typical data set, in this case for the Teflon particles. Except for the F-75 particles, all the other

TABLE III The particle-induced changes in contact angles and point of initial engulfment for the Teflon test particles. Values in italics indicate leachable, surface-active contaminants on particles

Teflon particle-induced changes in contact angles on reference PTFE film and point of initial engulfment				
Reference liquid	*Surface tension (dynes/cm)*	*Average initial contact angle (°)*	*Average final contact angle (°)*	*Engulfment (yes/no)*
Water	72.0	114, 116	116, 111	N
Glycerol	65.0	111, 110	*105, 103*	N
Formamide	58.0	105, 103	*85, 86*	N
Thiodiglycol	53.5	104, 106	*89, 89*	N
Methylene Iodide	49.0	85, 89	*82, 81*	N
1-bromo Naphthalene	45.0	81, 83	83, 80	N
Methyl Naphthalene	39.0	78, 80	*75, 72*	N
Dicyclohexyl	33.0	68, 69	*54, 55*	N
Hexadecane	27.5	47, 48	*20, 20*	N
Decane	24.0	35, 33	*25, 25*	N

particles behaved as could be predicted from their nominal critical surface tensions with regard to their spontaneous engulfment or exclusion from the various diagnostic liquids. Teflon particles, with nominal critical surface tension of about 18 dynes/cm, were insufficiently "wetted" by any of the diagnostic liquids to be engulfed into the liquid drops (see Tab. III). PE particles, with nominal critical surface tension in the mid-30's dynes/cm, were engulfed by methylnaphthalene, suggesting that the surface zone of these particles was probably oxidized during the preparative steps, as occurs with polyethylene cups for artificial hips [13]. Alumina and titanium particles, nominally of high critical surface tensions, were spontaneously "wetted" and engulfed by all the diagnostic liquids. F-75 particles, from cobalt-chromium-molybdenum alloy with a nominally high critical surface tension, were excluded from diagnostic liquids of high surface tension, but were engulfed by liquids having surface tensions of 33 dynes/cm or less. These wettability data show that the F-75 particles were coated with a low-surface-energy substance, probably similar to typical organic-polish-contaminated Co-Cr-Mo surfaces of fabricated medical devices [16, 17].

Infrared (IR) spectra for the Ti particles, dried from a distilled water slurry, showed no significant absorption bands above baseline, indicating that any water-leachable contaminant associated with these particles was present in very small quantities. For the Teflon particles, dried from a distilled water slurry, only the carbon-fluorine peaks around 1200 cm^{-1} were evident, confirming the identity of this polymer, but providing no additional compositional evidence about the endotoxins present. The F-75 particles, also dried from a distilled water slurry, did not show any elutable organic contamination in their IR spectra; recall that no significant *leachable* surface-active material was detected from the contact angle analyses. IR spectra for the alumina particles, dried from a distilled water slurry, had a broad peak centered at 3400 cm^{-1}, and sharper peaks at 1650 and 1375 cm^{-1}, indicating the presence of a hygroscopic salt (probably a carbonate) contamination phase. The PE particle spectrum revealed both the characteristic absorbance bands for hydrocarbons, two between 2800 and 2900 cm^{-1} and two around 1400 cm^{-1}, and a broad band at 3400 cm^{-1} consistent with partial surface oxidation of the PE particles.

DISCUSSION

Exposure to implantable medical device materials and any associated particles resulting from wear or biodeterioration can elicit an oxidative burst from phagocytes of the immune surveillance system. Reactive oxygen intermediates (ROI), which are usually generated during phagocytic events, might vary in amounts produced and have different material-dependent consequences for tissue compatibility and inflammation. Prior studies showed that both phagocytic and ROI responses are surface-contact and adhesion dependent [1]. Although it also has been shown that macrophages exhibit variable morphological changes upon contact with chemically-diverse substrata [18, 19], no prior work has actually discriminated between the separate and combined influences of large-area biomaterial substratum surface chemistry and small debris particles, which might be simultaneously present at an implant location. This research used a chemiluminescent (CL) assay, scanning electron micrographs (SEM), and energy dispersive x-ray spectra to document the ROI production and morphological alterations of monocytes interacting concurrently with characterized small particles of Teflon, polyethylene, Co-Cr-Mo alloy, titanium, and alumina on two characterized polystyrene substrata ("bacterial"-grade and "tissue culture"-grade) [7]. The relative ROI production associated with these interactions was monitored by an oxidative burst assay [2]. Considered together, the SEM and CL assay results showed significantly greater ($p < .001$) monocyte-derived macrophage spreading upon contact with higher-surface-energy substrata (GPTPS), which significantly suppressed ($p < .001$ in most cases; $< .05$ in all cases) the "oxidative burst" observed when lower-energy substrata of the same bulk composition (PS) were contacted. Concurrent presence of particulate debris had little additional influence. Slightly more ROI was produced in the presence of titanium or alumina particles ($p < .05$), relative to the no-particle controls.

Formation of oxygen-derived free radicals by various phagocytes (neutrophils, eosinophils, monocytes, and macrophages) is catalyzed by a membrane-bound enzymatic complex, the NADPH oxidase, which is dormant in resting cells, but becomes activated during surface contact, phagocytosis or following interaction of the cells with suitable soluble stimulants [3]. This respiratory burst oxidase of phagocytes catalyzes the one-electron reduction of molecular oxygen by NADPH to form superoxide anion (O_2^-), the precursor of a number of other reactive oxidants that function as microbicidal agents [4, 5].

In addition to the NADPH oxidase, several other enzymes in phagocytes are key to controlling contact-induced ROI production, including superoxide dismutase (SOD) [20, 21], catalase and myeloperoxidase (MPO) [22]. Superoxide is dismutated to hydrogen peroxide spontaneously or by SOD. Hydrogen peroxide is further reduced to hydroxy radical *via* the iron-catalyzed Haber-Weiss reaction, in which ferrous iron (Fe^{2+}) reduces H_2O_2 to form hydroxy radical ($\cdot OH$) and hydroxide ion (OH^-). The reactive oxygen intermediates produced by phagocytes are responsible for maintaining pathogen-free environments as a part of the normal host defense mechanism, but can also pathologically mediate cell injury during activation. Current evidence concerning the mechanisms of cell injury by activated oxygen species suggests that O_2^- and H_2O_2 injure cells as a result of the generation of more potent oxidizing species [23]. It is thought that O_2^- reduces a cellular source of ferric to ferrous iron, which reacts with H_2O_2 to produce the known more potent oxidizing species, hydroxy radical ($\cdot$OH), that can then initiate the peroxidative decomposition of the phospholipids of cellular membranes, damage the inner mitochondrial membrane, or oxidatively damage DNA [23].

The biological consequences of particulate wear debris associated with implants have been related to many factors of both the particulate and bulk material, including material composition, size, shape, surface area, density and surface topography [1, 24, 25]. The rate of production of the particulate, relative to the number of phagocytes present, also may determine the extent of the biological response.

One explanation for the clear division of results obtained here, between the reference PS and GPTPS substrata groups, is that the altered chemistry of the GPTPS surface, with a surface polarity notably higher and most similar to "tissue culture" polystryene (also produced by a GPT process), was more effective in sustaining intimate contact with the cell-surface membranes. Distortion of the plasma membranes by spreading and tension could have impaired the ability of the membrane-associated NADPH oxidase to catalyze ROI production. Another explanation is that ROI was produced and was either sequestered or dissipated within the cell/substratum pocket and, thus, was unavailable to the CL probe. Other methods do exist for measuring reactive oxygen intermediates [26–31] and should be pursued in future studies.

CONCLUSIONS

These data show that modulation of monocyte-derived macrophage production of reactive oxygen intermediates (ROI), including superoxide anion (O_2^-) and hydrogen peroxide (H_2O_2), was more dependent on the contact surface properties of the large unbroken substrata surfaces than on the types of particles these cells contacted and engulfed during the first hour of exposure. This was the case for monocyte-derived macrophages in contact either with reference or gas-plasma-treated polystyrene and any of 5 types of potential implant-related wear particles (titanium, cobalt-chromium-molybdenum alloy, Teflon, polyethylene, and alumina). Specifically, cell contact with gas-plasma-treated polystyrene (GPTPS) suppressed the human monocyte-derived macrophage release of superoxide anion and hydrogen peroxide to amounts approximately 40% of those produced by identical cell contact with reference polystyrene ($p < .05$). The concurrent morphological consequence of cell contact with GPTPS was spontaneous cell spreading to about four times the surface area noted for monocyte-derived macrophages attached to reference PS. Since particle engulfment was confirmed by microscopy and elemental analysis for the monocyte-derived macrophages in both instances, differential cell spreading did not impair the events of phagocytosis but did modulate reactive oxygen generation. Or, if ROI was generated and sequestered, differential cell spreading modulated detection of the ROI by the CL probe.

Titanium particles showed slight enhancement of ROI production by monocyte-derived macrophages in contact with both reference and GPTPS, associated with loss of cell viability. Endotoxin contamination of the test particles to amounts less than 50 EU/mL did not increase ROI production or cell spreading on nonengulfable substrata. Variations of surface hydrophobicity, hydrophilicity, or apparent critical surface tension of the particles were of minor influence on monocyte-derived macrophage ROI production and morphology. Much greater during the one-hour exposure period were the influences of modification of the unbroken polystyrene surface properties from low-to high-surface-energy.

Polystyrene substratum surface energy modifications produced by GPT in argon gas and in air, and in different GPT devices, resulted in increased cell spreading and decreased ROI production by contacting monocyte-derived macrophages. These results reflect the relatively permanent changes in the surface energy state of GPTPS, persisting for long times beyond the independently documented decay of GPT-induced surface free radicals [7, 13, 15]. Suppression of "inflammatory" ROI bursts, by GPT-induced surface energy and surface polarity increases, may explain superior implantation results found for GPT prosthetics [32-34]. Follow-up experiments with flat coupons of untreated and GPT implant materials (*e.g.*, Ti, F-75, PE) in the absence of particles are planned.

Acknowledgment

The authors are grateful to Dr. Thomas Nicotera of the Roswell Park Cancer Institute, Buffalo, NY, for expert advice on reactive oxygen intermediate species.

References

[1] Carter, L. C., *Handbook of Biomaterials Evaluation* (Taylor & Francis, Philadelphia, 1999), 2nd edn., Chap. 14, pp. 241–252.
[2] Kaplan, D. S., Picciolo, G. P. and Mueller, E. P., *U.S. Patent* **5**, 294, 541 (1994).
[3] Bellavite, P., *Free Radic. Biol. Med.* **4**, 225 (1988).
[4] Babior, B. M., *Environmental Health Perspectives* **102**, 53 (1994).
[5] Johnston, R. B., *Fed. Proceedings* **37**, 2759 (1978).
[6] Picciolo, G. L., Kaplan, D. S., Kapur, R., Rudolph, A. S., Dulcey, C. S. and Tamerius, J. D., *Cells and Materials* **6**, 291 (1996).
[7] Baier, R. E., Rittle, K. H. and Meyer, A. E.,*Principles of Cell Adhesion* (CRC Press, Boca Raton, 1995), Chap. 2, pp. 41–62.
[8] Clarke, I. C. and McKellop, H. A., *Handbook of Biomaterials Evaluation* (Macmillan, New York, 1986), Chap. 10, pp. 114–130.
[9] Flynn, H. E., Natiella, J. R., Meenaghan, M. A. and Carter, J. M., *J. Prosthetic Dent.* **48**, 82 (1982).
[10] Carter, L. C., Analysis of the Cellular Healing Response of the Chick Chorioallantoic Membrane to Implanted Poly(Glycolic) Acid. *Doctoral Thesis.* State University of New York at Buffalo (1993).
[11] Hensher, R., *Brit. J. Hosp. Med.* **53**, 455 (1995).
[12] Baier, R. E., Gott, V. L. and Feruse, A., *Trans. Amer. Soc. Artif. Int. Organs* **XVI**, 50 (1970).
[13] Jahan, M. S., Thomas, D. E., Banerjee, K., Trieu, H. H., Haggard, W. O. and Parr, J. E., *Radiation Phys. and Chem.* **51**, 593 (1998).
[14] Baier, R. E., *Adsorption of Microorganisms* (Wiley-Interscience, New York, 1980), pp. 59–104.
[15] Rittle, K. H., Influence of Surface Chemical Factors on Selective Cellular Retention. *Doctoral Thesis.* State University of New York at Buffalo (1991).
[16] Baier, R. E., Gott, V. L. and Dutton, R. E., *J. Biomed. Mater. Res.* **6**, 465 (1972).
[17] Carter, J. M., Flynn, H. E., Meenaghan, M. A., Natiella, J. R., Akers, C. K. and Baier, R. E., *J. Biomed. Mater. Res.* **15**, 843 (1981).
[18] Anderson, J. M., *Trans. Amer. Soc. Artif. Int. Organs* **34**, 101 (1988).
[19] Lohmann, C. H., Schwartz, Z., Koster, G., Jahn, U., Bochhorn, G. H., MacDougall, M. J., Casasola, D., Liu, Y., Sylvia, V. L., Dean, D. D. and Boyan, B. D., *Biomaterials* **21**, 551 (2000).
[20] McCord, J. M. and Fridovich, I. J., *Biol. Chem.* **244**, 6049 (1969).
[21] Salin, M. L. and McCord, J. M., *J. Clin. Invest.* **54**, 1005 (1974).
[22] Robinson, P. J. and Babcock, G. F., *A Guide for Research and Clinical Evaluation* (Wiley-Liss, New York, 1998).
[23] Farber, J. L., *Environmental Health Perspectives* **102**, 17 (1994).
[24] Shanbhag, A. S., Jacobs, J. J., Black, J., Galante, J. O. and Glant, T. T., *J. Biomed. Mater. Res.* **28**, 81 (1994).
[25] Voronov, I., Santerre, J. P., Hinek, A., Callahan, J. W., Sandhu, J. and Boynton, E. L., *J. Biomed. Mater. Res.* **39**, 40 (1998).
[26] Goth, L., *Clin. Chim. Acta* **143** (1991).

[27] Packer, L. and Glazer, A. N., *Methods in Enzymology* **186** (1990).
[28] Vowells, S. J., Sekhsaria, S., Malech, H. L., Shalit, M. and Fleisher, T. A., *J. Immunol. Methods* **178**, 89 (1995).
[29] Pick, E. and Mizel, D. J., *Immunol. Methods* **46**, 211 (1981).
[30] Miyasaka, C. K., Souza, J. A. A., Pires de Melo, M., Pithon Curi, T. C., Lajolo, F. M. and Curi, R., *Gen. Pharmac.* **31**, 37 (1998).
[31] Hosker, H. S., Kelly, C. and Corris, P. A., *Blood Rev.* **3**, 88 (1989).
[32] Baier, R. E. and Meyer, A. E., *J. Dent. Ed.* **52**, 788 (1988).
[33] Meyer, A. E., Baier, R. E., Natiella, J. R. and Meenaghan, M. A., *J. Oral Implantol.* **14**, 363 (1988).
[34] Baier, R. E., Natiella, J. R., Meyer, A. E. and Carter, J. M., *Tissue Integration in Oral and Maxillo-Facial Reconstruction* (Excerpta Medica, Amsterdam, 1986), pp. 13–40.

The Body's Response to Inadvertent Implants: Respirable Particles in Lung Tissues

R. BAIER, A. MEYER, E. AXELSON,
R. FORSBERG, M. KOZAK and P. NICKERSON

Industry/University Center for Biosurfaces, State University of New York at Buffalo, Buffalo, NY 14214-3007, USA

D. GLAVES-RAPP

Roswell Park Cancer Institute, Buffalo, NY, USA

Instillation of respirable glass fibers to rat lungs served as an *in vivo* model for the detection and evaluation of differential local biological responses to particulate matter in the deep lung. Three compositions of vitreous glass, stonewool, and refractory fiber materials (MMVF 10, HT, and RCF1a) were harvested with surrounding lung tissues and examined both histologically and by physical/chemical assays to correlate the observed differential dissolution events with specific biological responses associated with each material. Specimens at 2-days, 7-days, 30-days and 90-days post-instillation were compared from at least three rats for each condition and for phosphate-buffered-saline controls. HT fiber surface and bulk chemistry uniquely allowed direct histochemical visualization of fiber degradation steps by Prussian Blue staining, while multiple attenuated internal reflection infrared spectroscopy and energy-dispersive X-ray analysis of unfixed, fresh lung lobe slice surfaces revealed the concurrent biochemical changes. Insulation glass (MMVF 10) dissolved most quickly in extracellular compartments, as well as after phagocytosis of small fragments; stonewool (HT) was externally thinned by surrounding phagocytes and fragmented into shorter lengths engulfable by macrophages; refractory ceramic (RCF1a) resisted both external dissolution and macrophage uptake, becoming embedded in granulomatous nodules. It is clear from these results that the lung can process inadvertently respired particulates in different ways dependent on the specific compositions of the particles.

The animal model and analytical scheme reported here also show substantial promise for evaluating the effects of bioaerosols, and synergistic effects of respirable toxins with particulates, and consequences of dental aspirates into the lung.

INTRODUCTION

Respirable particulate matter, and especially asbestos, has a long history of significant health risk, including the causation of cancer [1]. If history is not to be repeated, the bioevaluation of alternative materials that may become environmental particles when used in construction and fabrication becomes of paramount importance. This study used physical–chemical and histochemical methods to determine the location, physiologic environment, tissue reaction to, and durability of 3 types of respirable fiberglass as a model for the pattern by which other particles, inadvertently delivered to the lungs or other body tissues, might be biologically processed. Particles that persist undissolved and resistant to cellular engulfment (phagocytic action by macrophages) have the potential to cause chronic health hazards [1–6]. The three types of fiberglass were selected based on prior observations of their significantly different dissolution rates *in vitro* in model lung fluid [1, 7].

Prior research has focused mainly upon *in vitro* correlates with physical persistence of fibers in the lung [1, 3, 8, 9], not upon characterization of the *in vivo* physiologic milieu and cellular components which determine whether fibers are dissolved or if they remain and present a potential biohazard [10–13]. Furthermore, live animals proved to be necessary for studies of the fate of glass fibers, after other reports showed that the behavior of particles *in vitro* does not necessarily predict their behavior in the animal [8–13].

The study reported here utilized intratracheal delivery of particles to rat lungs, tissue preservation, and evaluation of particles in tissues, two to ninety days after instillation. Other organs also were retrieved to determine whether phagocytes transported particles away from the lungs over time.

The results clearly showed composition-dependent differences in the paths of biological processing of these fibers, with the specific involvement of phagocytic cells that could externally digest and fragment a material (HT stonewool) otherwise insoluble in extracellular fluid.

Work currently in progress includes retrieval of lung macrophages and particles by lung lavage, 2 to 90 days after instillation. The interactions of living macrophages and glass fibers are being studied by localized pH indicators [14, 15], confocal microscopy, and immunohistopathology.

METHODS AND MATERIALS

Animal Model

Fisher 344/Charles River Rats (female) were selected, since this species/strain is the standard model in published studies on interaction of particles with the lung. Rats are of sufficient size to permit visual and tactile monitoring of intratracheal tube insertion into the oropharynx without magnification. Three main types of particles with different rates of dissolution *in vitro*, with known and suspected differential impacts on lung physiology [7] were instilled. Groups of at least 3 rats per particle type accommodated variation between individuals and provided scientifically analyzable and valid data. For each of the three particle types, there were groups of 3 test rats per each of 4 post-instillation times (2, 7, 30 and 90 days), for a total of 36 animals. A single series of control rats instilled with carrier fluid (phosphate – buffered saline) was used per each of the 4 sample times [12 rats]. One additional group of untreated rats served as a control group [3 rats]. Additional time points immediately after instillation were tested to develop the various animal tissue harvest, preservation, and analysis protocols [24 rats].

Vitreous Fibers

Three types of vitreous fibers were used: refractory ceramic fiber [RCF1a], insulation fiber [MMVF 10], and stonewool fiber [HT]. Fibers already sorted for their respirable size (averaging 1 micron diameter, 20 microns in length) were obtained from fiberglass manufacturers. Fiber sizes and general compositions were confirmed by scanning electron microscopy (SEM) and energy-dispersive X-ray (EDXray) analysis. Published dissolution constants for the three types of fiberglass are given in Table I.

For each group of 3 rats, a mass of fibers sufficient for instillation of the group was measured into a 4 ml vial and delivered to the animal

TABLE I Dissolution rates of fiberglasses used in rat lung instillation experiments

Fiberglass type	*Dissolution constant* [$ng/cm^2/h$]	*Literature reference*
MMVF 10	300	[1]
HT	59	[7]
RCF1a	3	[1]

surgery facility. Just prior to instillation, saline was added to the vial and the particle suspension was vortexed briefly to distribute the fibers evenly. The initial HT fiber dispersion was unstable but this was overcome by prior mild glow-discharge treatment [16], which did not compromise the relevance of the data to actual environmental exposures.

A dose of 1.2 mg of particles in 0.2 ml saline was instilled in each rat. Each instillation control rat received 0.2 ml saline intratracheally. Other control rats received no treatment.

Particle-instillation Procedure

A blunt-end, 18-gauge intubation needle was gently inserted into the trachea of the anesthetized rat. After insertion, the particle suspension was instilled into the trachea. Instillation ensured that a prescribed dose of particles was accurately delivered to the target organ in sufficient particle numbers with potential to cause discernable tissue reactions.

After designated post-instillation time periods (2–90 days), each animal was again anesthetized prior to perfusion, and then euthanized with an overdose of the original anesthetic prior to tissue harvesting. Blood samples were obtained to ascertain the relative populations of leukocytes (white blood cells): polymorphonucleocytes, lymphocytes, monocytes, eosinophils, and basophils. This type of analysis is called a "differential blood analysis" [17], whereby the percentages of the different types of leukocytes are determined. Significant shifts in blood cell populations can be signals of inflammatory or immune reactions, or reveal problems of infection or toxin contamination that would compromise the interpretations of the subsequent analyses.

After blood samples were obtained, saline was perfused into the rat's left ventricle to displace blood from the vascular system, most importantly from microvessels in the lungs and from other organs. After saline perfusion, one lung lobe was isolated, ligated, and excised for subsequent infrared spectroscopy and SEM/EDXray analysis. The remaining lung lobes then were perfused and chemically crosslinked ("fixed") with formalin. The lungs were inflated with formalin *via* the trachea prior to final dissection and immersion in formalin-filled containers. Thoracic lymph nodes and abdominal organs (spleen, kidney, liver) also were excised and fixed in formalin for later evaluation of fiber transport from the lungs.

The animals in this study were used in accordance with regulations and standards promulgated by the New York State Department of Health, the United States Department of Agriculture, the Public Health Service, Roswell Park Cancer Institute, and the University at Buffalo. Pain or discomfort to animals was limited to that which was unavoidable in the conduct of scientifically valid research. To the best of our knowledge, the studies did not unnecessarily duplicate any other in the published literature. The animal species, numbers and procedures used were the most appropriate for the investigation performed.

Analysis of Fibers and Surrounding Tissues

Lungs from rats with intratracheally-instilled glass fibers were harvested at 2, 7, 30 and 90 days post-instillation. Lung lobes from each animal were histologically processed and stained with hematoxylin and eosin (H&E), Masson Trichrome, and Prussian Blue reagents. The Prussian Blue stain yielded superb visualization of the HT fibers at various stages of their deterioration within the lung.

Fresh freeze-fractured lung lobes and fragments were examined by SEM and EDXray analysis, as-prepared and after glow-discharge-plasma processing for "relief" exposure of the inorganic fibers from their embedding organic tissues. Fresh lung lobe slices (saline-perfusion only prior to lung lobe excision; no paraffin embedment) were applied to the faces of germanium internal reflection crystals and dried under ambient laboratory conditions. The chemistries of lung tissue slices on the germanium plates were determined using multiple-attenuated-internal-reflection-infrared (MAIR-IR) spectroscopy. The IR spectra were analyzed for absorptions related to protein (1650 and 1540 cm^{-1}), lipid (1750 – 1700 cm^{-1}), and carbohydrate and silica moieties (1020 – 1080 cm^{-1}). Collagen protein was differentiated from glycoproteinaceous extracellular substances by the IR absorption band at 1240 cm^{-1}, attributed to the uniquely high hydroxyproline content of the collagen. Following MAIR-IR spectroscopy, the tissue slices were rehydrated and removed from the germanium plates. A "print" (residue) of the lung tissue remaining adhesively bonded on the plate was examined again by MAIR-IR and then by SEM and EDXray techniques.

RESULTS

All three types of vitreous fiber inclusions could be identified in H&E and Masson Trichrome-stained sections, 2 and 7 days post-instillation in deep lung tissue. They were mainly located within granulomas, collections of tightly-packed phagocytic cells surrounding the individual fibers. Figure 1a is a scanning electron microscopic [SEM] view of an unstained, freeze-fractured lung lobe, in a region where a refractory ceramic fiber is exposed near the surface. Figure 1b is a series of elemental maps for the major components of the fiber and surrounding tissue seen in the SEM view. It is noteworthy that calcium was *not* detectable in the RCF1a fiber. It was extremely difficult to resolve the RCF1a and MMVF 10 fibers in histological sections, even with the aid of stains, but the HT fiber type did yield positive results from direct fiber-staining attempts. Despite even greater problems of resolving micrometer-level details with infrared illumination, the application of Fourier Transform Infrared Microscopy [18] also has been useful in identifying the different chemistries of the embedded fibers, their surrounding granulomatous nodules, and the general lung tissue structures. Preliminary results from IR microscopic characterization of glass fibers in lung sections are addressed in the Discussion.

Surveys of stained sections, prepared from the complete range of lung samples and viewed by light microscopy, indicated that granulomatous nodules generally diminished in number, size, and staining intensity with increasing post-instillation time for the insulation fibers (MMVF 10) and stonewool (HT). Nodules persisted and may have even increased with time in the 30–90 day post-instillation cases for the refractory ceramic fibers (RCF1a). Quantitative image analysis is in progress, to examine the prospects that insoluble refractory fibers become protected from further cellular attack in the interstitial spaces by their connective tissue and scar capsules. In the cases examined here, by the 90-day observation time, most of the granulomas around the MMVF 10 and HT fibers were resolved and only few persistent fibers of these types were found embedded and integrated in flat tissue capsules. At the light microscopic level, the HT fragments also had lost most of their Prussian Blue staining intensity, by 90-days post-instillation.

Results of Microscopic Analyses

Results of light microscopy, SEM, and EDXray analyses are reported in this section.

(a)
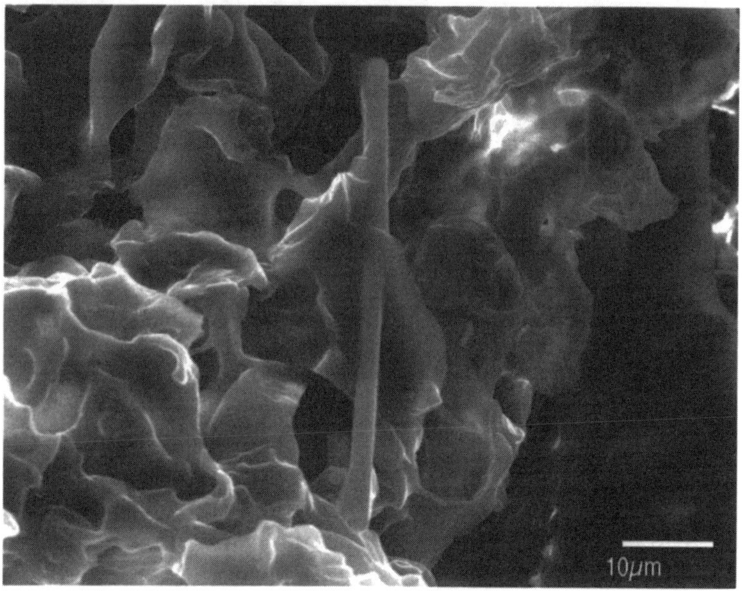

FIGURE 1 (a) Scanning electron micrograph for RCF1a fiber in rat lung for 30 days. (b) Elemental maps acquired by energy-dispersive X-ray analysis for constituents of fiber and lung tissue structures shown in SEM micrograph.

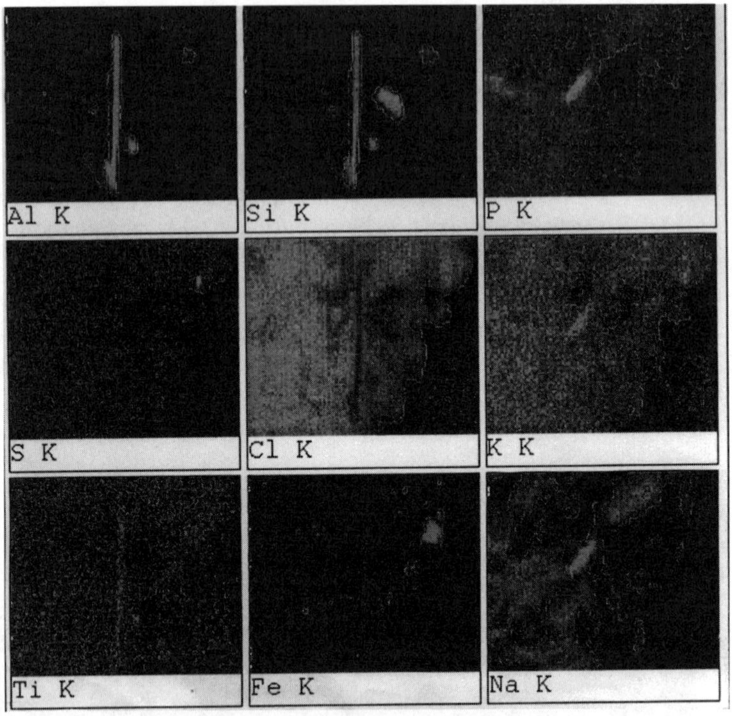

(b)

FIGURE 1 (Continued). (See Color Plate I).

The HT fibers uniquely gave a bright blue-purple reactive coloration when the tissue preparation was stained with Prussian Blue dye. This allowed rapid location of the small fibers and direct visualization of fiber breakdown. Light microscopic views, such as that shown in Figure 2, demonstrated that HT fiber breakdown often occurred in two steps: (1) localized fiber breakage or digestion at locations where closely approximated phagocytes were associated with selectively thinning fiber sites; and (2) generalized dissolution over time in a patchy fashion along the fiber. Broken segment lengths usually appeared in close proximity to the adjacent, surrounding reactive cells.

Figure 3 shows such fibers transferred from a fresh lung lobe slice to a smooth germanium internal reflection prism, in a "lung print" left behind when this tissue was gently peeled from the prism after multiple attenuated internal reflection infrared analysis. In Figure 3, the fibers

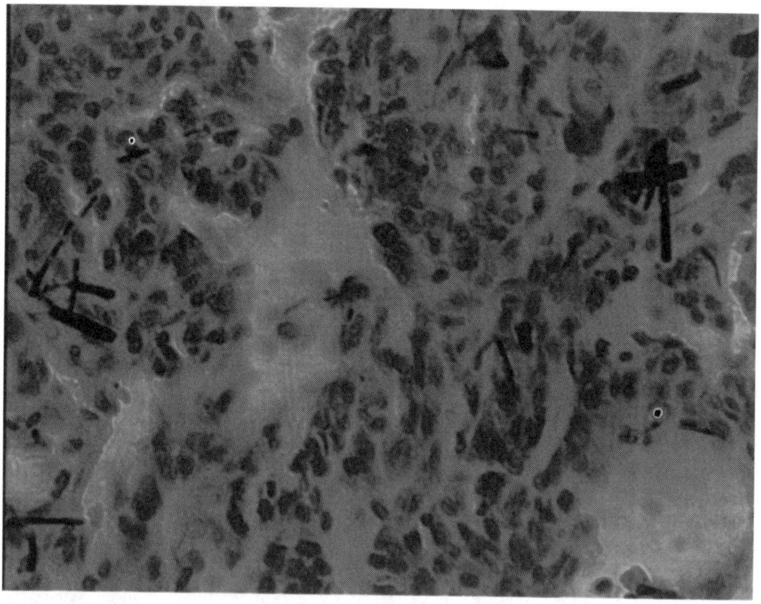

FIGURE 2 Light microscopic view of HT fibers in rat lung 2 days after instillation. Note irregular edges and ends of broken fiber segments in upper-left quadrant of this view. Average diameter of fibers seen here is 1 micrometer.

seen are from freshly-harvested lung tissue 7-days post-instillation with HT (stonewool). It is noteworthy that these fiber surfaces appeared, at only 7 days, already significantly micro-pitted and porous. In contrast, the exteriors of RCF1a refractory fibers, similarly examined even at the longer 30- and 90-day observation points, remained superficially smooth and optically dense.

Figure 4a is a scanning electron microscopic view of a freeze-fractured, glow-discharge-relieved deep lung tissue segment 7-days post-instillation of HT fiber. Such exposed HT fiber surfaces usually were seen to be pitted and porous. Figure 4b confirms the chemistry of HT fibers as published and compared with numerous other vitreous fiber types [7] by EDXray analysis. HT is uniquely rich in calcium as well as having the high amounts of aluminum and silicon characterizing refractory RCFla fibers [7].

Figure 5 is the elemental map for this same tissue region, where the co-locations of aluminum, silicon, and calcium within the HT fiber

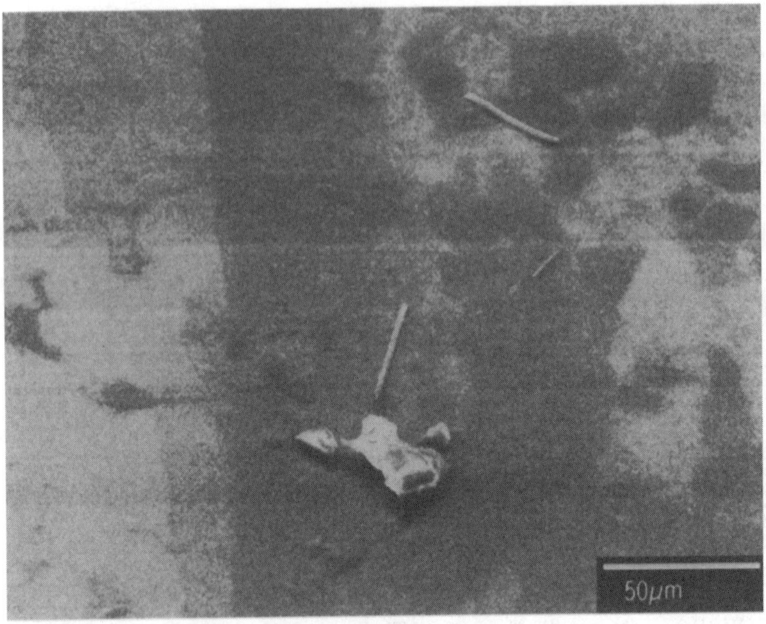

FIGURE 3 Scanning electron micrograph of HT fibers transferred from fresh lung slice to germanium internal reflection prism. Tissue slice was peeled from the prism after analysis by MAIR-IR spectroscopy. The IR spectrum of the residue seen here also was obtained.

body are obvious. However, the aluminum and silicon elemental signals are considerably less intense than obtained under the same conditions for similarly-sized RCF1a fibers (compare this 7-day result for HT with the 30-day result for RCF1a, Fig. 1b). These energy-dispersive X-ray results also independently confirmed the light-microscopic findings, where Prussian Blue staining showed rapid and substantial loss of apparent HT fiber mass in regions of high phagocytic cellular abundance.

Many of the 7-day specimen SEM views of HT fibers showed narrowing fiber diameters (necking) adjacent to the regions of closest cell membrane approximation, suggesting an active digestive process that did not require fiber internalization or engulfment [19].

Some scanning electron microscopic views of 30-day HT-instilled-lung specimens showed still-smooth fibers "torn" from adjacent tissue

(a)
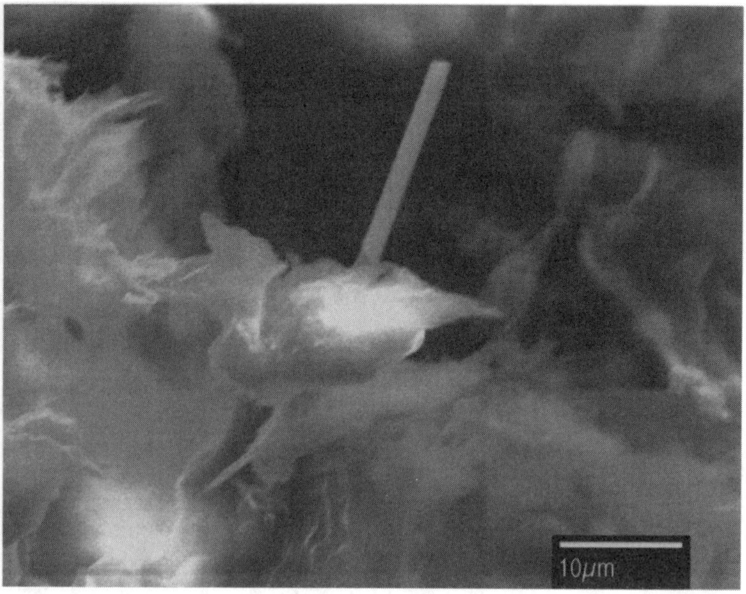

(b)
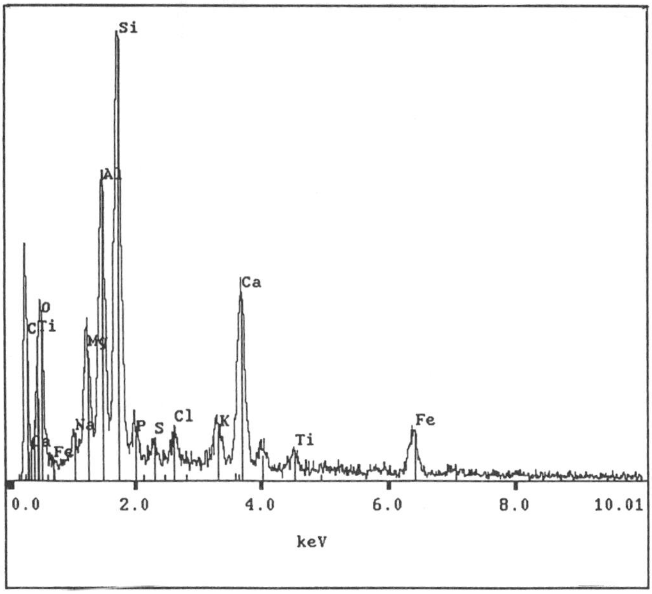

FIGURE 4 (a) Scanning electron micrograph for HT fiber in rat lung for 7 days. (b) Energy-dispersive X-ray spectrum of fiber and lung tissue shown in SEM micrograph.

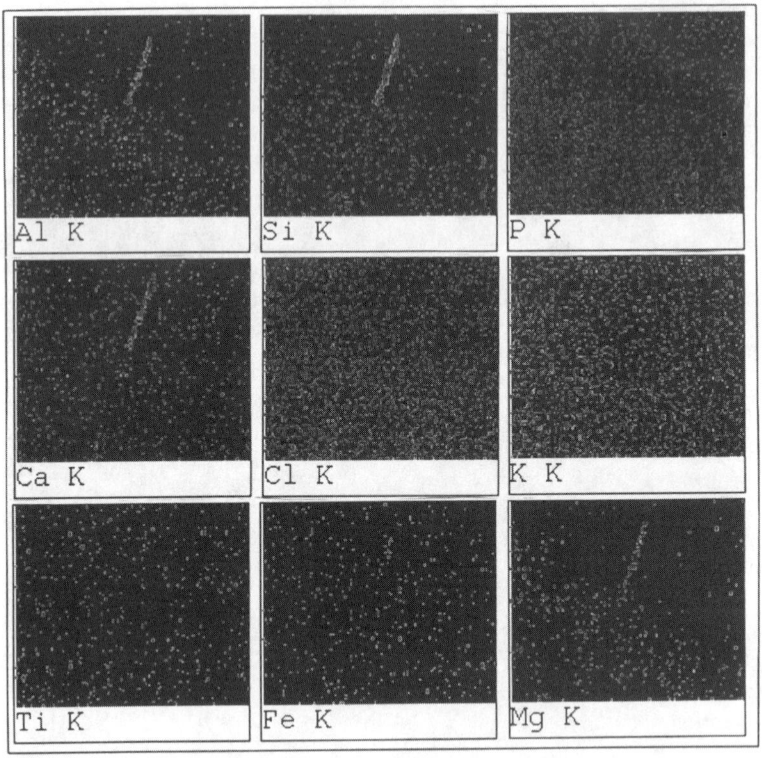

FIGURE 5 Elemental maps acquired by energy-dispersive X-ray analysis for constituents of fiber and lung tissue structures shown in Figure 4a. Compare with data obtained for refractory ceramic fiber RCF1a in tissue for 30 days (Fig. 1b). (See Color Plate II).

crypts. Such SEM views correlated with "scar" sites seen in the stained tissue sections where fiber-shaped voids were present as well as irregularly-shaped, still-present fiber fragments nearby. The longer-preserved HT fibers apparently had been protected from the scavenging phagocytes by collagen and fibroblasts of the usually dominant "foreign body response" of living tissues [20, 21]. Collagen was identified in these capsules by its characteristic blue–green color after Masson Trichrome staining of tissue sections. This result shows that general body fluids, extracellular matrix, or tissue chemistry away from reactive phagocytes, were not responsible for the degradation of HT fibers seen in the rat lungs examined in the current study.

Results of Infrared Spectroscopy

As illustrated in Figure 6, the removal of the lung slices from the faces of the analytical IR prisms correlated with a spectral loss of absorption bands typical of the collagen-based lung tissue scaffolding, leaving behind abundant glycoproteinaceous extracellular substances and variable amounts of lung surfactant (lipid).

Labels on Figure 6 designate the covalent chemical absorption bands characterizing the faces of fresh lung slices placed on germanium prisms, examined by multiple attenuated internal reflection infrared [MAIR-IR] spectroscopy. The spectra of as-placed segments (Fig. 6a) and the same specimens after peeling the bulk

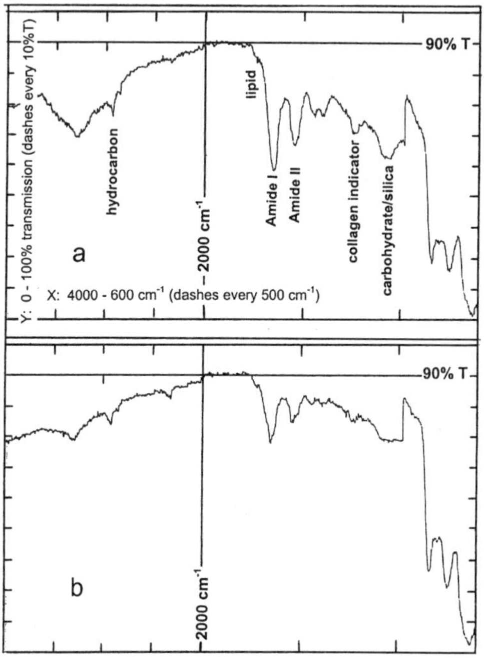

FIGURE 6 (a) Infrared spectrum, obtained by multiple-attenuated internal reflection technique, of rat lung tissue (HT, 2 days) on germanium internal reflection prism. (b) IR spectrum of residue on germanium prism after rehydrating and removing slices of lung tissue. Note that 100% of the lung surfactant (lipid component indicated by absorption between 1700 and 1750 cm^{-1}) remains as residue after the tissue is removed. Bands associated with protein and carbohydrate lost an average 61% of their intensities. Figure 2 is a SEM micrograph of the residue on the prism.

tissues from the prisms to leave behind "lung prints" always produced records of the sort noted in Figure 6b. These IR spectra reveal how removal of the bulk tissue selectively diminished the collagen-related absorption at about 1240 cm^{-1}, while leaving behind most of the lung surfactant phase with its lipid-related absorption at about 1730 cm^{-1}. Specific absorption bands for harvested lung lobe slices were compared with one another and from lung-to-lung and fiber-type-to-fiber-type throughout the 2-day to 90-day experimental period.

Figure 7 charts the most revealing of these comparisons, the ratio of protein to extracellular matrix materials (polysaccharide-rich) over time. The significantly greater and persisting abundance of protein in the refractory fiber RCFla-instilled lungs correlated with the formation and persistence of granulomas surrounding these dissolution-resistant fibers, as viewed by light microscopy of stained sections from the same lungs. It is helpful, here, to reinspect Figure 1a (a SEM view) and Figure 1b (a corresponding energy-dispersive X-ray elemental map) for fiber segments in a freeze-fractured and then glow-discharge-relieved lung specimen from a rat harboring RCF1a fibers for 30 days. Intense signals for aluminum, silicon and titanium characterize the fiber; weaker signals for phosphorus, sodium and iron characterize the cells of the surrounding granuloma. Fiber "blocking" of the

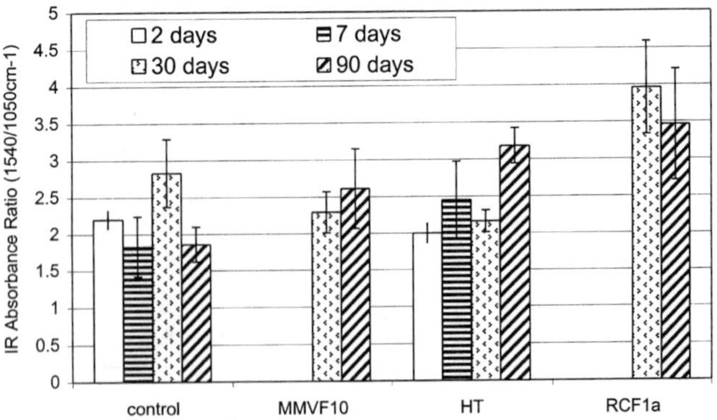

FIGURE 7 Chart of IR absorbance ratios of the Amide II protein absorbance (1540 cm^{-1}) and the carbohydrate absorbance (1050 cm^{-1}) of rat lung tissue after instillation of phosphate–buffered saline (control) and vitreous fibers. Data are presented as averages ± standard deviations.

background chlorine signal from the tissue in which it embedded attests to the retained electron density of the structure, despite 30 days of apparently aggressive attack by particle-digesting cells.

Adjacent areas of these same tissues, which appeared fiber-free at the scanning electron microscopic level, often indicated below-surface vitreous fiber presence by the energy-dispersive X-ray signals seen for the major fiber elements.

Infrared spectra showed similar increases and persistence for other macrophage-related absorptions in RCF1a-instilled lungs, while the control specimens displayed – by both infrared and microscopic methods – diminishing macrophage involvement by the 90-day observation time.

In these latter cases, although more analysis is required to confirm this finding, the IR spectra also indicated the persistence of silica-related absorptions in the originally fiber-instilled lungs. Export of the products of dissolution of glass from lung tissues was not as rapid a process as fiber disappearance in the rats.

TABLE II Results of differential blood analyses, showing average percentages of 5 types of leukocytes in blood of groups of test animals. Except where noted, $n = 3$ animals per group. [N: neutrophils, PMNs; L: lymphocytes; M: monocytes; EOS: eosinophils; BO: basophils]

Fiber type and	*Cells as percentage of Leukocytes present in blood*				
post-instillation time	*N*	*L*	*M*	*EOS*	*BO*
(PBS control)					
*2 days	17 ± 2	79 ± 1	2 ± < 1	2 ± 2	< 1
7 days	16 ± 4	80 ± 6	4 ± 2	< 1	< 1
92 days	22 ± 2	76 ± 1	< 1	2 ± 1	< 1
*$n = 2$					
MMFV 10					
91 days	12 ± 1	86 ± 1	1 ± 1	< 1	< 1
HT					
2 days	16 ± 3	83 ± 3	1 ± < 1	< 1	< 1
7 days	17 ± 4	80 ± 6	1 ± 1	2 ± 1	< 1
30 days	15 ± 4	83 + 4	1 ± < 1	1 ± < 1	< 1
90 days	10 ± 3	84 ± 3	5 ± 2	1 ± < 1	< 1
RCF1a					
30 days	16 ± 2	80 ± 2	2 ± < 1	1 ± < 1	< 1
92 days	19 ± 5	77 ± 6	3 ± 2	1 ± < 1	< 1

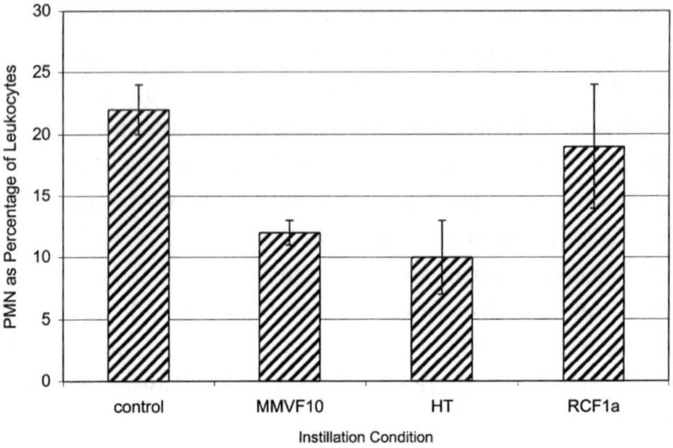

FIGURE 8 Percentage of PMN cells (polymorphonucleocytes, neutrophils) in leukocyte populations of blood samples from rats 90 days after instillation of phosphate–buffered saline (control) and vitreous fibers. Data are presented as averages ± standard deviations.

Results of Differential Blood Analyses

Results of the differential blood analyses are summarized in Table II. From Table II, it is noted that percentages of the different types of leukocytes were relatively consistent within any particular group of animals, among post-instillation time periods, and between fiber types. Lymphocyte levels were particularly consistent across all groups. Figure 8, however, highlights a trend noted in percentages of polymorphonucleocytes (PMNs, or neutrophils) 90 days after instillation. It is seen that animals instilled with MMVF 10 or with HT fibers had fewer neutrophils in their blood than animals instilled with RCF1a or only the phosphate-buffered saline.

DISCUSSION

Instilled fibers of all three types usually were located within granulomatous cell clusters formed within 2 days of fiber instillation. HT fibers, uniquely, were broken down most rapidly into smaller segments at the sites of phagocytic cells attached along their lengths. These fibers otherwise were too long to be engulfed and internally digested within the acidic granules of macrophages [1]. The observation that HT fibers not yet broken into smaller lengths showed local narrowing and loss of Prussian Blue-stainable substance mainly in regions with tightly attached phagocytes, is consistent with the finding by Carter [22] of external digestion of large particles by phagocytes converting to osteoclasts. Residual HT fragments were found to be depleted of constituents in proportion to the fibers' original elemental ratios, when examined by EDX-ray analysis. Early results from examination of the thoracic lymph nodes retrieved from HT-instilled rats confirm that fiber fragments were transported from the lung to the lymph nodes and possibly more peripheral locations.

Infrared imaging results obtained by colleagues at the National Institutes of Health, for representative lung specimens from this study, also identified the general lung tissue background to be collagen and glycoprotein-dominated "ground" substance, with some lipoidal surfactant. Granulomatous accretions had enhanced infrared absorptions in regions associated with carboxyl and nitroso groups, attributed to macrophages, as well as a small silicate band attributed to embedded glass fibers. Isolated glass fiber lengths protruding into lung vacuoles were surrounded by spread cells and had characteristic infrared spectra dominated by the silica absorption band and by macrophage-related bands; little collagen, glycoprotein, or fatty substance was present.

Infrared spectra of sequentially-harvested lung lobe slices, from rats living 2, 7, 30 and 90 days after glass fiber instillation, showed increasing ratios of absorption bands associated with granuloma formation only for RCF1a fibers at the final date. All specimens of fiber-instilled lungs continued to display silica absorption bands for the entire 90-day experimental periods, however. These findings show that the components of the rapid dissolution of the MMVF 10 insulation fiber and of the slower dissolved (shortened, then engulfed) HT stonewool remained predominantly within the lung tissue and were not exported to other organs in significant quantities. It is possible that dissolution-product-induced sequestration of PMNs within the lung, diminishing the numbers in circulation, accounts for the data displayed in Figure 8. Additional analyses of lymph nodes, liver, spleen, and kidney from the test animals are continuing, to investigate transport kinetics of instilled fibers or dissolution products from the lungs to these other organs.

Results of this study with respirable vitreous fibers of known, controlled compositions suggest that all particles entering the lungs, in spite of initial encounters with similar extracellular fluid, matrix, surfactant and cellular types, can induce composition-dependent differential responses from the active phagocytic cells that surround them. The types of observations made in various tissue sites have recently been reviewed [23] and examined from the perspective of hypersensitivity induction [24]. A growing body of literature focuses on particles cast as wear debris into internal body tissues [25–27], illustrating the need for more distinct discrimination of the cellular/particle interactions in many body sites.

Most intriguing is the observation by Carter and co-workers [19, 22] that giant cells with osteoclast-like (mineral-dissolving) properties are preferentially induced by calcium–phosphate and other "digestible" particulate debris while giant cells induced by polymethylmethacrylate particles, polypropylene particles, or even by mineral talc particles, display features characteristic only of classical "foreign body" inflammatory polykarya, which do *not* degrade extracellular substrata. It is relevant that the quicker "clearing" HT fibers investigated here were calcium-rich, and also contained some phosphate. This evidence, linking material compositional properties to the induction of specific biological "clearing" responses, should be taken into account when evaluating new prosthetic devices that may generate wear debris during their functional correction of deformities, or pathological states, in human recipients. More such evidence must be gathered regarding the "inadvertent implants" of occupationally- or environmentally-generated particles that collect in the lung, and in other body sites.

The infrared spectroscopy results reported here, revealing generalized retention of dissolved particulate silica throughout the lung bed, require confirmation and extension with newer histopathologic staining techniques to identify morphological changes that could forecast later functional changes. Blood specimens taken from experimental animals at the times of lung harvest, when inspected for changes in white cell "differential" counts, were within the normal values reported for rats [17]. Since the blood of the rats in this study remained within the published broad range of "normal" blood chemistry, it is unlikely that the results reported here were confounded by any infectious, toxic, or other pathologic process.

Yet, in comparison with the saline-instilled (control) rats in this study, an interesting pattern was seen that requires more studies to be verified. Whereas the saline-control and insoluble RCF1a fiber-instilled animals showed increases in circulating neutrophil abundance over the 90-day period, the more-soluble-fiber-instilled animals (MMVF10 and HT) showed decreases in neutrophil abundance, relative to control animals. This might be interpreted as preliminary evidence of an ongoing nontoxic but still "reactive" response of the lung tissue to dissolution products of fiberglass, while the saline control and insoluble particle-instilled lungs recovered more quickly from the initial ionic imbalance. Of equal interest and importance is the observation that the rat blood lymphocyte counts were essentially invariant over the whole experimental animal population, confirming and extending the findings that these animals remained immunologically uncompromised, throughout the study [17].

The influences of the body's surface-active substances should not be neglected in considering how small particles are dealt with physiologically. Here, the lung provides a possibly extreme example. The surface properties of the lung's inner envelope are generally attributed to an abundant lipid (dipalmitoyl phosphatidylcholine, DPPC)-dominated material called "pulmonary surfactant" [28]. Pulmonary surfactant preparations made by chloroform–methanol extraction of cell-free brochoalveolar lavages of lungs of freshly-killed calves improve lung mechanics significantly in both premature lambs and human infants [29], and display surface tensions as low as 42 mN/m. Our research group has performed preliminary Langmuir-Adam trough experiments on lung surfactant interactions with vitreous fibers. Lung surfactants concentrated at the air/water interface (*in vitro*) reduced the surface tension of pure water and salt-solutions to the same low values, and triggered the aqueous-phase engulfment (sinking) of the respirable glass fibers used here that otherwise were retained at the air/water interface. Similar results were obtained with asbestos fibers.

New techniques are required to make measurements within harvested lung tissues, macrophages from lung lavages, and additional harvested animal specimens to test hypotheses on the environment and fate of instilled vitreous fibers and other particle types unambiguously. Animal trials must be extended to examine consequences of respirable fiber inhalation *versus* intratracheal instillation. Extrapolation to events associated with similar particles impacting ocular tissues [30] must be done with care.

It is important that the required methods be advanced quickly. Alveolar macrophages undergo various morphological and surface changes upon contact with different particulate substrata, including glass, aluminum foil and cotton [31, 32], and similar changes have been noted from humans having occupational exposures to inorganic particles, including asbestos, silica, and coal [33].

Production of particulate wear debris is an inevitable consequence of the implantation of alloplastic materials into loaded human joints, sometimes with tragic sequelae as in the universal failure of certain TMJ (temporomandibular joint; jaw joint) implants fabricated from silicone, fluorocarbon, hydrocarbon, metallic, and mineral constituents [19, 34]. It is a regulatory necessity that the required database on particle-tissue interactions be significantly expanded. Similarly, since acute and chronic respiratory distress syndromes can develop from inadvertent contact, adhesion, and uptake of aerosolized diesel exhaust particulates and even dental drill particulates entering the lung [35], there is a strong demand for these data in environmental health fields, as well [36].

CONCLUSION

Particle contact-induced differentiation of a host's reacting cell populations can follow alternate paths with separate end points responsive to the specific surface chemistries of the contacting materials themselves. In all cases of particle intrusion to the body, it is critical to detect and decipher the relationships between the surface properties of the particles and the receiving biofluid/cells at the instant of effective contact. Independent documentation of the identity of those particles engulfed and modified by the cellular bodies also is important, to prevent attribution of negative consequences to an improperly identified material. It is possible, for example, that differentiation of blood monocytes into either macrophages, which engulf and digest particles internally, or osteoclast-like phagocytes, which can acid-digest nonengulfable particles externally, depends on the calcium–phosphorus contents of implanted mineral debris.

Acknowledgments

This work was supported by the Industry/University Cooperative Research Center for Biosurfaces (IUCB), and used respirable fiberglass particles donated by industry members of IUCB. We thank Dr. Pina Colarusso (National Heart, Lung, and Blood Institute; Laboratory of Kidney and Electrolyte Metabolism) for IR microscopic and synchrotron analyses of lung sections. We also are grateful to Dr. Bruce Holm and Dr. Edmund Egan of the University at Buffalo, who provided expert advice on lung surfactant and lung cell harvest.

References

[1] Eastes, W. and Hadley, J. G., *Inhal. Toxicol.* **8**, 323 (1996).
[2] Hesterberg, T., Miller, W., Hart, G., Bauer, J. and Hamilton, R., *J. Occup. Health Safety* **12**, 345 (1996).
[3] Yu, C. P., Dai, Y. T., Boymel, P. M., Zoitos, B. K., Oberdorster, G. and Utell, M. J., *Inhal. Toxicol.* **10**, 253 (1998).
[4] Hesterberg, T. W., Axten, C., McConnell, E. E., Oberdorster, G., Everitt, J., Miller, W. C., Chevalier, J., Chase, G. R. and Thevenaz, P., *Environ. Health Perspectives* **105**(Supp. 5), 1223 (1997).
[5] Hesterberg, T. W., Hart, G. A., Chevalier, J., Miller, W. C., Hamilton, R. D., Bauer, J. and Thevenaz, P., *Toxicol. Appl. Pharmacol.* **153**, 68 (1998).
[6] Kamstrup, O., Davis, J. M. G., Ellehauge, A. and Guldberg, M., *Ann. Occup. Hyg.* **42**, 191 (1998).
[7] Hesterberg, T. W., Chase, G., Axten, C., Miller, W. C., Musselman, R. P., Kamstrup, O., Hadley, J., Morscheidt, C., Bernstein, D. M. and Thevenaz, P., *Toxicol. Applied Pharmacol.* **151**, 262 (1998).
[8] Potter, R. M. and Mattson, S. M., *Glastech. Ber.* **64**, 16 (1991).
[9] Mattson, S. M., *Ann. Occup. Hyg.* **38**, 857 (1994).
[10] Eastes, W. and Hadley, J. G., *Inhal. Toxicol.* **7**, 179 (1995).
[11] Bernstein, D. M., Morscheidt, C., Grimm, H.-G., Thevenaz, P. and Teichert, U., *Inhal. Toxicol.* **8**, 345 (1996).
[12] Eastes, W., Morris, K. J., Morgan, A., Launder, K. A., Collier, C. G., Davis, J. A., Mattson, S. M. and Hadley, J. G., *Inhal. Toxicol.* **7**, 197 (1995).
[13] Mattson, S. M., *Environ. Health Perspectives* **102**(Supp. 5), 87 (1994).
[14] Eisner, D. A., Kenning, N. A., O'Neill, S. C., Pocock, G., Richards, C. D. and Valdeolmillos, M., *Pflugers Arch.* **413**, 553 (1989).
[15] Seksek, O., Henry-Toulme, N., Sureau, F. and Bolard, J., *Analytical Biochem.* **193**, 49 (1991).
[16] Baier, R. E. and Meyer, A. E., *Intl. J. Oral Maxillofac. Implants* **3**, 9 (1988).
[17] Harkness, J. E. and Wagner, J. E., *The Biology and Medicine of Rabbits and Rodents* (Lea & Febiger, Publishers, Philadelphia, 1989), p. 49.
[18] Paschalis, E. P., Boskey, A. L. and Nancollas, G. H., *Advances in Materials Sciences and Implant Orthopedic Surgery* (Kluwer Academic Publishers, Amsterdam, 1995), pp. 47–60.
[19] Carter, L. C., Carter, J. M., Nickerson, P. A., Baier, R. E., Wright, J. R. and Meenaghan, M. A., *J. Adhesion* **74**(1–4), 53–77 (2000).
[20] Coleman, D. L., King, R. N. and Andrade, J. D., *J. Biomed. Mater. Res.* **8**, 199 (1974).

[21] Baier, R. E., Meyer, A. E., Natiella, J. R., Natiella, R. R. and Carter, J. M., *J. Biomed. Mater. Res.* **18**, 337 (1984).
[22] Carter, L. C., Analysis of the Cellular Healing Response of the Chick Chorioallantoic Membrane to Implanted poly(Glycolic Acid). Doctoral Dissertation. State University of New York at Buffalo (1993).
[23] Carter, L. C., *Handbook of Biomaterials Evaluation* (Taylor & Francis, Philadelphia, 1999), Chap. 14, pp. 241–252.
[24] Merritt, K., *Handbook of Biomaterials Evaluation* (Taylor & Francis, Philadelphia, 1999), Chap. 18, pp. 291–300.
[25] Goodman, S. B., Lind, M., Song, Y. and Smith, R. L., *Clin. Orthopaed. Related Res.* **352**, 25 (1998).
[26] Santerre, J. P., Labow, R. S. and Boynton, E. L., *Canadian J. Surgery* **43**, 173 (2000).
[27] Brown, S. L., Silverman, B. G. and Berg, W. A., *Lancet* **350**, 1531 (1997).
[28] Holm, B. A. and Matalon, S., *Anesth. Analg.* **69**, 805 (1989).
[29] Notter, R. H., Egan, E. A. and Kwong, M. S., *Pediatr. Res.* **19**, 569 (1985).
[30] Stokholm, J., Norn, M. and Schneider, T., *Scand. J. Work Environ. Health* **8**, 185 (1982).
[31] Leake, E. S. and Wright, M. J., *J. Reticuloendothelial Society* **25**, 417 (1979).
[32] Leake, E. S., Wright, M. J. and Myrvik, Q. N., *J. Reticuloendothelial Society* **17**, 370 (1975).
[33] Takemura, T., Rom, W. N., Ferrans, V. J. and Crystal, R. G., *Am. Rev. Respir. Dis.* **140**, 1674 (1989).
[34] Baier, R., Axelson, E., Meyer, A., Carter, L., Kaplan, D., Picciolo, G. and Jahan, M., *J. Adhesion* **74**(1–4), 79–101 (2000).
[35] Mazzarella, M. A. and Flynn, D. D., *An Introduction to Experimental Aerobiology* (Wiley-Interscience, New York, 1969), Chap.18, pp. 437–462.
[36] Effros, R. M., *Inhalation Aerosols: Physical and Biological Basis for Therapy* (Marcel Dekker, Inc., New York, 1996), Chap. 5, pp. 139–154.

Elastic and Viscoelastic Contributions to Understanding Particle Adhesion

Measurement of the Adhesion of a Viscoelastic Sphere to a Flat Non-Compliant Substrate

MARK REITSMA and SIMON BIGGS

The Centre for Multiphase Processes, Department of Chemistry, The University of Newcastle, Callaghan, NSW 2308, Australia

VINCE S. J. CRAIG

The Centre for Multiphase Processes, Department of Chemical Engineering, The University of Newcastle, Callaghan, NSW 2308, Australia

The adhesion between a single polystyrene bead (radius, 27 μm) and a flat silica surface has been measured with an atomic force microscope as a function of two variables: (a) The maximum applied load and, (b) the loading time at a constant maximum applied load. Analysis of the results indicates significant plastic deformation of the bead under the action of the load forces. There is also evidence for time-dependent viscoelastic effects as a load is exerted on the bead. The contact zone of the polystyrene bead used for these experiments was examined using Scanning Electron Microscopy. The microscope images revealed a surface covered in small polymer beads with a radius of only 115 nm. In the contact zone these beads had undergone substantial and permanent deformation as a function of the applied load. Basic geometric analysis reveals that the large sphere is not contacting the flat surface under any load. The results presented here indicate the value of being able to measure adhesion using an atomic force microscope. The importance of being able to characterise the contact zone accurately is also highlighted.

INTRODUCTION

The adhesion of spheres to flat surfaces is of direct relevance to a range of processes both from the industrial viewpoint and in the natural world. For example, in the natural world adhesion plays a critical role in processes such as blood clotting, platelet binding, and leukocyte adhesion to cell walls [1]. In the technological world, the control of adhesion has important consequences in a wide variety of processes such as drug delivery, xerography, paints, and solids handling [2]. As a result, much effort has been devoted to gaining both a theoretical understanding and experimental insights into adhesion mechanisms.

The adhesion of any sphere to a surface will depend on the area of contact at separation [3]. However, theoretical determination of this area of contact is not easy. The application of contact mechanics to problems of this type has a long history. Early work utilised simple Hertzian deformation characteristics and assumed that the spheres were elastic indentors [4, 5]. In this approach, the adhesion between the surfaces is unsustainable and the compressive stresses in the contact region lead to the conclusion that, at detachment, no tensile load is required and the area of contact is zero. These theories were subsequently enhanced to include contributions from surface forces [6]. It can be shown that these surface forces may be sufficiently large as to exceed the elastic limit of the materials and allow plastic deformations. The major problem is to decide over what areas of the interacting surfaces these forces are large enough to have a real effect. For a material that is essentially elastic under a standard load regime, two main theories have been developed. These are the so-called JKR (Johnson-Kendall-Roberts) [7] and DMT (Derjaguin-Muller-Toporov) [8] theories. The JKR approach assumes that surface forces act in the contact region but are absent outside of it. This leads to the interesting result that the tensile stresses are infinite at the edge of the contact zone. The DMT theory does allow for the action of cohesive surface forces outside of the contact zone but it assumes that these forces do not alter the shape of the material in this zone from the Hertzian profile. This results in the adhesive stress being zero in the contact region and finite in this cohesive zone. The differences between these theories are significant and lead to very different predictions. In the DMT case, the contact area – load profile is essentially equivalent to the Hertzian case offset such that at any given load the contact area will be larger. The additional surface forces result in the retention of a finite area of contact under tensile loads up to the point where detachment occurs; at this point, the contact area goes continuously to zero. The JKR case is more complex under tensile loads. As a tensile load is applied, the contact area decreases towards a minimum, but finite, value. The retention of some contact area, even at the pull-off point under tension, is characteristic of the JKR approach.

It has been shown by Tabor [9] that these two theories apply to opposite ends of the same spectrum. The important parameter to consider is the ratio of the elastic deformation at pull-off (separation) to the range of the surface forces. For large compliant spheres, where the deformation may be substantial, this ratio is much larger than one. This is the region of application for the JKR theory. Alternatively, the DMT theory is applicable to small, stiff spheres where this ratio is less than one. Subsequently, the transition region between these two limits was investigated and complete solutions were developed [10]. A full description of the transition region between these two theories is the subject of much continuing research. In a recent paper, Johnson and Greenwood [11] presented a comprehensive adhesion map to describe the limits for each theory based on the deformation ratio and the normalised load (applied load/radius). It should be remembered, of course, that these theories and their developments are used to describe the contact of elastic bodies.

In reality, very few materials can be described as simple elastic spheres; this is especially true when we consider polymeric materials. The mechanics of contact may then be described as elastic, non-linear elastic (elasto-plastic), or plastic. In addition, time-dependent or visco-elastic effects may be important for any material under load. The type of contact mechanics that is applicable, and the extent of viscoelasticity, will be affected by the magnitude of any applied load and the time scale over which it is applied. As is clear from the above discussion, elastic contact mechanics have been extensively investigated. The plastic regime has also been the subject of investigation. Maugis and Pollock (MP) [12] developed the JKR theory to take into account plastic deformations of the interacting surfaces. The non-linear elastic or elasto-plastic regime is currently less-well described. Visco-elastic effects have also been the subject of recent research [13]. In these cases, the theories attempt to describe the change in the contact radius with time as a result of viscoelastic creep under load.

The development of new experimental techniques over the last decade has allowed a comprehensive investigation of the applicability of these theories. Clearly, the area of contact as a function of applied load is of critical importance. The contact between atomically-smooth mica surfaces has been probed using the Surface Forces Apparatus (SFA) [14]. The results of these investigations indicated a contact area – load relationship that was well modelled by the JKR relationship. These measurements were performed between two crossed hemicylinders of mica with radii in the order of 1 cm; the material properties of mica coupled with these large radii are clearly sufficient to access the JKR regime.

Other experimental approaches have involved the use of nanoindenters [15] and scanning electron microscopy (SEM) [16, 17]. The use of SEM micrographs is an interesting approach; direct images of the contact areas and contact line deformations can be obtained as a function of time. Such images have been reported both for compliant spheres on rigid substrates and for rigid spheres on compliant substrates. The results indicate that dependent upon the nature of the system and the size of the interacting species a full range of the possible contact mechanics responses can be probed. In addition, many viscoelastic creep effects are apparent in the data.

More recently, scanning probe microscopy (SPM) techniques have been extensively used to probe surface material properties on the nanometre length scale. The majority of this research has utilised the integral tips of the simple cantilever probes to investigate these surface properties [18]. Whilst the value of this research is undoubted, certain limitations of using these probes are apparent. In general, the radius of the tip is poorly controlled and, at contact, atomic scale defects can become very important. Since application of any theory relies upon the input of a radius, this uncertainty is critical. Also, the chemistry of the tips is limited. In most cases, commercial tips are constructed from one of silicon, silicon nitride, or tungsten. Some control over these issues can be gained by attaching individual spheres of known dimensions, in the micron-size range, to the end of the cantilever springs. This opens up a virtually unlimited set of possible interaction chemistries. As well, it introduces the possibility of probing geometric effects; for example, different size ratios of interacting spheres could be used. Systems that have been studied include tin [19], xerographic toner particles [20], polystyrene [21], cross-linked polydimethylsiloxane [22], glass [23], and metal oxides [24]. In most cases, the authors reported that the measured adhesion forces at pull-off were 2 to 3 orders of magnitude smaller than predicted from theory. This is attributed to a contact region that is dominated by surface asperities [19]. In all these early cases, the experiments were performed using springs that had a relatively weak spring constant. The load regime was such that, even for small-radius surface asperities, the yield points of the materials were not exceeded.

In a recent investigation, the effects of repeated loading and unloading cycles, (using large maximum loads) for a polystyrene microsphere on a mica surface were reported [21]. In this case, the adhesive forces recorded at pull-off were shown to be less than a factor of 3 different from the expected values using JKR theory. This improved correlation was attributed to the plastic deformation of surface asperities resulting in a contact area that was much closer to that for perfectly smooth surfaces. The loads applied were sufficient to mean that the small surface asperities were in the elasto-plastic or plastic regimes. Across the entire sphere radius the applied loads were still in the purely elastic regime.

In this paper, we extend this earlier work using a greater loading range. Again, the data presented are for a single polystyrene bead interacting with a non-compliant smooth surface. We will present direct evidence, from SEM images of the damage inflicted upon surface asperities as a result of plastic or elasto-plastic deformations. By using a variety of loading rates at a single maximum load we are also able to probe directly viscoelastic effects for the bead used here.

EXPERIMENTAL SECTION

Particle Preparation and Characterisation

The polystyrene spheres used here were prepared in-house using a standard suspension polymerisation process [25]. The resultant stable suspension of particles was cleaned by filtration through glass wool followed by dialysis against Millipore® water. The dialysis water was changed daily over a three-week period.

The particle size distribution was analysed using a Malvern Mastersizer S. The resultant size distribution is shown in Figure 1. It should be noted that the results show a bimodal distribution of

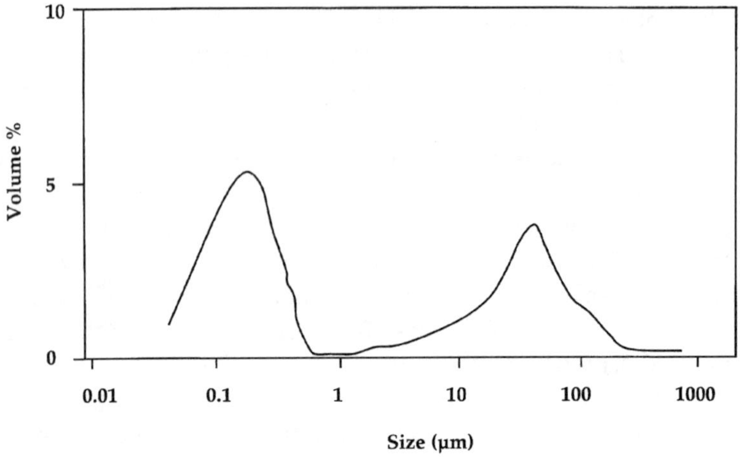

FIGURE 1 Particle size distribution plot for the polystyrene latex sample used in the adhesion measurements reported here.

particle sizes. This is not totally unexpected. The growth mechanism of particles, synthesised using suspension polymerisation, relies upon a progression from initial small polymer particles to large beads *via* an agglomeration and annealing process. The particle peak observed at a diameter of about 250 nm is, therefore, due to the presence of some unaggregated small spheres.

Probe Preparation

For ease of manipulation, and to minimise problems of glue contamination, beads with radii of between 10 and 20 μm are preferred. To facilitate preparation of the probe, a tungsten wire was etched to a fine point [26]. Using an optical microscope a single bead was selected from a drop of a diluted sample of the cleaned suspension. The wire was then used to remove the bead from this drop. Once removed, the bead quickly dried under the heat of the microscope lamp.

The isolated polymer bead was then glued to a single beam silicon cantilever (Digital Instruments, Inc.) using a small amount of epoxy resin. A scanning electron microscopy image of the colloid probe used in the studies reported here is shown in Figure 2.

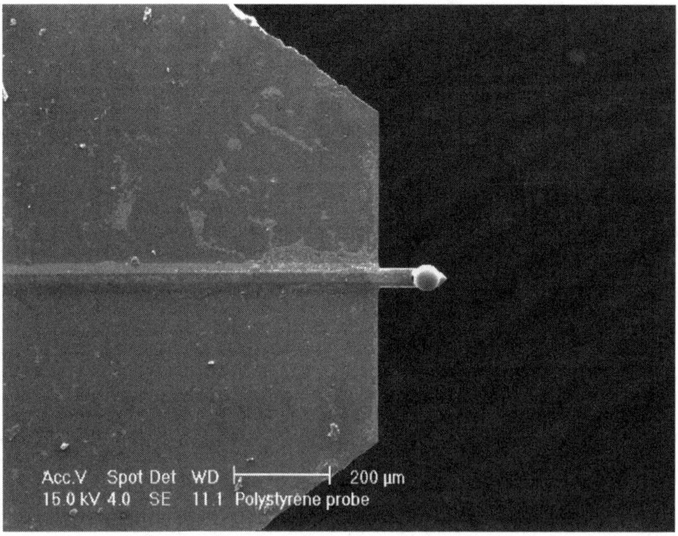

FIGURE 2 A scanning Electron Microscopy image of the colloid probe used in the measurements reported here. The probe consists of a 27.2 µm radius polystyrene bead attached to a single beam cantilever spring with a spring constant of 27 ± 1 N/m.

The spring constant of the cantilever used was determined using the technique of Cleveland *et al.* [27]. It was found to have a spring constant of 27 ± 1 N/m.

Silica Surface Preparation

The silica surfaces used in this investigation were prepared by oxidising a silicon wafer. The wafer was oxidised under an oxygen atmosphere at elevated temperatures using standard procedures. The resultant wafer had a surface oxide layer with a thickness of 140 ± 1 nm. The roughness of the surface, as determined using AFM images, was ± 2 nm over an area of 5 μm^2. The surface was cleaned by rinsing sequentially in ethanol, water and acetone followed by rapid drying under a stream of clean dry nitrogen.

SPM Force Measurements

Force-distance information was obtained from a Nanoscope® III AFM (Digital Instruments) which was operated in the "force mode".

In this mode the *X*-*Y* raster motion of the sample on the scanning piezoelectric crystal is suspended and the sample is moved towards and away from the cantilever in the *Z*-direction by the application of a saw-tooth voltage. In a typical experiment, the colloid probe was mounted in the commercial liquid cell (Digital Instruments). The surface used was a clean oxidised silicon wafer. AFM images indicated a surface roughness of less than ± 1 nm over an area of 1 μm^2. Prior to an experiment, the liquid cell was purged with a stream of clean dry nitrogen. During measurement, the cell was maintained under a dry nitrogen atmosphere.

Any experiment involves the silica surface being driven towards the polystyrene probe and deflections in the spring holding this probe being measured. Control of the scan size, scan rate and contact point between the surfaces allows control over the total contact time and the maximum applied load. The adhesive force is determined directly from these force-distance data as the surfaces are separated from one another.

RESULTS AND DISCUSSION

In Figure 3, data for the measured pull-off force as a function of the maximum applied load are given for the interaction between a 27.2 μm polystyrene (PS) sphere and a flat silica substrate. Two data sets are given in Figure 3, representing two consecutive loading cycles. The data in the first series show the effect of subjecting the bead to progressively increasing loads in the range from 3.5 μN to 95 μN. In this case, each run was performed after only a minimal time gap from the previous run (< 30 s). The effect of the increasing load is to cause an increase in the pull-off force for the bead from the surface. Immediately after application of the maximum load in Series 1, the applied load was reduced to < 5 μN and a second load sweep was performed. It is clear that reducing the load did not result in the complete loss of the increased adhesion generated during the loading cycle of Series 1. Again, increasing the load during Series 2 led to an increase in the adhesive pull-off force measured.

As a comparison, data are presented in Figure 4 for the measured adhesive pull-off force between a single 41 μm radius silica sphere

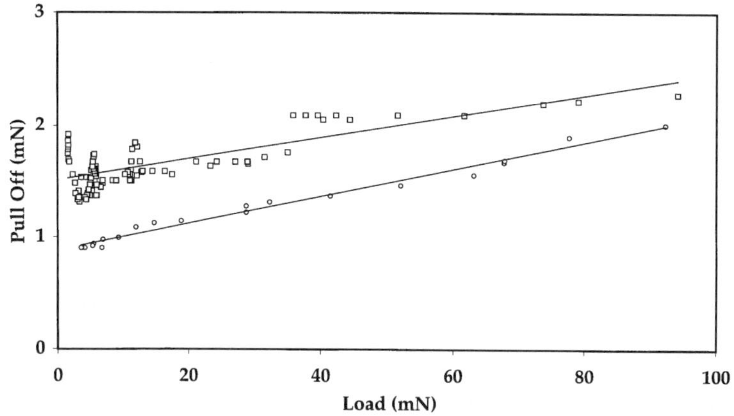

FIGURE 3 Data for the measured pull-off force *versus* applied external loading force between a 27.2 μm radius polystyrene sphere and a flat silica substrate. Two data sets are for two consecutive runs using increasing loads with the same bead and surface. The second run was started immediately the first had been completed. ○ First series, □ Second series.

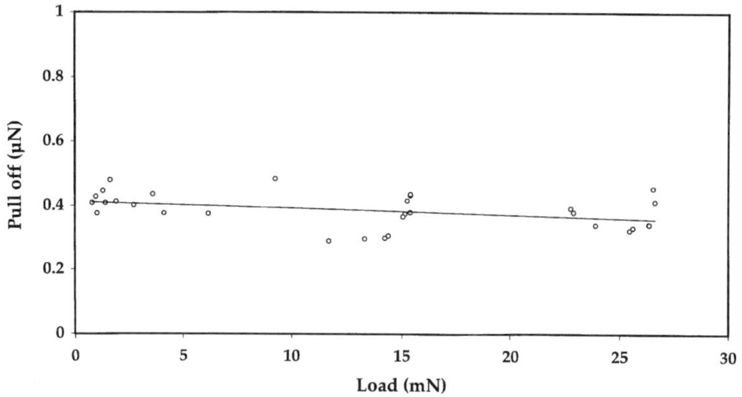

FIGURE 4 Data for the measured pull-off force *versus* applied external loading force between a 41 μm radius silica glass sphere and a flat silica substrate.

against the same silica flat. Two points are immediately apparent: the magnitude of the pull-off force is an order of magnitude smaller than in the PS case, and there is no increase in the pull-off force with increasing load.

Two important points are apparent from the load-pull-off data for the PS system. First, there must be some form of plastic or elasto-plastic deformation of the sphere since the pull-off increases as the applied load increases. Secondly, careful analysis of the data in Series 2 indicates the partial recovery of the sphere under the lowest loads. However, it is not clear from these data if this recovery, which is due to the removal of the load, is time-independent. However, the recovery is definitely indicative of elasto-plastic effects and it may also indicate visco-elasticity. This point will be discussed in more detail later.

For the silica – silica system the results are similar to those reported previously for non-compliant elastic solids [24]. The fact that the adhesive pull-off force is invariant with applied load indicates simple elastic contact between the surfaces. For a large elastic sphere in contact with a flat solid surface, the JKR theory [7] predicts that separation will occur when

$$F_{\text{pull-off}} = -\frac{3}{2}\pi R W_{12} \tag{1}$$

R is the probe radius and W_{12} is the adhesion work term. W_{12} is often approximated using $W_{12} = 2\sqrt{\gamma_1\gamma_2}$. The calculated value of the pull-off force from this equation, using a value of 78 mJ/m^2 for the silica surface energy [28], is 30 μN. This value for the surface energy is at the low end of the quoted scale for silica; values as high as 350 mJ/m^2 have been reported [29]. Therefore, the calculated value given here should be considered as a minimum. Notwithstanding this point, the value is two orders of magnitude greater than the recorded value from the colloid probe measurements. The most probable explanation for this result is that the contact zone is dominated by surface asperities. The role of surface asperities in reducing the magnitude of the adhesion is a well-documented phenomenon [19].

The PS-silica system can be analysed in a similar way. In this case, the predicted value of the pull-off force from JKR theory, using a value of 30 mJ/m^2 for the PS surface energy, is 12.4 μN. This is compared with values for the pull-off of between 1 and 2 μN measured using the colloid probe. Clearly, the measured values are much closer to the predicted values from this simple elastic theory. However, despite this close agreement, the data presented in Figure 3 indicate that this system has significant deviations from linear elastic behaviour.

The possibility of plastic deformation for the PS bead can be examined using the Maugis-Pollock (MP) [12] approach. In this theory, the conditions for the onset of elasto-plastic and full plastic behaviour are given in terms of the radius of the circle of contact: a_e and a_p, respectively.

$$a_e = \frac{4.90RY}{K} \tag{2}$$

$$a_p = \frac{60RY}{E} \tag{3}$$

in this simple theory, a dimensionless value, w^*, is defined as

$$w^* = \frac{W_{12}K^2}{RY^3} \tag{4}$$

where Y is the material yield strength and K is an elastic modulus term. Under zero applied load, when the value of w^* exceeds 5.2 the system can be assumed to be in the elasto-plastic regime. It will enter the plastic deformation zone when $w^* \geq 12000$.

The mechanical properties of polystyrene are well known [30]. Using values of $E = 2.55\,\text{GPa}$, $K = 4.3\,\text{GPa}$ and $W_{12} = 0.03\,\text{N/m}$ in the above equations, we can calculate that, for a bead of 27.2 μm radius, $w^* = 16$. Therefore, even at zero load we may expect this system to be in an elasto-plastic deformation regime.

This may possibly explain the rise in adhesion as a function of applied load. However, it should still be noted that the value of the adhesive pull-off forces at any applied load were a factor of between 5 and 10 times too small when using the simple JKR elastic theory. If we actually did have some elasto-plastic deformation in the bead, we would expect the contact area to be higher than predicted by the JKR approach and so the adhesive pull-off force should be greater than the value from the theory.

A possible explanation of why this result is obtained is to be found again in surface asperity contacts. An SEM micrograph of the bead used in this study is shown in Figure 5. Examination of this photo indicates the presence of significant surface debris. A larger scale image of the same bead is shown in Figure 6. Attention is drawn

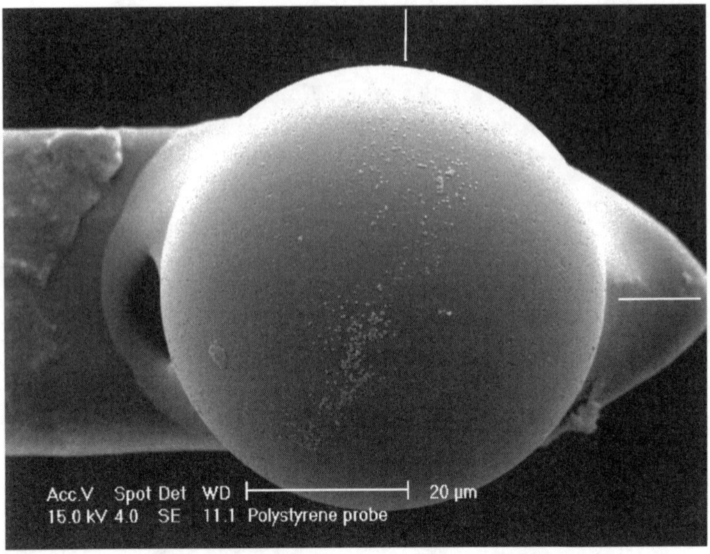

FIGURE 5 A Scanning Electron Microscopy image of the polystyrene bead used in the measurements reported here. Examination of the surface reveals the presence of significant adsorbed surface debris.

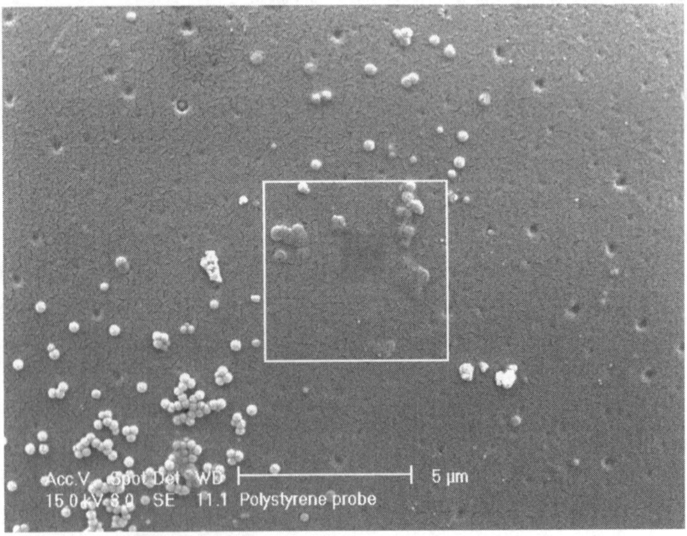

FIGURE 6 A Scanning Electron Microscopy image of the polystyrene bead showing the approximate region of contact, indicated by a white rectangle, on the surface of the polystyrene sphere at the conclusion of the loading cycles shown in Figure 3.

initially to the lower left hand corner of this image. Here, substantial numbers of small spherical particles can be seen. The sizes of these particles are approximately 230 nm. This is the size of the small residual contaminant particles that were present in our synthesised particle sample (*cf.* Fig. 1). It seems likely, therefore, that these particles are small polystyrene beads. Further evidence that these small spheres are polystyrene beads is gained from the presence of pits in the surface of the larger bead. These pits are approximately the same dimensions as the small particles and are probably positions from which, after the synthesis, partially-embedded small spheres have become dislodged. Remember that the big particles grow by agglomeration and fusion of the small spheres. In the top right corner of Figure 6 a small sphere that has become engulfed in the larger one can also be seen.

If we assume that these small spheres are indeed polystyrene beads, an interesting analysis of the contact zone can be performed. Examination of the central region of Figure 6 shows some small beads that appear to have been squashed. This region, highlighted by a white box, corresponds to the contact zone for the large sphere on the silica surface. Let us return once again to the simple JKR analysis of the expected pull-off force. However, in this case the calculation can be done assuming that the contact between the surfaces is through a single small sphere, radius = 115 nm. This results in a value for the adhesive pull-off force of 0.052 μN; our smallest measured value was approximately 1 μN. Thus, we would need around 20 of the small spheres to be contacting the silica surface at pull-off; in the SEM image there are around half this number. Two possibilities for the measured value can be postulated. In the first, there is some contribution to the overall adhesive force from the larger sphere. The second possibility is that the small spheres have undergone some plastic deformation resulting in a larger contact area and, hence, higher adhesion energy. A cursory inspection of the squashed beads in the contact zone suggests that this is the most likely explanation.

If we take the contact region, from Figure 6, to be a box with sides of 5 μm, a simple geometric analysis can be done. From Figure 7, we can see that such an analysis would result in a maximum height difference of 116 nm from the edge to the centre of this box when considering the surface of the larger sphere. The small spheres are

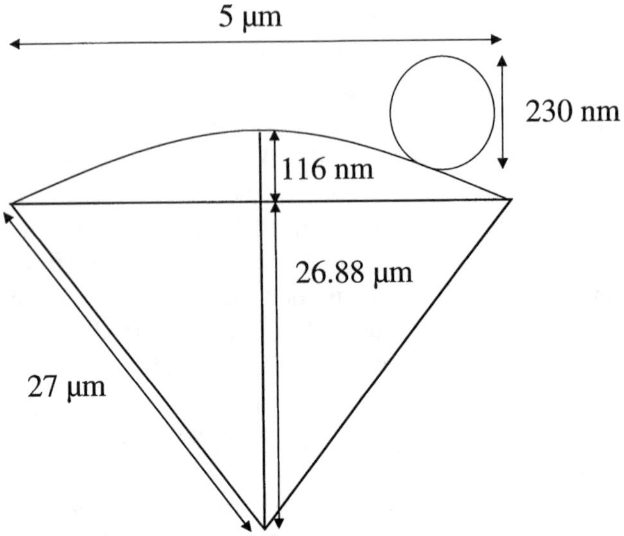

FIGURE 7 A schematic representation of the contact geometry of the sphere used in the measurements reported here. The height variation, from the edge to the centre of the curvature, across the width of the approximate contact zone (given in Fig. 6) is shown.

about twice this size. Therefore, even if the asperities were close to the edge of the box we would expect them to contact the flat surface well in advance of the large sphere surface. Indeed, a small sphere at the edge would have to deform by > 115 nm before the large sphere could have any contact. It is likely, therefore, that all of the observed adhesion is caused by these surface asperities.

A simple MP analysis using the small sphere radius predicts a value of $w^* = 3800$. Again, this is in the elasto-plastic regime. Thus, as we apply greater and greater loads, the permanent deformation will be expected to be continuously increasing and so the adhesion is seen to increase.

As we mentioned above, the data presented in Figure 3 appear to show some viscoelastic effects. After completion of the first loading run (Series 1) the system was allowed to rest for the standard 30 s. The first data point of Series 2 was collected at an applied load of 1 μN. Interestingly, this run resulted in an adhesion value that was approximately the same as the last point collected in Series 1 at 95 μN. Increasing the applied load was then seen to result initially in a reduction in the measured adhesion up to a load of about 5 μN. After this point, in general, the adhesion data were again seen to rise as a function of applied load, as for Series 1. It should be noted, however, that the data exhibit significant scatter up to a load of about 20 μN. All of these points indicate the presence of time-dependent effects in the measured data.

The initial reduction in adhesion observed in Series 2 is directly attributable to a partial recovery of the deformed contact zone of the bead. Clearly, all of the deformation experienced by the bead at the highest load in Series 1 is unsustainable at the low loads (< 5 μN) used initially in Series 2. However, the recovery is not instantaneous and, so, we see this apparent drop in the adhesion initially in Series 2. At loads of greater than 5 μN the adhesion begins to increase again with further increases in load. Since these values are larger than the corresponding adhesion values in Series 1, the relaxation process does not permit the full recovery of the bead structure. That is, plastic deformation has occurred. At all loads in Series 2 the measured pull-of adhesion values were higher than in Series 1. This indicates that at all applied loads in Series 1 the maximum adhesion value at that load had not been achieved. In other words, the creeping flow of the bead under that applied load was not complete in the timescale of the experiment. Recent advances in experimental methodology have driven an increased activity in the theoretical description of the contact mechanics for a viscoelastic material. This is a non-trivial problem. Unertl [31] has found that the maximum area of contact for a viscoelastic material under load can occur significantly after that load was applied. This appears to be the case here.

Further information about time-dependent viscoelastic effects in the system used here was found by measuring the adhesive pull-off force as a function of loading time for a single applied load. Results for this experiment are shown in Figure 8. These data indicate clearly that the adhesion increases with loading time. It should be remembered, of course, that the results given here are for a consecutive series of loading cycles on the same bead going from short to long times. Thus, the final load must contain some compounded information from the previous runs. Despite this, the data clearly highlight the difficulties of measuring adhesion forces for systems of this type. Clearly, to probe viscoelastic effects accurately using the probe microscopy technique

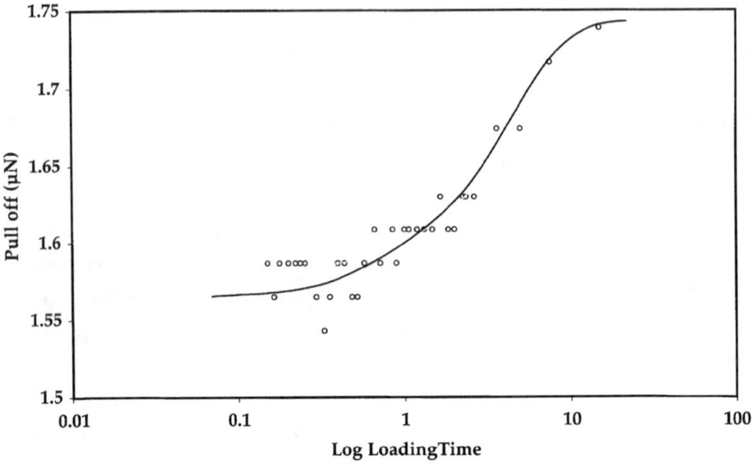

FIGURE 8 The measured pull-off force as a function of loading time at a constant applied load between a 27.2 µm polystyrene sphere and a flat silica substrate.

requires an improved methodology. This is the subject of ongoing research.

Although not planned as such, the fortuitous result of having adhering small particles on the surface of the larger bead has allowed the effects of small surface asperities to be probed in detail. Further experiments are under way in an attempt to control the contact zone better such that controlled amounts of asperity contacts might be introduced.

CONCLUSIONS

Data for the lift-off forces between a single 27 µm polystyrene bead and a flat silica surface have been obtained. Analysis of the contact zone indicated the presence of surface asperities. These asperities are attributed to the presence of small adsorbed polystyrene spheres ($R = 230$ nm). In the contact zone, there are around ten of these spheres. Under the loads applied here, all of these spheres were subjected to significant elasto-plastic deformations. The deformation of these asperities leads to an increasing adhesive pull-off force as a function of load. Control of the contact zone is extremely important in

AFM force measurements. Accurate analysis of the morphology of this zone can allow meaningful comparison of the measured data with established theories. The experimental data collected here also indicate the importance of time-dependent deformations due to viscoelastic creep. Longer contact times resulted in larger contact areas and higher pull-off forces.

Acknowledgment

The authors acknowledge the support of the ARC Special Research Centre for Multiphase Processes.

References

[1] Weeks, B. S., *International J. Molecular Medicine* **1**, 361 (1998).
[2] Mittal, K. L. (Ed.), *Particles on Surfaces 2: Detection, Adhesion, and Removal* (Plenum Press, New York, 1989).
[3] Visser, J., *Particulate Sci. Technol.* **13**, 169 (1995).
[4] Bradley, R. S., *Trans. Far. Soc.* **32**, 1088 (1936).
[5] Derjaguin, B. V., *Kolloid Z.* **69**, 155 (1934).
[6] Krupp, H., *Adv. Colloid Interface Sci.* **1**, 111 (1967).
[7] Johnson, K. L., Kendall, K. and Roberts, A. D., *Proc. R. Soc. London Ser A.* **324**, 301 (1971).
[8] Derjaguin, B. V., Muller, V. M. and Toporov, Yu. P., *J. Colloid Interface Sci.* **53**, 314 (1975).
[9] Tabor, D., *J. Colloid Interface Sci.* **58**, 2 (1977).
[10] Maugis, D., *Colloid Interface Sci.* **150**, 243 (1990).
[11] Johnson, K. L. and Greenwood, J. A., *J. Colloid Interface Sci.* **192**, 326 (1997).
[12] Maugis, D. and Pollock, H. M., *Acta. Metall.* **32**, 1323 (1984).
[13] Hui, C-Y., Baney, J. M. and Kramer, E. J., *Langmuir* **14**, 6570 (1998).
[14] Yoshizawa, H., Chen, Y-L. and Israelachvili, J. N., *J. Phys. Chem.* **97**, 4128 (1993).
[15] Pollock, H. M., *J. Phys. D.* **11**, 39 (1978).
[16] DeMejo, L. P., Rimai, D. S. and Bowen, R. C., *J. Adhesion Sci. Technol.* **2**, 331 (1988).
[17] Rimai, D. S., DeMejo, L. P. and Bowen, R. C., In: *Fundamentals of Adhesion and Interfaces* (VSP, Utrecht, The Netherlands, 1995), pp. 1–24.
[18] Thundat, T., Zheng, X-Y., Chen, G. Y., Sharp, S. L., Warmack, R. J. and Schowalter, L. J., *Appl. Phys. Lett.* **63**, 2150 (1993).
[19] Schaefer, D. M., Carpenter, M., Gady, B., Reifenberger, R., DeMejo, L. P. and Rimai, D. S., In: *Fundamentals of Adhesion and Interfaces.* (VSP, Utrecht, The Netherlands, 1995), pp. 35–48
[20] Ott, M. L. and Mizes, H. A., *Colloids Surfaces A.* **87**, 245 (1994).
[21] Biggs, S. and Spinks, G., *J. Adhesion Sci. Tecnnol.* **12**, 461 (1998).
[22] Chaudhury, M. K., Weaver, T., Hui, C. Y. and Kramer, E. J., *J. Appl. Phys.* **80**, 30 (1996).
[23] Schaefer, D. M., Carpenter, M., Gady, B., Reifenberger, R., DeMejo, L. P. and Rimai, D. S., *J. Adhesion Sci. Technol.* **9**, 1049 (1995).
[24] Mizes, H. A., *J. Adhesion* **51**, 155 (1995).

[25] Goodwin, J. W., Hearn, J., Ho, C. C. and Ottewill, R. H., *Colloid Polym. Sci.* **252**, 464 (1974)
[26] Method described in detail in Nanoscope III users manual, Appendix 1 (Digital Instruments, Santa Barbara, CA, USA).
[27] Cleveland, J. P., Manne, S., Bocek, D. and Hansma, P. K., *Rev. Sci. Instrum.* **64**, 3583 (1993).
[28] Schultz, J. and Nardin, M., In: *Modern Approaches to Wettability* (Plenum Press, New York, 1992), p. 82.
[29] Overbury, *Chem. Rev.* **75**, 555 (1975).
[30] Rimai, D. S., Moore, R. S., Bowen, R. C., Smith, V. K. and Woodgate, P. E., *J. Mater. Res.* **8**, 662 (1993).
[31] Unertl, W. N., In: *Microstructure and Tribology of Polymers*, Tsukruk, V. V. and Wahl, K. J., Eds. (ACS Books, Washington DC., 1999).

Surface Forces and the Adhesive Contact of Axisymmetric Elastic Bodies

A.-S. HUGUET and E. BARTHEL

Laboratoire CNRS/Saint-Gobain "Surface du Verre et Interfaces", 39, quai Lucien Lefranc, BP 135, F-93303 Aubervilliers Cedex, France

A synthetic approach to the problem of the adhesive contact of axisymmetric elastic bodies is proposed. A convenient and general formulation is thus obtained, which is shown to yield directly most of the useful models. In particular, the roles of the shape of the indenter on the one hand, and of the nature of the attractive interactions on the other hand are clearly separated. By nature, this approach can also be used in the case where the bodies are in interaction but not in contact. This results in a consistent treatment of long-range interactions and contact properties.

1. INTRODUCTION

The problem of the adhesive contact of elastic bodies is basically well understood. Historically, the first attempt, by Derjaguin, in his otherwise famous paper on the Derjaguin approximation [1], was quite disappointing: he found a pull-off force equal to $\pi w R$, where w is the adhesion energy and R the radius of the sphere. However, this result was at variance ("Daher ergibt sich ein zweimal so kleiner Wert [...]" [1]) with the just-discovered Derjaguin approximation, which describes the non-contact part of the same force curve and predicts a force twice as large at zero distance. This was the first attempt at matching long-range forces and pull-off force. Later, Derjaguin apparently reconciled his theories, since the DMT model [2] does predict a pull-off force equal to $2\pi wR$. As a result, continuity between the contact and non-contact parts of the force curve was restored (Fig. 1). There remained a question, however: the slope of the DMT curve at zero distance is zero, while a typical non-contact force curve is expected to exhibit a non-zero derivative at zero distance. Consequently, there apparently still is a discontinuity in the *derivative* of the force curve, a somewhat puzzling result. In addition, in the meantime, Johnson, Kendall and Roberts had put forward a model [3] giving a pull-off force equal to $3\pi wR/2$. The ensuing controversy was solved when, following Tabor [4], numerical computations [5–8] showed that the two models could be considered as limits in a continuous transition. The relevant parameter was shown to be the ratio of the gap between the surfaces just outside the contact zone and the range of the interactions (this ratio is denoted [5, 9] μ or λ). The JKR model applies when λ is much larger than unity. In this case, clearly, the non-contact part of the force curve is almost non-existent. The DMT model, on the contrary, applies when the range of the interactions is large ($\lambda \ll 1$), and matching the two parts of the force curve remains an open issue.

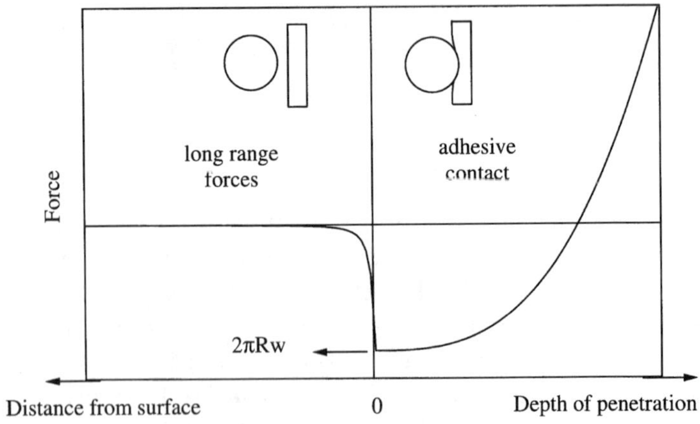

FIGURE 1 Typical matching problem for the non-contact (long-range forces) part of a force curve described by the Derjaguin approximation and the contact part of the same force curve described by the DMT model: although the force curve is continuous at zero distance (equal to $2\pi Rw$), the derivative is not (*cf.* also Fig. 6).

Simultaneously, the JKR model was studied through a rather different approach, in terms of fracture mechanics [10–12]. The culminating point of the fracture mechanics approach was reached in 1992, when Maugis proposed a Dugdale model for the adhesion of spheres [9]. This paper brought a number of fruitful ideas, allowed the first analytical description of the transition, and trigerred a number of additional contributions along this line [13–17].

Following these recent advances, we try to provide a compact general formulation of the adhesive contact of axisymmetric elastic bodies and also to describe the non-contact part of the experimental curves within the same theory.

Thus, we first give general expressions for the force and penetration assuming a given attractive stress distribution outside the contact zone and a given shape of the indenter. As a result of the adhesive process, two phenomena are simultaneously observed: compared with the non-adhesive case (Hertz [18]),

1. the contact zone increases (as in the JKR model);
2. there is an additional tensile force (as in the DMT model).

We discuss their relative weight in the general solution in terms of lateral extension of the attractive *interaction zone* c (Fig. 2). In this

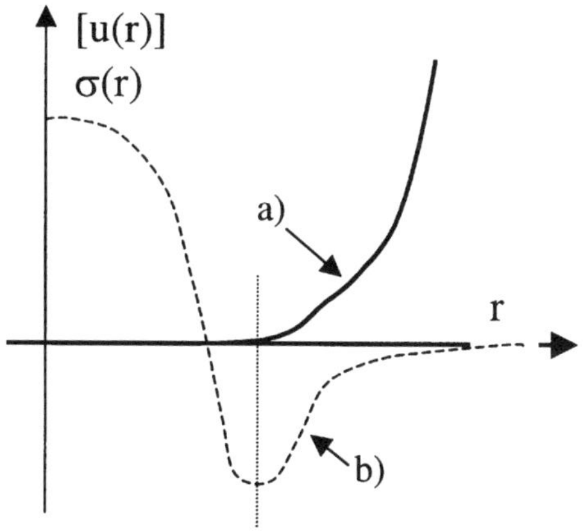

FIGURE 2 Typical gap profile (full line) and stress distribution (dashed) of an indenter in adhesive contact with a flat non-deformable surface. The radius of the contact zone is a, the radial extension of the interaction zone is c.

manner, the mechanics of the adhesive contact problem, and in particular of the JKR and DMT limits, can be obtained. Up to that point the description can be applied to an arbitrary type (in terms of range or reversibility, for example) of attractive interaction.

Then, we will assume a more specific description of the adhesive process in terms of interaction *potential* and, thus, of *adhesion energy*. Using the previous expressions for the mechanical response of the surface, we show in the general case that, when the interaction is short-ranged, the contact-zone-increase effect accounts for the full adhesion energy and, thus, obtain the generalization of the JKR model to bodies of arbitrary shapes. We also show that the linear elastic fracture mechanics approach (in terms of energy release rate) is easily derived from the previous expressions. In the case of long-range interactions, we show that the DMT limit is obtained only for paraboloidal bodies (which is the usual approximation for the sphere). The same limit is either basically useless or does not exist for other shapes. For intermediate cases, we propose a rationalization of the approximation schemes developed so far.

Finally, we outline the treatment of the non-contact part of the force curve and give an example of the consistent treatment of long-range and contact parts of an experimental force curve. This example is illustrative of the improvement obtained by the proposed modeling.

2. DESCRIPTION OF THE PROBLEM

2.1. Self-consistency

The adhesive contact between two bodies is a process which involves two phenomena:

1. the interaction between the bodies, through surface forces,
2. the mechanical deformation of the bodies.

The complexity of the problem comes from the fact that these phenomena are interdependent: the mechanical deformation will obviously depend upon the interaction and the interaction will depend upon the distance between the surfaces and, hence, upon the deformation of the bodies. It follows that a self-consistent solution is required.

Note, however, that these two phenomena are also at work when the surfaces interact without contact. In this case, however, one will usually consider that the surfaces are non-deformable – because mechanical instability of the experimental device will usually occur before the deformations are sizable – and the Derjaguin approximation [1] will be used. That this is not necessarily correct will be shown below.

2.2. Contact Zone and Interaction

The initial question one is faced with, for a given problem, is thus to describe both the mechanical behavior of the solids and their interaction. However, in doing so, one should keep in mind that the division of the phenomena between surface interaction and mechanical behaviour may in some cases be quite artificial.

More specifically, it is clear that the interaction potential between surfaces is always repulsive at small distance. This is due to Born repulsion between atoms, and ensures the stability of matter. Now, it is not trivial to get an accurate description of this repulsion. An *ad hoc* potential like the d^{-12} used in the Lennard-Jones potential is simply indicative of the trend. Actually, an approximation scheme can be devised, which retains the main feature of this repulsive part of the potential: its very steep variation with distance. Thus, a very small change in distance will induce a very large change in interaction. Or, conversely, the interaction can take on an arbitrary value – within the bounds of allowed values (maximum tensile stress and elastic limit) – with almost no change in separation. As a result, the problem can be reformulated in the following way: there is a zone in which the distance is prescribed and the stress is free. This is, of course, the contact zone. Thus, we have turned a prescription on the interaction into a prescription on the mechanics: the displacement is fixed in the contact zone.

As a result, there are, broadly speaking, two families of approaches:

1. one can prescribe the full interaction (attractive and repulsive parts) and, from the knowledge of the surface response to externally-applied stress, try and solve the self-consistent problem [5, 7, 8, 19]

or

2. one can exclude the very short range repulsive part of the interaction and work out the solution using mixed boundary conditions

for the mechanical side of the problem: the displacement is prescribed inside the contact zone and the stress outside [2, 3, 9, 16, 18, 20].

The interactions outside of the contact zone may be very diverse. The many types of interactions known in the field of surface forces may be operating – chemical interaction, electrostatic, electrical double-layer, meniscus force, ... – including the van der Waals force, which, although a body force, turns into a surface force when volume integration is performed on bodies with small curvature [21]. In addition, non-conservative forces, like viscous dissipation at the periphery of the contact zone, may also be considered [10]. In fact, the approach we present is independent of the nature of the interaction as long as it is a surface interaction, *i.e.*, it results in a surface stress distribution.

Once these questions have been sorted out, we cannot proceed until the description of the mechanical behaviour of the surface under given boundary conditions is known.

3. SURFACE ELASTICITY: USEFUL RESULTS

This technical section deals with the response of an elastic flat surface: the aim is to formulate general expressions for the total force, displacement and elastic energy in terms of an auxiliary function, g. We also show how g depends upon the surface stress distribution, or upon the normal surface displacement.

The present formulation relies upon results for the description of the elastic response of a flat surface known in the literature [22–29]. We have tried, in the Appendices, to provide a complete and "minimal" (in term of complexity) derivation of those results which are necessary here by consistently using the Fourier transform method (as suggested by Landau [30], page 26, note 1). Although it differs in a number of details, this approach is, thus, essentially Sneddon's [22]. We believe, however, that the present derivation side-steps several more difficult points in Sneddon's texts. In addition, it directly provides some necessary relations (Eqs. (3.10) and (3.11)) usually not reported in the literature.

3.1. Surface Displacement

Let us assume linear elastic behaviour, a frictionless contact and an axisymmetrical geometry. In particular, we are interested in the surface displacement, $u_z(r)$, induced by a normal stress distribution at the surface, $F_z(r)$.

We will be using the result that:

$$u_z(r) = \frac{2}{\pi}\frac{1}{\mathcal{K}}\int_0^r \frac{g(t)}{(r^2 - t^2)^{1/2}}\,dt, \tag{3.1}$$

where the g function thus introduced is

$$g(t) = \int_t^{+\infty} \frac{sF_z(s)}{(s^2 - t^2)^{1/2}}\,ds, \tag{3.2}$$

and

$$\mathcal{K} = \frac{E}{2(1 - \nu^2)}. \tag{3.3}$$

Note the factor 3/8 with the frequent definition

$$K = \frac{4E}{3(1 - \nu^2)} \tag{3.4}$$

which is linked to the case of the spherical indenter, and the factor 1/2 with the usual definition

$$E^\star = \frac{E}{1 - \nu^2}. \tag{3.5}$$

The important fact here is that according to Eq. (3.1), the surface displacement at r is obtained through the datum of the auxiliary function $g(s)$ for $s < r$ only. Conversely, $g(r)$ is known from the datum of $F_z(s)$ for $s > r$ only (Eq. (3.2)). This property will prove invaluable in the context of mixed boundary conditions which appears in the adhesive contact problem (Section 4).

3.2. Surface Stress

One of the main features of Eqs. (3.1) and (3.2) is that they can be inverted. In the present derivation (Appendices), which is based on the successive use of Hankel and cosine transforms, this property stems from the fact that each of these transforms is invertible. Explicitly, one gets:

$$g(t) = \mathcal{K}\frac{\partial}{\partial t}\int_0^t dr \frac{ru_z(r)}{(t^2 - r^2)^{1/2}} \tag{3.6}$$

$$= \mathcal{K}\left[u_z(r=0) + t\int_0^t dr \frac{u_z'(r)}{(t^2 - r^2)^{1/2}}\right], \tag{3.7}$$

and

$$sF_z(s) = -\frac{2}{\pi}\frac{d}{ds}\int_s^{+\infty} dt \frac{tg(t)}{(t^2 - s^2)^{1/2}} \tag{3.8}$$

or

$$F_z(s) = \frac{2}{\pi}\left[\frac{\Delta g(s_0)\Theta(s_0 - s)}{(s_0^2 - s^2)^{1/2}} + \int_s^{+\infty} dt \frac{g'(t)}{(t^2 - s^2)^{1/2}}\right] \tag{3.9}$$

Equations (3.7) and (3.9) were obtained by integrating by parts. We also provided for the possibility that g be discontinuous by introducing the jump $\Delta g(s_0)$ of g at s_0. Equivalent relations were introduced by Sneddon under the denomination of Abel's relation [31].

Here again, Eqs. (3.6) or (3.7) express $g(t)$ from the datum of $u(r)$ for $t > r$ only, while the surface stress at s is obtained through the datum of the auxiliary function $g(t)$ for $s < t$ only (Eqs. (3.8) and (3.9)).

Note that this is just the general form of the results Maugis used in his paper [9]; although, since these equations do not exist as such in Sneddon's papers, he had to build them up, in a special case, by linear superposition of two particular cases [31, 32].

3.3. Force and Energy

Computing the total force, F, applied to one of the bodies through the interface as a function of the auxiliary function, g, we obtain

$$F = 4\int_0^{+\infty} ds\, g(s). \tag{3.10}$$

Similarly, the total mechanical energy, $\mathcal{E}$, is

$$\mathcal{E} = \frac{2}{\mathcal{K}} \int_0^{+\infty} ds\, g^2(s). \tag{3.11}$$

Following our approach (Appendix B), these relations are again obtained through general properties of the Hankel and cosine transforms.

4. GENERAL EQUATIONS FOR THE AXISYMMETRICAL CONTACT

We now further investigate the properties of the g function. We show that, in the contact problem, g can be calculated from the boundary conditions – *i.e.*, the shape of the indenter and the nature of the interactions – from Eqs. (3.1) and (3.2) and their inverses. Further, we also show that, in addition to the total force and elastic energy (Section 3.3), the *penetration* of the indenter can also be expressed as a function of g, thus completing the general solution to the contact problem.

Let us consider an axisymmetrical deformable body with convex shape $h(r) > 0$ (with $h(0) = 0$) in contact with a perfectly rigid flat plane. We assume there is a contact zone of radius a. Outside the contact zone, we suppose there is a normal stress distribution, $p(r)$, which acts on the bodies. No specification as to the form and origin of this stress distribution is given. In particular, no reference to the adhesive mechanism is used for now.

The boundary conditions are, thus:

$$u_z(r) = \delta - h(r) \quad \text{when } r < a, \tag{4.1}$$

$$F_z(r) = -p(r) \quad \text{when } r > a. \tag{4.2}$$

Since the interaction outside the contact zone is assumed to be attractive, $p > 0$. Inward displacement, u_z, is counted positive. Similarly, the penetration, δ, will be counted positive if interpenetration would result, were the bodies non-deformable.

The solution to the adhesive contact problem then proceeds as follows: from the boundary condition (4.1) [resp., (4.2)] and Eq. (3.7) [resp., (3.2)], the g function can be calculated for $r < a$ [resp., $r > a$]. Matching these two parts of the function g for continuity (which results from the continuity of stress at a) gives an equation from which the penetration is obtained. Then, the total force is calculated through (3.10). Of course, all the other quantities in the problem can also be derived from g, such as the stress distribution inside the contact zone (Eq. (3.8)), the surface displacement outside the contact zone (Eq. (3.1)) and the total elastic energy (Eq. (3.11)).

4.1. A Singular Case: The Flat Punch

As a first example, we show how to calculate the solution for the adhesionless ($p(r)=0$ for $r > a$) rigid flat punch of radius a indenting the plane down to a penetration δ_{fp}, a well-known result due to Boussinesq [20]. We here have $h=0$. Let us determine the g function from the boundary conditions. We use Eqs. (3.7) and (3.2) from which we get:

$$g(t) = \mathcal{K}\delta_{fp} \quad \text{when } t < a, \tag{4.3}$$

$$g(t) = 0 \quad \text{when } t > a. \tag{4.4}$$

As a result, from Eq. (3.10), we obtain the relation between force and displacement

$$F_{fp} = 4\mathcal{K}a\delta_{fp}. \tag{4.5}$$

In addition, from Eqs. (3.1) and (3.9), we obtain the information complementary to the boundary conditions, *i.e.*, the displacement outside the contact zone and the stress distribution inside:

$$u_{z,fp}(r, z=0) = \frac{2}{\pi}\delta_{fp} \arcsin\left(\frac{a}{r}\right) \quad \text{for } r > a, \tag{4.6}$$

$$F_{z,fp}(r, z=0) = \frac{2}{\pi}\frac{\mathcal{K}\delta_{fp}}{(a^2 - r^2)^{1/2}} \quad \text{for } r < a. \tag{4.7}$$

4.2. General Solution for Non-singular Indenter Shapes

Let us now assume a regular shape for the indenter (*i.e.*, with a continuous derivative). Under the external loading and the adhesive interaction, the indenter penetrates the plane and the contact radius reaches a value a. We want to calculate expressions for the penetration, δ, and the force, F, as a function of a. Let us first calculate g. Denoting

$$\delta_0(t) = \frac{\partial}{\partial t}\int_0^t \frac{rh(r)dr}{(t^2 - r^2)^{1/2}} = t\int_0^t \frac{h'(r)dr}{(t^2 - r^2)^{1/2}}, \tag{4.8}$$

from Eqs. (3.7) and (3.2) we have

$$g(t) = \mathcal{K}(\delta - \delta_0(t)) \quad \text{when } t < a, \tag{4.9}$$

$$g(t) = -\int_t^{+\infty} \frac{sp(s)ds}{(s^2 - t^2)^{1/2}} \quad \text{when } t > a. \tag{4.10}$$

4.2.1. Penetration

We now calculate the penetration, δ. If the shape of the indenting body, h, and its derivative, h', are continuous at a, then we have continuity of the normal stress at a and, thus, continuity of g, which allows us to determine δ in the following way.

Let us first consider the case of the non-adhesive contact. We know from Eq. (4.10) that $g(a) = 0$. Thus, the penetration is

$$\delta = \delta_0(a). \tag{4.11}$$

We conclude that the function $\delta_0(t)$, which only depends upon the shape of the indenter, is the penetration which is necessary to obtain a radius of contact, a, in the absence of attractive interaction [31]. The function δ_0 and all the subsequently-introduced functions are explicitly given for the case of the cone and the paraboloid in the Table.

TABLE Values of the various expressions describing the adhesionless contact for two different indenter shapes

	$h(t)$	$\delta_0(t)$	$\phi_0(t)$	$a\delta_0(a)-\phi_0(a)$	F_0
Cone	αt	$\pi/2\ \alpha t$	$\pi/4\ \alpha t^2$	$\pi/4\ \alpha a^2$	$\pi\mathcal{K}\alpha a^2$
Paraboloid	$t^2/2R$	t^2/R	$t^3/3R$	$2a^3/3R$	$8\mathcal{K}a^3/3R$

Similarly, in the case of the adhesive contact, the penetration, δ, is simply

$$\delta = \delta_0(a) + \frac{1}{\mathcal{K}} g(a). \tag{4.12}$$

Equation (4.12) means that attractive interactions reduce the penetration necessary to reach a contact radius, a, by a distance, $g(a)/\mathcal{K}$ (note that $g(a)$ is negative by definition). This increase of the contact zone due to the adhesive interaction – for which we obtain a general formulation – is the central concept in the JKR approach [3]. It clearly appears as a competition between the interaction, as weighted in $g(a)$, and the rigidity of the surface expressed by $\mathcal{K}$. We, thus, observe that some increase of the contact zone will be observed for any stress distribution, although $g(a)$ clearly emphasizes the contributions close to the contact zone.

4.2.2. Force

From Eqs. (3.10), (4.9) and (4.10), the total force is expressed as

$$F(a) = F_0(a) + 4ag(a) + 4\int_a^{+\infty} ds\, sp(s) \arcsin\frac{a}{s} - 2\pi\int_a^{+\infty} ds\, sp(s), \tag{4.13}$$

where F_0 is the force necessary for the adhesionless indenter to reach a contact radius a and reads

$$F_0(a) = 4\mathcal{K}\{a\delta_0(a) - \phi_0(a)\}, \tag{4.14}$$

with

$$\phi_0(t) = \int_0^t ds\, \delta_0(s). \tag{4.15}$$

We can picture this Hertz-like repulsive contribution as the force required to push a flat punch down to a penetration $\delta = \delta_0(a)$ minus a (positive) quantity which accounts for the reduction in resistance due to the upward sloping shape.

In order to describe the other contributions, which arise as a result of the attractive interaction, let us start from the adhesionless punch indenting the plane with a penetration δ (Fig. 3a). The radius of contact is a_{ini}, and the force is the bare Hertz contribution $F_0(a_{\text{ini}})$. If we switch an attractive interaction on, then an additional surface displacement occurs (Fig. 3b), Eq. (3.1), which increases the contact radius to a. However, this additional displacement has to be cancelled inside the contact zone, due to the boundary condition Eq. (4.1). Thus, the stress distribution inside the contact zone rearranges, so that the boundary conditions be fulfilled inside the contact zone while preserving the continuity of the stress distribution at a. Thus, the final stress distribution inside the contact but close to the contact line is tensile (Fig. 2).

As a result, the Hertz contribution increases to $F_0(a)$. In addition, a tensile stress distribution develops inside the contact zone, which is determined by the final radius of contact and the external stress

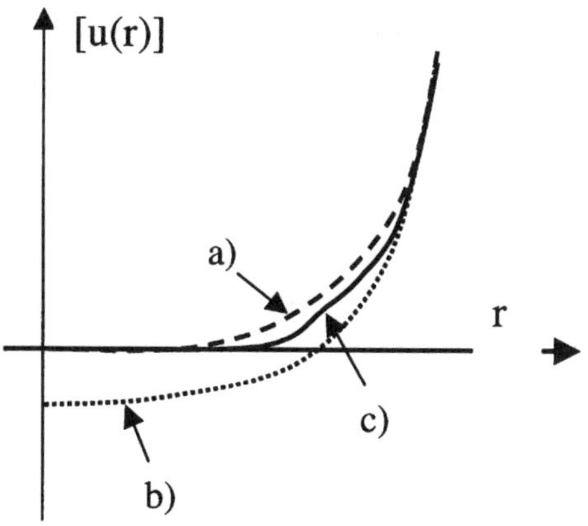

FIGURE 3 Gap profile of the indenter in contact with a flat non-deformable surface. (a) in the absence of attractive interaction; (b) with the additional displacement due to the attractive interactions alone; (c) with the additional displacement due to both the attractive interactions and the boundary conditions inside the contact zone.

distribution, and leads to the next two terms. Finally, the total force, F, breaks down into four components (Eq. (4.13)). The first term, $F_0(a)$, in Eq. (4.13) is the usual Hertz repulsive (*i.e.*, positive) component of the adhesionless punch with the final contact radius a. Thus, the increase in the contact zone has induced an increase in the repulsive Hertz term from $F_0(a_{\text{ini}})$ to $F_0(a)$. The second and third terms, F_{incr} and F_{add} form the contribution from the attractive stress distribution inside the contact zone. Note that F_{incr} (tensile) results from the global displacement, $g(a)/\mathcal{K}$ (flat punch displacement), inside the contact zone, and reflects the contact zone increase effect. It is partially offset by F_{add} (compressive), which results from the cancellation of the surface displacement inside the contact zone due to the stress distribution outside. The last term, F_{ext} (negative), is the contribution from the stress distribution outside the contact zone.

We are also in a position to discuss the relative weight of the adhesive contributions in terms of the radial extension of the surface stress distribution and compare with the standard theories. If this extension is large compared with the radius of contact, examination of the various expressions shows that F_{ext} dominates and F_{incr} and F_{add} cancel at first order. Thus, the attractive stress is essentially *outside* the contact zone, does not bring about substantial increase of the contact zone but simply acts as an externally-applied force. This situation is reminiscent of the DMT model. On the contrary, if the range is small, F_{incr} dominates and the other two interaction terms cancel at first order. Thus, the contact zone increase effect is dominant, and the attractive stress is essentially *inside* the contact zone. In this case, the picture which emerges from Eqs. (4.12) and (4.13) is the additional flat-punch retraction (*cf.* Section 4.1 with a displacement $g(a)/\mathcal{K}$) central to the JKR theory, and the contact is completely described by the specification of $g(a)$.

As an illustration, let us consider the case where the attractive stress is a constant inside the interaction zone, as assumed by Maugis [9]. We assume a radius of contact and an interaction amplitude equal to unity. Figure 4 displays the ratio of the absolute value of the various interaction force terms in this case as a function of the extension of the interaction zone, c.

Thus, we have given general expression for the two effects of attractive interactions:

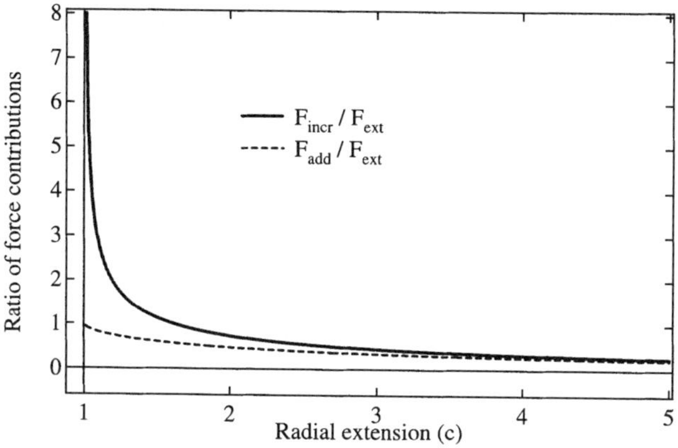

FIGURE 4 Absolute value of the relative magnitude of the three force contributions due to the attractive interaction as a function of the radial extension of the interaction zone, c, in the Maugis model [9]. The stress amplitude, σ_0, and the radius of contact a are chosen equal to unity. Note that F_{add} is positive, while F_{incr} and F_{ext} are negative. At small radial extension, F_{incr} dominates, while F_{add} and F_{ext} cancel (JKR limit). At large radial extension, F_{ext} dominate, while F_{incr} and F_{add} cancel (DMT limit).

1. by a wetting-like effect, they tend to increase the contact radius at a given penetration
2. they contribute an overall attractive force, F_{ext}.

The general behaviour results from the sum of these effects.

4.2.3. The Gap

Since the interaction between the contacting bodies will depend upon (at least) the local value of the gap, the self-consistent treatment of the problem requires the knowledge of the shape of the gap outside the contact zone. Thus, in this more technical subsection, we consider the deformations outside the contact zone in more detail.

Let us define the gap between the surfaces as

$$[u(r)] = u_z(r) - \delta + h(r). \tag{4.16}$$

Naturally, $[u(r)]$ is zero inside the contact zone. Outside the contact zone, the only non-trivial contribution comes from the deformation

$u_z(r)$, $r > a$. This deformation is the sum of the adhesionless term

$$u_{z,0}(r) = \frac{2}{\pi}\int_0^a \frac{\delta_0(a) - \delta_0(t)}{(r^2 - t^2)^{1/2}}\, dt, \tag{4.17}$$

and the term due to the adhesive interactions

$$u_{z,\mathrm{int}}(r) = \frac{g(r)}{\mathcal{K}} - \frac{2}{\pi}\frac{1}{\mathcal{K}}\int_a^r \arcsin\left(\frac{t}{r}\right) g'(t)\,dt. \tag{4.18}$$

In particular, we observe that

$$\frac{\partial u_{z,\mathrm{int}}(r)}{\partial r} = \frac{2}{\pi \mathcal{K} r}\int_a^r \frac{t g'(t)}{(r^2 - t^2)^{1/2}}\, dt. \tag{4.19}$$

Thus, as expected, if g is differentiable with a continuous derivative (which implies the same regularity for F_z), then the gap profile is regular.

Note, however, that, when $g'(t)$ is increasingly peaked and tends to $g'(t) = \Delta g(a)\delta(t - a)$, that is to say when g tends to a function with a discontinuity $\Delta g(a)$ at a, we do tend to the singular JKR flat-punch deformation, $u_{z,fp}$ (Eq. (4.6)), with displacement $g(a)/\mathcal{K}$. Thus, as is now well known, the singular JKR case appears as a limit within a non-singular model.

Up to now, we have given and investigated the general contact equations for a body with arbitrary shape and an arbitrary stress distribution outside the contact zone. We have identified the contributions of two rather different effects: for a given penetration, the part of the stress distribution far away from contact essentially leads to an overall additional attractive force, while no additional deformation close to the contact zone is incurred; the part of the stress distribution close to the contact zone essentially leads to an increase of the contact zone by a spreading-like effect described by the ratio $g(a)/\mathcal{K}$.

5. SELF-CONSISTENT APPROACH

We now tackle the self-consistent part of the approach. Up to now, no hypothesis as to the nature of the stress distribution outside the contact zone was made. Let us now assume that the stress outside the contact zone is derived from some interaction potential between the surfaces, $V([u])$. An exact self-consistent approach, thus, requires that everywhere outside the contact zone,

$$F_z(r) = -\frac{\partial V}{\partial [u]}([u(r)]); \tag{5.1}$$

Now, as we have just seen, $[u(r)]$ also depends upon $F_z(s)$ in a non-trivial manner. This exact formulation is, thus, difficult to use directly, except for numerical calculations.

Let us rather introduce a weaker (approximate) formulation of the self-consistent approach. The value of the potential for a zero surface separation is some non-zero adhesion energy, w. From this bare definition of the adhesion energy, we obtain the condition:

$$w = \int_a^{+\infty} p(s) \frac{\partial [u]}{\partial s} ds. \tag{5.2}$$

This approximate self-consistency equation will turn out to be quite sufficient to specify the stress distribution to first order outside the contact zone and, thus, describe the adhesive contact.

We first examine the consequences of Eq. (5.2) in the case where the interactions are short-ranged.

5.1. Short Range Interactions – The General JKR Model

If the interaction range is small, the radial extension of the interaction will also be small, g' is peaked around a and, from Eq. (4.19), the dominant term in Eq. (5.2) will be the $\partial u_{z,\mathrm{int}}/\partial s$ term. Thus,

$$w \simeq -\frac{2}{\pi}\frac{1}{\mathcal{K}} \int_a^{+\infty} dt\, tg'(t) \int_t^{+\infty} \frac{p(s)ds}{s(s^2 - t^2)^{1/2}} \simeq -\frac{1}{a} \int_a^{+\infty} g'(t)g(t), \tag{5.3}$$

where the s and t factors in the integrals were replaced by a to first order since $p(s)$ decreases rapidly outside the contact zone. As a result, in the case of short-range interactions, the self-consistency equation boils down to

$$w = \frac{g(a)^2}{\pi \mathcal{K} a}. \tag{5.4}$$

Thus, $g(a)$, which has been shown above to describe completely the adhesive contact in the present case, is determined as a function of w. The generalization of the JKR equations to an arbitrary shape of indenter clearly appears.

5.1.1. Penetration and Force

As a result of Eq. (5.4),

$$g(a) = -(\pi w a \mathcal{K})^{1/2} \tag{5.5}$$

and, taking into account the considerations of Section 4.2.2, the penetration and force are then easily derived from their adhesionless counterparts through Eqs. (4.12) and (4.14):

$$\delta = \delta_0(a) - \left(\frac{\pi w}{\mathcal{K}}\right)^{1/2} a^{1/2} \tag{5.6}$$

$$F = F_0(a) - 4(\pi \mathcal{K} w)^{1/2} a^{3/2}. \tag{5.7}$$

In particular, we clearly observe that the adhesive process is independent of the shape of the indenter.

5.1.2. Fracture Mechanics Approach

Maugis and Barquins have shown that the generalization of the JKR approach can be obtained under the viewpoint of fracture mechanics [10–12]. They have used expansions of the Sneddon expressions to calculate the adhesion-induced stress intensity factor. We now show that the present formulation of the problem also allows a discussion of the same issue directly through the energy release rate. Following these authors, we assume that the interactions are infinitely short-ranged, so that

$$g(a_-) < 0 \tag{5.8}$$

$$g(r) = 0 \quad \text{for } r > a, \tag{5.9}$$

that is to say, there is an attractive stress distribution characterized by $g(a)$ inside the contact zone, but the attractive stress distribution outside the contact zone has been shrunk to zero. Then, from Eq. (3.11), the mechanical energy, $\mathcal{E}$, is

$$\mathcal{E} = \frac{2}{\mathcal{K}} \int_0^a ds\, g^2(s). \tag{5.10}$$

5.1.3. Equilibrium

From a fracture mechanics viewpoint, the contact radius, a, is a thermodynamic quantity, and equilibrium is obtained if, for an elementary variation in contact area, dA, the variation in mechanical energy, $d\mathcal{E} = GdA$, is equal to the variation of the interfacial energy, $w\,dA$, or, stated otherwise, if the adhesion energy, here rather acting as a surface tension, is equal to the energy release rate, G.

Now, Eq. (4.9) shows that $g(t)$ is independent of a for $t < a$. Thus, at constant penetration, δ,

$$G = \frac{d\mathcal{E}}{dA} = \frac{1}{2\pi a}\frac{d\mathcal{E}}{da} = \frac{g^2(a)}{\pi \mathcal{K} a}, \tag{5.11}$$

from which we again obtain Eq. (5.4).

5.1.4. Stability

Maugis has extensively studied the stability of the equilibrium in various cases. This point is all the more significant as instability means contact rupture: hence, the pull-off force is determined by the stability condition. The system is stable if

$$\frac{d^2\mathcal{E}}{dA^2} > 0 \tag{5.12}$$

Most noteworthy, the stability depends upon the loading conditions, *i.e.*, if in Eq. (5.12) the derivative is taken at constant penetration (fixed grip) or constant force (fixed load). To that purpose, adequate expressions for G are directly obtained from Eqs. (4.12) and (4.13):

$$g(a) = \mathcal{K}\{\delta - \delta_0(a)\} = \frac{F(a) - F_0(a)}{4a}, \tag{5.13}$$

and

$$G = \frac{\mathcal{K}(\delta - \delta_0(a))^2}{\pi a} = \frac{(F(a) - F_0(a))^2}{16\pi a^3 \mathcal{K}}. \tag{5.14}$$

This last expression for the energy release rate often occurs in Maugis' papers. From these latter expressions, the general stability conditions can be calculated as:

$$\frac{\partial \delta_0}{\partial a} > \frac{-g(a)}{2a\mathcal{K}} \tag{5.15}$$

or

$$\frac{\partial \delta}{\partial a} = 0 \tag{5.16}$$

at fixed grip and

$$\frac{\partial F_0}{\partial a} > -6g(a) \tag{5.17}$$

or

$$\frac{\partial F}{\partial a} = 0 \tag{5.18}$$

at fixed load.

As a result, these expressions show how the shape of the indenter determines the value of the contact radius for which contact break-up occurs and, thus, the pull-off force.

To conclude this subsection, let us consider again Derjaguin's initial paper [1]. Assuming energy balance between the Hertz elastic energy, $2\mathcal{K}a^5/5R^2$, and the interfacial energy, $\pi w a^2$, Derjaguin implicitly assumed that a variation of contact radius leads to a variation of interfacial energy proportional to the full adhesion energy, w, that is to say that the interactions are very short-ranged, and the displacement right outside the contact zone is very steep since it spans the full decay length of the interaction potential. This concept, central to the JKR theory, is inconsistent with a bare Hertz deformation.

5.2. Intermediate Range Interactions

In the case of intermediate range interactions, the results will depend upon the balance of the long-range and short-range parts of the interaction. Thus, exact results require that the interaction be precisely described. However, we have shown that for reasonably well-behaved interaction potentials, good approximations are obtained when, in addition to the adhesion energy, w, the interaction is simply described by a typical decay length [16]. We will, thus, now describe the approximate analytical models of the JKR-DMT transition one can obtain along this line of thought.

In 1992, Maugis showed, using a constant stress approximation for the attractive interaction (Dugdale model), that an analytical model could be presented for the contact of spheres [9]. The results show that the JKR-to-DMT transition occurs because $g(a)$ decreases to zero while F_{ext} increases to $2\pi wR$ as the range of the interactions increases to infinity (and their amplitude goes to zero, so as to keep w finite).

After Maugis, various approximate models have been developed. One chooses some reasonable form of the stress distribution outside the contact zone: the cases studied thus far were low order polynomials: constant, linear, and quadratic [16]. The distribution, however, is characterized by a radial extension, c, and an amplitude, σ_0. The solution to the contact problem can be completely calculated analytically, as shown above. We, thus, obtain analytical expressions for the force, the penetration, the gap and also for the self-consistency Eq. (5.2) in terms of a, c and σ_0. As a result, for each value of the contact radius, a, and the interaction amplitude, σ_0, the radial extension, c, can be calculated numerically from the self-consistency equation, and the numerical values of the force and penetration determined.

Thus, force curves can be calculated for various values of the interaction amplitude, σ_0. Note that σ_0 and w typically specify the range of the interactions which is of the order w/σ_0. Such approximation schemes have been shown to reproduce closely the results of extensive numerical calculations for definite interaction potentials, at a much lower computational cost [15, 16].

Another kind of approximation has been introduced recently by Johnson and Greenwood [15]. Instead of using a polynomial stress distribution in the attractive interaction zone, they introduced an ellipsoidal distribution, which lends itself more easily to analytical calculations. Actually, we can observe from Eq. (3.9) that this form of interaction stems from a low-order polynomial approximation *on the g function*. In the above mentionned instance [15], one has

$$g(t) \propto (c^2 - t^2) \quad \text{for } a < t < c. \tag{5.19}$$

Thus, one can anticipate a whole class of such approximations, with similar behaviour.

5.3. Long Range Interactions

The long-range interactions seem easy to tackle. In this case, we assume that the interaction zone spreads out far away from the contact zone and, as a result of the finite value of the adhesion energy, the amplitude of the interaction is vanishing. Thus, in the self-consistency equation, Eq. (5.2), the gap is essentially determined by the shape of the indenter, $h(r)$; the additional deformations due to the contact, which typically spread out only as a far as a few times the contact radius, are negligible.

In the discussion, we will keep to integer order indenter shape. The best known case is the case of the paraboloid. There, we have

$$w = \int_a^{+\infty} p(s)\frac{s}{R}\,ds. \tag{5.20}$$

As a result, as defined in Eq. (3.2) $g(a)$ goes to zero; that is to say, there is no increase of the contact zone. If we compare the adhesion energy, w, and F_{ext}, we directly obtain that

$$w = \frac{1}{2\pi R}F_{\text{ext}}, \tag{5.21}$$

which is the DMT result [2]. Typically, the DMT case is known to apply when a sizeable meniscus of condensed liquid surrounds the contact zone.

If the shape of the indenter is described by a higher-order power law, then $g(a)$ also goes to zero but, in addition, F_{ext} also becomes small compared with w: only the smaller-ranged part contributes, and this goes to zero.

In contrast, in the case of the cone, we observe that $g(a) \equiv w$ while F_{ext} goes to infinity!

We conclude that, *except in the case of the paraboloid*, it is physically impossible to assume that the interaction zone spreads out far away from the contact zone: in the case of higher-order shapes, the gap slopes up too fast, while, in the case of the cone, the contact zone grows up too fast. In these cases, therefore, we must resort to the intermediate range description when the JKR case does not apply. Note that in the bi-dimensional case (2D), a similar phenomenon has been found when considering the contact of a parabolic indenter: the pull-off force goes to zero when the interaction zone spreads out far from contact, and no "2D-DMT" limit is observed for a parabolic shape [33].

These results do not reduce the relevance of the DMT limit. First of all, the paraboloidal case is *the* significant case, since any axisymmetric non-conformal asperity will adequately be described for all relevant purposes as a paraboloid. Moreover, the existence of this limit is also probably the reason why approximate schemes like those mentioned in Section 5.2 are efficient.

6. INTERACTING SURFACES WITHOUT CONTACT

The assessment of adhesion energies from pull-off forces thus only depends, to first order, upon the range of the relevant interactions. We emphasize the fact that the description of the interacting surfaces before contact, or after contact rupture, should be consistent with the assumptions made for the description of the contact. Thus, if the long-range interactions are measured, their contributions to the total pull-off force may be calculated and the remaining part ascribed to short-range forces, which cannot be measured directly.

Assuming reversible processes throughout, as we did, the non-contact part of the force curve is described by simply specifying the stress distribution

$$F_z(r) = -p(r), \tag{6.1}$$

and the surface deformations computed through Eqs. (3.2) and (3.1). As a result, one obtains a continuous curve taking into account interactions and deformations with and without contact. A good (theoretical) example of such a full force curve can be found in Ref. [8].

Thus, if some assumption is made regarding the interaction between surfaces for the contact part of the curve, one can check whether this assumption is consistent with the measured non-contact part of the curve. Parts of the full curve may be inaccessible, however, due to instability, either plain physical instability, or more often experimental instability, due to the finite stiffness of the measuring system.

In the absence of instability at contact rupture, the contact radius will go to zero, and assuming a regular enough interaction potential, it appears that the force curve will be continuous with a continuous derivative at zero contact radius. Thus, the zero slope in the DMT model at zero contact radius mentioned in the Introduction is due to the fact that the dominant force term, F_{ext}, is a constant equal to $2\pi Rw$. Under consistent assumptions, it will still essentially be constant for some distance after contact rupture, so that the slope after contact rupture is indeed also zero. Thus, if the interaction is such that the slope is not zero after contact rupture, one should conclude that the DMT model is not adequate to describe the contact, and that a more elaborate model like Maugis' or others should be used.

We will now elaborate on the necessary consistency between the treatment of contact and non-contact parts of the force curve using an experimental example. Figure 5 displays the scaled results of the full (backward) force curve measured between silica surfaces in dry air, as obtained with a very rigid surface forces apparatus [34]. The linear non-contact part of the curve allows the long-range interactions to be described as meniscus-induced. Thus, a Maugis model appeared adequate. However, it turned out that it was impossible to fit consistently the contact and non-contact data with a Maugis model because, for the relevant values of the parameters ($\lambda = 0.3$), no force jump appears in the Maugis model. It, thus, became clear that an additional much shorter-ranged force also played a role in the interaction. A proper model taking into account two very different length scales was devised [35], which allowed a good fit of the data, showing that the long-range interaction (the meniscus force, described as in the Maugis model by a constant stress zone, with a λ

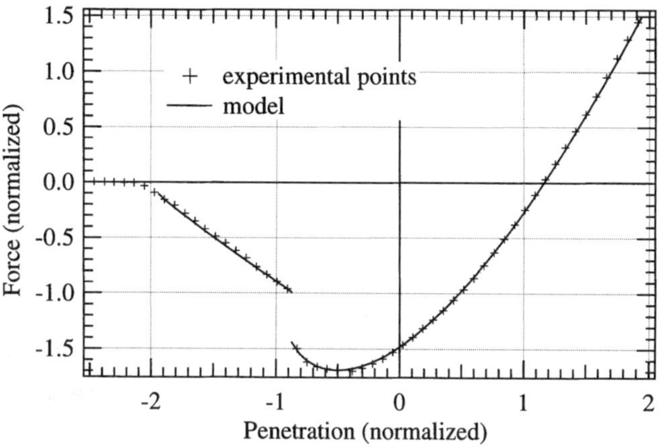

FIGURE 5 Normalized force *vs.* normalized penetration for a silica/silica contact in dry air. The experimental points are from Ref. [34], the model is described in Section 6 and Ref. [35].

close to 0.3) accounts for about two-thirds and the short-range interaction (described similarly, but with a very large λ arbitrarily taken as 10) for one-third of the total adhesion.

This type of data treatment, therefore, allows some estimate of the very-short-range forces to be obtained through the comparison of long-range forces and pull-off force. We have also tried to use the same approach in the very different context of ultra-high vacuum force measurements between a metal tip and oxide substrates [36].

7. CONCLUSION

We have shown that the Sneddon approach to the elastic response of a flat surface to axisymmetric loadings is convenient to describe the adhesive contact of bodies. It allows general and useful expressions to be derived easily, and highlights the generic features of the various theories. In addition, the useful extension to the case where the surfaces interact but are not in contact is readily obtained.

We have, thus, introduced the generalized JKR model under various aspects. Several approximate models for the intermediate cases have been introduced and classified. The DMT limit has been shown to depend crucially upon the shape of the indenter.

References

[1] Derjaguin, B. V., *Kolloid Z* **69**, 155 (1934).
[2] Derjaguin, B. V., Muller, V. M. and Toporov, Yu. P., *J. Collid Interface Sci.* **53**, 314 (1975).
[3] Johnson, K. L., Kendall, K. and Roberts, A. D., *Proc. Roy. Soc. London A* **324**, 301 (1971).
[4] Tabor, D., *J. Colloids Interface Sci.* **58**, 2 (1977).
[5] Muller, V. M., Yushenko, V. S. and Derjaguin, B. V., *J. Colloids Interface Sci.* **77**, 91 (1980).
[6] Muller, V. M., Yushchenko, V. S. and Derjaguin, B. V., *J. Colloids Interface Sci.* **92**, 92 (1982).
[7] Attard, P. and Parker, J. L., *Phys. Rev. A* **46**, 7959 (1992).
[8] Greenwood, J. A., *Proc. R. Soc. Lond. A* **453**, 1277 (1997).
[9] Maugis, D., *J. Colloids Interface Sci.* **150**, 243 (1992).
[10] Maugis, D. and Barquins, M., *J. Phys. D: Appl. Phys.* **11**, 1989 (1978).
[11] Barquins, M. and Maugis, D., *J. Mech. Theor. Appl.* **1**, 331 (1982).
[12] Maugis, D., *J. Adhesion Sci. Tec.* **1**, 105 (1987).
[13] Johnson, K. L., *Langmuir* **12**, 4510 (1996).
[14] Johnson, K. L. and Greenwood, J. A., *J. Colloid Interface Sci.* **192**, 326 (1997).
[15] Greenwood, J. A. and Johnson, K. L., *J. Phys. D: Appl. Phys.* **31**, 3279 (1998).
[16] Barthel, E., *J. Colloid Interface Sci.* **200**, 7 (1998).
[17] Barthel, E., *Thin Solid Films* **330**, 27 (1998).
[18] Hertz, H., *J. reine und angewandte Mathematik* **92**, 156 (1882).
[19] Hughes, B. D. and White, L. R., *Q. Jl Mech. Appl. Math.* **32**, 445 (1979).
[20] Boussinesq, J., *Application des Potentiels àl'Etude de l'Equilibre et du Mouvement des Solides Elastiques* (Gauthier–Villars, Paris, 1885).
[21] Hamaker, H. C., *Physica* **4**, 1058 (1937).
[22] Sneddon, I. N., *Fourier Transform* (McGraw-Hill, New York, 1951).
[23] Muki, R., "Progress in solid mechanics", In: Sneddon, I. N. and Hill, R. Eds., *Progress in Solid Mechanics*, **I** (North Holland, Amsterdam, 1960), Chap. VIII, p. 399.
[24] Spence, D. A., *Proc. Roy. Soc. A* **305**, 55 (1968).
[25] Gladwell, G. M. L., *Contact Problems in Linear Elasticity* (Sijthoff and Noordhoff, Alphen aan den Rijn, 1980).
[26] Johnson, K. L., *Contact Mechanics* (Cambridge University Press, Cambridge, 1985).
[27] Fogden, A. and White, L. R., *J. Colloid Interface Sci.* **138**, 414 (1990).
[28] Hills, D. A. J., Nowell, D. and Sackfield, A., *Mechanics of Elastic Contacts* (Butterworth-Heinemann, 1993).
[29] Basire, C. and Fretigny, C., *C. R. Acad. Sci. Paris* **326**, 323 (1998).
[30] Landau, L. D. and Lifsitz, E. M., *Theory of Elasticity*, 2nd edn. (Pergamon Press, Oxford, 1970).
[31] Sneddon, I. N., *Int. J. Eng. Sci.* **3**, 47 (1965).
[32] Lowengrub, M. and Sneddon, I. N., *Int. J. Eng. Sci.* **3**, 451 (1965).
[33] Baney, J. M. and Hui, C. Y., *Adhes. Sci. Techol.* **11**, 393 (1997).
[34] Barthel, E., Lin, X. Y. and Loubet, J. L., *J. Colloid Interface Sci.* **177**, 401 (1996).
[35] Barthel, E., *Colloids and Surfaces A* **149**, 99 (1999).
[36] Sounilhac, S., Barthel, E. and Creuzet, F., *J. Appl. Phys.* **85**, 222 (1999).

APPENDIX A

We describe the linear elastic response of a flat solid submitted to an axisymmetric normal stress distribution, F_z. Let us start from the mechanical equilibrium condition

$$\mathrm{div}(\bar{\bar{\sigma}}) + \bar{F} = 0, \tag{A1}$$

and the linear elasticity relation

$$\bar{\bar{\sigma}} = \frac{\nu E}{(1-2\nu)(1+\nu)} \mathrm{Tr}(\bar{\bar{\varepsilon}})\bar{\bar{\delta}} + \frac{E}{(1+\nu)}\bar{\bar{\varepsilon}}, \tag{A2}$$

where $\bar{\bar{\sigma}}$ is the stress tensor, $\bar{\bar{\varepsilon}}$ the strain tensor, $\bar{F}$ the external stress distribution, $\bar{\bar{\delta}}$ the unit tensor and E and ν the Young's modulus and Poisson's ratio. We will write down an equation for the displacement field, u, knowing that the strain tensor is the symmetric part of the gradient of the displacement field.

We now take the axisymmetric nature of the problem into account by resorting to the form of Fourier transform which arises in such conditions, *i.e.*, the Hankel transform. Indeed, in the angular integration of the Fourier transform, one comes across the quantity

$$\int_0^{2\pi} \exp(ikr\cos(\theta))d\theta, \tag{A3}$$

which directly leads to the use of Bessel functions like

$$J_0(r) = \frac{1}{\pi}\int_0^{\pi} \cos(r\cos(\theta))d\theta \tag{A4}$$

and Hankel transforms. In addition, we take into account the tensorial nature of the fields we are looking for by noting that the normal components, u_z, and, F_z, of the displacement and force fields transform as scalars under a rotation around the z axis, while the tangential components u_r and F_r transform as vectors. As a result, we will selectively use the Hankel transforms of order 0 and 1 as follows:

$$\tilde{u}_{z,0}(k,z) = \int_0^{+\infty} dr r\, J_0(kr) u_z(r,z), \tag{A5}$$

$$\tilde{F}_{z,0}(k,z) = \int_0^{+\infty} dr r\, J_0(kr) F_z(r,z), \tag{A6}$$

$$\tilde{u}_{r,1}(k,z) = \int_0^{+\infty} dr r\, J_1(kr) u_r(r,z), \tag{A7}$$

$$\tilde{F}_{r,1}(k,z) = \int_0^{+\infty} drr\, J_1(kr) F_r(r,z), \tag{A8}$$

where J_1 is the first-order Bessel function

$$J_1(r) = \frac{1}{\pi}\int_0^{\pi} \sin(r\cos(\theta))\cos(\theta)d\theta. \tag{A9}$$

From the expressions for the divergence and gradient in cylindrical coordinates, and the properties of Hankel transforms, one obtains the following equilibrium conditions on the displacement field

$$\left(-\alpha^2 k^2 + \frac{\partial^2}{\partial z^2}\right)\tilde{u}_{r,1}(k,z) - \beta k \frac{\partial}{\partial z}\tilde{u}_{z,0}(k,z) = -\frac{2(1+\nu)}{E}\tilde{F}_{r,1}(k,z), \tag{A10}$$

$$\beta k \frac{\partial}{\partial z}\tilde{u}_{r,1}(k,z) + \left(\alpha^2 \frac{\partial^2}{\partial z^2} - k^2\right)\tilde{u}_{z,0}(k,z) = -\frac{2(1+\nu)}{E}\tilde{F}_{z,0}(k,z), \tag{A11}$$

with

$$\alpha^2 = \frac{2(1-\nu)}{1-2\nu} \tag{A12}$$

and

$$\beta = \frac{1}{1-2\nu}. \tag{A13}$$

If we now assume that the tangential component, F_r, of the force field is zero, Eq. (A10) becomes

$$\mathcal{O}_1 \tilde{u}_{r,1}(k,z) - \mathcal{O}_2 \tilde{u}_{z,0}(k,z) = 0, \tag{A14}$$

where the operators are defined as

$$\mathcal{O}_1 = -\alpha^2 k^2 + \frac{\partial^2}{\partial z^2}, \tag{A15}$$

$$\mathcal{O}_2 = \beta k \frac{\partial}{\partial z}. \tag{A16}$$

Now $\mathcal{O}_1$ and $\mathcal{O}_2$ commute, so that one can introduce a potential function, $G(k,z)$, such that, from Eq. (A14),

$$\tilde{u}_{z,0}(k,z) = \mathcal{O}_1 G(k,z), \tag{A17}$$

$$\tilde{u}_{r,1}(k,z) = \mathcal{O}_2 G(k,z). \tag{A18}$$

Then Eq. (A14) is automatically fulfilled and since

$$\beta = \alpha^2 - 1, \tag{A19}$$

Equation (A11) becomes

$$\left(\frac{\partial^2}{\partial z^2} - k^2\right)^2 G(k,z) = -\frac{(1+\nu)(1-2\nu)}{(1-\nu)E}\tilde{F}_{z,0}(k,z), \tag{A20}$$

which provides a solution for a solid submitted to an axisymmetric distribution of normal stress.

1. Elastic Surface Loaded Axisymmetrically

Let us now examine the response of a flat elastic surface submitted to a normal stress distribution. We now introduce the boundary conditions relevant for the surface. The plane of the surface will be chosen as $z=0$. All fields and, thus, the potential $G(k,z)$, vanish above the surface ($z > 0$). We assume that the external forces are described by the *surface* distribution, $\tilde{F}_{z,0}(k)\delta(z)$. We are now essentially looking for $G_{\text{sing.}}(k,z)$, the singular part of $G(k,z)$, which will give rise to the δ function on the right-hand side of Eq. (A20). Given the fact that $G(k,z)$ is zero in the upper half-space, a cosine transform (on the negative part of the z axis) is in order. Introducing the cosine transform of $G_{\text{sing.}}(k,z)$ (denoted $\tilde{G}_{\text{sing.}}(k,\xi)$), the solution to Eq. (A20) is obtained as

$$\tilde{G}_{\text{sing.}}(k,\xi) = -\frac{(1+\nu)(1-2\nu)}{(1-\nu)E}\tilde{F}_{z,0}(k)\frac{1}{2\pi}\frac{1}{(\xi^2+k^2)^2}, \tag{A21}$$

from which, by inverse cosine transform, one gets

$$G_{\text{sing.}}(k,z) = -\frac{(1+\nu)(1-2\nu)}{2(1-\nu)E}\tilde{F}_{z,0}(k)\frac{1}{k^2}\left(\frac{1}{k} - z\right)\exp(kz) \tag{A22}$$

for $z < 0$. By definition, $G_{\text{sing.}}(k, z) = 0$ for $z \geq 0$. This expression provides a solution of the mechanical equilibrium equation, which expresses the balance of forces, when a surface is subjected to a surface distribution, $F_z(r)$, of normal forces. It is not, however, a solution for a *free* surface, for the surface shear, in this solution, is not zero. Therefore, we now build the zero-surface shear solution.

2. Frictionless Elastic Surface Loaded Axisymmetrically

The general solution to Eq. (A20) is made up of the singular part we have just calculated plus a regular term of the form $(A + Bz)\exp(-kz) + (C + Dz)\exp(+kz)$, where A, B, C and D depend only upon k. Keeping only the term $G_{\text{reg.}}(k, z) = (C + Dz)\exp(+kz)$ which vanishes when z goes to $-\infty$, the general potential is now

$$G(k, z) = G_{\text{sing.}}(k, z) + G_{\text{reg.}}(k, z). \tag{A23}$$

The boundary conditions on the free surface are, using the obvious notation:

$$\sigma_{zz,\text{reg.}}(k, z)|_{z=0} = 0 \tag{A24}$$

$$\sigma_{rz,\text{reg.}}(k, z)|_{z=0} + \sigma_{rz,\text{sing.}}(k, z)|_{z=0} = 0. \tag{A25}$$

Applying the Hankel transform of order 0 to Eq. (A24) and the Hankel transform of order 1 to Eq. (A25), again in agreement with their tensorial nature, one gets

$$\nu\beta k\tilde{u}_{r,1,\text{reg.}}(k, z) + \frac{\alpha^2}{2}\frac{\partial}{\partial z}\tilde{u}_{z,0,\text{reg.}}(k, z) = 0 \tag{A26}$$

$$\frac{\partial}{\partial z}\tilde{u}_{r,1}(k, z) - k\tilde{u}_{z,0}(k, z) = 0. \tag{A27}$$

From the definitions of $G(k, z)$, $\tilde{u}_{r,1}(k, z)$ and $\tilde{u}_{z,0}(k, z)$ (Eqs. (A23), (A17) and (A18)) one obtains a system of two linear equations in C and D, from which the final solution for a frictionless axisymmetrically loaded elastic surface is obtained as:

$$\tilde{u}_{r,1}(k,z) = \left(\frac{1+\nu}{E}\left(\frac{1-2\nu}{k}+z\right)\tilde{F}_{z,0}(k)\right)\exp(kz) \tag{A28}$$

$$\tilde{u}_{z,0}(k,z) = \left(\frac{2(1+\nu)}{E}\left(\frac{1-\nu}{k}-\frac{z}{2}\right)\tilde{F}_{z,0}(k)\right)\exp(kz). \tag{A29}$$

For contact problems, the important general result is that, at the surface ($z=0$),

$$\tilde{u}_{z,0}(k) = \frac{2(1-\nu^2)}{E}\frac{\tilde{F}_{z,0}(k)}{k}, \tag{A30}$$

where $\tilde{u}_{z,0}(k)$ [resp. $\tilde{F}_{z,0}(k)$] denotes $\tilde{u}_{z,0}(k,z=0)$ [resp. $\tilde{F}_{z,0}(k,z=0)$].

This expression is clearly the equivalent of Landau's $1/r$ Green's function in real space [30]. Since $k\tilde{u}_{z,0}(k)$ is essentially the transform of the gradient of $u_z(r)$, this relation simply states that the normal compliance of the surface is $\mathcal{K}^{-1}$.

APPENDIX B: USEFUL FORMS OF THE GENERAL RESULT AND ITS RELEVANCE TO MIXED BOUNDARY CONDITIONS

As a result of Eq. (A30) and using the inverse Hankel transform (the form of which is identical to the direct transform), one can express the surface displacement as:

$$u_z(r) = \frac{2(1-\nu^2)}{E}\int_0^{+\infty}\tilde{F}_{z,0}(k)J_0(kr)dk. \tag{B1}$$

Now, with the change of variable $t = r\cos\theta$ in Eq. (A4), one obtains the following expression for the Bessel function:

$$J_0(kr) = \frac{2}{\pi}\int_0^r \frac{\cos(kt)dt}{(r^2-t^2)^{1/2}}, \tag{B2}$$

that is to say, the Bessel function of order 0 is also the cosine transform of the function

$$\frac{\Theta(r-t)}{(r^2-t^2)^{1/2}}, \tag{B3}$$

where Θ is the usual Heaviside step function.

Indeed, let us now introduce the cosine transform of the quantity $\tilde{F}_{z,0}(k)$, which is itself the Hankel transform of the axisymmetric surface stress distribution:

$$g(t) = \int_0^{+\infty} \tilde{F}_{z,0}(k)\cos(kt), \tag{B4}$$

one readily gets Eqs. (3.1) and (3.2), by using, for instance, Parseval's relation:

$$u_z(r) = \frac{4}{\pi}\frac{(1-\nu^2)}{E}\int_0^r \frac{g(t)}{(r^2-t^2)^{1/2}}dt,$$

and similarly,

$$g(t) = \int_t^{+\infty} \frac{sF_z(s)}{(s^2-t^2)^{1/2}}ds.$$

This trick in the inverse Hankel transform allows one to dispense altogether with dual integral equations [22] and Erdelyi-Kober operators [27]. Now the adequation of this form of solution to mixed boundary conditions becomes obvious since $u_z(r)$ depends *only* on the values of $g(t)$ smaller than r, while $F_z(s)$ depends *only* on the values of $g(t)$ larger than s.

In addition, Eqs. (3.1) and (3.2) can both be inverted through the use of inverse Hankel and Fourier transforms, leading to the inverse relations (3.6), (3.7), (3.8) and (3.9).

Let us introduce two additional useful properties of this representation. The total force exerted on the surface F can be easily expressed as (Eq. (3.10))

$$F = 2\pi\int_0^{+\infty} drrF_z(r) = 2\pi\tilde{F}_{z,0}(k=0) = 2\pi\frac{2}{\pi}\int_0^{+\infty} dsg(s),$$

where the second equality stems from the Hankel transform and the last from cosine transform.

Similarly, the total mechanical energy, $\mathcal{E}$, can be calculated using Parseval's relation for the Hankel and the cosine transform (Eq. (3.11))

$$\begin{aligned}\mathcal{E} &= \pi \int_0^{+\infty} drrF_z(r)u_z(r) = \pi \int_0^{+\infty} dkk\tilde{F}_{z,0}(k)\tilde{u}_{z,0}(k) \\ &= \frac{\pi}{\mathcal{K}} \int_0^{+\infty} dk\tilde{F}_{z,0}(k)^2 = \frac{2}{\mathcal{K}} \int_0^{+\infty} dsg^2(s),\end{aligned}$$

where the second equality comes from the Parseval's relation for the Hankel's transform, the third from the equilibrium relation Eq. (A30), and the fourth from Parseval's relation for cosine transform.

Finite Element Modeling of Particle Adhesion: A Surface Energy Formalism

DAVID J. QUESNEL

Mechanical Engineering Department, University of Rochester, Rochester NY 14627-0132, USA

DONALD S. RIMAI

NexPress Solutions, LLC, Rochester NY 14653-6402, USA

The adhesion of particles is modeled with finite element analysis using an energy approach comparable with that used in the JKR formalism. The strain energy of a cylindrically symmetric system, comprising a particle adhering to a surface with a fixed contact size, is computed as a function of contact size and then added to an energy term that is linearly proportional to the contact patch area. These computations also include contributions from the potential energy of a body force comparable with that which might be applied by a centrifuge. The results show regions of stability (adhesion) where a local energy minimum exists and regions of release where separation of the particle from the surface leads to a continuous decrease in the energy of the system. The effect of the deformation of the particle is included implicitly as a result of the FEM which provides details of the strains and stresses within the system. Discussion concentrates on the physical meaning of the behaviors and the significance of JKR-like theories that use an effective surface energy to represent electrostatic and van der Waals contributions to the adhesion. Modeling the effects of surface roughness of particles and the plastic deformation of particles through an effective surface energy is considered.

INTRODUCTION

The adhesion of particles is well known to be the direct result of the electrical interactions between molecules. These can be either electrodynamic, such as in the case of van der Waals interactions between electrically neutral species, or electrostatic, as in the case of direct attraction between oppositely-signed charged species. Some researchers favor explanations such as the charged-patch model [1, 2] dominated by electrostatic behavior while others believe the electrostatic attraction is primarily a far-field effect with the near-field behavior controlled by van der Waals interactions [3–6]. Computations have been made that include both contributions to explain the spatial dependence of the force measured between a particle mounted on an AFM probe and a surface [7].

The charged-patch model of particle adhesion and the uniformly charged approach both focus on the distribution of charge on the surface of the particle and the attractive forces that are developed as a result of the electric charge interactions. Alternatively, the Lifshitz–Hamaker approach [8–10] is taken to include the van der Waals contribution. These two approaches have generated a dichotomy of thought regarding particle adhesion. Those favoring the Lifshitz–Hamaker approach tend to view adhesion in terms of surface energy contributions, such as those discussed in the JKR model [11]. Alternatively, those favoring the electrostatic models tend to view adhesion and particle detachment in terms of balancing the attractive and applied (detachment) forces. Both the electrostatic and van der Waals contributions include action at a distance, namely the direct attraction of opposite sign charge *via* the $1/r^2$ law or the attraction of induced dipoles that lead to the $1/r^6$ dependence, sometimes broadly referred to as the London dispersion force [12]. While it has been shown for simple geometries [13] that the van der Waals forces, as modeled by the Lennard–Jones potential, can give rise to an apparent surface energy, the detailed mechanics of the attraction cannot be explained by a surface energy alone [14, 15]. Rather, the spatial rearrangement of the atoms that occur as a result of the interaction forces must be included. This rearrangement of atoms gives rise to larger forces than would be expected for van der Waals interactions between static atomic groupings and leads to a hysteresis in the load-displacement behavior [16, 17].

It seems clear that the use of the surface energy paradigm is a simplification of the physics described above, with the intent that the important aspects of the adhesive behavior would be captured using the surface energy representation. Indeed, the wide-spread acceptance of the JKR theory [11] suggests that much of the behavior of the particle and surface interaction has been captured with the use of an effective surface energy. In the larger perspective, the use of the surface energy under the rubric of an energy method is well known to represent the energetics of a situation without regard to the detailed mechanisms involved. Indeed, this is one of the beauties of the energy approach to mechanics problems, in that the messy mechanistic treatment can sometimes be avoided. The downside of the energy approach is that its result provides no information as to the mechanism or its details.

Experimental approaches to the study of particle adhesion have shown that the mechanical properties of the particles play a substantial role in the apparent adhesion to surfaces [18]. Rimai *et al.*, have shown that the relationship between contact-patch size and particle size varies depending on the mechanical properties of the particles and substrates. Theoretical treatments that involve plasticity [19] have also shown that the permanent deformation that occurs as a result of electrostatic and van der Waals interactions leads to a reduction in the stored strain energy of the system. Since this stored energy is recovered upon removal of an elastic particle, the plasticity serves to enhance the apparent adhesion by reducing the elastic energy that can be recovered during particle removal. Analyses of this type are difficult to perform for the commercially-important irregular geometries, hence, the benefit of finite element methods. In this paper, we use finite element methods on simple geometries to establish the overall approach. In subsequent papers, it is our intent to use more complex geometries and surface textures that more closely model the interaction of irregularly-shaped particles and particles with surface addenda.

In order to lend credibility to the finite element approach to the study of adhesion from a mechanistic point of view, the present paper takes a critical approach by examining the implications of the use of an effective surface energy. This approach has two direct benefits. First, it allows an independent assessment of the assumptions and analytical procedures of the JKR theory by computing the strain energies associated with the deformation process by an independent means. Second, it allows us to examine the influence of changes in how the loads are applied, such as the use of body forces rather than equipollent loading in the far field. This is of particular significance when the results are to be compared with particle experiments where forces are applied *via* centrifugation. On a more far-reaching scale, this approach provides a mechanism to apply the JKR formalism to particles of irregular geometry and particles where large scale deformations can occur. In addition, the effect of additional long-range interactions on particle adhesion can be determined. Such long-range interactions are not allowed in the JKR formalism, which assumes that all interactions occur strictly within the area of contact.

ANALYTICAL BACKGROUND

In the field of particle adhesion, the dominant theory is that first proposed in 1971 by Johnson, Kendall and Roberts (JKR) [11]. It has withstood the test of time and has gained wide acceptance, having been found to predict adhesional behavior accurately under a variety of circumstances [20]. Earlier theoretical works by Derjaguin *et al.* (DMT) [21], while providing similar phenomenology, do not represent the experimental data as well and are drifting out of favor. Subsequent work that expands on the JKR theory by incorporating plasticity, such as that of Maugis and Pollock (MP) [22], or that allow deformations of larger degree [23, 24], have yet to gain the degree of overall acceptance of the JKR theory. There are numerous works which leverage the initial application of the energy balance between elastic strain energy and surface energy first proposed by Johnson *et al.* [11] and the interested reader is referred to broader reviews in the area of contact mechanics [25].

Beautiful in its simplicity, the JKR formalism takes a continuum mechanical approach to describing the adhesion of particles to other particles and to substrates. Specifically, the JKR theory calculates an equilibrium contact radius, a, from the sum of the elastically stored energy, the energy associated with the surface forces, and the mechanical potential energy of the externally-applied load. The resulting relationship is given by

$$a^3 = \frac{R}{K}\left\{P + 3w_A\pi R + \left[6w_A\pi RP + (3w_A\pi R)^2\right]^{1/2}\right\} \tag{1}$$

where R, P, and w_A represent the particle radius, the applied force on the particle, and the work of adhesion between the particle and substrate, respectively, and K is an effective stiffness related to the Young's moduli and Poisson's ratios of the system. For the case where one material is considered rigid relative to the other, the effective stiffness is dominated by the modulus and Poisson's ratio of the more compliant material. Consequently, from the experimental point of view, a hard spherical particle with an assortment of compliant substrates provides the same equilibrium contact patch dimensions as the more difficult-to-fabricate assortment of compliant spherical particles on rigid substrates.

Examining the above expression provides specific relations between contact and particle radii and the applied forces. For example, if $P = 0$, the zero load contact radius is given by

$$a_0 = \left[\frac{9w_A\pi(1-\nu^2)}{2E}\right]^{1/3} R^{2/3}. \tag{2}$$

When $P > 0$ so that the particle is pressed into the surface, a increases. Alternatively, when $P < 0$, so that the load is pulling the particle from the surface, a decreases. Johnson *et al.* [11] have argued that, because the solutions to the JKR equation must be real, separation must occur when

$$P = -\frac{3}{2}w_A\pi R, \tag{3}$$

which corresponds to a finite contact radius at separation given by $a_s \approx 0.63a_0$. There are no constant load solutions for $a < 0.63a_0$.

Implicit in the JKR formalism are several basic assumptions, including that of small strains, as required by the analytical methods, and the assumption that all interactions occur within the zone of contact. This latter assumption is a consequence of the use of a surface energy and suggests that the forces resulting from changes in surface energy must make themselves felt near to the points or areas of contact. However, the requirement that interactions vanish at the circumference of the contact zone implies an infinite stress at that location, resulting in locally large strains. The implications of these assumptions have been discussed in detail by Tabor [26]. The assumption regarding force location is not consistent with the action-at-a-distance concept usually associated with electrostatic and electrodynamic interactions.

The stress singularity that occurs at the edge of the contact patch has been shown to have the same general properties as the singularity at the tip a sharp crack, leading to the application of fracture mechanics [27] to the problem of adhesion. This allows a substantial expertise in the area of fracture mechanics to be applied to the adhesion problem and the interested reader is advised to seek out references in this area as well [28 – 30]. The earliest fracture mechanics work goes back to the energy balance of Griffith [31] and his use of the stress analysis of Inglis [32]. Early in his work, Griffith qualitatively argued that the elastic strain energy liberated by the growth of a crack could be used to balance the work needed to create new surfaces based on thermodynamic arguments. However, he did not publish his theory until he could provide a mathematical justification, which, as it turns out, relied on the work of Inglis [32] published seven years earlier. Although somewhat more difficult to see, the tendency of an adhesive joint to spread by wetting of one surface by another has the same mechanics. The energy liberated by the wetting of a curved surface by a planar surface is used to deform both bodies until the rate of energy liberation by advancement of the contact area just balances the rate of storage of energy in the strain fields of the bodies. When external loads are applied to separate the adherends, the problem becomes a fracture problem. In the area of fracture mechanics, it is well known that surface energy alone does not control the macroscopic fracture behavior. Rather, an effective surface energy that is many times the actual surface energy is needed to rationalize and explain the observed behavior. The reason for the additional energy is that far-field deformations occur during the growth of a crack. This deformation consumes energy far in excess of the intrinsic surface energy and dominates the overall fracture behavior, particularly when plastic deformation occurs. Even so, however, a simple scalar number is used to characterize fracture behavior. This is called the critical crack driving force or the critical stress intensity but, in the final analysis, it is equivalent to an effective surface energy. With the same physics controlling the removal of particles, the critical surface energy per unit distance of crack advance will exceed that attributable to surface energies by substantial margins.

Since effective surface energies control both fracture and adhesion, then particle adhesion can be thought of as a fracture mechanics problem with specific geometric features that enable stresses to be developed without external loads. The question of whether or not a particle adheres to a surface can be discussed in terms of whether or not it is energetically favorable for a crack to propagate along the particle-substrate interface when loads are applied. In that sense, a change in the JKR contact patch size with applied load is merely a manifestation of the Griffith's criterion. The JKR separation condition expresses a loss of stability where a decrease in contact patch size reduces the overall system energy for all loading conditions.

Although the JKR model does accurately describe numerous aspects of particle adhesion, it also raises many questions. For example, why does a model that assumes small strains work despite the fact that the strains can be quite large near the edge of the contact zone? Why does separation of the particle from the substrate occur at a specific finite radius? Are there differences in the stresses (and, therefore, the removal forces) if the load encompasses the entire particle (as would be exerted by an ultracentrifuge), as opposed to a point (*e.g.*, the particle attached to an AFM cantilever). In this study, preliminary results addressing these questions using finite element modeling are presented. The successes of these methods allow us to look forward to the use of the FEM to answer questions regarding what happens when the materials respond either plastically or hyperelastically.

COMPUTATIONAL DETAILS

The JKR formalism of particle adhesion was modeled as a large-strain, linear-elasticity problem using ANSYS 5.3, a commercial finite element modeling package. In this modeling, a particle, having an elastic modulus in the range of 5 MPa and a Poisson's ratio of 0.495, was adhered to a rigid substrate for a specific contact radius by judicious application of boundary conditions normal to the surface of the substrate. Situations with larger and smaller contact radii were modeled by changing the number of nodes that were required to be in normal contact with the rigid surface. Motion at the constrained nodes in the plane of the rigid surface was allowed, providing for lateral expansion and contraction of the two contacting surfaces. It was assumed, and subsequently verified, that the stresses would be localized within the half of the particle actually nearest the substrate, thereby allowing us to concentrate the nodes of the finite element package within that region. Computations were performed with arbitrary dimensions, thereby enabling the results to be scaled for any size particle.

Each calculation began by applying boundary conditions to the undeformed axisymmetric representation of the lower half of the particle. Displacement boundary conditions pulled the nodes on the surface of the spherical particle down into a common plane representing the rigid substrate while not restricting radial motions. Body force boundary conditions were applied as accelerations corresponding to specific numbers of "g" forces as would occur in a centrifuge. The boundary conditions were ramped in 10 substeps with converged intermediate solutions using the large deformation (non-linear geometry) option. Output files were queried for stress tensors, strain tensors, total loads at the displacement boundary conditions, total strain energy of the particle, radius of the contact patch, and motion of the center of the particle. Motion of the center of mass was computed from the strained locations of the nodes.

By varying the specific nodes bonded to the surface and the magnitude of the body force, the total energy of the system is computed as a function of the equilibrium contact patch dimension. Following the JKR model, this total energy was obtained as a sum of the strain energy of the system, the surface energy of the system, and the mechanical potential of the exterior forces.

RESULTS

Figure 1 shows a plot of the total energy (assuming no externally-applied load) as a function of the diameter of the contact patch for a 5 μm radius particle in contact with the rigid substrate (assuming $w_A = 0.17\,\text{J/m}^2$ and $E = 4\,\text{MPa}$). The equilibrium contact radius, as determined by the location of the minimum in the total energy, is found to be at $a = 2.0\,\mu\text{m}$. Experimental data by Rimai and DeMejo [20] report a contact radius for similar-size glass particles on a

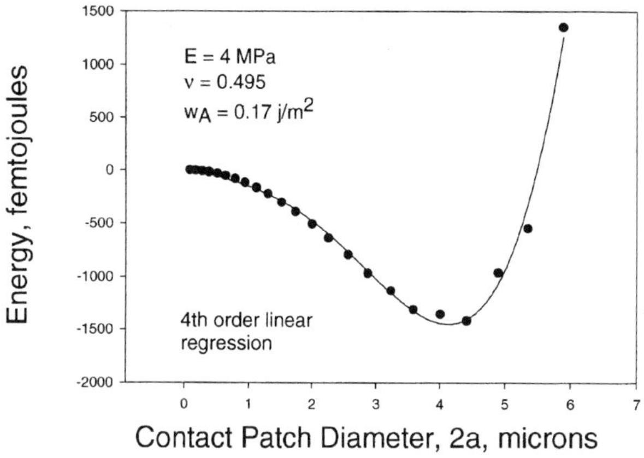

FIGURE 1 Energy as a function of contact patch diameter for a spherical particle on a rigid substrate. Work of adhesion between particle and substrate is shown along with properties of the spherical particle.

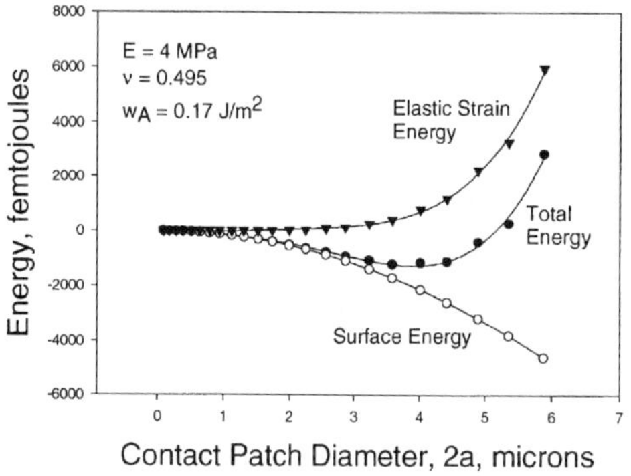

FIGURE 2 Energy as a function of contact-patch size showing surface energy and elastic energy contributions compared with total system energy.

polyurethane substrate ($E \approx 4\,\text{MPa}$) of approximately 2 μm, in good agreement with the finite element results.

Figure 2 presents the elastic strain energy and the surface energy contributions to the total energy as a function of contact patch diameter for the same conditions of $E = 4\,\text{MPa}$ and $w_A = 0.17\,\text{J/m}^2$ discussed above. The total energy is shown as well. It is clear from Figure 2 that the reason for the contact-patch-size-dependent minimum in energy, and, thus, a stable value of the contact patch, is the competition between the decrease in energy due to the surface wetting and the increase in system energy stored in elastic strain.

As previously indicated, the JKR model predicts that particle-substrate separation occurs during the application of a negative load when the contact radius decreases to approximately $0.63a_0$. The behavior obtained with finite element modeling suggests a similar behavior, which is illustrated in Figure 3. Here we note that the negative of the slope of the energy *vs.* contact patch size represents a generalized restoring force akin to the crack-driving force in fracture mechanics. The positive slope shown in Figure 3 for increases in contact patch size represents a restoring force that is negative, characterizing a compression in the contact zone, that resists displacing the system to larger contact sizes. That is, the system resists an increase in contact patch size due to compressive loads because the energy of the system rises when the strain energy contributions exceed the energy reduction associated with the surface energy term.

When attempts are made to pull the particle off and, thus, decrease the contact patch size, Figure 3 suggests a local tension that increases at first but then reaches a maximum value at $\sim 0.63a_0$. With further decreases in contact radius, the derivative, or generalized restoring force, decreases smoothly from its maximum to zero. This implies that the restoring force exerted by the substrate on the particle, as a result

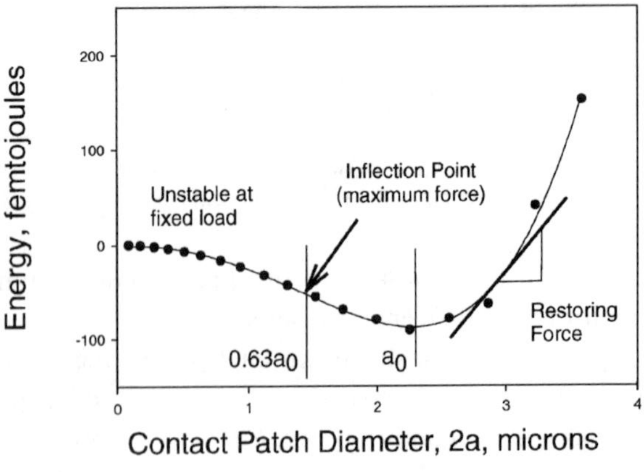

FIGURE 3 Energy *vs.* contact-patch diameter illustrating the generalized restoring force that results from excursions in contact-patch size. The instability at $\sim 0.63a_0$ is shown with an arrow.

of the balance between surface energy reduction and elastic strain energy increase, is a smooth continuous function with a local maximum. Naturally, this translates into a force normal to the substrate surface during various stages of removal that reaches a maximum and then decreases to zero if the particle is not allowed to respond to the forces, a situation akin to displacement control. In load control, however, the system has no solutions for values of $a < 0.63a_0$. The JKR model appeals to the need for a positive real value of the square root in Eq. (1), that is, a real solution, to explain why no solutions are available for $a < 0.63a_0$ and why the particle must detach. Here it is seen that, for load control, the restoring force that is generated by the deflected shape in terms of strain and surface energy contributions is everywhere less than the force applied to get it into this position, leading to an unstable separation of the particle from the surface.

Whereas the schematic representations of the energy *vs.* contact-patch size behaviors examined thus far help us conceptualize the physics that generate the restoring forces, any energy contributions associated with applied loads must be included as well. This energy due to loading is the potential energy of the load and it represents the ability of the applied loads to do work if and when the system deforms to a new geometry. Figure 4 shows the total energy of the particle-substrate system with the inclusion of the mechanical

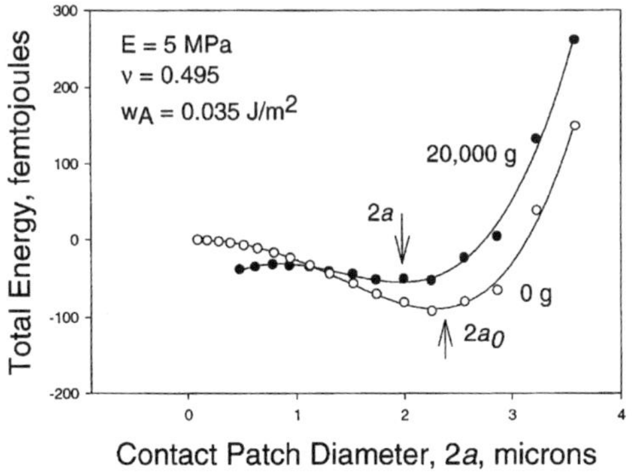

FIGURE 4 Comparison of total system energies at 0 g and 20,000 g accelerations, including the potential energy of the mechanical forces due to acceleration. Note the work of adhesion is about five times smaller, leading to a smaller value of a_0.

potential energy of an exterior body force. This represents a load-control situation where a centrifuge is used to apply tensile loading to the particle by applying 20,000 g of acceleration. In these calculations, the interaction energy is substantially smaller because we have changed the parameters E and w_A to match physical experiments where particle removal can be achieved by centrifugation. The physical parameters of the 5-micron particle are given in Figure 4, which illustrates that the minimum energy position has shifted. As shown in Figure 4, increasing the loading to the equivalent of 20,000 g causes the energy curve to tilt. This causes the local minimum to shift to the left, leading to a smaller equilibrium contact patch, designated by the symbol $2a$ as compared with the initial contact patch size of $2a_0$.

Figure 5 shows the individual contributions to the total system energy. As before, the elastic strain energy increases with contact-patch size and the surface energy contribution decreases with contact-patch size. In addition, there is a mechanical potential energy term resulting from the acceleration. This mechanical potential energy of

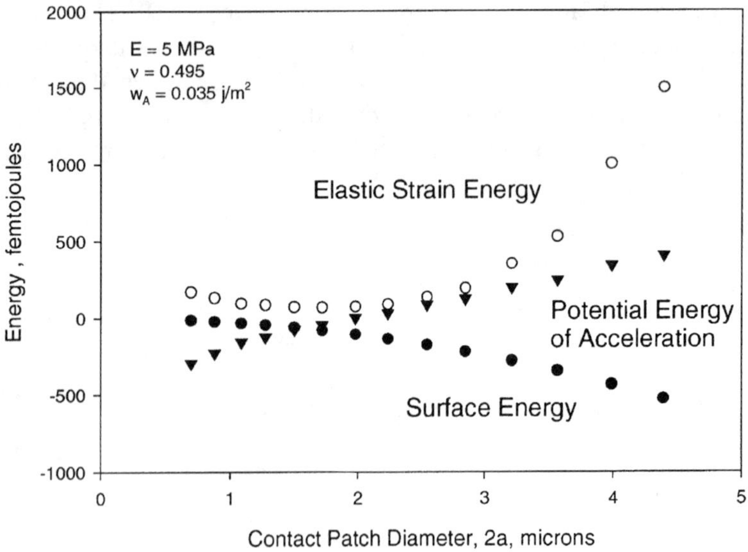

FIGURE 5 Energy components for the system shown in Figure 4. Note that the potential energy of the acceleration is nearly linear with contact-patch size. See text for implications of this linearity. The applied force is $P = -1.284\,\mu N$ where negative sign indicates tension.

the body force caused by acceleration increases approximately linearly with contact-patch size. Referring once again to Figure 4, if one imagines a marble settling into the local energy minimum like a ball in a bowl, as the mechanical potential is applied, the linearity makes the bowl appear to tilt. Larger values cause even more tilt until a point is reached where the marble rolls out of the local minimum and down the energy curve. At this point, there is a critical value of the contact-patch size which is unstable against further displacements to smaller contact-patch sizes. This is the physical reason why there are no load solutions for contact patches sizes less then $\sim 0.63a_0$. It is energetically favorable for the system to separate and there is no local minimum in energy.

Conceptualizing the problem in this way, it can also be observed that there is a local energy maximum created by the mechanical potential term. The change in energy from the local minimum to the local maximum is a measure of the energy which must be supplied to remove the particle while it is being centrifuged at 20,000 g. It provides a measure of the stability of the system against fluctuations in the critical parameters that control adhesion. To remove the particle, the acceleration (or other applied forces) must be increased until the energy change needed to decrease the contact size either vanishes or can be supplied by random motion of the particle, perhaps as available from Brownian motion of the surrounding medium.

Figures 6a – d show a series of plots of total energy *vs.* contact patch size for accelerations of 0, 5000 g, 20,000 g and 50,000 g, respectively. The four graphs of Figure 6 are on the same scale for ease of comparison. As the force due to the acceleration is increased, the graphs tilt progressively in a way that produces smaller contact patch sizes and decreased stability. At the largest acceleration of 50,000 g, the lowest energy position has clearly slid off the graph to the left indicating a zero contact-patch size is appropriate.

Figure 7 shows a parametric plot of the total energy expressions given in the original JKR theory development [11] as a function of P_c, the critical load for particle removal sometimes referred to as the pull-off, separation, or detachment force. At $P = 0$, the curve shows a distinct minimum at $2a_0$ and, as expected, gradually changes shape until, at $P = 1.2P_c$, a curve is generated that has a negative slope throughout, comparable with Figure 6d. Unexpectedly, the case for $P = 1.0P_c$ clearly shows a negative slope at all values of contact-patch

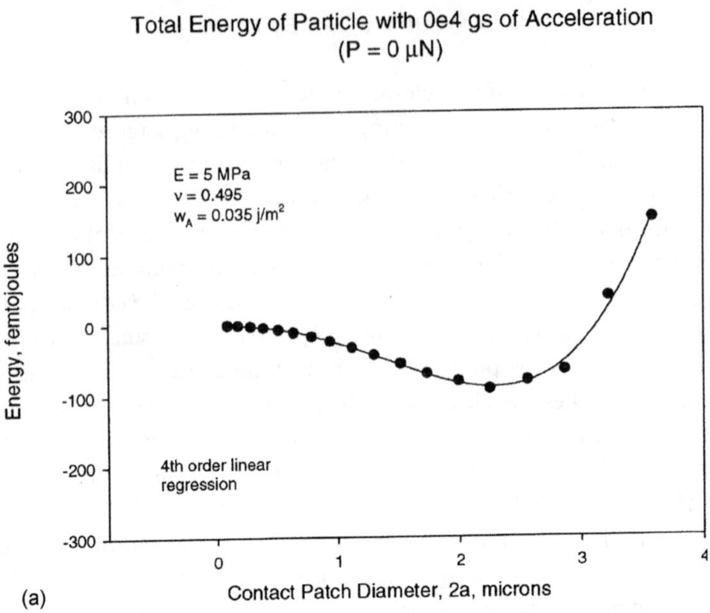

(a)

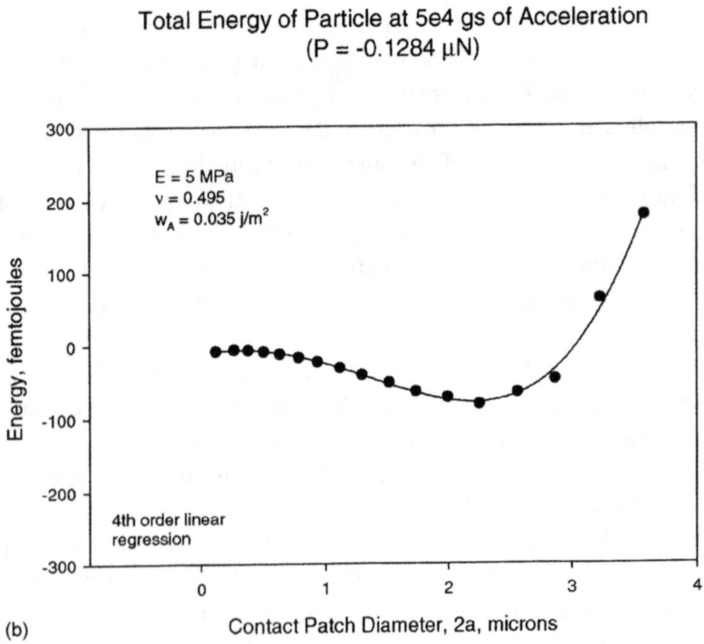

(b)

FIGURE 6 Comparison of energy *vs.* contact patch size for various applied accelerations that might be applied in an ultra-centrifuge. (a) 0 g; (b) 5,000 g; (c) 20,000 g; (d) 50,000 g. Note that the energy minimum shifts left and disappears as the acceleration increases.

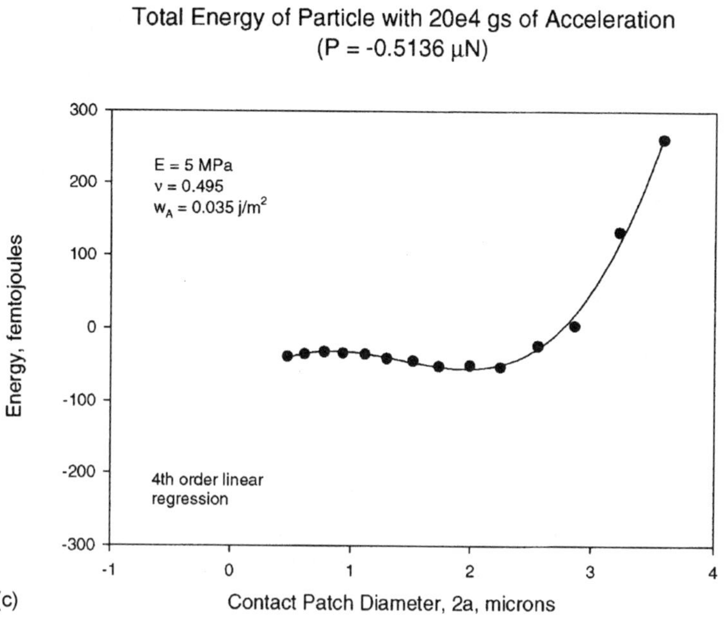
Total Energy of Particle with 20e4 gs of Acceleration
(P = -0.5136 μN)
E = 5 MPa
ν = 0.495
w_A = 0.035 j/m²
4th order linear regression
Energy, femtojoules
300
200
100
0
-100
-200
-300
-1
0
1
2
3
4
Contact Patch Diameter, 2a, microns

(c)

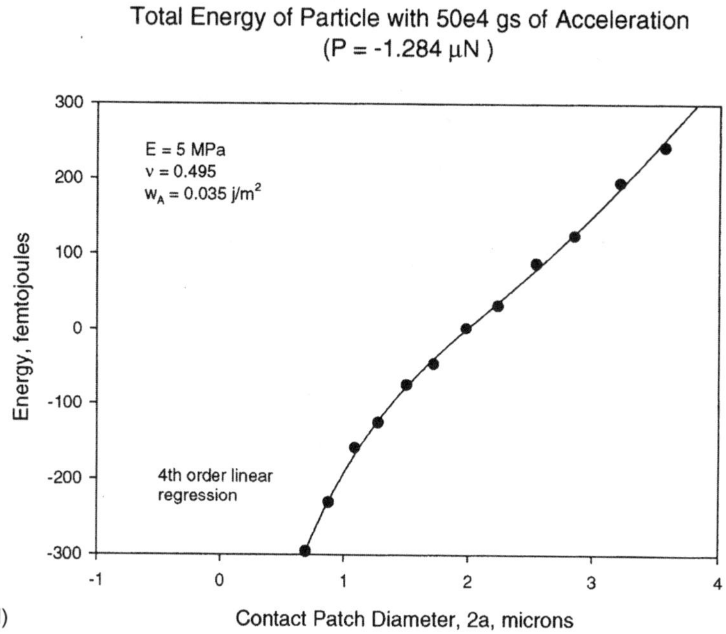
Total Energy of Particle with 50e4 gs of Acceleration
(P = -1.284 μN)
E = 5 MPa
ν = 0.495
w_A = 0.035 j/m²
4th order linear regression
Energy, femtojoules
300
200
100
0
-100
-200
-300
-1
0
1
2
3
4
Contact Patch Diameter, 2a, microns

(d)

FIGURE 6 (Continued).

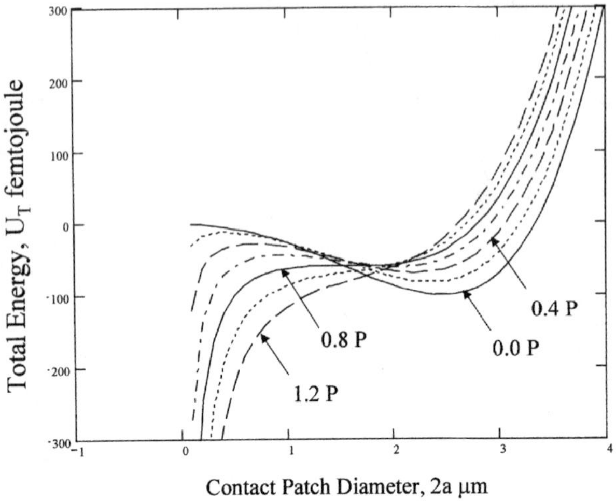

FIGURE 7 Parametric plots of total energy expressions from JKR theory development [11] as a function of the critical load needed for particle removal, P_c.

size, rather than the level behavior anticipated. The level behavior seems to occur at smaller values closer to $0.8P_c$ instead. The present authors do not have an explanation for this observation. Overall, however, the results of the finite element analysis and the results obtained analytically by Johnson, Kendall and Roberts [11] are in reasonable qualitative agreement in light of the differences in loading methods used in the two analyses.

In addition to the contact radii and energies, the stresses and strains were also calculated using finite element modeling. The results are very similar to the analytical results obtained from Maugis and Pollock [19], suggesting that the finite element analysis is capable of solving problems of this type in a reliable fashion. The method may now be applied to irregular geometries, hollow particles, biological cell structures, particles with surface roughness, particles with coatings and property gradients, and particles with distinct surface charges.

CONCLUSIONS

Many features and assumptions in the JKR adhesion formalism can be modeled and tested using finite element methods. Finite element analysis is able to illustrate the physics behind the predictions of the JKR theory, and helps shed light on the implications of the assumptions used in that theory. Further work into topics such as the consequences of the large stresses encountered at the contact zone, as well as the effects of yielding and hyperelasticity, is envisioned. This method also allows for the possibility of examining the consequences of assuming the existence of surface energy in other cases where analytical complexity precludes traditional analysis methods, such as irregular geometries, geometries with gradients in properties, or situations where localized space charges cause electrostatic effects.

References

[1] Hays, D. A. and Wayman, W. H., *J. Imag. Sci.* **33**, 160 (1989).
[2] Hays, D. A., *J. Adhesion* **51**, 41 (1995).
[3] Gady, B., Schleef, D., Reifenberger, R., Rimai, D. and DeMejo, L. P., *Phys. Rev.* **B53**, 8065 (1996).
[4] Gady, B., Schleef, D., Reifenberger, R. and Rimai, D. S., *J. Adhesion* **67**, 291 (1998).
[5] Gady, B., Reifenberger, R. and Rimai, D. S., *J. Appl. Phys.* **84**, 319 (1998).
[6] Gady, B., Reifenberger, R., Rimai, D. S. and DeMejo, L. P., *Langmuir* **13**, 2533 (1997).
[7] Toikka, G., Spinks, G. M. and Brown, H. R., *Proc. 22nd Annual Meeting of the Adhesion Society*, Speth, D. R. Ed. (Adhesion Society, Blacksburg, Virginia, 1999), pp. 14–16.
[8] Lifshitz, E. M., *Sov. Phys. JEPT* **2**, 73 (1956).
[9] Hamaker, H. C., *Physica* **4**, 1058 (1937).
[10] Krupp, H., *Adv. Colloid Interface Sci.* **1**, 111(1967).
[11] Johnson, K. L., Kendall, K. and Roberts, A. D., *Proc. R. Soc. London Ser. A* **324**, 301 (1971).
[12] Zimon, A. D., *Adhesion of Dust and Powder*, 2nd edn., translated from Russian by Johnston, R. K. (Plenum Pub. Corp., New York, 1982).
[13] Quesnel, D. J., Rimai, D. S. and DeMejo, L. P., *J. Adhesion Sci. Technol.* **9**, 1015 (1995).
[14] Burnham, N. A., Colton, R. J. and Pollock, H. M., *Nanotechnology* **4**, 64 (1993).
[15] Pollock, H. M., Burnham, N. A. and Colton, R. J., *J. Adhesion* **51**, 71 (1995).
[16] Quesnel, D. J., Rimai, D. S. and DeMejo, L. P., *J. Adhesion* **51**, 49 (1995).
[17] Quesnel, D. J., Rimai, D. S. and DeMejo, L. P., *J. Adhesion* **67**, 235 (1998).
[18] Rimai, D. S., DeMejo, L. P. and Bowen, R. C., *J. Adhesion* **51**, 139 (1995).
[19] Maugis, D. and Pollock, H. M., *Acta Metall.* **32**, 1323 (1984).
[20] Rimai, D. S. and DeMejo, L. P., *Annu. Rev. Mater. Sci.* **26**, 21 (1996).
[21] Derjaguin, B. V., Muller, V. M. and Toporov, Yu. P., *J. Colloid Interface Sci.* **53**, 314 (1975).
[22] Maugis, D. and Pollock, H. M., *Acta Metall.* **32**, 1323 (1984).
[23] Maugis, D., *Langmuir* **11**, 679 (1995).
[24] Barthel, E., *Proc. 21st Annual Meeting of the Adhesion Society*, Dickie, R. A. Ed. (Adhesion Society, Blacksburg, Virginia, 1998), pp. 255–257.

[25] Johnson, K. L., *Contact Mechanics* (Cambridge University Press, Cambridge, 1987).
[26] Tabor, D., *J. Colloid Interface Sci.* **58**, 2 (1977).
[27] Maugis, D., *J. Colloid Interface Sci.* **150**, 243 (1992).
[28] Broek, D., *Elementary Engineering Fracture Mechanics*, 3rd edn. (Martinus Nijhoff Pub., Boston, MA, 1982).
[29] Lawn, B., *Fracture of Brittle Solids*, 2nd edn. (Cambridge University Press, Cambridge, MA, 1993).
[30] Anderson, T. L., *Fracture Mechanics Fundamentals and Applications* (CRC Press, Boca Raton, FL, 1991).
[31] Griffith, A. A., *Phil. Trans. Ser. A* **221**, 163 (1920).
[32] Inglis, C. E., *Trans. Inst. Naval Architects*, **55**, 219 (1913).

Creep Effects in Nanometer-scale Contacts to Viscoelastic Materials: A Status Report

W. N. UNERTL

Department of Physics and Laboratory for Surface Science and Technology, University of Maine, Orono, ME 04469, USA

Effects of creep on the behavior of nanometer-scale contacts to viscoelastic materials are described from the viewpoint of the contact mechanics theory developed by Ting. The two most important effects are: (1) The time at which maximum contact area and maximum deformation occur can be delayed substantially from the time of maximum applied load. (2) The deformation at separation is related to the loss tangent. These long-range effects due to creep are distinct from the much shorter-range crack tip effects induced by adhesion at the periphery of the contact and associated with the names Barquins and Maugis. Consideration of relevant time scales reveals that creep effects are expected to dominate in SFM-scale contacts for a wide range of compliant viscoelastic materials. Guidelines for selection of optimal experimental parameters for nanometer-scale studies are presented. The need for a comprehensive theory is emphasized.

1. INTRODUCTION AND BACKGROUND

There is increasing need for quantitative measurements of properties of materials with nanometer-scale resolution. Applications include materials characterization [1, 2], microelectromechanical systems and tribology [3 – 7], and biological systems [8]. At present, the scanning force microscope (SFM) [1] is the predominant method for such measurements although other techniques are promising [9]. In SFM, the indenter is a sharp tip that is contacted to the sample and the mechanical response measured. The edge of the contact is frequently pictured as the tip of a crack [10, 11] so that any increase or decrease in contact size is equivalent to the closing or opening of a crack at the contact periphery. The SFM tip shape is usually modeled as a paraboloid of revolution $f(r) = r^2/2R_o$ where $f(r)$ is the height at radius r and R_o is the radius of curvature at the point of contact [12, 13]. Figure 1a compares a parabolic profile with a typical dimension for SFM tips ($R_o = 50$ nm) with spherical profiles of radii 50 nm. Cylindrical coordinates (r, z) are used. Notation used to describe the contacts is shown in Figure 1b for a rigid parabolic probe in contact with an initially-flat, perfectly-elastic substrate under time-dependent

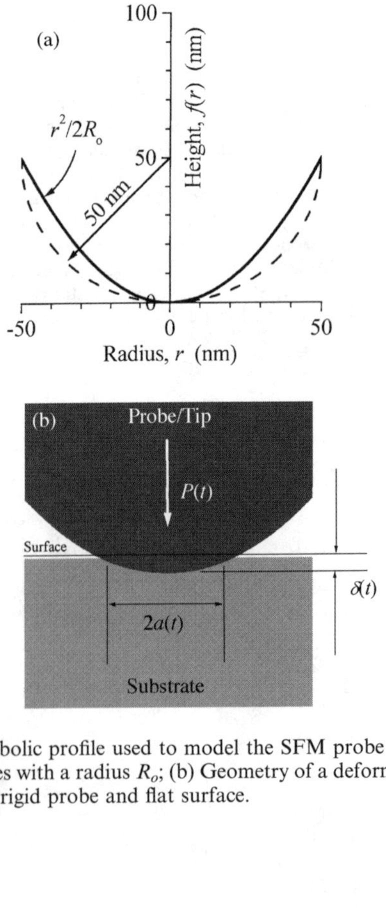

FIGURE 1 (a) Parabolic profile used to model the SFM probe (solid line) compared with a spherical profiles with a radius R_o; (b) Geometry of a deformable contact between an axially-symmetric, rigid probe and flat surface.

load $P(t)$. The radius of the circular contact is $a(t)$ and the rate at which it changes is $V = da/dt$; $\delta(P, t)$ is the deformation along the symmetry axis. Mechanical properties such as modulus, yield strength, and work of adhesion are extracted from the SFM data using continuum mechanics models from the field of contact mechanics [14, 15]. These models include adhesion and are very well developed for contacts between elastic materials [16].

Unfortunately, the appropriate contact mechanics analysis is not yet completed for viscoelastic materials. However, Johnson [17] has recently identified two limiting regimes, each with a characteristic response time. In this chapter, these regimes will be called the *crack regime* and the *creep regime* and their characteristic times the crack relaxation time, τ_{crack}, and the creep relaxation time, τ_{creep}. In the creep regime, bulk deformations due primarily to the Hertz contact pressure are dominant and the effects of adhesion are neglected. Since the contact radius, a, characterizes the range of creep deformations,

$$\tau_{\text{creep}} \approx a/\langle V \rangle \tag{1}$$

where $\langle V \rangle$ is the average value of V. In the late 1960's, Ting [18, 19] solved the creep problem for cases where adhesion can be neglected. More recently, Wahl, Stepnowski and Unertl reported creep effects in SFM studies of 1,2 polybutadiene [20, 21]. The primary purpose of this article is to describe the application of the Ting theory to nanometer-scale contacts.

In the crack regime, viscoelastic response is limited to a narrow zone at the periphery of the contact and long-range deformations are treated elastically; *i.e.*, creep effects are completely ignored. The theoretical description of the crack regime was developed by Barquins and Maugis [22, 23]. They incorporated viscoelastic response into the fracture mechanics approach to contact mechanics. The crack length, l, provides a measure of the range of crack tip effects so that

$$\tau_{\text{crack}} \approx l/\langle V \rangle. \tag{2}$$

Barquins and Maugis validated their model experimentally for macroscopic contacts to polyurethane [10, 22, 23]. Basire and Fretigny [24] used SFM to study the engulfment of the SFM probe into a styrene-butadiene copolymer in the absence of any applied load. This process is driven entirely by adhesion forces [25].

Since $l \leq a$, $\tau_{crack} \leq \tau_{creep}$ [17]. For SFM-scale contacts to compliant materials the difference is expected to be large, so that $\tau_{crack} \ll \tau_{creep}$ [26]. Thus, in SFM, one expects the response of the contact to be dominated by the regime whose relaxation time is closest to the relaxation time of the viscoelastic material under study. Furthermore, both τ_{crack} and τ_{creep} can be varied over wide ranges by increasing or decreasing the total contact time, $t_{contact}$ and, thus, changing $\langle V \rangle$.

The experimental range of $t_{contact}$ accessible in quantitative SFM experiments is fairly large, typically from a fraction of a 1 ms to about 1000 s. The longest contact times are limited by thermal drift and creep of the piezoelectric elements. The shortest times are limited by cantilever response times and roll-off frequencies of electronic filters. Measurements can also be affected by mechanical resonances and frequency-dependent phase shifts of the instrument. These instrumental limitations, and others [8, 12], must be quantitatively understood before reliable measurements of mechanical properties can be made. Contact radii are typically in the range $1\,\text{nm} \lesssim a \lesssim 100\,\text{nm}$ for the probe tips and loads normally used in SFM experiments. Thus, the average rate of change of the contact radius $\langle V \rangle \equiv (da/dt)$ lies in the range $0.001\,\text{nm/s} \lesssim \langle V \rangle \lesssim 100\,\mu\text{m/s}$. This overlaps the range of some macroscopic contact experiments, which typically have $\langle V \rangle \gtrsim 1\,\mu\text{m/s}$ [23].

The major goal of this paper is to provide an overview of current understanding of the role that creep plays in the formation of contacts with nanometer-scale dimensions. The primary theoretical formulation of viscoelastic contact mechanics, due to Ting, is reviewed.

Quantitative examples based on simple mechanical models are used to illustrate the creep effects expected in dynamic SFM loading experiments. Some of these results have been presented previously [20, 21]. The major limitation of the Ting theory is its neglect of adhesion. Approaches to include adhesion, at least approximately, are discussed. Considerations for design of SFM experiments to emphasize either the creep or crack regimes are presented. Much of the material also applies to ultra-low-load indentation experiments like those of Asif, Wahl and Colton [9].

2. CONTACT MECHANICS FOR VISCOELASTIC MATERIALS IN THE ABSENCE OF ADHESION

Viscoelastic effects become important whenever contact dimensions change in a time interval that is comparable with a characteristic relaxation time, τ, of the viscoelastic material. Ting [18], following earlier work by Lee and Radok [27] and Graham [28], obtained a general solution to the equilibrium Hertz contact problem for the case of a *rigid axisymmetric* probe with shape $f(r)$ and an *isotropic* substrate with *linear viscoelastic* response. This response will be referred to as creep. He assumed a linear viscoelastic stress–strain ($\sigma_{ij} - \varepsilon_{ij}$) relation of the form

$$\sigma_{ij}(t) = \int_{0^-}^{t} [2G(t-\tau)\partial\varepsilon_{ij}(\tau)/\partial\tau + \delta_{ij}\lambda(t-\tau)\partial\varepsilon_{kk}(\tau)/\partial\tau]d\tau \qquad (3)$$

where $G(t)$ and $\lambda(t)$ are time-dependent relaxation moduli. $G(t)$ is the relaxation modulus in shear. The bulk modulus $K(t)$, the Poisson's ratio, $\nu(t)$, and the Young's modulus, $E(t)$, are all time dependent and related to $G(t)$ and $\lambda(t)$ by $K(t) \equiv \lambda + (2/3)G$, $\nu(t) \equiv (1/2)(3K - 2G)/(3K + G)$ and $E(t) \equiv 9KG/(3K + G)$ [29]. The lower limit 0^- allows for discontinuous jumps in stresses and strains at $t = 0$. Ting obtained explicit expressions for the contact radius, $a(t)$, and the penetration, $\delta(t)$, of the tip of the indenter assuming the following boundary conditions at $z = 0$:

$$\left.\begin{array}{ll} \sigma_{zz} = \sigma_{rz} = 0 & r > a(t) \\ u_z(r,t) = \delta(t) - f(r)H(t) \quad \text{and} \quad \sigma_{rz} = 0 & r \leq a(t) \end{array}\right\} \qquad (4)$$

where $u_z(r, t)$ are the vertical displacements under the indenter in the z-direction and $H(t)$ is the Heavyside step function. These boundary conditions, Eq. (4), assume the stress component in the interface to be zero. This condition can be met either if the contact is frictionless or if the probe and substates have identical mechanical properties, requirements that are probably seldom met in real contacts. However, at least in the case of elastic contacts, failure to satisfy this condition leads to errors of only a few percent in contact radius and pull-off force [14, 30]. The z-axis coincides with the symmetry axis of the indenter, is directed into the sample, and has its origin at the undeformed surface of the sample. These boundary conditions assume no adhesion and no friction at the interface between the indenter and sample. The solutions can be expressed in terms of two functions, $\phi(t)$ and $\psi(t)$, whose Laplace transforms are related by $s\hat{\psi}(s) = 1/s\hat{\phi}(s) = s\hat{G}(s)/[1 - s\hat{\nu}(s)]$. If the Poisson's ratio, ν, is assumed constant, $\psi(t)$ has the same form as $G(t)$, the relaxation modulus in shear, and $\phi(t)$ has the same form as the creep compliance in shear [18, 29].

The specific form of the solutions depends on whether the contact radius, $a(t)$, increasing or decreasing and whether it is larger or smaller than its value at $t = 0$. Figure 2 shows the three cases of interest here. The initial contact radius is a_o. Non-zero a_o can occur two ways. (1) The contact can reach equilibrium following application of a load in the distant past. (2) The viscoelastic response has a perfectly elastic component as is the case for the Maxwell and standard solid models. In case I (Fig. 2), $a(t)$ is increasing ($t \leq t_{\max}$) and the solution confirms the earlier result of Lee and Radok [27] which was based on the Boltzmann principle of correspondence [29]. Specifically, for

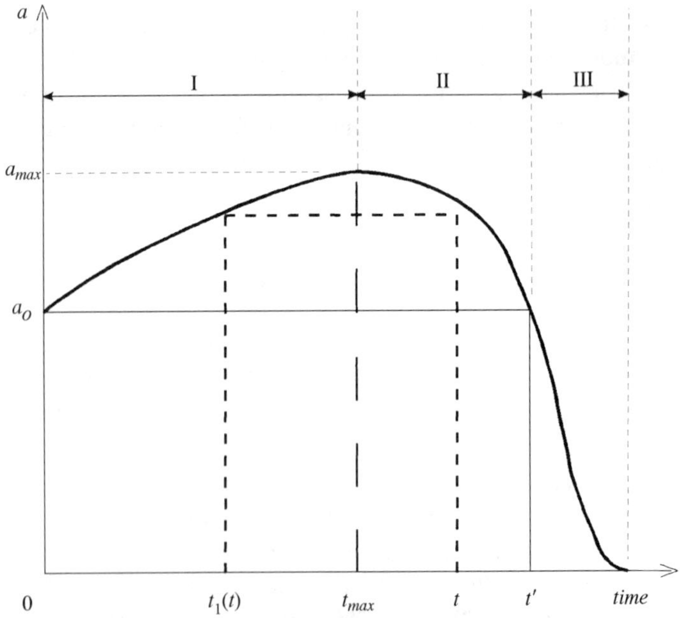

FIGURE 2 Variation of contact radius with contact time.

a parabolic indenter with $f(r) = r^2/2R_o$:

$$\left.\begin{aligned} P(t) &= \frac{8}{3R_o}\int_{0^-}^{t}\psi(t-\tau)\frac{\partial a^3(\tau)}{\partial\tau}d\tau \\ \delta_{\mathrm{I}}(t) &= a^2(t)/R_o \end{aligned}\right\} \quad \text{if } da(t)/dt \geq 0. \tag{5}$$

If the applied load is known, $a(t)$ is determined from Eq. (5) using Laplace transforms [18].

Case II is the interval between $t_{\max}$ and t', where a begins to decrease but is still larger than its initial value, a_o.

$$\left.\begin{aligned} P_{\mathrm{II}}(t) &= \frac{8}{3R_o}\int_{0^-}^{t_1(t)}\psi(t-\tau)\frac{\partial a_{\mathrm{I}}^3(\tau)}{\partial\tau}d\tau \\ \delta_{\mathrm{II}}(t) &= \delta_{\mathrm{I}}(t) - \int_{t_{\mathrm{peak}}}^{t}\phi(t-\tau)\frac{\partial}{\partial\tau}\int_{t_1(\tau)}^{\tau}\psi(\tau-\eta)\frac{\partial\delta_{\mathrm{I}}(\eta)}{\partial\eta}d\eta d\tau \end{aligned}\right\} \quad \text{if } a(t_{\mathrm{peak}}) \geq a(t) \geq a_o \tag{6}$$

where $\partial\delta_{\mathrm{I}}/\partial t$ is evaluated assuming Eq. (5) is valid beyond $t_{\max}$. These equations have two surprising features. First, the value of $a(t)$ at any t in the interval $t_{\max} < t < t'$ is determined entirely by the history of the contact *prior* to $t_1(t)$, the time at which a initially reached the same radius as at t. Second, the value of $\delta_{\mathrm{II}}(t)$ depends only on the history between $t_1(t)$ and t and is independent on the behavior *prior* to $t_1(t)$.

Finally, case III describes the behavior after a has decreased below a_o; *i.e.*, for $t \geq t'$,

$$\left.\begin{aligned} P_{\mathrm{III}}(t) &= \frac{8}{3R_o}\phi(t)a_{\mathrm{I}}^3(t) \\ \delta_{\mathrm{III}}(t) &= \delta_{\mathrm{II}}(t') - \int_{t'^+}^{t}\phi(t-\tau)\frac{\partial}{\partial\tau}\int_{0^+}^{\tau}\psi(\tau-\eta)\frac{\partial\delta_{\mathrm{I}}(\eta)}{\partial\eta}d\eta d\tau \end{aligned}\right\} \quad \text{if } a(t) \leq a_o \tag{7}$$

where a_{I}, δ_{I} and δ_{II} are evaluated assuming Eqs. (5) and (6) are valid beyond $t_{\max}$. Equations (5)–(7) scale with R_o in the same way that the elastic Hertz results do; *i.e.*, $a \propto R_o^{1/3}$ and $\delta \propto R_o^{-1/3}$.

3. EXAMPLES

We now use the three simple mechanical models shown in Figure 3 to illustrate the effects of creep on the formation and rupture of

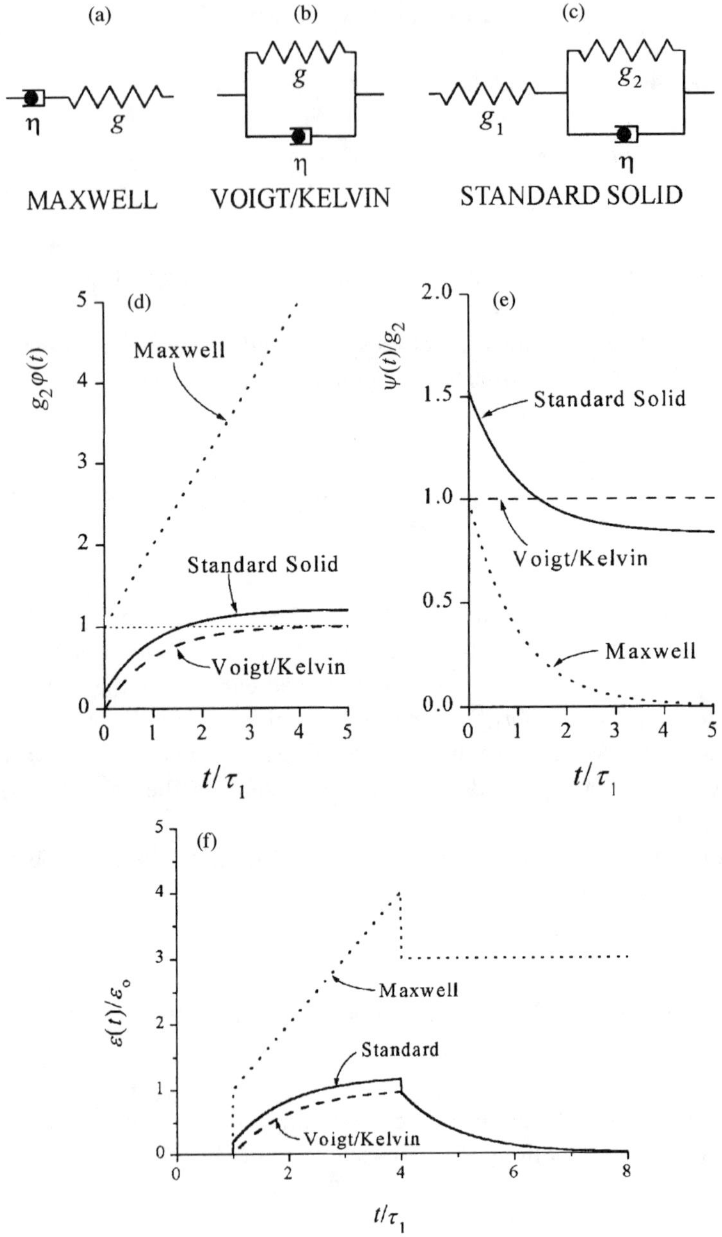

FIGURE 3 Simple mechanical models of linear viscoelastic response. (a) Maxwell model; (b) Voigt/Kelvin model; (c) Standard solid model; (d) Creep compliance functions; (e) Stress relaxation functions. (f) Strain following a step increase of stress at t/τ_1 and a step decrease at 4 t/τ_1.

contacts with nanometer-scale dimensions [26, 31]. The elastic elements are Hookeian springs each with shear modulus g. Each viscous element is a Newtonian dashpot with viscosity η. These mechanical models do not quantitatively describe actual polymers. However, they do capture the major features of the polymer response and, therefore, provide simple models that are useful for semi-quantitative analysis of data and for the design and interpretation of experiments. It is in this spirit that they are used here. Unless stated otherwise, specific examples shown below use the parameter values $g = g_2 = 2\,\text{MPa}$, $g_1 = 10\,\text{MPa}$, $\eta = 10\,\text{MPa}\cdot\text{s}$, with corresponding relaxation times (see below) of $\tau_M = \tau_V = \tau_1 = 50\,\text{s}$ and $\tau_2 = 8.33\,\text{s}$.

The Maxwell model (Fig. 3a) consists of a spring and dashpot in series. Figures 1d and 3e show the creep compliance and stress relaxation functions

$$\left.\begin{aligned} \phi_M(t) &= (1 + t/\tau_M)/g \\ \psi_M(t) &= g\exp(-t/\tau_M) \end{aligned}\right\} \tag{8}$$

where $\tau_M \equiv \eta/g$ is the characteristic relaxation time following a step change in strain. The Maxwell model is fluid-like if the viscosity is small and solid-like if it is large. Figure 3f shows the strain response of the Maxwell model subjected to a sudden step increase in stress by σ_o at $t/\tau_1 = 1$ followed by a sudden decrease by $-\sigma_o$ at $t/\tau_1 = 4$. The instantaneous elastic response of the spring is followed by a linear response of the dashpot. When the stress is removed, the spring instantly relaxes but the dashpot remains extended; the Maxwell model can exhibit permanent deformation.

The Voigt/Kelvin model (Fig. 3b) has a spring and dashpot connected in parallel. The creep compliance and stress relaxation functions are

$$\left.\begin{aligned} \phi_V(t) &= (1 - \mathrm{e}^{-t/\tau_V})/g \\ \psi_V(t) &= g(1 + \tau_V\delta(t)) \end{aligned}\right\} \tag{9}$$

where $\tau_V \equiv \eta/g$ is the characteristic relaxation time following a step change in stress. Equations (9) are plotted in Figures 3d and 3e. The dashpot prevents any instantaneous elastic response. Unlike the Maxwell model, there is no permanent deformation (Fig. 3f).

The Maxwell model is a better approximation to the stress relaxation of polymers while the Voigt/Kelvin models better approximates creep response [31].

The standard solid model (Fig. 3c) is a three-parameter model that gives a better overall approximation to the response of polymers to changes in stress and strain. It consists of a Voigt/Kelvin model with parameters g_2 and η in series with an elastic spring g_1. The creep compliance and stress relaxation functions are

$$\left.\begin{aligned}\phi(t) &= \frac{1}{g_2}\left[\left(\frac{\tau_1}{\tau_1-\tau_2}\right) - \exp(-t/\tau_2)\right]\\ \psi(t) &= g_2\frac{(\tau_1-\tau_2)}{\tau_1}\left[1+\left(\tfrac{\tau_1-\tau_2}{\tau_1}\right)\exp(-t/\tau_1)\right]\end{aligned}\right\} \quad (10)$$

where $\tau_1 \equiv \eta/g_2$ is the relaxation time following a step change in stress and $\tau_2 \equiv \eta/(g_1+g_2)$ is the relaxation time following a step change in strain; τ_1 is always greater than τ_2. Equations (10) are plotted in Figures 3d and 3e. There is an instantaneous elastic response due to g_1.

If the loading varies cyclically at frequency ω as in dynamical mechanical testing, the response is characterized by a complex modulus and phase lag. For example, the mechanical response to a strain and stress

$$\begin{aligned}\varepsilon(t) &= \varepsilon_o \exp(i\omega t)\\ \sigma(t) &= \sigma_o \exp i(\omega t+\delta)\end{aligned} \quad (11)$$

with amplitudes ε_o and σ_o is given by

$$G(\omega) = (\sigma_o/\varepsilon_o)\exp i\delta = G_1(\omega) + iG_2(\omega) \quad (12)$$

and

$$\tan\delta(\omega) = G_2(\omega)/G_1(\omega) \quad (13)$$

where G_1 is the storage modulus and G_2 is the loss modulus. $G_1(\omega)$ and $G_2(\omega)$ are related to $G(t)$ by

$$G_1(\omega) = \omega\int_0^\infty G(t)\sin\omega t\,dt \quad \text{and} \quad G_2(\omega) = \omega\int_0^\infty G(t)\cos\omega t\,dt. \quad (14)$$

Time scales of experimental creep data are converted into frequency scales using the relationship $\omega = 1/t$ [29].

The fraction of the maximum energy stored that is lost per cycle is $\Delta W/W = 2\pi \sin\delta$ which has its maximum value at ω_o, the frequency at which $\tan\delta$ is maximum. Figure 4 illustrates the typical frequency dependence of $G_1(\omega)$, $G_2(\omega)$ and $\tan\delta$ for the standard solid model with $g_1 = 1$ GPa, $g_2 = 1$ MPa and $\eta = 100$ MPa · s. For these parameters, $\tau_1 = 100$ s, $\tau_2 = 0.1$ s and $\omega_o = 0.316\,\mathrm{s}^{-1}$. This model has properties similar to a styrene-butadiene random co-polymer with glass transition temperature near room temperature [24, 32]. G_1 is smallest at low ω where the polymer has rubbery response and largest at high ω where the polymer has glassy response. At high frequency, the dashpot cannot respond and the system responds like an elastic spring with modulus $G_1(\omega \to \infty) \to g_1$. At low frequency, the dashpot can fully relax and the system again responds elastically but with modulus $G_1(\omega \to 0) \to g_1 g_2/(g_1 + g_2) \approx g_2$ if $g_2 \ll g_1$ as in the example. $\tan\delta$ peaks at $\omega_o = 1/\sqrt{\tau_1 \tau_2}$ where G_1 is beginning to increase rapidly.

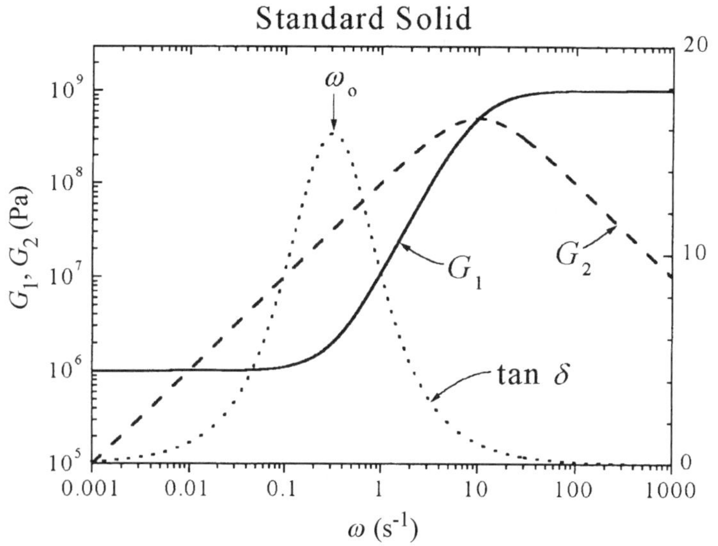

FIGURE 4 Frequency dependence of the storage modulus, G_1, loss modulus, G_2, and $\tan\delta$ for the standard solid model with $g_1 = 1$ GPa, $g_2 = 1$ MPa and $\eta = 100$ MPa · s ($\tau_1 = 100$ s, $\tau_2 = 0.1$ s and $\omega_o = 0.316\,\mathrm{s}^{-1}$).

Real polymers may have multiple peaks in tanδ. These peaks are frequently associated with excitation of particular molecular motions of the polymer, the glass transition temperature, or the melting temperature.

For the standard model, Figure 5 shows the interrelationships between the relaxation times, τ_1 and τ_2, the frequency, ω_o, and tanδ_{max}. The solid lines show the dependence of the maximum value of tanδ on τ_1 and τ_2 for tanδ_{max}=0.001, 0.01, 0.1, 1, 10 and 100; *i.e.*, from nearly elastic to very dissipative materials. The diagonal dashed lines show the values of ω_o. The ranges of τ_1, τ_2 and ω_o were selected to span that generally accessible to SFM instruments. When reading the plot, it is important to keep in mind that $\tau_1 > \tau_2$, always. Highly

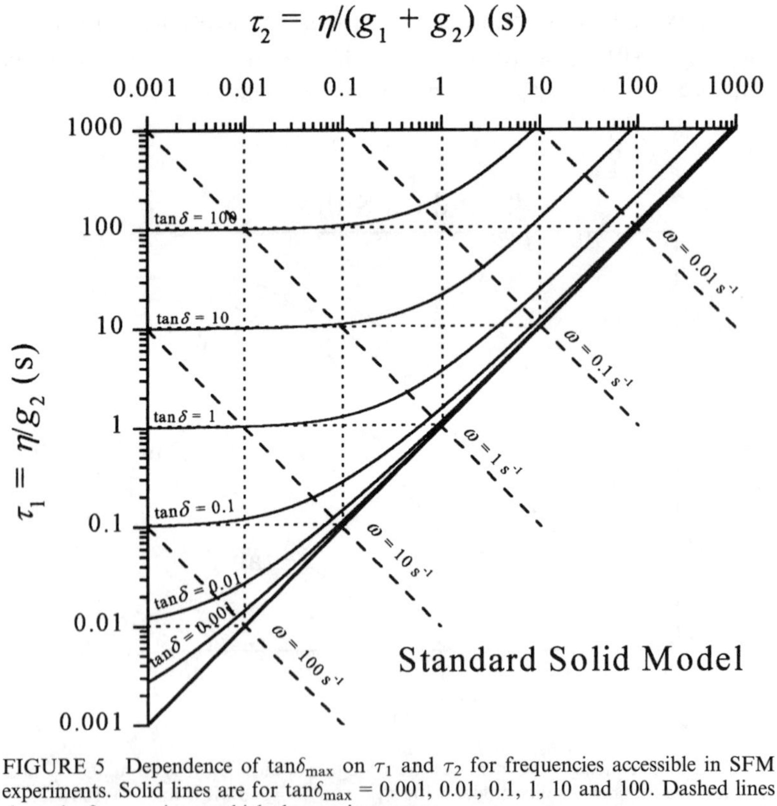

FIGURE 5 Dependence of tanδ_{max} on τ_1 and τ_2 for frequencies accessible in SFM experiments. Solid lines are for tanδ_{max} = 0.001, 0.01, 0.1, 1, 10 and 100. Dashed lines show the frequencies at which the maxima occur.

dissipative materials occur toward the upper left corner and more elastic materials toward the lower right diagonal edge. This type of plot is useful in selecting parameters to match the standard model to a particular set of data. For example, if τ_1 and ω_o can be estimated, then τ_1 and $\tan\delta$ can be determined directly from Figure 5. In the case of 1,2-polybutadiene [20], shear modulation measurements suggest $\tau_1 \approx 50\,\text{s}$ and $\tan\delta \approx 1$ so that $\tau_2 \approx 20\,\text{s}$ and $\omega_o \approx 0.03\,\text{s}^{-1}$.

In the following subsections, these mechanical models are used to demonstrate the ways creep can affect nanometer-scale contacts. The response is calculated using a time-dependent applied load (Fig. 6a) selected to be similar to the load variation in a typical SFM force-distance measurement. The total time of contact between the probe and substrate is t_{contact} and the time at which maximum load, P_{max}, occurs is yt_{contact} where $0 < y < 1$. Typically in SFM, $y \approx 0.2 - 0.3$. The unloading time is $t_{\text{unload}} = (1 - y)t_{\text{contact}}$. The slopes of the loading and unloading curves can be varied independently which is not the usual case in SFM experiments. The probe shape is assumed to be a paraboloid of revolution as in Figure 1 with $R_o = 50\,\text{nm}$. For this probe shape and linear loading, the contact radius and deformation

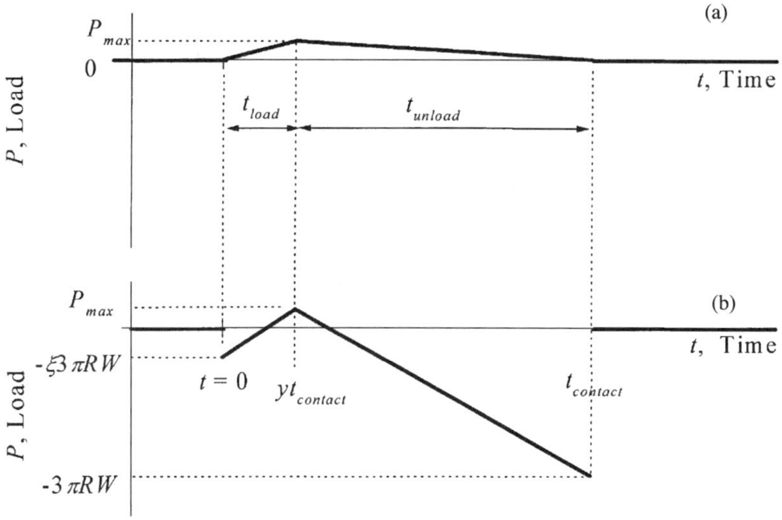

FIGURE 6 (a) Time-dependent linear loading ramp; (b) Time-dependent loading ramp with jump-to-contact and pull-off behavior.

scale as $a(t) \propto P_{\max}^{1/3} R_o^{1/3}$ and $\delta(t) \propto P_{\max}^{2/3} R_o^{-1/3}$. Thus, all the specific results presented below for $P_{\max} = 2\,\text{nN}$ and $R_o = 50\,\text{nm}$ are easily extended to any other combination of $P_{\max}$ and R_o.

The choice of probe shape bears additional comment. A parabolic probe is frequently assumed for model calculations primarily for computational convenience. It is also the one choice made in all the commonly-quoted results from contact mechanics (Hertz, JKR, DMT, Barquins-Maugis, *etc.*) and, therefore, facilitates comparisons between viscoelastic and elastic materials. Additionally, a paraboloid is a good approximation to a sphere as long as the contact radius is much smaller than the sphere's radius ($a \ll R_o$). As an example, Figure 1a compares a parabola and sphere both with radius R_o = 50 nm. However, as soon as a reaches $R_o/10$ or $R_o/20$, the paraboloid ceases to be a good approximation to a sphere. This condition is usually exceeded in SFM-scale experiments on compliant materials. Additionally, real SFM probes are generally neither parabolic nor spherical, except within about the first nanometer from the end [13]. Thus, for very compliant materials, which have $\delta \gg R_o$ even in the absence of an applied load [24, 33], the shape of the probe far from the end must be included if quantitative comparisons between experiment and theory are to be made. A conical probe shape may be a better choice [24, 33]. Incorrect choice of probe shape can have serious consequences. For example, in Hertzian contacts δ scales as $P^{2/3}$ for a parabolic probe, but as $P^{1/2}$ for a conical probe.

3.1. Maxwell and Voigt/Kelvin Models

We now use the Maxwell and Voigt/Kelvin models to illustrate some of the important characteristics of creep. These models are simple enough that the Ting equations [Eqs. (5)–(7)] can be solved analytically. In nearly all other cases, analytical solutions are not possible. Specific calculations below use $g = 2\,\text{MPa}$, $\eta = 100\,\text{MPa} \cdot \text{s}$ and $\tau_M = \tau_V = 50\,\text{s}$. These values are similar to those measured experimentally for 1,2-polybutadiene freshly cast from toluene solution [20, 21]. These models have behavior that brackets that expected for real materials.

Figure 7 shows the major features of the Ting solution for the Voigt/Kelvin model (Fig. 7a) and the Maxwell model (Fig. 7b). The contact radius, a, is plotted as a function of t/t_{contact} for various

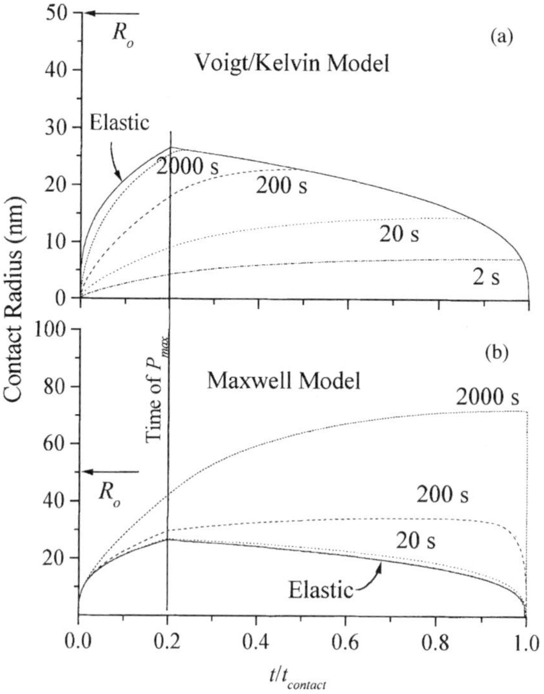

FIGURE 7 Solutions to the Ting model for contact of a rigid parabolic probe to a viscoelastic substrate described by (a) the Voigt/Kelvin model and (b) the Maxwell model.

t_{contact}. Each curve is labeled by its value of t_{contact}. The maximum load occurs at $y=0.2$. Also shown for reference is the response of an elastic material with the same g (solid line). The most striking feature of both models shown in Figure 7 is that the time of maximum contact radius, $t_{\max}$, does *not* coincide with the time of maximum load ($t=yt_{\text{contact}}$) as it does for elastic materials. Instead, maximum radius always occurs *after* the maximum load. This delay occurs because the substrate continues to respond by creep even during the unloading cycle.

Consider first the Voigt/Kelvin model (Fig. 7a). For finite contact times, the contact radius increases more slowly than in the elastic limit and its maximum value is always smaller. As t_{contact} goes to values much shorter than τ, the response of the dashpot increasingly lags the applied load because $\tan\delta=\tau_V\omega_o \sim \tau_V/t_{\text{contact}}$. The contact becomes increasingly rigid, the maximum value attained by a becomes smaller and smaller, and t_{max} shifts closer and closer to the end of the contact (*e.g.*, $t_{max} \rightarrow t_{contact}$) as shown Figure 8. Specifically,

$$t_{max} = -\tau_V \ln \left[\frac{y}{y - 1 + \exp(y t_{contact}/\tau_V)} \right], \tag{15}$$

i.e., t_{max} is a function of only (τ_V, y) and is independent of both P_o and R_o. The ordinate in Figure 8 is the fractional position of the peak in the unloading part of the cycle $(t_{max} - y t_{contact})$ normalized to the total unloading time $(t_{contact} - y t_{contact})$; it varies between zero and unity. For the Voigt/Kelvin model, y has little influence on the relative time in the unloading cycle at which maximum radius is reached. The total possible range of $0 \leq y \leq 1$ is shaded in dark gray in the figure. As $t_{contact}$ becomes much longer than τ_V, the dashpot is more responsive, and t_{max} decreases toward the time of maximum load; *i.e.*, the contact becomes more and more elastic. The most rapid variation of t_{max} occurs for $t_{contact}$ in the range $0.05\tau_V \lesssim t_{contact} \lesssim 5\tau_V$.

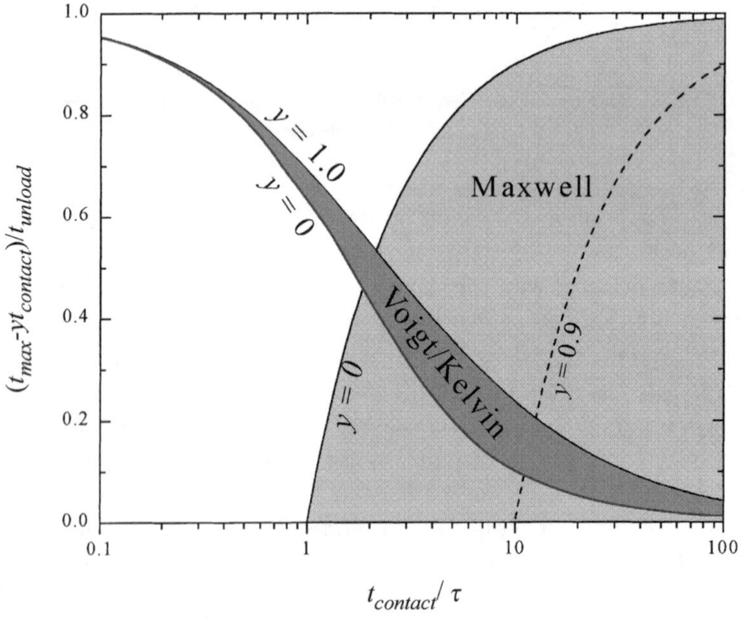

FIGURE 8 Relative time of the maximum contact radius in the unloading cycle as a function of $t_{contact}/\tau$ for the Maxwell and Voigt/Kelvin models. Adhesion is neglected.

In contrast to the Voigt/Kelvin model, the contact radius for the Maxwell model (Fig. 7b) is always larger than the elastic limit. In fact, $a \gtrsim R_o$ once $t_{\text{contact}} > \tau_M$ and this makes the more fluid-like Maxwell model more sensitive to the assumptions about the probe shape for a given (g, η). If $t_{\text{contact}} \leq \tau_M$, a has no delayed maximum. But as $t_{\text{contact}} \gg \tau_M$, the dashpot has more and more time to respond, the maximum reached by a becomes larger and larger, and t_{peak} moves closer to t_{contact} according to

$$\frac{t_{\max}}{t_{\text{contact}}} = 1 - \frac{\tau_M}{t_{\text{contact}}}; \tag{16}$$

i.e., the relative position of $t_{\max}$ is independent of y and depends only on τ_M. The relative location of $a_{\max}$ in the unloading cycle is compared with that of the Voigt /Kelvin model in Figure 8. The area shaded light gray indicates the full range of y values.

Experimentally it is desirable to have the maximum contact area occur in the middle of the unloading cycle since this region is isolated from the times of maximum loading and rupture where the maximum might be more difficult to distinguish. Examination of Figure 8 shows that this is generally possible if the maximum load occurs as soon as possible ($y \approx 0$) and if $t_{\text{contact}} \approx \tau$.

3.2. Standard Solid Model

The standard solid model is more representative of contacts to polymeric solids and exhibits behavior intermediate between the Maxwell and Voigt/Kelvin models. Johnson used the standard model to illustrate the time and load dependence of a from the viewpoint of the Ting model [17]. Typical response behavior is shown in Figure 9. The radius is confined between two limiting cases. For $t_{\text{contact}} \ll \tau_1$, the response approaches that of an elastic Hertzian contact of stiffness, g_1. For $t_{\text{contact}} \gg \tau_1$, it approaches that of an elastic Hertzian contact of stiffness $g_1 g_2/(g_1 + g_2)$. At intermediate t_{contact}, the behavior varies continuously between these limits. The contact radius, a, has the same delayed maximum behavior found for the simpler models. The relative time in the unloading cycle at which the maximum contact radius occurs is shown in

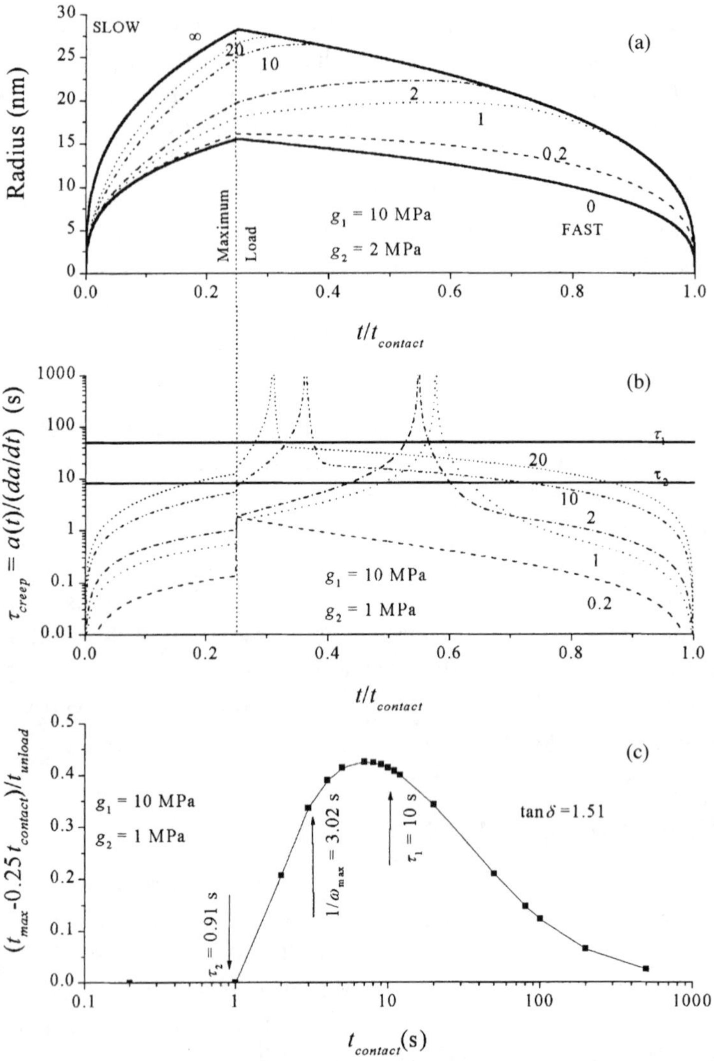

FIGURE 9 (a) Variation of the contact radius, a, for the standard solid model for $t_{\text{contact}}/\tau_1 = 0$, 0.2, 1.0, 2.0, 10, 20, ∞. The loading cycle is as in Figure 1a with $y = 0.25$ and $\eta = 10\,\text{MPa}\cdot\text{s}$; (b) $a(t)/(da/dt)$ for the same parameters as in (a); (c) Time in unloading cycle at which maximum contact radius occurs as a function of t_{contact}.

Figure 9c. The largest $t_{\max}$ occurs for $t_{\text{contact}} \approx \tau_1$. This time is close to $1/\omega_o$, the time at which energy dissipation has its maximum value.

Unfortunately, SFM measurements cannot directly determine the contact radius. Therefore, it is more important to examine the effects of creep on the deformation, δ, because this parameter can, in principle at least, be determined by SFM. Figure 10a shows P *vs.* δ obtained

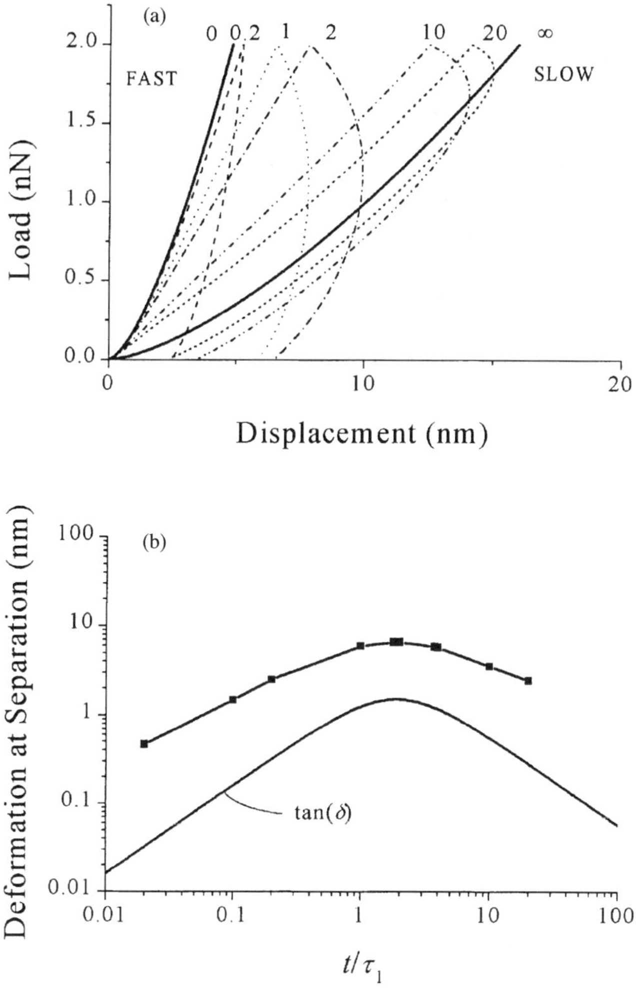

FIGURE 10 (a) Variation of the deformation, δ, with load for the standard solid model for various contact times, t_{contact}. Parameters are the same as in Figure 9a; (b) Variation of the deformation at separation with t_{contact} for the parameters in Figure 9a. Also plotted is tanδ where the frequency-to-time conversion has been made using $t=\pi/\omega$.

for the same parameters as in Figure 9a. The solid lines indicate the limiting elastic cases, which are reversible. The loading portion of the deformation curves shift continuously between these limits as $t_{contact}$ increases. As is the case for a, δ continues to increase after the load starts to decrease, *i.e.*, the probe tip continues to penetrate deeper into the substrate even though the load is decreasing. The maximum penetration is reached slightly after maximum a is reached. Separation of the probe and substrate at the end of loading (*i.e.*, $t = t_{contact}$ and $a = 0$) occurs well before δ can return to zero. The deformation at separation is plotted in Figure 10b and is largest when $t_{max} \approx 2\tau_1$. Also plotted in the figure is the loss tangent where the frequency has been converted to time assuming $t = \pi/\omega$ rather than the standard $t = 1/\omega$. Clearly, the maximum deformation at separation is correlated with the maximum energy dissipation and its shape, as a function of $t_{contact}$, is very similar to that of $\tan\delta$. The magnitude of the maximum deformation at separation is large, nearly 10 nm in this case. This large effect should be easily observable with SFM or other ultra low-load indention instruments.

3.3. Implications for Experiments on Viscoelastic Materials

These simple models provide a phenomenological framework for design and interpretation of experiments involving nanometer-scale contacts to viscoelastic materials. For example, the following general observations can be made:

1. The maxima attained by the contact radius, a, and deformation, δ, always increase as the contact time $t_{contact}$ increases.
2. The ranges on both a and can δ are bounded unless the polymer behaves more like a Maxwell fluid. The upper bound is set by the elastic response determined by the limiting value of the modulus at low frequencies. The lower bound is set by the elastic response as determined by the high frequency limit of the modulus.
3. Both a and δ attain their maximum values during the unloading for a large range of contact times. The maximum delay occurs when the contact time is approximately equal to the inverse of ω_o, the frequency at which $\tan\delta$ is maximum.

4. CREEP IN VISCOELASTIC CONTACTS WITH ADHESION

The discussion of creep given above completely ignores adhesion. Unfortunately, the contact mechanics problem for a viscoelastic material that includes both adhesion and creep has not yet been solved. Certainly, adhesion will increase the contact area, just as it does for elastic materials.

The boundary conditions on $z=0$, as given by Eq. (4), must be modified to include contributions to the radial pressure distribution, $p(r)$, from adhesion forces [16]. One possibility is to use

$$\sigma_{zz} = -p(r) = -\frac{\partial w(D)}{\partial D(r)} \quad \text{and} \quad \sigma_{rz} = 0. \tag{17}$$

everywhere at the interface, where $w(D)$ is the free energy of interaction between planar surfaces separated by D and might be approximated by Lennard-Jones or Dugdale potentials. This is the approach used by Hughes and White [34] and Barthel [16] to obtain a general solution to the equivalent elastic problem. Equation (17) is the Derjaguin approximation and is satisfied for SFM-scale contacts [12]. The adhesion energy, W, is related to w by [16]

$$W = -\int_0^{-\infty} \sigma(D)dD = \int_0^{-\infty} \frac{\partial w(D)}{\partial D} dD. \tag{18}$$

An alternative is to maintain the mixed boundary conditions of Eq. (4), but with $p(r)$ specified only for $r > a$:

$$\sigma_{zz} = 0 \rightarrow \sigma_{zz} = -p(r) \quad \text{for } r > a(t). \tag{19}$$

This approach has the advantage that only the attractive portion of w must be known. On the other hand, it assumes the spacing between the probe and substrate is constant everywhere inside the contact, usually estimated to be an interatomic spacing.

In the absence of an adequate theory, Tirrell and co-workers [35] have suggested an *ad hoc* approach to incorporate the effects of adhesion into the Ting model. In analogy with the Johnson, Kendall and Roberts (JKR) [36] theory of adhesive elastic contacts, they simply replace the load in Eq. (5) with an effective load based on the JKR result, *i.e.*,

$$P(t) \rightarrow P_{\mathrm{eff}}(t) \equiv P(t) + 3\pi W R_o + \sqrt{6\pi W R_o P(t) + (3\pi W R_o)^2}. \quad (20)$$

This substitution is expected to be valid only as long as the contact radius is increasing, since Eq. (5) applies only in this case. Equation (20) was found to give a good description of the increase in area of contacts between 0.7 – 1.2 mm diameter spheres of diblock copolymers of poly(ethylene)-poly(ethylene-propylene) [35]. In particular, the values of W extracted with this analysis were in excellent agreement with values determined by contact angle measurements [37].

Unertl extended this approach to include cases for which a is decreasing [26]. He used the Derjaguin-Muller-Toporov (DMT) limit $P_{\mathrm{eff}}(t) = P(t) + 2\pi R_o W$ (see Fig. 6b), rather than the JKR limit, for analytical simplicity. In the case of the Maxwell model, the only effect is the expected increase in the contact radius. More importantly for the present discussion, there was no effect on the time lag at which the contact area reaches it maximum value. In the case of the Voigt/Kelvin model, the maximum value of the contact radius also increases. However, unlike the Maxwell model, t_{max} shifts toward shorter delay times, but by no more than ten percent. These results suggest that the major conclusions of the preceding section, particularly those shown in Figures 8 and 9c, will not be significantly altered by inclusion of adhesion. Specifically, inclusion of DMT adhesion appears to make only minor modifications to the Ting analysis for SFM-scale experiments.

A model proposed recently by Hui, Baney and Kramer incorporates adhesion and viscoelasticity [38]. The key feature of this model is the explicit realization that the cohesive zone at the crack tip must be finite in extent. This means that stresses are finite everywhere and, consequently, the rate of energy flow into the cohesive zone is dependent on the crack speed, da/dt. That is, unlike the case of elastic materials, the stress intensity factor $K(t)$ must be time dependent. Viscoelastic effects are described using the approach of Yang [39], which restricts the results to monotonically increasing contact area.

The contact area and deformation are expressed as

$$a^3(t) = \frac{3R}{4}\left\{\frac{3R\pi\phi_o^2K_I^2(t)}{2} + \phi(t)*P(t) + \phi_oK_I(t)\sqrt{\left[\frac{3\pi R\phi_oK_I(t)}{2}\right]^2 + 3\pi R[\phi(t)*P(t)]}\right\} \tag{21}$$

and

$$\delta(t) = \frac{a^2(t)}{R} - \frac{K_I(t)\sqrt{\pi a(t)}}{\psi_o} \tag{22}$$

where ϕ_o is the short time compliance, ψ_o is the short time relaxation, $K_I(t)$ is the time-dependent Mode I stress intensity factor, and $\phi(t)*P(t) = \int_{0^-}^{t}\phi(t-\tau)(\partial P(\tau)/\partial\tau)d\tau$. $K_I(t)$ is eliminated from (21) and (22) using Schapery's solution for a closing crack [40], in which the cohesive forces are described by a Dugdale potential and the viscoelastic response by a simple creep compliance function

$$\phi(t) = \phi_o + \phi_1 t^m \tag{23}$$

where and ϕ_1 and m are constants. The long-time behavior of this model is fluid like. The rate of change of contact radius is given by

$$\frac{da(t)}{dt} = \frac{\pi K_I^2(t)}{8\sigma_o^2}\left[\frac{\phi_1}{\phi_o\lambda(Z)}\sqrt{\frac{\pi}{4}}\frac{\Gamma(m+1)}{\Gamma(m+3/2)}\right]^{\frac{1}{m}}\frac{1}{[1+\lambda(Z)]^2} \tag{24}$$

where

$$\lambda(Z) = \frac{2m+1}{4W(m+1)}\phi_oK_I^2(t)\left\{1+\sqrt{1-\frac{8Wm(m+1)}{\phi_oK_I^2(t)(2m+1)^2}}\right\} - 1. \tag{25}$$

For a load-controlled experiment, $P(t)$ is specified and Eqs. (24) and (21) or (22) are solved simultaneously to eliminate K_I and obtain the time variation of a and δ. For a displacement-controlled experiment, $\delta(t)$ is specified and Eqs. (24) and (21) or (22) are solved simultaneously to obtain the time variation of a and P.

Hui, Baney and Kramer show explicitly that Eq. (20), used by Tirrell and coworkers, is a misapplication of the correspondence principle and cannot be correct. In numerical simulations, they obtain reasonable fits to the data of Ref. [35] and also show that Eq. (20) overestimates $a(t)$.

The model used by Hui, Baney and Kramer requires a detailed knowledge of the local failure and bonding processes at the crack tip. This leads them to suggest that adhesive properties of contacts to viscoelastic materials are not best characterized by the work of adhesion, W, but rather by the stress intensity factor $K_I(t)$. The relationship between $a(t)$ and K_I, Eq. (24), is complex since $a(t)$ can depend on the entire history of K_I.

5. TIME SCALES IN SFM EXPERIMENTS

As discussed in the introduction, the creep and crack regimes can be distinguished by their relaxation times, τ_{crack} and τ_{creep}. In this section, the expected magnitudes of these relaxation times are estimated and the experimental regimes in which one is emphasized over the other are determined. Estimation of τ_{crack} for a viscoelastic material requires an estimate of the crack length, l, which can be obtained using the Dugdale interaction model [17, 41, 42], as follows.

The Dugdale model provides a more realistic description of the deformations near the contact periphery than does the JKR model. The pressures in the contact are compared for the two models in Figure 11 with the JKR model on the right and the Dugdale model on the left. The Dugdale interaction is also compared with a Lennard-Jones interaction in the insert on the left. The Dugdale interaction approximates the actual probe-substrate interaction force per unit area as a square well of depth, σ_o, and range, h_o. The work of adhesion is $W = \sigma_o h_o$.

In the JKR limit, the pressure in an adhesive contact between elastic bodies has two components, shown schematically on the right hand side of Figure 11. They arise from the external load and from the adhesion [14]. Superimposed on the figure (heavy lines) are the outlines of the substrate and parabolic tip as in Figure 1b. The Hertzian contribution from the external load (dashed line),

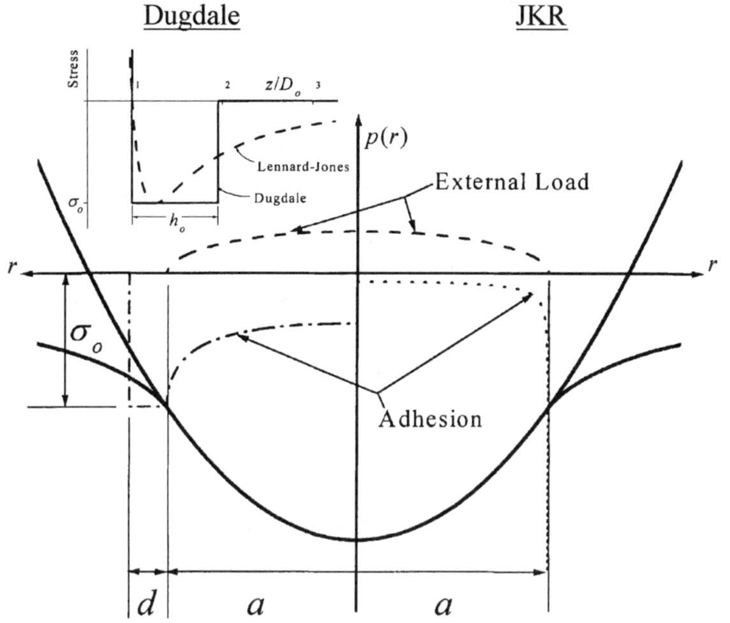

FIGURE 11 Comparison of the Dugdale and JKR models.

$P_{\text{external}}(r) \propto \sqrt{1-(r^2/a^2)}$, has the same form in both JKR and Dugdale models. It is largest in the center and falls to zero at the edges. In the JKR model, the adhesive contribution (dotted line) results in an additional contribution to the pressure, $p_{\text{JKR}}(r) \propto -1/\sqrt{1-(r^2/a^2)}$ where a now depends on the adhesion and, consequently, so does p_{external}. p_{JKR} is compressive and diverges strongly and unphysically at the periphery ($r=a$). This non-physical divergence is eliminated in the Dugdale model (dotdash line) where [10]

$$p_{\text{Dugdale}}(r) \propto \begin{cases} -\frac{2\sigma_o}{\pi}\tan^{-1}\sqrt{\frac{m^2-1}{1+(r/a)^2}} \text{ if } r<a \\ -\sigma_o \text{ if } a \le r \le a+d \end{cases}$$

where $m=(a+d)/a$. In both JKR and Dugdale models, it is clear that the dominant effects of adhesion occur at the contact periphery.

The lateral distance, d, over which the Dugdale force acts outside the contact defines the Dugdale zone. For a parabolic probe, d is determined from [10]

$$h_o = \frac{4Wa}{h_o \pi E^*}\left[1 - m + \sqrt{m^2-1}\cos^{-1}\left(\frac{1}{m}\right)\right] + \frac{a^2}{\pi R_o}\left[\sqrt{m^2-1} + (m^2-2)\cos^{-1}\left(\frac{1}{m}\right)\right] \tag{26}$$

where $E^* \equiv \left[\left(1+\nu_1^2\right)/E_1 + \left(1+\nu_2^2\right)/E_2\right]^{-1}$ is the effective modulus and the subscripts refer to the probe and substrate. In the limit $d \ll a$,

$$d \approx \pi E^* h_o^2 / 4W \tag{27}$$

and is independent of R_o. (The equivalent expression given in Ref. [17] is incorrect.)

Creep and crack response are very different at the SFM-scale because l and a are very different. To show this, first consider a crack between elastic materials. Greenwood and Johnson [42], using the approach of Schapery [43], showed that the crack length, l, can be estimated as $l \approx E^* h_o^2/2W$. Comparing this with Eq. (27) shows that $l \approx 2d/\pi$. Thus, d is a reasonable estimate of the crack length and, in the following, we assume $l = d$. For viscoelastic materials, Greenwood and Johnson further showed that the same result holds but with E^* replaced $1/\phi(\tau_{\text{crack}})$. Thus, d/a is in the range $E^*_\infty h_o^2/2aW \leq d/a \leq E^*_o h_o^2/2aW$ where E^*_∞ is the limiting modulus for a slow moving crack ($\tau_{\text{crack}} \to \infty$) and $E^*_o \gg E^*_\infty$ is the limiting value for a fast moving crack ($\tau_{\text{crack}} \to 0$). Combining Eqs. (1) and (2), and using the JKR radius at pull-off ($a_{\text{pull-off}} = \sqrt[3]{9\pi W R_o^2/8E^*}$) as a convenient lower bound to a, yields

$$\tau_{\text{crack}}/\tau_{\text{creep}} < h_o^2\left(\pi E_o^{*^2}/3W^2 R_o\right)^{2/3} \tag{28}$$

Equation (28) is plotted as a function of E^*_o in Figure 12 for the limiting combinations of W and R_o expected in SFM measurements. SFM probe tips have nominal radii in the range $10\,\text{nm} < R_o < 100\,\text{nm}$ and $10\,\text{mJ/m}^2 < W < 100\,\text{mJ/m}^2$ includes the range of W typically encountered in studies of polymer adhesion [44]. This analysis shows that $\tau_{\text{creep}} > 100\tau_{\text{crack}}$ for all materials with $E^*_o \leq 100\,\text{MPa}$, *i.e.*, the time scales are very different with τ_{creep} being significantly longer.

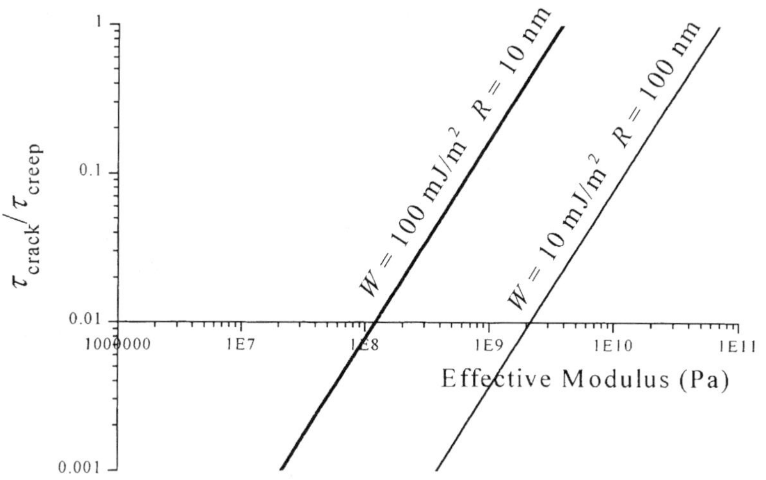

FIGURE 12 $\tau_{\text{crack}}/\tau_{\text{creep}}$ *vs.* E_o^*.

Since τ_{crack} and τ_{creep} are very different, it should be possible to study them separately. To do this, the SFM instrument must be able to satisfy two conditions. First, the characteristic relaxation time, τ, of the material under study must lie within the accessible range of experimental contact times, t_{contact}. Second, the experimental parameters must be adjusted so that either τ_{crack} or τ_{creep} is close to the characteristic relaxation time, τ of the sample under study. Using the approximations described above, the following expressions are obtained for *bounding* values of τ_{crack} and τ_{creep}

$$\tau_{\text{creep}} \geq a_{\text{pull-off}}/\langle V\rangle = \sqrt[3]{9\pi W R_o^2/8\langle V\rangle^3 E_o^*} \tag{29}$$

and

$$\tau_{\text{crack}} \leq d/V = \pi E_o^* h_o^2/4VW. \tag{30}$$

Note that τ_{crack} is independent of probe radius but $\tau_{\text{creep}} \propto R_o^{2/3}$. The bands of limiting values of Eqs. (29) and (30) that are shown in Figure 13 as a function of E_o^* were calculated for $\langle V\rangle = 100\,\text{nm/s}$ and $10\,\text{mJ/m}^2 < W < 100\,\text{mJ/m}^2$, and $10\,\text{nm} < R_o < 100\,\text{nm}$. For this range of parameters, the entire creep regime always lies above the band labeled "creep" and the crack regime lies below the band labeled "crack". Equivalent results for other choices of $\langle V\rangle$ are obtained by simply

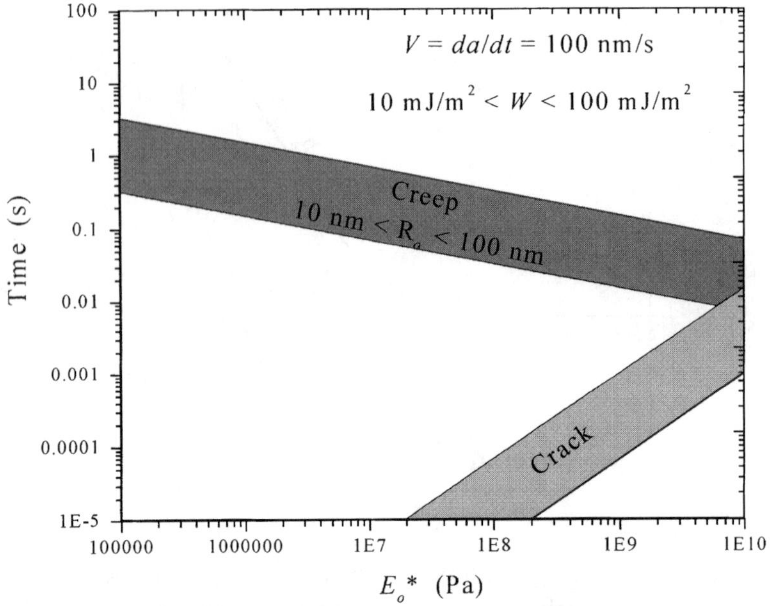

FIGURE 13 Creep and crack regimes accessible in SFM experiments for $\langle V\rangle =$ 100 nm/s.

shifting both bands vertically by $(100/\langle V\rangle)$ where $\langle V\rangle$ is measured in nm/s. For the example of Figure 13, creep effects are seen to dominate the SFM response for polymers with effective moduli below about 100 MPa and relaxation times in the range $0.1\ \text{s} \lesssim \tau \lesssim 10\ \text{s}$. Obviously, t_{contact} must lie in the same range as τ, which is the range used for the majority of force-distance curves reported in the literature. In order to bring the crack regime into this experimental range, V would have to be reduced by 3 to 4 orders of magnitude. SFM experiments have been done in this range by Basire and Fretigny [24] who measured the engulfment of an SFM probe into a styrene-butadiene random copolymer under zero load conditions. In this case, the increase in contact area is driven solely by surface forces and opposed by deformation of the substrate.

Tapping or intermittent contact mode [45] measurements involve very short contact times ($t_{\text{contact}} < 10^{-5}$ s or less) with very high $\langle V\rangle \gtrsim 10\ \mu\text{m/s}$. Again, creep effects will tend to dominate the measured response if the viscoelastic material also has $\tau \sim t_{\text{contact}}$.

In general, the creep regime is expected to dominate SFM measurements on compliant materials with $E_o^* \lesssim 100\,\text{MPa}$. For materials with higher moduli, the creep and crack regimes lie closer together and experiments are likely to be influenced by both. The requirement for low V and short $t_{contact}$ makes it difficult for SFM to operate in the crack regime for most compliant materials. Unfortunately, this is the regime with the most complete theoretical model [11].

The discussion given above is presented in terms of average speeds, $\langle V \rangle$. V cannot be controlled in actual SFM instruments and it varies over a very wide range. Thus, in a real contact, the behavior will not be as simple as indicated by the arguments given above. This is illustrated in Figure 9b for the standard model where the instantaneous value of $\tau_{creep}(t) = a(t)/(da/dt)$ is plotted for the $a(t)$ results shown in Figure 9a. The relaxation times, τ_1 and τ_2, for the mechanical model are indicated by the solid lines. $\tau_{creep}(t)$ changes by over five orders of magnitude. It is very short during the initial stage of contact formation and in the final stages of pull-off so that the assumption $a \gg l$ may no longer be valid. At the point of maximum contact area, $\tau_{creep}(t)$ exceeds 1000 s. Its average value ranges between 5 s and 33 s as $t_{contact}$ increases from $0.2\tau_1$ to $20\tau_1$. Even for the examples with the largest time lag to reach a_{max}, $\tau_{creep}(t)$ is near τ_1 and τ_2 for a relatively small fraction of the total contact time.

6. SUMMARY

The theory developed by Ting [18, 19] has been applied to describe the contributions of creep to the response of nanometer-scale contacts during loading and unloading. This theory neglects the effects of adhesion but can be solved analytically for several simple models. None the less, it gives valuable insights into the behavior of small contacts. The major result is that creep can cause the contact area, a, and deformation, δ, to reach their maximum values well *after* the maximum load has been applied to the contact. Results obtained using simple mechanical models to describe the response of the viscoelastic materials provide useful guidance on how to optimize experiments to study creep. The maximum deformation at the instant the contact is broken is large and correlates well with the time dependence of the loss function, tanδ. A typical SFM instrument should be able to study creep processes in materials whose characteristic relaxation times are in the range from roughly a millisecond up to a few hundred seconds.

Since the characteristic relaxation times of creep and crack regimes differ substantially for SFM experiments, the contact times of force-distance curves can be adjusted so that the response is by dominated one or the other.

Quantitative characterization of compliant viscoelastic materials requires that force *vs.* distance curves be carried out over as wide a range of contact times as possible. This observation should be particularly relevant for many biological materials where experimental contact times have been limited. Additionally, it suggests that studies of pull-off behavior that have traditionally relied on rather short contact times will need to be re-evaluated.

Most of the models described here are obviously too simple to be used for quantitative analysis of experimental data. However, most important differences are expected to be qualitative and the major conclusions reached using them will remain correct. The more rigorous analysis recently put forth by Hui, Baney and Kramer [38] correctly includes adhesion and creep, but only applies to cases where the contact size is increasing. It has not yet been applied to nanometer-size contacts. Once such an analysis becomes available, more complex systems such as layered films and living cells can be attacked quantitatively.

Acknowledgements

The author acknowledges financial support from the Department of Energy, the Office of Naval Research, the Paper Surface Science Program of the University of Maine, and the Maine Science and Technology Foundation. Discussions with K. L. Johnson and K. J. Wahl have also been invaluable.

References

[1] Burnham, N. A. and Colton, R. J., In: *Scanning Tunneling Microscopy and Spectroscopy*, Bonnell, D. A. Ed. (VCH Publishers, Inc., New York, 1993), p. 191.

[2] Hues, S. M., Draper, C. F. and Colton, R. J., *J. Vac. Sci. Technol.* **B12**, 2211 (1994).

[3] *Fundamentals of Friction: Macroscopic and Microscopic Processes*, Singer, I. L. and Pollock, H. M. Eds. (Kluwer Academic Publishers, Dordrecht, 1992).
[4] Colton, R. J., *Langmuir* **12**, 4574 (1996).
[5] Landman, U., Luedtke, W. D., Burnham, N. A. and Colton, R. J., *Science* **248**, 454 (1990).
[6] *Tribology Issues and Opportunities in MEMS*, Bhushan, B. Ed. (Kluwer Academic Publishers, Dordrecht, 1998).
[7] Mate, C. M., *Trib. Lett.* **4**, 119 (1998).
[8] You, H. X. and Yu, L., *Methods Cell Sci.* **21**, 1 (1999).
[9] Asif, S. A. A., Wahl, K. J. and Colton, R. J., *Rev. Sci. Instrum.* (1999).
[10] Maugis, D., *J. Colloid Interface Sci.* **150**, 243 (1992).
[11] Maugis, D. and Barquins, M., *J. Phys. D* **11**, 1989 (1978).
[12] Unertl, W. N., *J. Vac. Sci. Technol. A* **17**, 1779 (1999).
[13] Carpick, R. W., Agraït, N., Ogletree, D. F. and Salmeron, M., *J. Vac. Sci. Technol. B* **14**, 1289 (1996).
[14] Johnson, K. L., *Contact Mechanics* (Cambridge University Press, Cambridge, 1987).
[15] Savkoor, A. R., In: *Microscopic Aspects of Adhesion and Lubrication*, Georges, J. M. Ed. (Elsevier Sci. Publ. Co., Amsterdam, 1981), p. 279.
[16] Barthel, E., *J. Colloid Interface Sci.* **200**, 7 (1998).
[17] Johnson, K. L., In: *Microstructure and Microtribology of Polymer Surfaces*, Tsukruk, V. V. and Wahl, K. J. Eds. (American Chemical Society Books, Washington, D.C., 2000), p. 24.
[18] Ting, T. C. T., *J. Appl. Mech.* **33**, 845 (1966).
[19] Ting, T. C. T., *J. Appl. Mech.* **35**, 248 (1968).
[20] Wahl, K. J., Stepnowski, S. V. and Unertl, W. N., *Tribology Lett.* **5**, 103 (1998).
[21] Wahl, K. J. and Unertl, W. N., In: *Tribology Issues and Opportunities in MEMS*, Bhushan, B. Ed. (Kluwer Academic Publishers, 1999).
[22] Barquins, M. and Maugis, D., *J. Adhesion* **13**, 53 (1981).
[23] Barquins, M., *J. Adhesion* **14**, 63 (1982).
[24] Basire, C. and Fretigny, C., *C. R. Acad. Sci. Paris, Series IIb* **325**, 211 (1997).
[25] Kovacs, G. J. and Vincett, P. S., *Thin Solid Films* **111**, 65 (1984).
[26] Unertl, W. N., In: *Microstructure and Microtribology of Polymer Surfaces*, Tsukruk, V. V. and Wahl, K. J. Eds. (American Chemical Society Books, Washington, D.C., 2000), p. 66.
[27] Lee, E. H. and Radok, J. R. M., *J. Appl. Mech.* **27**, 438 (1960).
[28] Graham, G. A. C., *Int. J. Engng. Sci.* **3**, 27 (1965).
[29] Findley, W. N., Lai, J. S. and Onarian, K., *Creep and Relaxation of Nonlinear Viscoelastic Materials* (Dover Publications, New York, 1976).
[30] Savkoor, A. R., In: *Microscopic Aspects of Adhesion and Lubrication*, Georges, J. M. Ed. (Elsevier Science, Amsterdam, 1981), p. 279.
[31] Young, R. J. and Lovell, P. A., *Introduction to Polymers*, 2nd edn. (Chapman & Hall, London, 1991).
[32] Richard, J., *Polymer* **33**, 562 (1992).
[33] Domke, J. and Radmacher, M., *Langmuir* **14**, 3320 (1998).
[34] Hughes, B. D. and White, L. R., *Q. J. Mech. Appl. Math.* **32**, 445 (1979).
[35] Falsafi, A., Deprez, P., Bates, F. S. and Tirrell, M., *J. Rheol.* **41**, 1349 (1997).
[36] Johnson, K. L., Kendall, K. and Roberts, A. D., *Proc. R. Soc. London A* **324**, 301 (1971).
[37] Zhao, W., Rafailovich, M. H., Sokolov, J., Fetters, L. J., Plano, R., Sanyal, M. K., Sinha, S. K. and Sauer, B. B., *Phys. Rev. Lett.* **70**, 1453 (1993).
[38] Hui, C. Y., Baney, J. M. and Kramer, E. J., *Langmuir* **14**, 6570 (1998).
[39] Yang, W. H., *J. Appl. Mech.* **35**, 379 (1968).
[40] Schapery, R. A., *Int. J. Fracture* **39**, 163 (1989).
[41] Kim, K. S., McMeeking, R. M. and Johnson, K. L., *J. Mech. Phys. Solids* **46**, 243 (1998).

[42] Greenwood, J. A. and Johnson, K. L., *Phil. Mag.* **43**, 697 (1981).
[43] Schapery, R. A., *Int. J. Fracture* **11**, 369 (1975).
[44] Chaudhury, M. K., *Materials Sci. Engin. R* **16**, 97 (1996).
[45] Behrend, O. P., Oulevey, F., Gourdon, D., Dupas, E., Kulik, A. J., Gremaud, G. and Burnham, N. A., *Appl. Phys. A* **6**, S219 (1998).

Particle-Surface Interactions That Influence Adhesion

Experiments and Engineering Models of Microparticle Impact and Deposition

RAYMOND M. BRACH, PATRICK F. DUNN and XINYU LI

Particle Dynamics Laboratory, Department of Aerospace and Mechanical Engineering, University of Notre Dame, Notre Dame, IN 46556, USA

This article summarizes, reviews and consolidates some of the research work done by the authors over recent years. It covers a wide variety of topics related to the experimental and analytical investigations of the impact of microparticles with flat surfaces in the presence of adhesion and frictional forces. Over 180 experiments were conducted under vacuum conditions to study the effects of particle size, shape, incident translational and rotational velocities, and substrate surface roughness on the oblique impact response of the particle. Analytical models of the impact process were developed, including an algebraic, rigid-body model and a numerical simulation that can be used to predict rebound and capture conditions and to model the forces and displacements that occur during the contact duration. These models were validated using experimental results. Overall, the article covers impact conditions ranging from the more idealized case of a microsphere impacting a molecularly-smooth surface to the more realistic and complex situation of a biological microparticle impacting a typical indoor-room surface.

1. INTRODUCTION

The following article is a summary of combined experimental and analytical work done in order to study and model the process of impact of microparticles. It begins with a discussion of some of the classical experimental results already published in the literature. Then, it displays and examines more recent experimental results of the oblique impact of microspheres against ultrasmooth and rough surfaces. Data from the impact of nonsmooth particles (bioaerosols) are also presented.

Two analytical models of the planar oblique impact process in the presence of adhesion are summarized and discussed. One is an algebraic model based on rigid body[1] impact theory and which uses coefficients to represent the impact process. Another is based on the Hertzian elastic model in the contact region but which also takes into account an adhesion force as well as distinct representation of material and adhesion dissipation. The latter model, referred to as a dynamic simulation, is based directly on Newton's differential equations of motion and its solution is through numerical integration.

In addition to the impact models, another model is developed. It is a set of empirical equations that represent the kinematic coefficient of restitution of the rigid body model as a function of initial normal velocity. These equations when combined with the rigid body equations allow specific behavior, including capture, to be modeled for applications.

Finally, several analyses are carried out that relate to the impact process of microspheres. One studies the influence of surface roughness on the measured values of the coefficient of restitution and the frictional, coefficient for oblique collisions. The analysis shows that small unknown variations in the slope of the surface can lead to biased coefficient values computed from experimental measurements. A second analysis uses the rigid body impact model with Monte Carlo methods to investigate both the effects of surface roughness and the impact process. A third analysis is a sensitivity study of the different factors such as physical constants of the impact process and shows which are the most significant in controlling capture of a particle due to adhesion.

2. EXPERIMENTS ON MICROPARTICLE IMPACT

Many definitive experimental studies of microparticle impact onto surfaces exist, including the pioneering work of Dahneke [1], as well as

[1]The term *rigid body* here denotes the presence of rotational inertia (in contrast to a point mass) and does not imply inflexibility.

the work of Wang *et al.* [2], Wall *et al.* [13] and Dunn *et al.* [4, 5]. Experiments can be categorized into those limited to normal impacts and those including oblique impacts. Both involve the effects of adhesion but only the oblique collisions display effects of friction. The results of the microparticle/substrate surface impact experiments of Dunn *et al.* [4, 5] and Li *et al.* [5] are summarized in this section. In all, 187 different experimental combinations were conducted that included metal and glass microspheres and lycopodium spores, 5 different ranges of microsphere diameters, 11 different surface types, 14 different surface angles and about 14 different ranges of initial microparticle velocities. Each experiment consisted of approximately 40 individual impact events. All experiments were conducted under vacuum and neutral charge conditions in order to obviate the effects of additional forces acting near or during surface contact. The approach taken was first to investigate the most idealized impact conditions of microsphere impact with a molecularly-smooth, planar surface. Then, additional factors such as a wider particle-size distribution, a rougher substrate surface and a more complex particle surface were introduced in steps in order to understand each of their effects on impact and capture.

2.1. Equipment and Methods

This section describes the basic experimental facility used for the experiments, which was developed by Caylor [6] and described in detail by Dunn *et al.* [4]. The physical attributes of the microparticles and surfaces are described later when the results are presented. The system primarily consisted of a vacuum test cell (at 10^{-4} Torr), particle dispenser and target surface. In these experiments, the microparticles were dispensed using a neutral-charge particle dispenser (NPD) and in all cases the target surface was electrically grounded. The microparticles were placed on the bottom dispenser plate of the NPD. A rotor underneath the plate periodically contacted the plate and vibrated it, causing the microspheres to fall through a hole at its center. To control the falling particles, a hypodermic needle was connected to the hole. As a result, the microspheres were directed downward in a straight trajectory to the target surface. Once a particle was ejected from the NPD, it was accelerated by gravity to the target surface. The vertical distance between the dispenser and the target surface varied from 0.01 m to 1.0 m, providing a velocity range from 0.44 m/s to 4.4 m/s.

For oblique impact experiments in which the target surface was inclined at an angle with respect to the incident particle beam, a particle trajectory imaging system (PTIS) was used to record the microparticles incident and rebound trajectories, from which the velocity components were determined. This set-up is shown in Figure 1. The PTIS was comprised of an argon-ion laser, beam chopper, plano-convex lens, CCD camera and video recorder. The PTIS generated a pulsed laser light sheet that illuminated the individual particle as it approached and rebounded from the surface. The trajectories were processed to obtain the particle's incident and rebound angles and speeds. For this PTIS setup, an argon-ion laser beam (operated nominally at 1W) passed through a collimator to control the beam width to provide as narrow a light sheet as possible. The laser beam was then directed through a spinning disk with 10 evenly-spaced slots to produce a pulsed laser beam. Depending on the

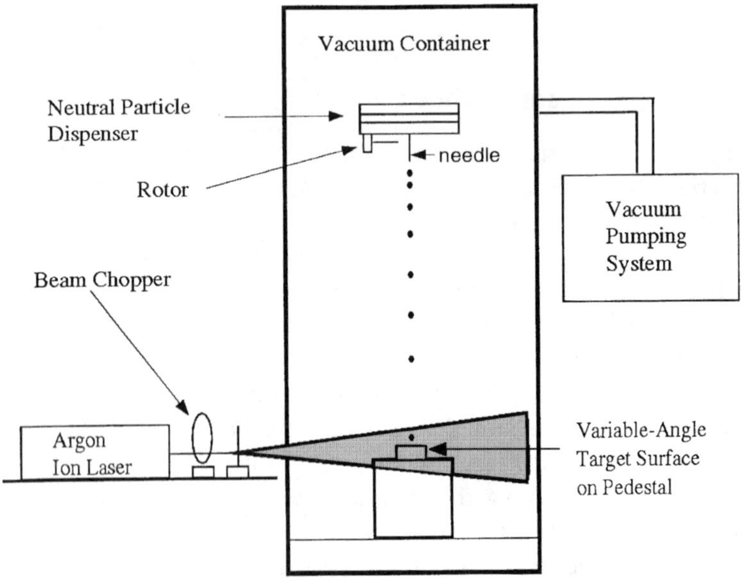

FIGURE 1 Schematic layout of the experimental equipment used for oblique impact experiments.

angular velocity of the disk, the pulsed frequency could be varied to obtain the desired track length. The chopped beam then passed through a plano-convex lens that formed a pulsed light sheet aligned in a vertical plane above the target surface. In preparing an experiment, another laser beam was sent through the hypodermic needle to make sure this laser beam was at the center of the pulsed light sheet. Video data were taken through a porthole located at the side of the vacuum chamber using a CCD camera and a video cassette recorder. The camera was installed approximately 90° to the light sheet. The image was enlarged as much as possible to get a clear trajectory. Based on the frame rate, the field of view, and the strobe frequency, the system could measure particle velocities ranging from about 0.1 to 30 m/s.

For normal impact experiments using microspheres, a two-dimensional phase Doppler particle analyzer (PDPA) was used in the 90° side-scatter mode to determine each microsphere's incident and rebound vertical velocity components and its diameter. The top view of this set-up is shown in Figure 2. The probe volume was positioned about 1 mm above the surface. Frequency shifting was used to discriminate between the incoming and outgoing particles. The particle size was determined from the Doppler signal based on the phase

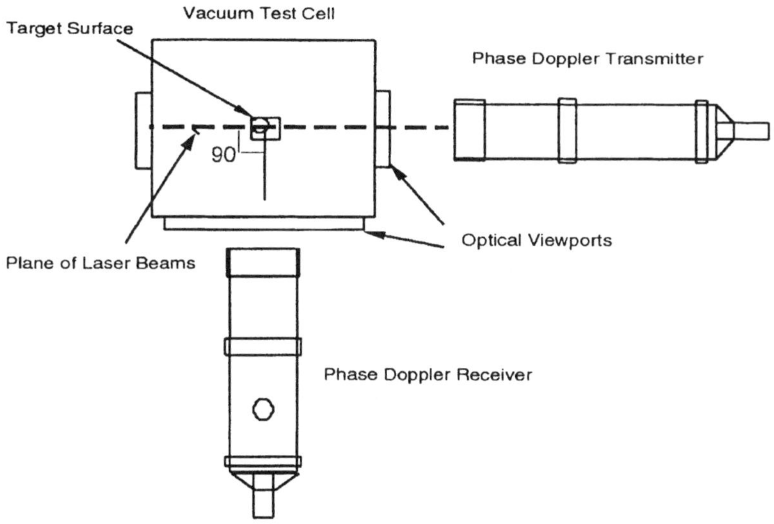

FIGURE 2 Top view of experimental set-up used for normal impact experiments.

differences between the three detectors. The system's laser was an air-cooled argon ion laser; the output power could be varied from 20 to 300 mW to obtain satisfactory scattered light intensity over the wavelength range between 454 and 514 nm. The raw PDPA data (including the incident and rebound velocity component values, particle diameter and time of acquisition) was stored in a data file. After the experiment, the data were processed to identify the incident and rebound normal velocity components for the *same* microsphere. This was accomplished by using an initial estimate of the vertical distance, h, between the laser probe volume and the target surface to initiate a pair-searching algorithm. Separate experiments were conducted to verify that the distance between the laser probe volume and the surface had no effect on pair identification. Finally, the paired velocity data corresponding to the position of laser probe volume were corrected by $\pm\sqrt{2gh}$ to give the velocity components at the surface. In this manner, the various parameters of interest could be computed for each individual impact event and then, subsequently, averaged over 40 individual events for an experiment.

Details of an uncertainty analysis of the above process can be found in Caylor [6] and in Dunn *et al.* [4]. The measurement and finite sampling uncertainties are combined to provide estimates of the true mean values at 95% confidence for each parameter. In the following presentation of the experimental data, the sample mean value of a quantity is plotted with an error bar. Each data point represents the sample mean value of approximately 40 *individual* impact events. An error bar designates the range within which the true mean value lies with respect to the sample mean value within a 95% probability.

2.2. Experimental Parameters

There are three experimental parameters that are used to characterize the impact event. These are the coefficient of restitution, e, the impulse ratio, μ, and the normalized translational kinetic energy loss, T_L. Referring to Figure 3, these parameters are defined as follows:

$$e = -V_n/v_n \tag{1}$$

$$\mu = (V_t - v_t)/(V_n - v_n) \tag{2}$$

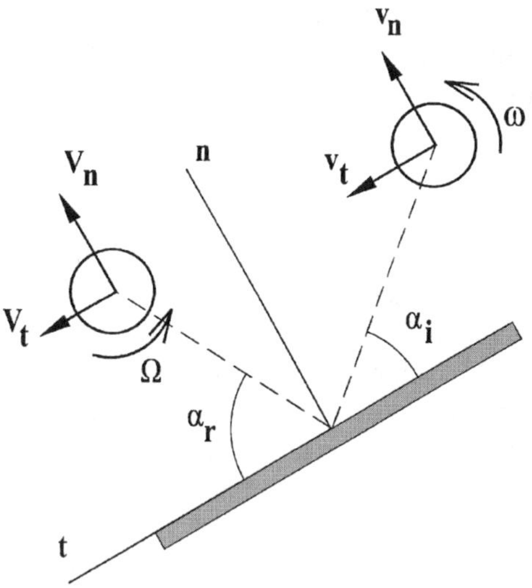

FIGURE 3 Variables and coordinates used for planar impact measurements.

$$T_L = (v^2 - V^2)/v^2 \tag{3}$$

Their derivations and physical interpretations will be presented in the section describing the analytical models. The following experimental results are presented in their context.

2.3. Normal Impact Results

Dahneke [1, 8] studied the capture velocity of 1.27 μm-diameter polystyrene latex spheres. In these experiments, it was found that a particle with an incoming velocity less than a critical velocity would stick to the surface, otherwise, if it had a larger velocity, it would rebound from the target surface. He also suggested that the elastic flattening during contact is very important. Dahneke [8] measured the particle velocity before and after collision directly. The velocity was determined by measuring the time for a particle to pass through two parallel laser beams. Dahneke found that the coefficient of restitution, defined as the ratio of rebound velocity over the incoming velocity, increased with the incoming velocity just above the critical (capture) velocity, reached an asymptotic level for some range of velocities and then finally decreased at higher incoming velocities.

Paw U [9] obtained the speed at impact for normal incidence above which ragweed pollen and lycopodium spores rebound from glass and the leaves of various plants. The results suggested that the geometry and material properties of the particle are more important than those of the surface to determine the critical velocity.

By using a high speed camera system, Rogers and Reed [10] measured the capture velocities for three different kinds of particles (copper, glass, steel) impacting onto surfaces. The results showed that the capture velocity is inversely related to the particle size. Although plastic deformation during contact was a focal point of the study, little or no evidence of plastic deformation was given.

With the help of laser Doppler velocimetry, Wall *et al.* [3] measured the velocities of incoming and rebounding ammonium fluorescein spheres having different diameters (2.58, 3.44, 4.90 and 6.89 μm). The velocity range was from about 1 m/s, near the capture threshold, up to 100 m/s. It was found that capture velocity decreases with a power-law dependence on particle size rather than the elastic flattening model proposed by Dahneke [7]. Plastic deformation was used in this study as the sole mechanism of energy dissipation. Yet, no observable evidence of plastic deformation was cited.

Dunn *et al.* [4] reported the experimental results of both normal and oblique impact of microspheres on different surfaces. A phase Doppler particle analyzer and a particle trajectory imaging system were applied to measure the particle velocity. It was found that, for an incidence velocity larger than 3 m/s, the coefficient of restitution was almost constant. This was consistent with the results of Dahneke [8] and Wall *et al.* [3]. When the velocity was less than 3 m/s, the coefficient of restitution decreased rapidly. Results also showed that the material properties affected the impact process. No capture velocities were obtained from the experiments.

Li *et al.* [5] reported the results of the normal impact of polydisperse stainless steel microspheres with molecularly-smooth silicon surfaces at velocities near the capture velocity. These experimental results were compared with a dynamic simulation model, which was used to predict the capture velocity. The predicted capture velocity was consistent with experimental observations.

In summary, for normal impact, experiments show that the capture velocity is related to the particle size, initial velocity, surface geometry and material properties. Yet, because there are no experiments in which the capture velocity has been measured directly, no direct verification of any theoretical models of the capture velocity has been made.

2.3.1. The Effect of Particle Diameter

Experiments that reveal that the most "classic" results are those involving the impact of microspheres onto a molecularly-smooth substrate surface. One set of experiments in this category used two kinds of Type 316 stainless steel microspheres, each with a different size range (SST65, 10 to 65 μm, $d_{10}=49$ μm and SST125, 60 to 125 μm, $d_{10}=75$ μm).[2] The target surface was a molecularly-smooth (to within 5 Angstroms) [1, 0, 0] plane silicon crystal. As shown in Figure 4 (SST65 solid squares; SST125 open squares), both cases display a similar trend; at higher initial velocities (above about 0.75 m/s), the coefficient of restitution is roughly constant. At lower initial velocities, the coefficient of restitution decreases as the initial velocity decreases.

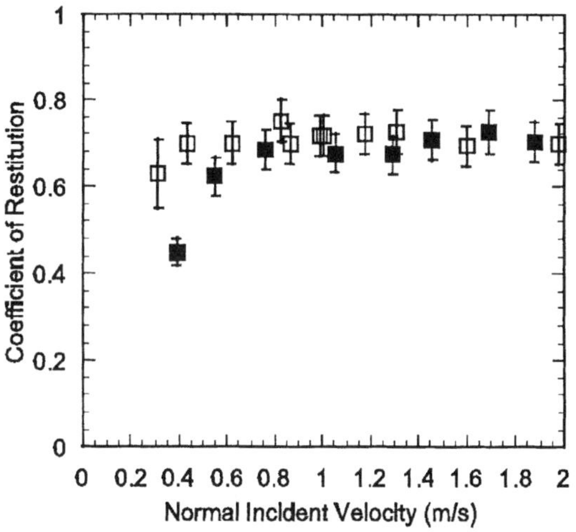

FIGURE 4 Coefficient of restitution from the normal impact of SST microspheres with diameters of d_{10} = 49 μm (solid squares) and d_{10} = 75 μm (open squares) for a molecularly-smooth, planar silicon crystal surface.

[2] d_{10} is the mean of the distribution of diameter.

This effect is more pronounced for smaller diameter particles indicating that adhesion effects are more significant for smaller diameter particles. Such trends with particle diameter agree with findings of Wall *et al.* [3].

For normal impact, the normal component is the total velocity. Thus, $T_L = 1 - e^2$. The normalized energy loss varying with the normal velocities is shown in Figure 5. It can be seen that, when incoming velocity gets smaller, the normalized kinetic energy loss becomes larger. This implies that, due to the effects of adhesion, there is more energy dissipation of the initial kinetic energy for impacts occurring at lower normal velocities. As a result, the microparticle will be captured if the incoming velocity becomes much smaller. The experiments showed that the microspheres started to pile up on the substrate surface (*i.e.*, were captured by the surface) at about 0.2 to 0.3 m/s. Because of the range of microsphere sizes, it was impossible to measure directly an exact capture velocity from the present experimental setup.

In summary, effects of adhesion manifest themselves at lower incident normal velocities, resulting in a decrease of the coefficient of restitution and an increase in the normalized translational kinetic energy loss. This effect occurs at relatively higher incident normal

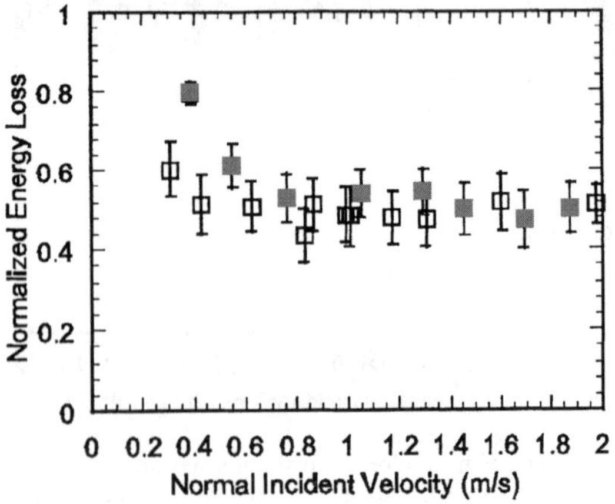

FIGURE 5 Energy loss from the normal impact of SST microspheres with diameters of $d_{10} = 49\,\mu m$ (solid squares) and $d_{10} = 75\,\mu m$ (open squares) for a molecularly-smooth, planar silicon crystal surface.

velocities for smaller diameter microspheres. This inherently leads to a lower capture velocity for larger diameter microspheres.

2.4. Oblique Impact Results

Broom [11] used glass spheres impacting on aluminum to study the adhesion of particles in filters. The nominal impact angles were 90° and 45°. A high speed camera system was used to take pictures of the particle trajectory and, further, to deduce the velocity. The results showed that the capture velocity was smaller for oblique impact, which implied that the particle rebounded more easily from the oblique surface than from the normal surface. It was also found that the nature of the impact surface is important; a polished surface exhibited an efficiency of capture higher than a rough surface.

Aylor and Ferrandino [12] studied the sticking probability of ragweed pollen and lycopodium spores impacting on glass cylinders and wheat stems. It was found that the speed for onset of rebound of a ragweed pollen was about 1.7 times as great as the critical speed for lycopodium spores. Changes in wheat stem characteristics as the plant aged also produced a measurable effect on this sticking efficiency. The experiments showed that the coefficient of restitution might not be constant for all impacts angles, but increases when the angle of impact becomes more oblique.

Wang *et al.* [2] investigated the adhesion efficiency of particles on a cylinder. Their measurements of particle rebound as a function of the position angle on the cylinder showed that rebound increases rapidly with angle. It was suggested that the tangential velocity component caused particle bounce, which possibly occurs through interaction with surface roughness.

Buttle *et al.* [13] conducted experiments with glass spheres impacting onto an aluminum substrate at 90° (normal), 50° and 29° respectively. The velocity was measured using laser Doppler velocimetry. The value of coefficient of restitution increased from around 0.5 at normal and 50° impact to 0.68 at 29° impact. This was considered to be the result of frictional force reduction and rotation of particle at very oblique impact.

Using the particle image technique, Dunn *et al.* [5] studied the effects of impact caused by the surface material properties and roughness, microsphere spin, particle size and electrical charge. It was found that surface roughness significantly biased the experimental results for very shallow (glancing) angles of incidence, yielding an unrealistically high value for the coefficient of restitution. Through examination of the microsphere's impulse ratio, four regions of different surface contact mechanics can be identified (in order of decreasing incident angle): a "rolling region" in which the microspheres are rolling without sliding by the end of contact, a "transition region" in which some of the microspheres slide throughout contact and some end up only rolling, a "sliding region" in which the microspheres all are sliding at the end of contact and the impulse ratio value is constant and, lastly, a region characterized by changes from a constant impulse ratio value. The initial spin of the particle also increased the uncertainty of the impulse ratio for a particular incidence angle.

In summary, the results of oblique impact experiments suggest that the capture velocity will not be a constant for all the impacting angles; the more oblique the impact angle, the smaller the capture velocity. The introduction of a friction force in addition to an adhesion force clearly makes the impact process more complex. Most of the data were acquired with the target surfaces oriented at various angles with respect to the incident particle trajectory. The cases examined ranged from the more idealized case of a microsphere impacting a molecularly-smooth surface to the more realistic and complex situation of a biological microparticle impacting a typical indoor-room surface. In the following, the oblique impact results are presented in the context of illustrating the effects of various parameters on the impact process.

2.4.1. The Effect of Surface Roughness

One of the first questions to be addressed is how the roughness of the substrate surface affects the microparticle's impact response. This is examined experimentally through comparison of various cases of "rough" substrate surfaces with that of the molecularly-smooth substrate surface. A base case for comparison is chosen as a relatively-monodisperse stainless steel microsphere (SST76, $d_{10} \approx 62\,\mu\mathrm{m}$) impacting a molecularly-smooth substrate surface ([1, 0, 0] plane silicon crystal).

The results of the base case for a nominal incident normal velocity of 1.66 m/s are shown in Figure 6, in which the solid symbols denote the smooth-surface base case and the open symbols the "rough"

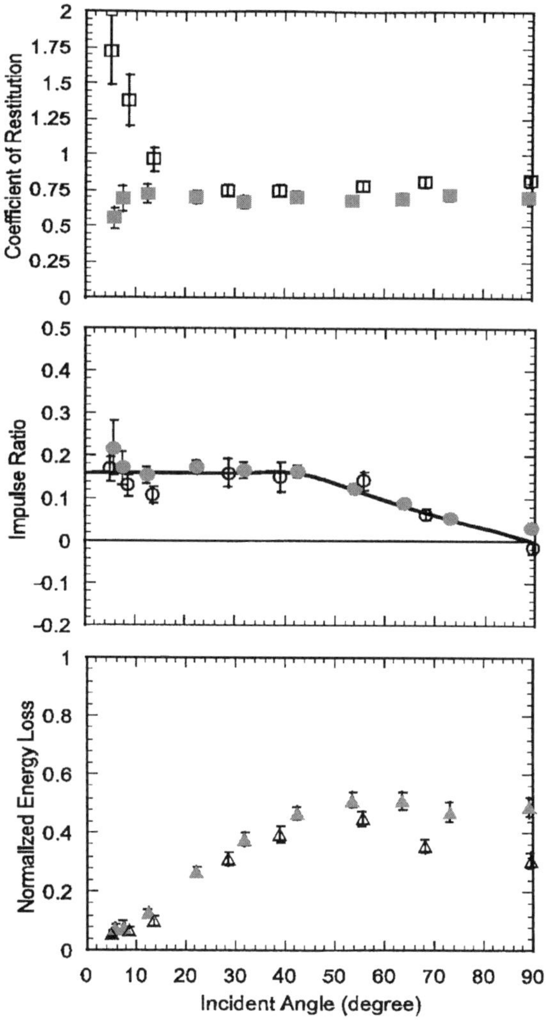

FIGURE 6 Comparisons of e, μ and T_L for a nominal incident normal velocity of 1.66 m/s where the solid symbols are from the smooth-surface (molecularly-smooth, planar silicon crystal) and the open symbols the "rough" surface case (back side of the same silicon surface). The solid curve is for the impulse ratio under the conditions of Coulomb friction and no initial angular velocity.

surface case. As shown in the figure, the coefficient of restitution is relatively constant ($e \approx 0.70$) with incident angle, except at incident angles of approximately 10° or less, where it decreases to $e \approx 0.50$. The impulse ratio curve exhibits a classic rolling-throughout-contact-duration behavior [15] from 90° down to approximately 50°, where it transitions to a constant value of $\mu \approx 0.15$. It is presumed that this value corresponds to that of the coefficient of friction between stainless steel and silicon. Below approximately 15°, the impulse ratio rises slightly. (The reason for this will be discussed in the section on analytical models.) The normalized translational kinetic energy loss is maximum at the higher incident angles, where most of the loss occurs due to overcoming surface adhesion.

The comparative rough substrate case consists of the same particles and the "rough" back-side of a silicon crystal. The results in Figure 6 are denoted by the open squares. It is seen that for incident angles larger than 20° the coefficient of restitution for the rough surface was slightly larger than that of the smooth surface. At lower incident angles (less than about 20°), values of the coefficient for the rough surface were much higher than those for the smooth surface. In fact, coefficient of restitution values exceed unity, which is not physically possible. Because of surface roughness, the "true" surface normal direction differs from the "nominal" normal and biases the value of e calculated using the nominal normal velocity components.[3] This is explained in detail later in Section 4. As far as the impulse ratio is concerned, for the rough surface the impulse ratio was not higher than that of the smooth surface, which implied that the rough surface might not result in a larger friction coefficient as expected. Because the adhesion force may contribute to the friction, if the adhesion force was smaller for rough surface, the friction may be smaller for the rough surface also. The normalized energy loss was less for the rough surface when the incident angles were higher than 50°, which might indicate that adhesion dissipation is less significant for rough surface impact. Whether or not the adhesion dissipation force could cause such an obvious influence on the normalized energy loss is still open to question.

---

[3]This is analogous to viewing the surface of the earth as flat when modeling meteorite impacts. On a broad plain or plateau the normal direction is close to a radial line but in a mountainous region the actual normal can differ greatly.

Additional cases were run to compare smooth with rough surface impact behavior. Shown in Figure 7 are the results of another comparison, in which the smooth surface is a "first-surface" mirror and

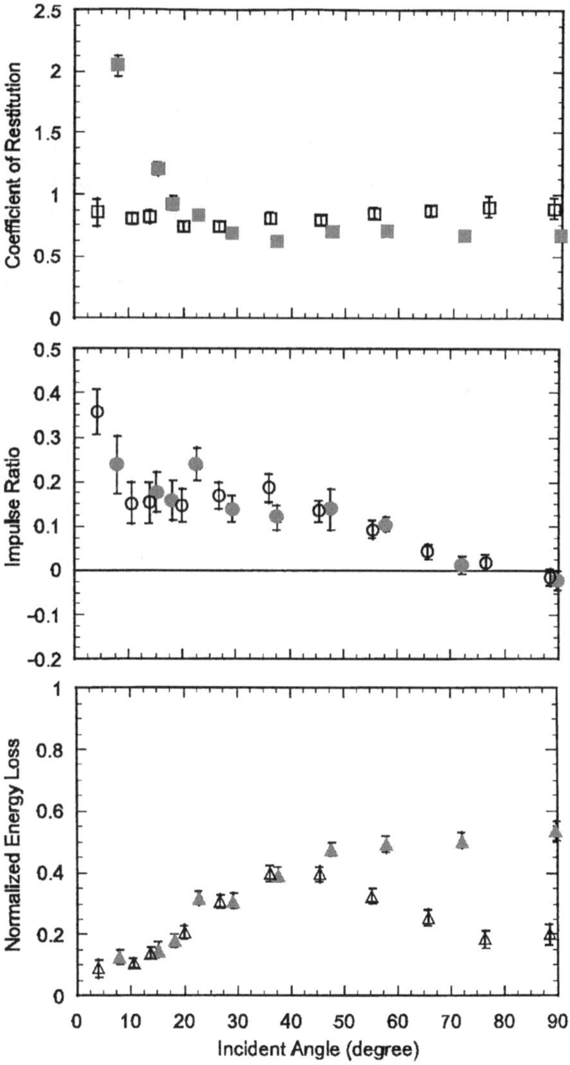

FIGURE 7 Comparisons of e, μ and T_L for a nominal incident normal velocity of 1.66 m/s where the solid symbols are from the smooth-surface (a "first-surface" mirror) and the open symbols the "rough" surface case (back side of the mirror).

the rough surface is the back-side of the same material. The comparison reveals some similar trends and some differences from what was previously observed. The coefficient of restitution values here are higher than before ($e \approx 0.70-0.90$ *versus* $e \approx 0.70-0.75$). Measured values of e for the rough case again increase to beyond unity at the lower incidence angles. The impulse ratio behavior is similar. The normalized translational kinetic energy loss, however, shows the opposite trend for the second rough-*versus*-smooth surface case, in that the rough surface case exhibits higher losses at high incidence angles. This can be explained by noting that this energy loss varies as $(1-e^2)$ at high incidence angles. Consequently, because e is lower energy loss will be higher.

To summarize, surface roughness can cause the measured value of the coefficient of restitution to be greater than unity at very low incidence angles (see Section 4) and change the amount of translational energy loss at higher incidence angles.

2.4.2. The Effects of Incident Rotational and Translational Velocities

Another consideration in characterizing a microparticle's impact response is to identify how the microparticle's incident rotational and translational velocities influence its capture or rebound. Incident translational velocities are changed easily by varying the spacing between the particle dispenser and the target surface. Rotational velocities, however, cannot be controlled, cannot be measured and possibly can vary systematically from one spacing to another. For this set of experiments, 11 different cases were studied. Relatively monodisperse stainless steel microspheres (SST65) were used along with the molecularly-smooth substrate surface ([1, 0, 0] plane silicon crystal). Results for e, μ and T_L *versus* incident angle are presented in Figures 8–10, for 3 of the 11 incident velocity cases examined. The incident velocity ranges were from 1.59 to 1.69 m/s for the 1.60 m/s case; from 1.01 to 1.09 m/s for the 1.05 m/s case; and from 0.39 to 0.49 m/s for the 0.45 m/s case. In Figure 8, the coefficient of restitution decreased slightly with the decreasing nominal incident velocity from 1.60 m/s to 0.45 m/s. For the same nominal incident velocity, the coefficient of restitution decreased slightly with decreasing incident

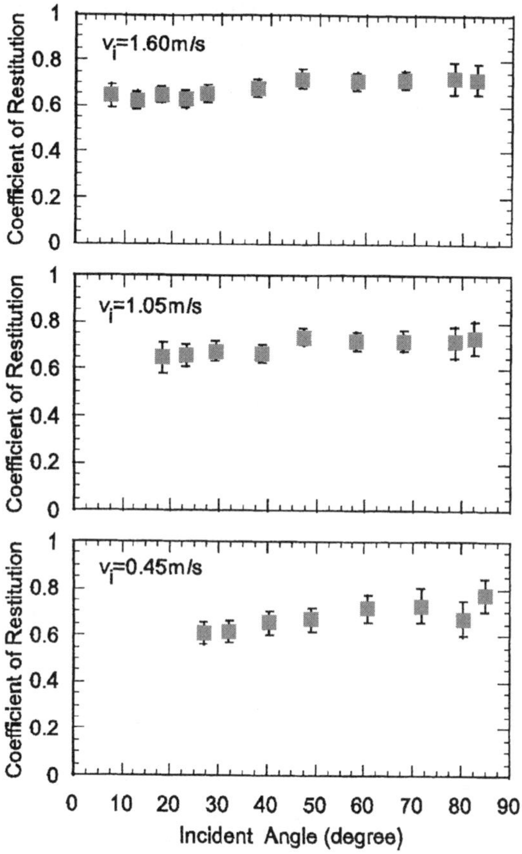

FIGURE 8 Comparisons of the coefficient of restitution, *e*, for stainless steel microspheres (SST65) with a molecularly-smooth, planar silicon crystal surface for 3 different ranges of total initial velocities; values indicated are nominal values.

angle. For the low incident velocity case ($v = 0.45$ m/s), the decreasing trend was more significant. For the higher incident velocity case ($v = 1.60$ m/s), the decrease in the coefficient of restitution was slight and the trend was not consistent at very shallow incident angles. The average value changed from 0.70 to 0.65 with the incident angle decreasing from 85° to 25°. Because the incident normal velocity decreases with decreasing incident angle (because the total velocity is held constant), the coefficient of restitution consequently decreases with the decreasing of the normal velocity component. For some cases,

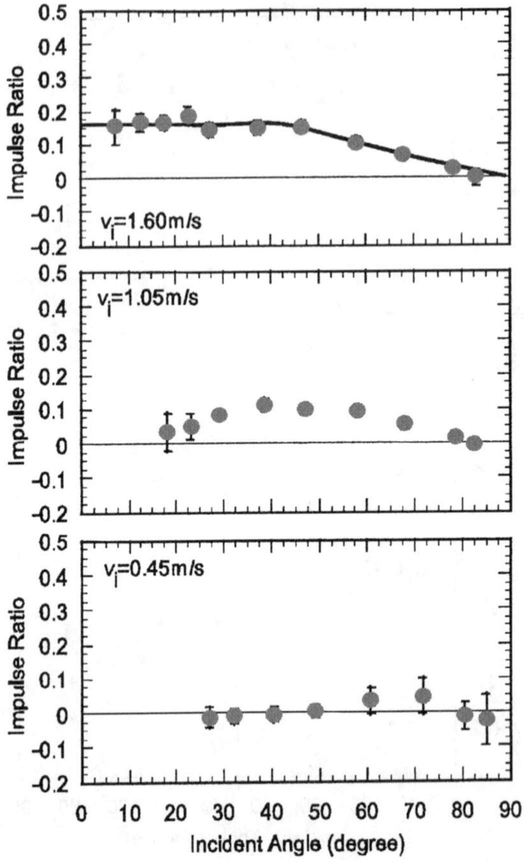

FIGURE 9 Comparisons of the impulse ratio, μ, for stainless steel microspheres (SST65) with a molecularly-smooth, planar silicon crystal surface for 3 different ranges of total initial velocities; values indicated are nominal values. The solid curve is the impulse ratio for coulomd friction from the rigid body model.

the normal velocity component was lower than the capture velocity of normal impact ($\approx 0.2-0.3$ m/s for SST76), yet no capture was observed and the coefficient of restitution was still rather high. This implies that, for oblique impact, the capture velocity is different from that for normal impact.

The impulse ratio behavior is shown in Figure 9. For $v = 1.60$ m/s, in the region of incident angles from 90° to 50°, the impulse ratio increases with decreasing incident angle. This is the "rolling region",

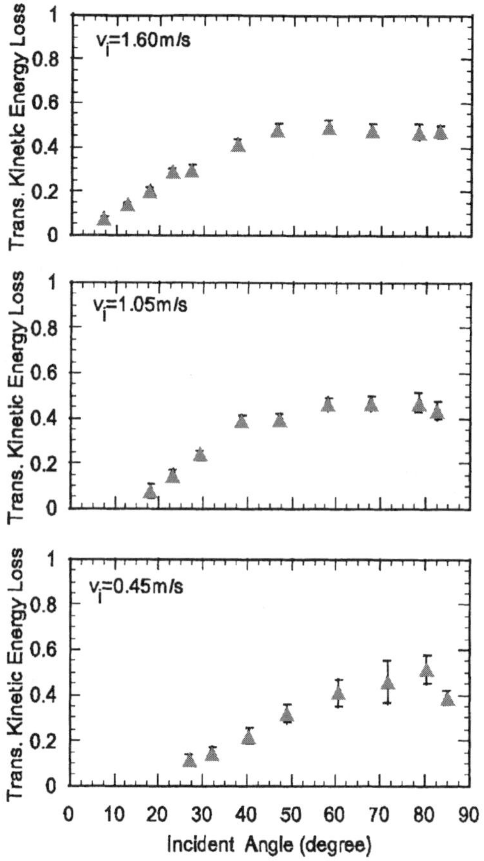

FIGURE 10 Comparison of the normalized energy loss, T_L, for stainless steel microspheres (SST65) with a molecularly-smooth, planar silicon crystal surface for 3 different ranges of total initial velocities; values indicated are nominal values.

where the microspheres are rolling without sliding on the surface at the end of contact. The observation was supported by the computational results of the rigid body model [5]. Starting from 50° down to 10°, the impulse ratio changes insignificantly and is basically constant ($\mu \approx 0.15$). This phenomenon was observed by Dunn *et al.* [5] and this region is designated as the "sliding region", where the microspheres are sliding (and rolling) on the surface at the end of contact. For this situation, a constant impulse ratio is interpreted as the sliding coefficient of friction according to the Amontons–Coulomb law.

For the $v = 1.05\,\text{m/s}$ and $v = 0.45\,\text{m/s}$ cases, the impulse ratio still increases when the incident angle decreases from 85° but the region of increasing impulse ratio becomes narrower. After that, the impulse ratio no longer is constant but decreases with decreasing incident angle. Some values even are less than or close to zero, which implies a zero-friction contact. The reason for the decreasing of impulse ratio with the decreasing of initial impact velocity is not clear yet. However, a dynamic simulation of the process reveals that these seemingly low impulse ratio values can be accounted for by assuming that the microspheres have a significant rotational velocity component prior to impact. The normalized translational kinetic energy loss is plotted in Figure 10. For the higher incident velocity case of $v = 1.60\,\text{m/s}$, the loss is approximately constant from 85° down to 50°. From the energy loss equation, when the incident angle is large and μ is small, $T_L \approx (1 - e^2)$. Values of the coefficient of restitution change very slightly in this region so small differences should be expected in T_L. With the incident angles decreasing from 50°, the normalized energy loss decreases consistently to near zero. Based on rigid body mechanics, under an irrotational initial condition, η is equal to $\tan^{-1}\,\alpha_i$ where α_i is the incident angle. When the incident angle decreases, η^2 increases faster than η, and both are larger than unity for $\alpha_i < 45°$. Therefore, T_L almost always decreases for small incident angles. For $v = 1.05\,\text{m/s}$, the constant normalized energy loss region becomes narrower (60° to 85°). This is even more apparent for the $v = 0.45\,\text{m/s}$ case in which the normalized energy loss decreases starting from 80°. In summary, the magnitudes of the incident velocities (translational and rotational) of the microspheres and the incident angle play a significant role on the particle's impact response.

2.4.3. Experimental Measurements of the Effect of Particle Shape

Most real microparticle impacts involve non-spherical particles. The effect of microparticle shape can be ascertained though comparison of spherical and non-spherical microparticles using the same substrate surface. To study the effects of microparticle surface roughness, *lycopodium* spores were used with molecularly-smooth silicon crystal substrate. *Lycopodium* spores are common bioaerosols. A surface profile of a spore reveals that the surface is irregular and full of cavities. Comparisons of irregular particle surfaces (lycopodium spores, solid symbols) and smooth microsphere surfaces (stainless steel microspheres, SST76, open symbols) with a silicon substrate surface are shown in Figure 11. For the SST76, the coefficient of restitution is approximately constant from 90° down to 10°, where the value of e then decreases with decreasing incident angle. When the incident angle decreased from 90° to 55°, the magnitude of the coefficient of restitution in both cases is approximately the same, and the value is nearly constant. Both the material and the surface irregularity differences appear to have little effect on the impact response in this region. Below this region, the experimental results for these two cases show some differences. The coefficient of restitution for the *lycopodium* spores decreased from about 0.70 down to less than 0.55 when the incident angle decreased from 48° to 38°. Paw U [9] estimated that for 20 – 40 μm diameter *lycopodium* spores the capture velocity was 1.30 m/s when impacting American elm leaves. In the present experiments, the normal incoming velocity component ranged from about 1.10 m/s to 0.91 m/s for incident angles within 48° to 38°. Thus, based on this finding, one would expect to observe a decrease of the coefficient of restitution in this range that would continue as the incident angle is decreased further. However, for incident angles less than 38°, the coefficient of restitution for the *lycopodium* spores actually increases. This behavior has been observed before by Dunn *et al.* [5] with microspheres impacting a rough surface. Their analysis shows that surface roughness causes a fictitious increase in the measured value of the coefficient of restitution. The surface roughness of a *lycopodium* spore likewise may cause an increase of the coefficient of restitution as the incident angle is decreased. Other factors, such as multi-collisions and/or non-central collisions due to the irregular *lycopodium* spore surface can also contribute to this increase. No definite conclusions can be drawn.

The trends of the impulse ratio with changing incident angle also show the effect of microparticle surface geometry, friction and initial conditions. For the SST76 case, the impulse ratio increases consistently as the incident angles decrease from 90° down to about 50°. Then, it remains approximately constant from 50° to 10°. It increases again when the incident angles become smaller (below about 10°).

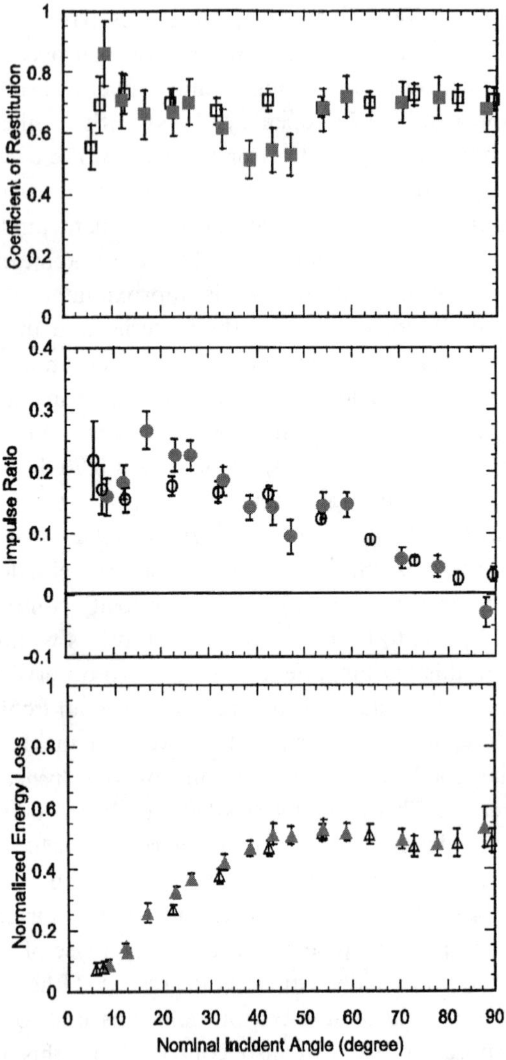

FIGURE 11 Comparisons of the impulse response from a silicon substrate surface of *lycopodium* spores (solid symbols) and stainless steel microspheres (SST76, open symbols).

As before, the increasing of the impulse ratio from 90° down to an incident angle of about 50° is explained by the microparticle rolling when it leaves the substrate surface at the end of contact. Below 50°, a constant impulse ratio occurs because of sliding throughout contact.

Below about 10°, the increase of the impulse ratio is suspected to be caused by adhesion. Adhesion increases the normal contact force and, therefore, the frictional force. The impulse ratio for the *lycopodium* spore behaves somewhat differently. There is no clear constant region. The impulse ratio increases as the incident angle decreases from 90° to about 20°. Then it decreases as the incident angle becomes less than 15°. The comparison of the normalized kinetic energy loss shows no substantial difference between these two cases. It seems that the differences in the microparticle and surface materials has no effect on the normalized kinetic energy loss under the present experimental conditions. From this observation, we learn that if we want to examine the effects of materials, surface geometry, *etc.*, on the impact response, the normalized energy loss, T_L, may not be the most sensitive parameter.

In summary, the microparticle's surface roughness complicates the behavior of e and μ. The coefficient of restitution at low incident angles exhibits trends similar to those observed for the rough substrate surface case, in which e rises with decreasing incident angle to values of unity or greater.

2.4.4. The Combined Effects of Particle Shape and Surface Roughness

Perhaps the most complex impact situation is that of a non-spherical microparticle contacting a typical rough indoor-room surface. Through comparison with the previously established base cases, these effects now are examined. Oblique impact experiments of *lycopodium* spores with a commercial *Formica* surface were conducted to explore the roughness effects of both substrate and particle. Profilometer scans of the *Formica* surface revealed an average surface roughness height of approximately 3.6 μm. The impact results are compared with those of SST76 microspheres impacting the same type of *Formica* surface and plotted in Figure 12. It can be seen that, for the same incident angles, the coefficient of restitution values for the stainless steel microsphere case (SST76, open squares) are significantly higher than those for the spore case (solid symbols), although the trends are similar. (Values of $e > 1$ occur as earlier in the rough-surface data in Figs. 6 and 7 and are explained in Section 4.) The coefficient of restitution increases with

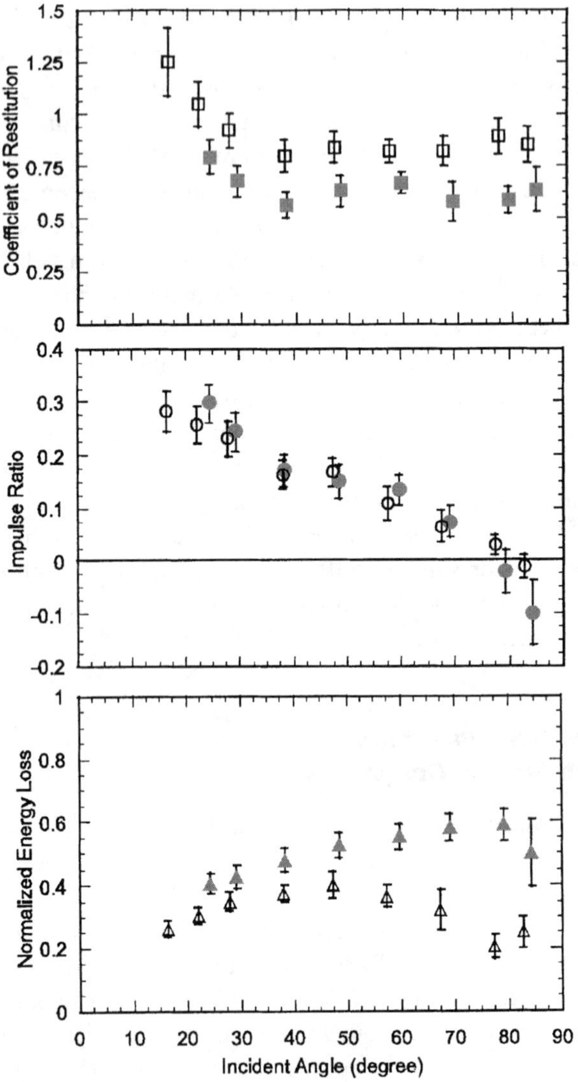

FIGURE 12 Comparisons of the impulse response from a commercial *Formica* surface of *lycopodium* spores (solid symbols) and stainless steel microspheres (SST76, open symbols).

decreasing incident angle for both cases. The measured values of the coefficient of restitution for the SST76 case exceed unity when the incident angle is less than 20°. This observation agrees with those reported by Dunn *et al.* [5]. Differences between the *Formica* surface and the aforementioned silicon surface cause a remarkable difference in the coefficient of restitution. These may be explained by a conjunction of factors, such as the substrate and particle surface roughness and material properties. Under the present experimental set-up, only those events having both incident and rebound velocities were observed and analyzed. The observed increase of the coefficient of restitution at low incident angles does not necessarily mean that there was no capture in the experiments. A statistical analysis of *lycopodium* spore oblique impact onto surfaces is presented later and examines some of those effects.

The impulse ratio for these two cases is roughly the same. The surface roughness *of Formica* may be a dominant contributor to friction. If so, the difference of the microparticle surface geometry for a *lycopodium* spore and a SST76 microsphere is not very important, as far as the impulse ratio is concerned.

The normalized kinetic energy loss of the *lycopodium* spores onto *Formica* is higher, especially at higher incident angles. This is not surprising because it has already been seen that the *lycopodium* spores give a lower coefficient of restitution. For high incident angles, $T_L \approx (1 - e^2)$ and the normalized kinetic energy loss for *lycopodium* spores should be higher. For incident angles less than 45°, the difference of T_L between these two cases becomes smaller although the difference in the coefficient of restitution still is significant.

In summary, both particle and surface roughness significantly affect the impact process, the degree of which depends upon the specific materials involved. For low incident angles, both particle and surface roughness lead to apparent increases in the coefficient of restitution. The impulse ratio never achieves a constant value. Normalized translational kinetic energy losses are higher at high incident angles for the case of combined particle and surface roughness.

3. ANALYTICAL MODELS OF THE IMPACT PROCESS WITH ADHESION

The analytical models developed for the process of microparticle impact with a substrate surface are presented in this section. Each of the analytical models is compared with the experimental results just presented.

The problem approached here is to investigate the impact mechanics of a microparticle just before, during and just after it collides with a surface in the presence of adhesion. The investigation specifically includes the effects of molecular attraction in the form of van der Waals force over the contact surface. Provisions were made to include other microforces [17], such as electrostatic and image forces; but this is not covered here. The problem includes determination of the rebound velocities of a microsphere that approaches a surface with arbitrary initial velocities (including angular velocities) and relating the impact process to the physical properties of the materials and to the adhesion force. The conditions under which the particle does not rebound (that is, it *attaches* or is *captured*) are of particular interest.

Two distinct but complementary models for the planar oblique impact of microspheres were developed. Both are derived directly from Newton's laws of motion. The first is an algebraic model based on the principles of rigid body impact and is referred to as the rigid body model. (The term *rigid body* refers not to a lack of flexibility of the microparticle, but rather that it is not treated as a point mass). The second is referred to as a dynamic simulation because it is based on the integration of ordinary differential equations of motion. Actually, there are two versions of the simulation. One is 2-dimensional or planar and the other is 3-dimensional. The two-dimensional simulation is described here; the three-dimensional dynamic simulation is an extension of the same concepts and will be described in a future technical paper. The rigid body model is algebraic and relatively simple. It imbeds the physical properties of the microsphere and surface (substrate) and impact process non-linearities into 3 coefficients. A major difference between the models is that the rigid body model deals only in impulses and changes in momentum, whereas the simulation deals with displacements and forces. The simulation uses Hertzian theory to model the contact mechanics normal to the surface. It introduces a unique approach to modeling of the van der Waals force as a tensile ring force around the dynamic periphery of the compressive Hertzian contact area. Both models are *quasi-static* in the sense that they ignore wave motion and model impact motion as a half-cycle of compressive vibration. Both models include tangential friction effects and can handle oblique impacts. The rigid body frictional model is completely general in the sense that tangential resistance to motion is from an impulse, without specifying the time-varying nature of the frictional force. The frictional force used in the dynamic simulation is more akin to Coulomb-type friction, but its formulation allows some versatility and permits changes as to how the friction force is developed. Both models have a common feature in that they distinguish between energy lost due to material deformation and energy lost in the adhesion process.

There are some basic, underlying assumptions to the dynamic simulation and rigid body model that deserve mention. The form taken by the simulation equations is a direct consequence not only of the assumptions mentioned above that energy dissipation due to adhesion and the particle-substrate materials are independent but, also, that plastic deformation plays an insignificant role in the dynamic contact process. The latter assumption is made due to the extremely short contact durations of microparticle impact and the consequent extremely high strain rates, in the order of 10^5 or higher. Current understanding is that plastic deformation develops through the process of dislocations and cannot develop at rates corresponding to such short time intervals. Moreover, objects whose dimensions are the order of magnitude of the microparticles are known to be less ductile and possess higher strength than like objects of much larger dimensions.

Studies were carried out related to the above models to augment or supplement their use. For example, a set of empirical equations was developed that can be used to represent the observed experimental behavior of the coefficient of restitution of the rigid body model for specific applications (specific materials, initial velocities, *etc.*). A unique feature of these equations is that the capture velocity can be determined from data taken near but not at capture conditions. Another auxiliary study is a Monte Carlo implementation of the rigid body model. This allows analysis of the impact process where one or more of the process parameters and/or initial conditions possess statistical variations. Another independent, but related, study that used the models is a sensitivity analysis of the impact process using methods from the field of design of experiments. All of the above are described and summarized in the following.

3.1. Rigid Body Impact Model

The simpler of the two analytical models is an algebraic model based on Newton's laws in the form of impulse and momentum and follows from some of the work of Brach [15]. The model is remarkably simple yet surprisingly powerful. Figure 13 shows the basic geometry of a sphere in contact with a surface with a normal coordinate, n, positive away from the surface, and a tangential coordinate, t, along the surface. P_n and P_t are the normal and tangential components of the impulse over the full contact duration. A contact couple impulse, M, is illustrated but has been shown to be negligible by Brach *et al.* [16] and is omitted from the model. Notation is such that initial conditions (velocities, angles, *etc.*) are designated by lower case symbols and final conditions by upper case or capitals. A complete derivation of the model equations is given in Brach and Dunn [17]. A summary of the results is presented here. In addition to its simplicity, the model has two distinct applications to the study of microsphere impact. It can be used to analyze and interpret experimental data simply by plotting measured response data against the response predicted by the model. By using experimental data to represent the model's coefficients, the second use is as a predictive model for specific applications.

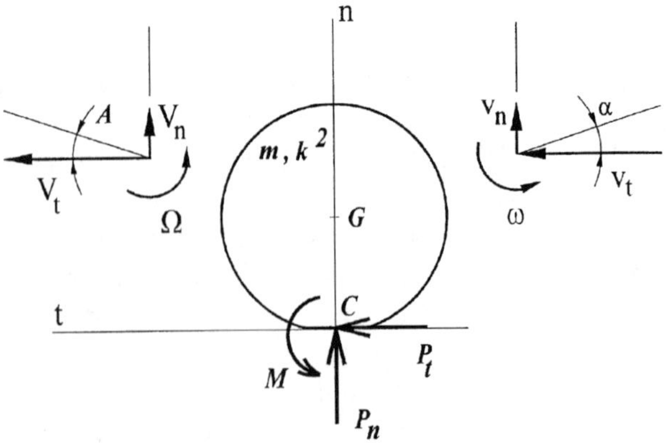

FIGURE 13 Free body diagram, variables and coordinates for the rigid body model of a microsphere impact.

3.1.1. Coefficients

Several coefficients are defined and used in the model. These are a coefficient of restitution in the absence of adhesion, R, an overall coefficient of restitution, e, an impulse ratio coefficient, μ, and an adhesion dissipation coefficient, ρ. The coefficient e is the commonly encountered kinematic coefficient of restitution defined as the ratio of the final to initial normal contact velocities as in Eq. (1). The impulse ratio is defined as the ratio of the tangential and normal components of the contact impulse (and leads to Eq. (2)):

$$\mu = P_t / P_n \tag{4}$$

The adhesion coefficient, ρ, is defined as the negative of the ratio of the (normal) impulse due to adhesion during rebound, P_A^R, to the (normal) impulse generated by deformation during approach, P_D^A, that is,

$$P_A^R = -\rho P_D^A \tag{5}$$

With the assumption that dissipation due to adhesion predominates during rebound, ρ gives a measure of the energy dissipation due to adhesion. The coefficients are not independent; e, R and ρ are related in a way that will be demonstrated in the next sections. The impulse ratio is related to the frictional drag that exists over the contact surface and is discussed shortly.

3.1.2. Solution Equations

The equations that provide the final conditions from known initial conditions are referred to as the solution equations. These are as follows.

$$V_n = -e\, v_n \tag{6}$$

$$V_t = v_t - \mu(1 + e)\, v_n \tag{7}$$

$$\Omega = \omega + 5\mu(1 + e)\, v_n / 2\eta \tag{8}$$

$$P_D = -m(1 + R)\, v_n \tag{9}$$

$$P_A^R = \rho R\, m v_n \tag{10}$$

Ω and ω are the final and initial angular velocities, respectively. An important auxiliary expression is for the tangential contact velocity, V_{Ct}, at point C. If this becomes zero at any time during the impact, the mode of motion of the sphere changes from sliding and rolling to rolling alone without sliding. This velocity is $V_{Ct} = V_t - r\Omega$ and can be expressed as

$$V_{Ct} = v_t - r\omega - 7\mu(1+e)/2\, v_n \tag{11}$$

The impulse ratio, μ_0, just large enough to cause rolling without sliding, or $V_{Ct} = 0$ is

$$\mu_0 = 2\eta/7(1+e) \tag{12}$$

where

$$\eta = (v_t - r\omega)/v_n \tag{13}$$

The impulse ratio represents tangential resistance to motion (frictional drag).

3.1.3. Energy Equations and Coefficients of Restitution, R and e

One of the most important aspects of a collision analysis is the energy loss. An expression for the energy lost during the collision can be found from the solution equations by subtracting the final kinetic energy from the initial using the solution equations to express final velocities in terms of the initial velocities and the coefficients. The energy loss, T_L, normalized to the initial translational kinetic energy, is:

$$2T_L/mv^2 = [(1-e^2) + 2\mu(1+e)\eta - \mu^2(1+e)^2]/(1+\eta^2), \tag{14}$$

Note that the value of impulse ratio, μ, is bounded by the level of friction in general, or friction coefficient when appropriate. For example, for a coefficient of friction f, $|\mu| \leq f$. In general, the work of an impulse is given by,

$$W = \int F dx = \int F(\tau)\dot{x} d\tau = \int \nu(\tau) dp \tag{15}$$

Consider a normal impact. The gain of kinetic energy during rebound is $e^2(1/2mv_n^2)$. Using Eq. (15) the work of the normal impulse during rebound is $1/2R^2(1-\rho)mv_n^2$. Recall that R is the coefficient of restitution in the absence of adhesion. Similarly, the work, W_A, of P_A^R, the rebound adhesion impulse is $-1/2R^2\rho(1-\rho)mv_n^2$. Equating the kinetic energy gain to the work of the total rebound impulse gives the fundamental relationship

$$e^2 = R^2(1-\rho)^2 \tag{16}$$

Figure 14 is a schematic diagram showing the trends of the impact coefficients as the initial normal velocity changes. It shows that as the initial velocity gets smaller, if ρ reaches unity, e becomes 0. This occurs at the capture velocity, $v_n = v_c$, where the particle does not rebound.

For $\rho = 0$, the work of the body deformation impulse and the work (energy dissipated) of the adhesion impulse are R^2 and zero, respectively. For attachment, $\rho = 1$ and the rebound work term is zero. This gives $e^2 = 0$, as expected, but the works of the body and adhesion impulses should be equal and opposite, not zero. A more direct approach to work-energy can be developed. The energy loss in a collision can be written as the energy lost due to material dissipation, $(1-R^2)1/2mv_n^2$, and the energy lost due to the work done by

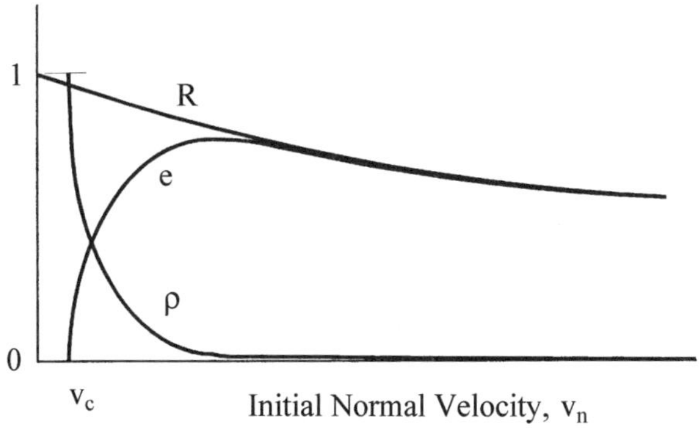

FIGURE 14 Trends of the restitution and adhesion coefficients for a microparticle impact.

adhesion, W_A. The total impact energy loss is:

$$T_L = (1 - e^2)1/2\,mv_n^2 = (1 - R^2)1/2\,mv_n^2 + W_A \tag{17}$$

The term including $1 - R^2$ is the energy lost due to dissipation in the sphere and W_A is the energy loss due to the work done by the adhesion force during rebound. Using Eqs. (16) and (17), the work done by the adhesion impulse (see Eqs. (5) and (10)) can be written as,

$$W_A = mv_n^2 R^2 \rho(2 - \rho)/z \tag{18}$$

For no adhesion, $\rho = 0$ and $W_A = 0$; for attachment (capture) $\rho = 1$ and the work of adhesion is $W_A = R^2\,mv_n^2/z$. Figures 15 and 16 show examples of the form of $e(v_n)$ for microsphere impact and illustrate the effects of adhesion on rebound as just discussed. The data in Figure 15 are direct experimental measurements [19] fit to empirical curves (discussed later) and the data in Figure 16 are from a dynamic impact simulation [25], similarly fit and described in the next section. Both illustrate the trend toward capture.

3.1.4. Impulse Ratio Coefficient, μ

Two distinct effects can be observed through examination of the impulse ratio. One is the effect of initial angular velocities and is most evident at high incident angles, near normal impact. The other is the

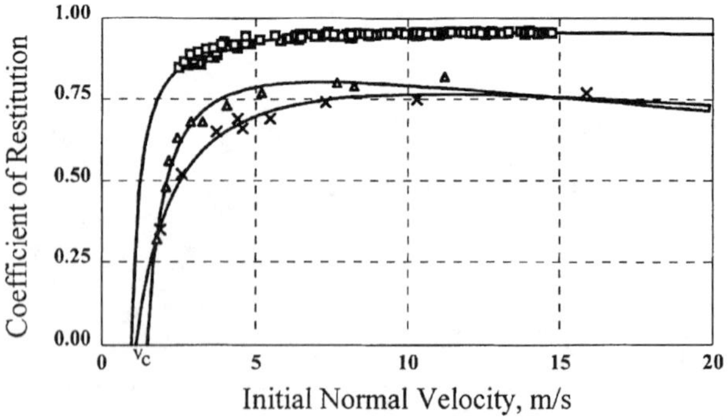

FIGURE 15 Experimental data from normal impacts: polystyrene latex particles on a quartz surface, □ ($v_c = 0.967$) [8]; ammonium fluorescein particles on molybdenum, Δ ($v_c = 1.47$), and mica, x ($v_c = 1.10$), surfaces [3].

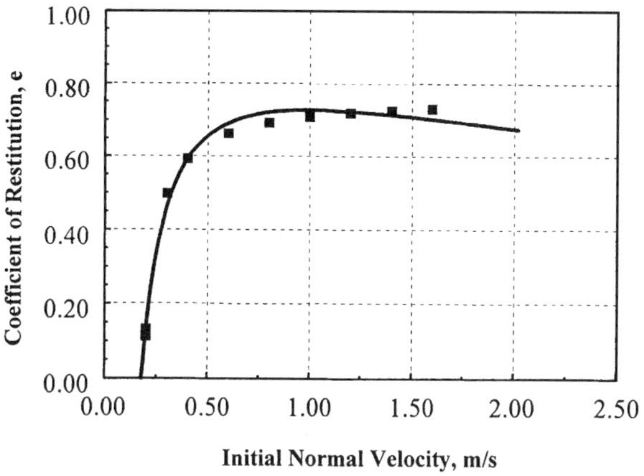

FIGURE 16 Simulation results based on experimental measurements, ■, of stainless steel microspheres impacting a silicon substrate, fit to the empirical equations; capture velocity is 0.18 m/s.

effect of adhesion causing a normal force and consequent frictional force for oblique collisions at very low initial normal velocities (low initial normal momentum).

All of the oblique impact experiments discussed here use the procedure where microspheres are dispensed vertically under gravity toward a target surface at a fixed distance from the dispenser. The surface is given a sequence of different orientation angles to produce different oblique angles of incidence, α. For a constant friction coefficient, f, and when the microspheres are not rotating as they fall ($\omega = 0$), the impulse ratio will follow the corresponding solid curves in Figure 17. For normal incidence, $\alpha = 90°$, and an initial angular velocity of $\omega = 0$, no tangential force and impulse will develop so $P_t = 0$, $\mu = 0$ and the result is the point $(\mu, \alpha) = (0, 90°)$. If $\omega \neq 0$, then $P_n \neq 0$ and then μ at 90° will be above or below the abscissa, depending on the sign of ω. Such effects of initial angular velocities were common; see Figures 6, 7, 11 and 12.

As α gets smaller the initial normal velocity component, the initial normal momentum and the normal impulse, P_n, approach zero and the consequent friction goes to zero. In the presence of adhesion, however, the normal force depends both upon the initial normal momentum and the adhesion force. As the angle of incidence grows small, adhesion pulls the particle toward the surface, creating an area

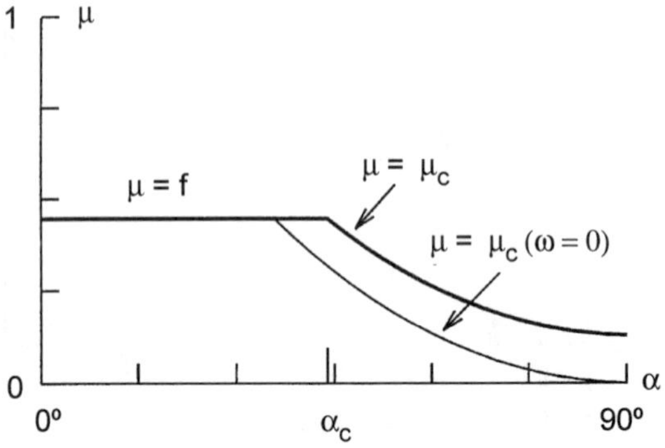

FIGURE 17 Idealized impulse ratio as a function of the angle of incidence for Coulomb friction as predicted by the rigid body model.

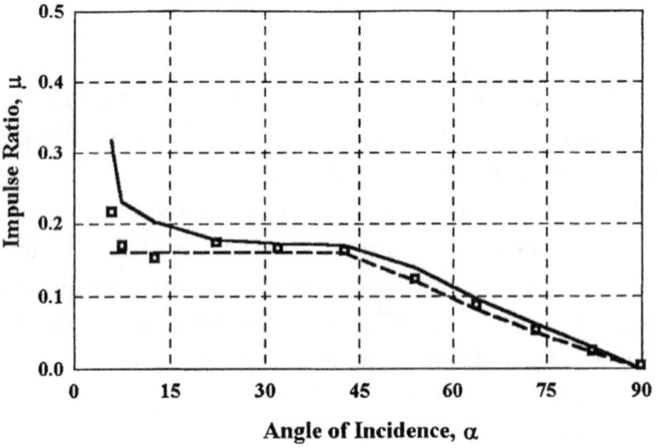

FIGURE 18 Impulse ratio as the angle of incidence changes for oblique collisions of stainless steel microspheres against a Si surface; dashed curve is from the rigid body model and the solid curve is from the simulation.

of normal compressive stress that supports friction. So, in the presence of adhesion, a frictional force and tangential impulse remain even as the initial normal momentum goes to zero. Consequently, the impulse ratio, $\mu = P_t / P_n$, grows as $\alpha \to 0$. This is reflected both in the experimental data and the simulation model and is shown in Figure 18. This upward trend of the impulse ratio has not been reported before. So, the rise in μ as $\alpha \to 0$ and $v_n \to 0$ for $v_t \neq 0$ is attributable to the presence and significance of adhesion.

3.2. Dynamic Simulation Model

A summary now is presented for the 2-dimensional simulation. A full derivation is contained in Brach and Dunn [17]. Figure 19 shows a free body diagram of a microsphere with an undeformed radius, r, whose mass center has elastic coordinates n and t and that rotates with angle θ with angular velocity $\omega = \dot{\theta}$. The normal displacement, n, of the mass center of a sphere is governed by Newton's second law. Combining Hertzian theory and the assumptions discussed above gives:

$$m\ddot{n} = \sqrt{r}K\,n^{3/2} - \sqrt{r}K\,n^{3/2}\,c_H\,\dot{n} - 2\pi a f_0 - 2\pi a f_0\,c_A\,\dot{n} \tag{19}$$

Of the terms on the right hand side, the first is the classical Hertzian restoring force with stiffness parameter K. The second term is a dissipation term corresponding to the Hertzian force and represents dissipation in the materials. The third term represents idealized adhesion attraction as a conservative, circumferential (line) force. The last term adds dissipation due to adhesion. The quantities c_A and c_H are the damping coefficients for the adhesion and Hertzian damping, respectively. The terms in Eq. (19) that model the adhesion force and

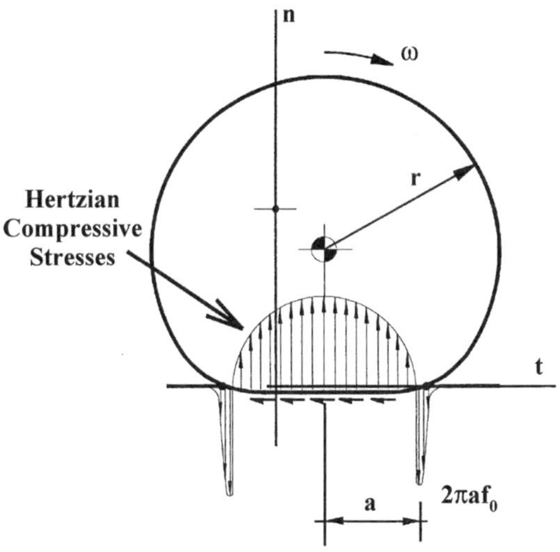

FIGURE 19 Idealized representation of the contact stresses showing the Hertzian compressive stress with a hemispherical distribution and adhesion represented as a ring stress around the periphery of the contact area.

dissipation may seem unorthodox. They specifically are chosen to provide a hysteretic form [17] for the adhesion force as is determined by measurement [23]. The contact radius, a, is related by Hertzian theory such that $a^2 = r\,n$. Hertzian stiffness, K, is given by:

$$K = 4/3\pi(k_1 + k_2)^2 \tag{20}$$

$$k_i = (1 - \nu_i^2)/\pi E_i \tag{21}$$

$$r = R_S\, r_S/(R_S + r_S) \tag{22}$$

Radii are the undeformed radius of the microsphere, R_S, and the local radius of curvature of the surface (substrate), r_S. For a flat surface, $r_S \to \infty$.

The damping terms can be modified by defining nondimensional dissipation parameters. The strength of the Hertzian dissipation rests in the magnitude of the constant c_H. A nondimensional parameter, ζ_H, is defined such that

$$\zeta_H = rc_H/T_H = c_H(R^3 K v_n^{1/2}/m)^{5/2} \tag{23}$$

where T is the period of contact predicted by Hertzian theory. Similarly, a nondimensional dissipation parameter is defined for the adhesion term as

$$\zeta_A = rc_A/T_H = c_A(R^3 K v_n^{1/2}/m)^{5/2} \tag{24}$$

For ζ_A and ζ_H to be constants it is necessary to use an appropriate nominal value of the initial normal velocity, v_n, for normalization; in fact, the capture velocity can be used, if known. Using these dissipation coefficients, the final form of Eq. (19) then becomes

$$\begin{aligned} m\ddot{n} = {} & \sqrt{r}K\,n^{3/2}\left[1 - \zeta_H(m/R^3 K v_n^{1/2})^{5/2}\,\dot{n}\right] \\ & - 2\pi\sqrt{rn}\,f_0\left[1 + \zeta_A(m/R^3 K v_n^{1/2})^{5/2}\,\dot{n}\right] = F_n(\tau) \end{aligned} \tag{25}$$

The model of the adhesion force used here is not derived from basic principles but rather represents an idealization of the force as observed over the past by others such as Johnson and Pollock [19], Fichman and Pnueli [20], Johnson *et al.* [21] and others, and as proposed by Brach and Dunn [17].

The tangential equation of motion is:

$$m\ddot{t} = F_t(\tau) \tag{26}$$

Tangential motion over the contact surface is considered to be either sliding or not sliding, so

$$F_t = -fF_n(\tau)\operatorname{sign}(\dot{t} - R\dot{\theta}), \;\; \dot{t} - R\dot{\theta} \neq 0 \tag{27}$$

where f is a frictional parameter (possibly a function) and

$$F_t = 0, \;\; \dot{t} - R\dot{\theta} = 0 \tag{28}$$

The third and final equation of motion governs angular motion and is:

$$I\ddot{\theta} = RF_t(\tau) \tag{29}$$

where I is the moment of inertia of the sphere.

3.2.1. Estimation of Dynamic Simulation Parameters

The parameters, or constants, that appear in the equations of the simulation model represent physical quantities such as mass and radius. Some of these are well known and are relatively easy to determine such as the Young's modulus, E, the coefficient of friction, f, and Poisson's ratio, ν. Other constants, such as the strength, f_0, of the adhesion ring force and damping, ζ_A, are not as easy to determine and could even be functions of v_n. Some of these are discussed in the following.

Assuming that f_0 remains relatively constant during an impact, its value can be estimated from equilibrium conditions. Using JKR theory, Li *et al.* [5] show that f_0 is related to the Dupré surface energy[4] constant, w_A, Hertzian radius, r, and stiffness, K, by

$$f_0 = (9Kr\,w_A^2/2\pi)^{1/3} \tag{30}$$

[4]The Dupré surface energy, w_A, has units of J/m^2 and actually is an energy density. In accordance with common usage, however, it will be referred to here as energy.

The procedure for determining the values of the two damping coefficients is through the use of experimental data taken at low and high initial normal velocities. For a high initial velocity (when adhesion dissipation is negligible), ζ_A is set to a value of zero and ζ_H is chosen by matching the experimentally-measured coefficient of restitution. With that value of ζ_H a value of ζ_A is then found by again matching the coefficient of restitution, but now at a low initial velocity. By repeating this process, a pair of nominal values of ζ_A and ζ_H is found that matches the coefficient of restitution over the desirable range of initial velocities and for specific materials. There seems to be very little information in the current literature concerning the mathematical nature of dissipation due to adhesion. For all of the simulation results, it is assumed that no significant adhesion dissipation occurs during establishment of adhesion, that is, during approach, so that $\zeta_A = 0$ for $\dot{n} < 0$. The simulation allows this to be changed if desirable.

The Hertzian stiffness, K, Eq. (20), is well defined for classical contact problems without adhesion. However, for the combined loading of inertial compression and adhesion, Hertzian theory must be augmented. The presence of adhesion makes a sphere deform somewhat more than in the absence of adhesion. This has been investigated by Brach *et al.* [22] and resulted in a reduced stiffness given by

$$K_R = K - \frac{5}{2}\left(\frac{5}{9}\right)^{5/2} \frac{w_A \pi R^2}{a_e^3} \tag{31}$$

where a_e is the static equilibrium value of the contact radius. The reduced stiffness should be used in place of K in applications of Eqs. (19) and (25).

3.2.2. Typical Results of the Dynamic Simulation

Some typical results are shown in Figure 20, taken from Li *et al.* [5]. This shows the normal contact force as it varies with time. These results are for the parameters of ammonium fluorescein spheres with a molybdenum substrate [3]. Compressive forces are shown as positive

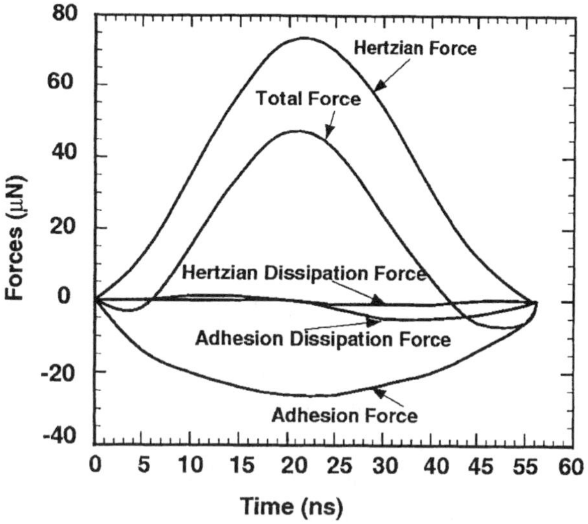

FIGURE 20 Typical simulation results showing the variations with time of the normal contact force. The case matches experimental results [3] for an ammonium fluorescein microsphere of diameter $d = 4.9\ \mu m$ impacting a molybdenum surface at 5.2 m/s. A value of $f_0 = 8.90$ N/m is used.

and the presence of adhesion is reflected by the negative dips to the total force near the beginning and end of contact. It is interesting to note that, for the conditions represented here, the dissipation forces are relatively small compared with the Hertzian and adhesion force.

Comparison of simulation results with experimental data are, of course, a major priority. This is not as easy at it sounds, however. The main reason is that while the simulation is capable of producing displacements, velocities, forces, *etc*., as functions of time during contact, direct measurements of these quantities are virtually impossible for microparticles with present-day instrumentation and techniques. Only initial and final mass center velocities are measurable. This means that when parameters such as mass and radius are changed in the simulation, only their *effects* on rebound velocities and energy loss from observable data can be compared. Angular rotation and angular velocity of a microsphere, likewise, are impossible to measure. Such rotational velocities generally will exist due to the presence of frictional drag over the contact surfaces and may exist initially due to prior impacts and during dispensing. Angular velocities and their changes are impossible to measure not only during contact but both before and after. Yet it is possible to assess the influence of angular velocities. For example, Dunn *et al.* [5] measured the approach and rebound velocities of stainless steel microspheres impacting a ultra flat silicon surface. The points in Figure 21 show kinetic energy loss based on the measured mass center velocities before and after impact. The two curves in the figure are from a corresponding simulation where the solid curve shows the true energy loss and the dashed curve shows the results of the dynamic simulation where the rotational kinetic energy has been ignored. Since the experimental data more closely match the dashed curve, it can be concluded that the rotational kinetic energy for microsphere impact can be significant. In fact, for an incident angle near 50° the final angular velocity from the simulation is about 2.3×10^4 rad/s. This should not be surprising since the final angular velocity, $\Omega = V_t/r$, where r is a radius of the order of 10×10^{-6} m. Another aspect of interest is the contact durations. Simulations of stainless steel microspheres and a silicon surface produce contact durations ranging from 300 to 700 ns. Based on half of this duration for approach and half for rebound, this implies strain rates as high as 10^5 to 10^6 s.

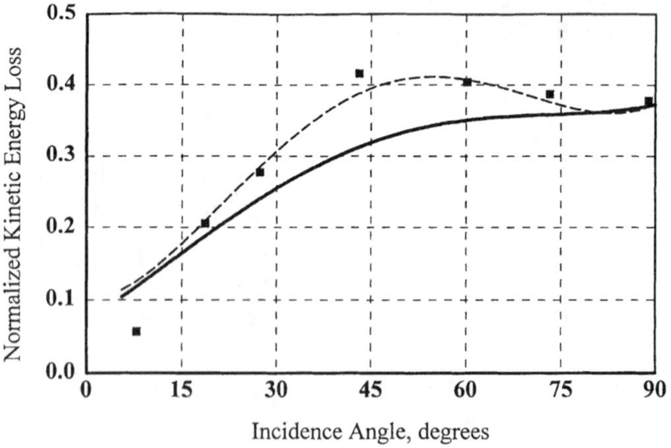

FIGURE 21 Points are experimental values of normalized kinetic energy loss for stainless steel microsphere impacts on a Si surface; solid and dashed curves are from the simulation where the dashed curved ignores energy due to the final angular velocity.

3.2.3. *Applications of the Dynamic Simulation, Rolling Dissipation*

One of the applications of the microsphere impact simulation was a study of the effects of rotational dissipation during contact. When rotation (rolling) of a microsphere takes place during contact and in the presence of adhesion, the leading edge of the contact area is continually establishing new contact with the surface and the trailing edge likewise is breaking contact (peeling). Since the adhesion process is not reversible, energy is lost. Consequently, rolling[5] in the presence of adhesion should cause a dissipation couple or moment over the contact area (see M in Fig. 13). A study was carried out by Brach *et al.* [16] that used the simulation and an independent mathematical analyses of the rolling contact problem to assess the effect of rotational dissipation. Their results show that the magnitude of rolling deformation and adhesion bond peeling couples are direct functions of the contact radius. Because the contact radius generally is relatively small, the effect of rolling dissipation can be neglected. This is why the moment impulse, M, in Figure 13 was neglected.

3.3. Empirical Equations

The rigid body model is based on the impulse and momentum equations of Newton's laws. The normal rebound of a microsphere is modeled simply as $V_n = -e\,v_n$, where the coefficient of restitution is treated as a constant in the collision problem. In actuality, e is a physical quantity that depends nonlinearly on the initial velocity, that is, $e = e(v_n)$. The other constants, R and ρ, also play a role as defined earlier. Actual behavior for specific materials, initial velocity ranges, *etc.*, is reflected in the model through $e(v_n)$. A series of empirical equations was suggested by Dunn *et al.* [4] and developed by Brach and Dunn [18] in which the constants are determined from data (experimental and/or analytical) through the use of fitting procedures.

---

[5]Note that rolling can occur simultaneously with sliding or in the absence of sliding. The latter is referred to as *pure rolling*.

A feature of these equations is that they can be used directly with the rigid body model and essentially tailor the model to specific applications. Another, and very important feature, is that they allow determination of the capture velocity using impact data from experiments where capture never actually occurs. In fact, it is difficult to measure directly the capture velocities since they represents a limiting behavior. When capture does occur, particles often collect on the surface and can hinder impact measurements.

The most convenient and effective form of the empirical equations is:

$$R = \frac{k_1^p}{k_2^p + |v_n|^p} \tag{32}$$

and

$$\rho = \frac{\kappa_1^q}{\kappa_2^q + |v_n - v_c|^q} \tag{33}$$

where the constants k, κ, p, q and v_c are determined from experimental data. All of these quantities have a functional dependence on the parameters of the impact and adhesion processes such as particle size, material properties, *etc*. Note that the k's and κ's have units of velocity.[6] For the simple case of $k_1 = k_2 = k$ and $\kappa_1 = \kappa_2 = \kappa$ and recognizing that $e = R(1 - \rho)$ from rigid body theory (see Eq. (16)), the overall coefficient of restitution becomes:

$$e = e(v_n) = \left[\frac{k^p}{k^p + |v_n|^p}\right]\left[1 - \frac{\kappa^q}{\kappa^q + |v_n - v_c|^q}\right] \tag{34}$$

These equations have been applied above, such as in Figures 15 and 16 in fitting of experimental and simulation data and for the determination of the capture velocity. Fitting of the equations and determination of the constants can be done in many ways, such as by least squares. Most data analysis software have such routines.

Combined with the empirical equations $e(R, \rho, v_n, v_c)$, the rigid body model is a versatile algebraic model of the impact process in

[6]Variations of these equations are possible and the equations can also be defined in terms of nondimensional velocities, v_n/k and/or v_n/κ.

the presence of adhesion. It first is used to study the effects of changes in the incident angle in order to study surface roughness. It then is used in a Monte Carlo analysis where the parameters are given statistical distributions. Examples of this are now discussed.

4. ANALYSES AND APPLICATIONS

4.1. Surface Roughness Effects

Consider a microsphere with a nominal angle of incidence, α, striking a surface with some waviness as shown in Figure 22. The microsphere strikes the surface at a point where the true normal is n' and the true angle is $\alpha+\phi$. The impact parameters, e and μ, characterize the mechanical processes normal and tangential to the true surface. However, their experimentally *measured* values are computed using velocity components measured relative to the nominal surface, that is, $e_m = -V_n / v_n$ and $\mu_m = (V_t - v_t) / (V_n - v_n)$. Expressions for the nominal velocity components, V_n, V_t, v_n and v_t, can be expressed in terms of the true velocity components, V_n', V_t', v_n' and v_t', using transformation equations. If this is done, the above two equations become, respectively,

$$e_m = \frac{e \cos\phi - V_t' \sin\phi / v_n'}{\cos\phi - v_t' \sin\phi / v_n'} \tag{35}$$

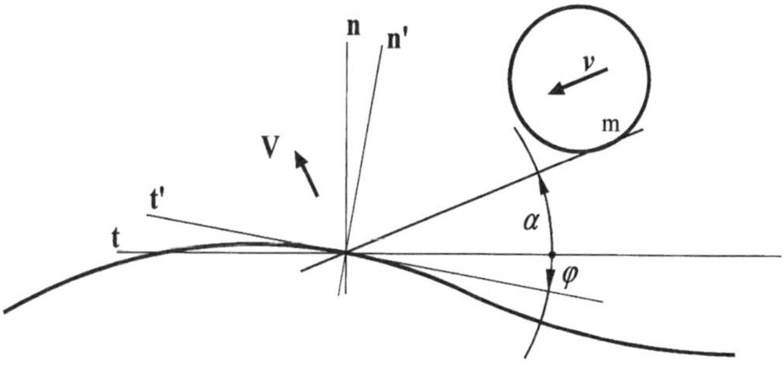

FIGURE 22 Apparent incidence angle, α, and normal coordinate, n, compared with true angle, $\alpha+\phi$, and normal coordinate, n', for variations in surface flatness.

and

$$\mu_m = \frac{\mu - \tan\phi}{1 + \mu\tan\phi} \tag{36}$$

where the subscript m denotes the measured values.

4.1.1. Shallow Angles of Incidence, $\alpha \approx 0$

It was noted earlier that for rough surfaces and shallow angles of incidence measured values of e exceed 1 (see Figs. 6, 7 and 12, for example). This can be explained using Eq. (35). Consider the case where both angles α and ϕ are small, such that $\cos\phi \approx 1$ and $\sin\phi \approx \phi$. Additionally, for a relatively low friction coefficient, f, sliding usually continues throughout the contact duration and $f\alpha \ll 1$. Under these conditions,

$$e_m \approx \left[1 + \frac{\phi}{\alpha}\right]\left[e + \frac{\phi}{\phi + \alpha}\right] \tag{37}$$

and

$$\mu_m \approx \mu - \phi, \quad \mu = \pm f \tag{38}$$

where $\tan\phi \approx \phi$ and $\mu\phi \ll 1$. The effect of waviness with a positive slope is to give a higher coefficient of restitution and give a negative bias to the impulse ratio. Note that for $\phi \approx \alpha$, e_m can exceed $2e$. So, although conservation of energy requires that $0 \le e \le 1$, measured values can exceed 1, or even 2.

4.1.2. Normal Incidence, $\alpha = \pi/2$

Normal incidence implies that $\alpha = \pi/2$ and $v_t = 0$. For normal and near normal impacts, rebound is under the condition of pure rolling, that is, $\mu = \mu_0$. However, if the microsphere has a high initial spin, or with any initial spin and low friction, it still will be sliding at separation. For a waviness with a small slope and the condition of rolling, the experimental coefficient of restitution becomes:

$$e_m \approx e - \phi^2 + (1+e)\,\mu_c\,\phi \tag{39}$$

The critical impulse ratio is:

$$\mu_0 = \frac{2}{7(1+e)}\frac{v_t' - r\omega}{v_n'} \tag{40}$$

When $\omega = 0$, $(1+e)\,\mu_c \approx -2\phi/7$ and $e_m \approx e - 9\phi^2/7$. When $\omega \neq 0$, a bias with the sign of ω can exist. For rolling at separation, the expression for the experimental value, μ_m, of the impulse ratio for normal incidence is

$$\mu_m \approx -\phi\left(1 - \frac{2/7}{1+e}\right) + \left(\frac{2/7}{1+e}\right)\frac{r\omega}{v_n'} \tag{41}$$

The angular velocity adds the potential for a bias to μ_m; otherwise, with $\omega = 0$, the measured impulse ratio is directly related to the slope, ϕ.

4.2. Monte Carlo Analysis

A computer program has been developed that allows various statistical distributions in arbitrary combinations to be assigned to process parameters. Written in the Microsoft Quick Basic language, the program permits a choice for each process parameter to be assigned a uniform, normal (Gaussian), standardized normal, lognormal or user-specified distribution. Random variables are generated through the use of the random number generator supplied with the Quick Basic language and the different distributions are generated through the use of the central limit theorem.

This program was applied to model the results of microsphere (SST76) impact with a rough surface (*Formica*). The actual microsphere diameter distribution (found to be normal) and the incident velocity distribution, both determined from direct measurements using the PDPA system, were inputs to the code. In addition, the empirically-determined local surface angle distribution (the distribution of ϕ) was another input. For each Monte Carlo calculation, a randomly-selected value of ϕ was added to the incident angle, α, to yield the true surface angle, $\alpha+\phi$. Because there was no equivalent molecularly-smooth *Formica* surface available for comparison, it was assumed that such a surface behaved similarly to the molecularly-smooth silicon case. This required that the coefficient of restitution values needed to be adjusted to compensate for initial differences due to the different material properties of *Formica* and silicon. This was accomplished by simply subtracting from the coefficient of restitution values for each of the two cases the respective values at 85°; this is referred to as a relative coefficient of restitution.

The results of the Monte Carlo calculations are presented in Figure 23. The dashed line in the figure is a fit to the base case data

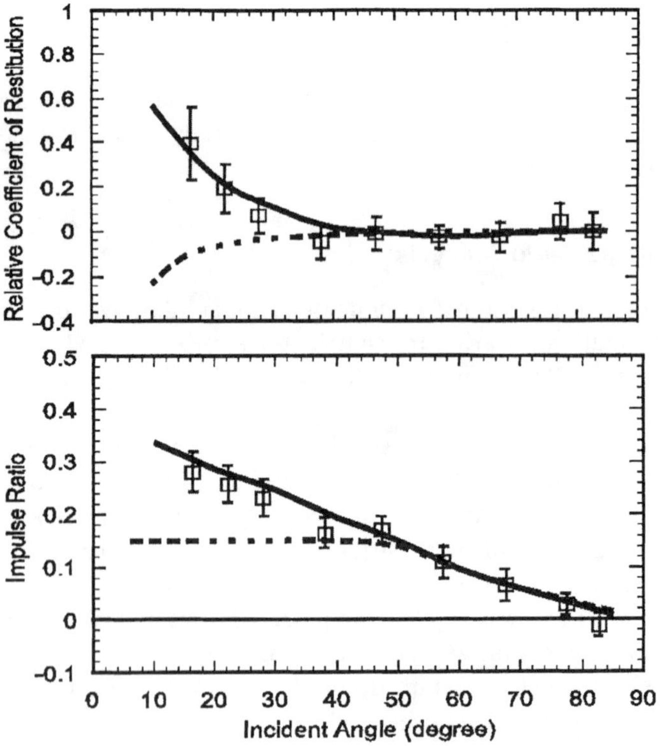

FIGURE 23 Monte Carlo method used with the rigid body model for a random variation in surface angle and particle diameter (solid curves) compared with experiments □ (with error bars) and deterministic rigid body results (dashed curves).

corresponding to the molecularly-smooth surface. The data are presented in the figure as open squares (see also Fig. 12) and the Monte Carlo results as solid lines. It is seen that the calculations follow the data very well. From approximately 90° down to 45°, the molecularly-smooth and rough surface cases are similar. Below 45°, they differ. As the incident angle is decreased, the relative coefficient of restitution for the rough surface increases toward unity, whereas, for the molecularly-smooth surface it decreases. Values of the impulse ratio for the rough case never become constant with decreasing incident angle, whereas, for the molecularly-smooth case, they become constant and equal to ≈ 0.15.

Thus, by considering the microparticle diameter and substrate local-incident-angle distributions, the Monte Carlo method can successfully predict the impact results of rough surfaces. This shows that the departure of experimental data from the ideal rigid body model can be attributed to surface roughness.

4.3. Sensitivity Analysis

The sensitivity of the impact process to variations in process parameters can be assessed using Monte Carlo techniques. However, other methods also are available. In particular, it is possible to use the methods of the design of experiments, DOE (see, for example, Guttman *et al.* [24]) to assess and rank which process parameters are the most significant. In particular, a study by Brach *et al.* [25] was carried out to determine which parameters are the most critical in influencing the capture velocity. The procedure used in the study contains the following steps:

1. define the dependent variable of interest, called the *response*, here the capture velocity,
2. define the parameters of interest, called *factors*,
3. establish nominal (typical) values of each factor as well as upper and lower *levels* of each factor,
4. determine the capture velocity for combinations of factors,
5. calculate the effect that each factor and factor interaction has on the response using DOE methods,
6. assess and rank the significance of the factors and their interactions.

The five factors chosen for this study are lettered A – E; they are: A, the Hertzian stiffness, K; B, the Dupré surface energy, w_A; C, microsphere radius, r; D, the damping constant, ζ_A, associated with adhesion energy dissipation and E, the damping constant, ζ_H, associated with material energy dissipation. Nominal values of the factors were chosen to correspond to stainless steel microspheres and the ultrasmooth silicon surface. These were chosen because of the availability of experimental data for these materials. The DOE method requires that each factor be assigned low and high values, (−/+). This was done in two different ways, resulting in two sets of results of the sensitivity study. The first set of upper and lower levels was treated as a process in estimating a realistic uncertainty in determination of the factors. The second procedure was simply to take ± 5% of the nominal values for the upper and lower levels. Values of the capture velocity did not have to be determined for all possible

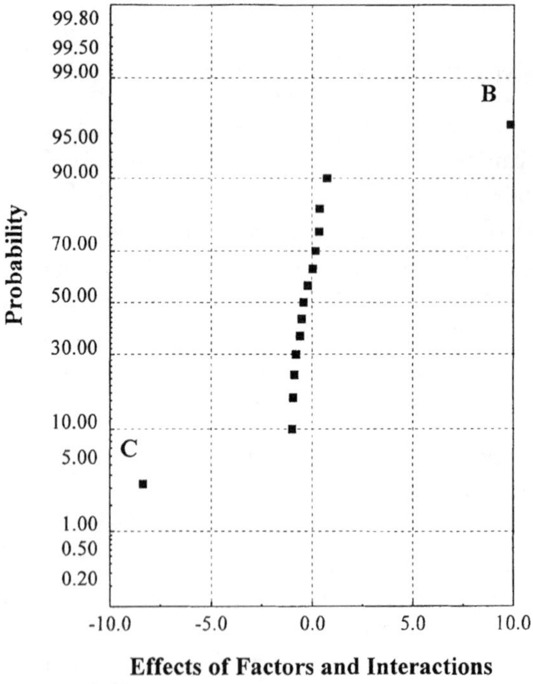

FIGURE 24 Effects of factors and factor interactions for *realistic* variations.

combinations of factor levels. This is possible through the use of a fractional factorial design.

The collision responses were calculated using the dynamic simulation model and the capture velocity values were found using the empirical fit equations. Figure 24 shows the results in the form of a normal probability plot of the effects and interactions from the analysis using levels based on factor uncertainty. It is quite clear that factors C, microsphere radius, and B, Dupré surface energy, stand out. This implies that these two variables control the capture process. The effect of radius is negative, indicating that the larger the radius, the lesser a tendency for capture conditions to exist and *vice versa*. The surface energy effect is positive, so the greater the Dupré surface energy, the more likely for capture to occur. The second part of the sensitivity study used *uniform* upper and lower levels and produced effects shown in Figure 25. Here, the influence of other factors begins

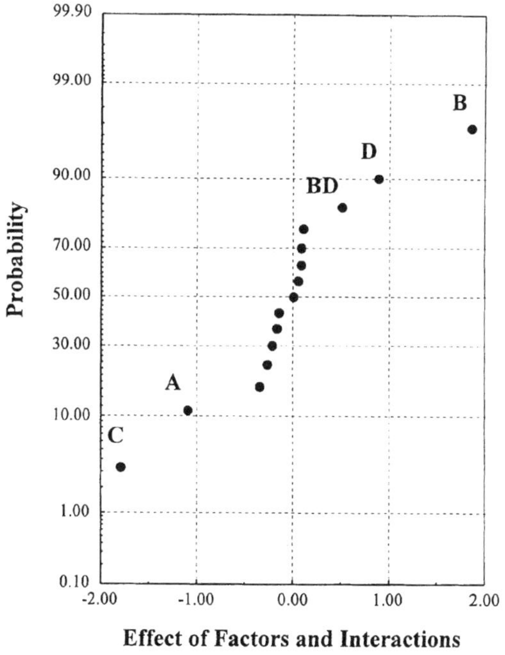

FIGURE 25 Effects of factors and factor interactions for *uniform* variations.

to appear. Again, factors C and B have the greatest influence, but factors A, Hertzian stiffness, and D, adhesion damping, show significance. The BD interaction, which is an interaction between Dupré surface energy and adhesion damping, also shows significance.

By and large, the results indicate that small particles with high adhesion energy are the most likely to be captured. Such a conclusion may seem to be rather obvious. Yet it does indicate that the other quantities such as the mechanical stiffness and material dissipation play secondary roles. Furthermore, such a conclusion also points out that the models being used (to generate the response values for the sensitivity analysis) agree with the intuition built up from observation of behavior of the impact-adhesion process. These results also agree with the observation made earlier of the relatively small magnitude of the dissipation forces from the simulation (see Fig. 20).

5. DISCUSSION

This section presents a discussion some of the more significant experimental and analytical results covered in this paper.

6. EXPERIMENTS

The experimental results reported herein are based on normal and oblique impacts of either microspheres and microparticles with either molecularly-smooth and rough surfaces. The experiments conducted using microspheres and molecularly-smooth surfaces served as an "ideal" base case with which other results were compared. All experiments were conducted in a vacuum chamber (see Fig. 1) and with charge-free surfaces and particles. Ideal, base-case conditions were chosen so that the van der Waals force acting at the contact surface was the only additional force due to the smaller size of the particle.

Normal impact experiments confirmed what others have found, that, as initial velocities decrease, there is a rapid decrease in the coefficient of restitution due to adhesion effects ending in capture at low velocities (for example, see Figs. 15 and 16). Capture velocities decreased as particle size increased (see Fig. 4) because the effect of van der Waals force diminishes with increasing particle size.

Oblique impact experiments show that additional complexities occur when both friction and microsphere rotation are present. In the case of normal impacts, capture would usually manifest itself by a buildup of particles on the target surface. This did not happen for oblique impacts for two reasons. The first is because captured particles can roll away from the (tilted) target zone. But, perhaps more importantly, the conditions of capture for oblique impact were found to be different from that for normal impacts. Particles that were captured with a certain initial velocity during normal impacts were not captured with the same normal velocity (component) during oblique impacts. In fact, capture never was observed for oblique impacts (although this may have been because experimental initial velocities never were low enough). This was despite the fact that for oblique collisions friction combines with the van der Waals force to increase the impact energy loss. The presence of friction, however, does more than reduce the tangential velocity; it also causes a microsphere to take on a significant change in angular velocity. In fact, the final rotational velocities were determined to reach the order of 10^5 rad/s; kinetic energy associated with the (final) rotational velocity of the microspheres was found by modeling to be a significant portion of the overall final energy (see Fig. 21).

Collisions at small, or shallow, incident angles were found to be sensitive to surface roughness. When the approach angle was of the order of about 10° or less, surface roughness (variations in local surface slope) caused confounding of results. Under these conditions, coefficients of restitution calculated from experimental normal velocity measurements are significantly affected by differences between the actual, local normal direction and the target's nominal normal direction. Using the nominal instead of the actual normal angles in computation gave values of the coefficient of restitution greater than one (see Figs. 6 and 7, for example), a physical impossibility. This phenomenon was modeled analytically, verifying that roughness can lead to anomalous coefficient values if the actual normal surface direction is not used.

The effects of friction on impact were not measured directly because, depending on initial conditions (including initial angular velocities), various combinations of sliding and rolling occur throughout the contact duration. Tangential effects were measured using the ratio of the change in tangential momentum to the change in normal momentum. This equals the ratio of the tangential to normal impulses, or the impulse ratio. Under a wide range of conditions, good agreement between the experimental impulse ratios and their theoretical counterparts from Coulomb's law was obtained (see Figs. 6b and 9a). This was not always the case, however (see Figs. 9b and 9c), especially when both rough surfaces and nonspherical particles were used.

7. MODELING

A significant departure was made in this research with respect to conventional assumptions of how energy is lost in the contact process. Many prior studies and models attribute the large majority of energy loss during microparticle impact to plastic deformation of the materials. Here, the view is taken that plastic deformation during impact is minimal, but that a significant amount of energy is lost due to the hysteretic nature of the adhesion process. The drop in the coefficient of restitution (as the initial velocity decreases) and capture, when it occurs (see Figs. 15 and 16), is attributed primarily to the effects of adhesion, not to plastic deformation.

Two analytical models of the impact process have been derived and applied extensively. One is a rigid body model (so named because it uses *rigid body* dynamics, not *point mass* dynamics). It is based on the principles of impulse and momentum and uses coefficients such as the coefficient of restitution, the impulse ratio and the adhesion coefficient. A fundamental assumption of the model is that the majority of contact energy loss (due to adhesion) occurs during rebound as opposed to approach. This allows the definition of an adhesion coefficient. In conjunction with the rigid body model, a set of empirical equations was developed that models the behavior of the impact coefficients as the initial velocity changes. The empirical equations are devised in such a way that they automatically determine a unique value of the capture velocity without direct measurements of the capture process itself (see Fig. 15).

A second analytical model simulates elastic behavior in the contact region using Hertzian theory and a new way of modeling the adhesion force, namely through the use of an attraction force with a ring geometry around the periphery of the contact region. Two distinct sources of dissipation are included, one related to strain and strain rate in the materials and the other due to adhesion dissipation. With the exception of the coefficients of these dissipation forces, all of the coefficients of the differential equation terms are determined directly from physical properties, including the adhesion force.

The dissipation coefficients were determined from experimental data for specific materials; this is because an analytical model for adhesion dissipation does not exist. For the sphere sizes and materials corresponding to the experiments, simulations showed that the peak adhesion force was approximately 1/3 of the peak Hertzian force and the corresponding peak adhesion dissipation force was about 1/5 of the peak adhesion force (see Fig. 20).

One of the primary uses of the rigid body model was to allow a comparison of the experimental values of the impact coefficients with impact behavior in the absence of adhesion. For example, the coefficient of restitution in the absence of adhesion typically approaches unity as the initial normal velocity approaches zero. In the presence of adhesion, however, the restitution coefficient approaches zero and the coefficient of adhesion approaches unity at the capture velocity. In general, for oblique impact and Coulomb's law of friction, the impulse ratio has a certain behavior as the angle of incidence changes (see Figs. 6 and 9); with some notable exceptions, this behavior was found applicable to microsphere impacts. An exception was for rough surfaces (see Fig. 12b) and some anomalous results for smooth surfaces (see Fig. 9). A Monte Carlo simulation, based on the rigid body model and using statistical distributions for the particle diameters, initial velocity and (rough) surface angles showed excellent agreement with corresponding data (see Fig. 23).

Finally, the simulation model was used to determine which physical parameters have the greatest influence on the capture velocity. Among the various physical parameters (such as Hertzian stiffness, adhesion dissipation, *etc.*) that influence the contact impact process, the microsphere radius and the Dupré surface energy were found to have the greatest influence on capture.

NOTATION

a	contact radius, m
e	impact coefficient; coefficient of restitution, velocities normal to the surface, rigid body model
c_A, c_H	coefficients of adhesion and Hertzian damping terms, simulation model, s/m
E	Young's modulus (modulus of elasticity), N/m^2
F	force, N
f	Coulomb friction coefficient
f_0	magnitude of adhesion ring force, simulation model, N/m
I	moment of inertia of sphere, m^{2}kg
K	Hertzian stiffness, N/m^2
k_i	empirical constant, empirical equations, m/s
K_R	reduced stiffness, N/m^2
m	mass, kg
P	Impulse, N-s
p	exponent, empirical equations
q	exponent, empirical equations
R	coefficient of restitution in the absence of adhesion, rigid body model
R_S	undeformed radius of microsphere, m
r	Hertzian radius of contacting bodies, m
r_S	radius of surface (substrate), m
T_H	period of Hertzian impact contact, s
T	kinetic energy, J
T_L	kinetic energy loss (initial minus final), J
V, v	final and initial velocity, respectively, rigid body impact model, m/s
w	Dupré surface energy constant (also, specific work of adhesion), J/m^2
W_A	work of the surface adhesion force (and impulse), J
$x, \dot{x}$	displacement, m, velocity, m/s
α	angle of incidence
η	ratio of initial tangential to normal contact point velocity, rigid body model
ϕ	angle
κ	empirical constant, empirical equations, m/s

μ	impact coefficient; ratio of normal and tangential impulse components, rigid body model
μ_0	critical impulse ratio; value of μ for zero final relative tangential contact velocity, rigid body model
ν	Poisson's ratio
Ω, ω	final and initial angular velocity, respectively, rigid body model
ρ	impact coefficient; ratio of adhesion impulse during rebound and elastic (deformation) impulse during approach
τ	time, simulation model, s
ζ_A, ζ_H	nondimensional coefficients of adhesion and Hertzian damping terms, simulation model

Subscripts

c	capture, empirical equations
m	measured, surface roughness analysis
n, t	normal, tangential components
e	equilibrium

Superscripts

A	approach phase of an impact, rigid body model
D	deformation
R	rebound phase of an impact, rigid body model

References

[1] Dahneke, B., "The Capture of Aerosol Particles by Surfaces", *J. Colloid Interface Sci.* **37**, 342–353 (1971).

[2] Wang, H. C. and John, W., "Dynamic Adhesion of Particles Impacting a Cylinder", In: *Particles on Surfaces, 1: Detection, Adhesion and Removal*, Mittal, K. L. Ed. (Plenum, New York, 1988), pp. 211–224.

[3] Wall, S., John, W., Wang, H.-C. and Goren, S. L., "Measurement of Kinetic Energy Loss for Particles Impact Surfaces", *Aerosol Science and Technology* **12**, 926–946 (1990).

[4] Dunn, P. F., Brach, R. M. and Caylor, M. J., "Experiments on the Low Velocity Impact of Microspheres with Planar Surfaces", *Aerosol Science and Technology* **23**, 80–95 (1995).

[5] Li, X., Dunn, P. F. and Brach, R. M., "Experimental and Numerical Studies on Microsphere Normal Impact with Surfaces", *J. Aerosol Sci.* **30**, 439–449 (1999).

[6] Caylor, M. J., "The Impact of Electrically Charged Microspheres with Planar Surfaces Under Vacuum Conditions", *Ph.D. Dissertation*, University of Notre Dame (1993).
[7] Dahneke, B., "Measurements of the Bouncing of Small Latex Spheres", *J. Colloid Interface Sci.* **45**, 584–590 (1973).
[8] Dahneke, B., "Further Measurements of the Bouncing of Small Latex Spheres", *J. Colloid Interface Sci.* **51**, 58–65 (1975).
[9] Paw U, K. T., "The Rebound of Particles from Natural Surfaces", *J. Colloid Interface Sci.* **93**, 442–452 (1983).
[10] Rogers, L. N. and Reed, J., "The Adhesion of Particles Undergoing an Elastic–Plastic Impact with a Surface", *J. Phys. D: Appl. Phys.* **17**, 677–689 (1984).
[11] Broom, G. P.,"Adhesion of Particles in Fibrous Air Filters", *Filtration and Separation* **16**, 661–669 (1979).
[12] Aylor, D. E. and Ferrandino, F. J., "Rebound of Pollen and Spores during Deposition on Cylinders by Inertial Impaction", *Atmos. Environ.* **19**, 803–806 (1985).
[13] Buttle, D. J., Martin, S. R. and Scruby, C. B., *Harwell Laboratory Report AERE-R13711*, Oxfordshire, UK (1989), pp. 1–30.
[14] Dunn, P. F., Brach, R. M. and Janson, G. G., "Surface Contact Mechanics during Oblique Impact of Microspheres with Planar Surfaces", *Aerosol Science and Technology* **25**, 445–465 (1996).
[15] Brach, R. M., *Mechanical Impact Dynamics* John Wiley, New York (1991).
[16] Brach, R. M., Dunn, P. F. and Cheng, W., "Rotational Dissipation During Microsphere Impact", *Aerosol Science* **10**, 1321–1329 (1999).
[17] Brach, R. M. and Dunn, P. F., "Macrodynamics of Microparticles", *Aerosol Science and Technology* **23**, 51–71 (1995).
[18] Brach, R. M. and Dunn, P. F., "Models for Rebound and Capture in Oblique Microparticle Impacts", *Aerosol Science and Technology* **29**, 379–388 (1998).
[19] Johnson, K. L. and Pollack, H. M., "The Role of Adhesion in the Impact of Elastic Spheres", *Midwest Mechanics Seminar Series*, University of Notre Dame (1993).
[20] Fichman, M. and Pnueli, D., "Sufficient Conditions for Small Particles to Hold Together Because of Adhesion Forces", *J. Applied Mech.* **52**, 105–108 (1985).
[21] Johnson, K. L., Kendall, K. and Roberts, A. D., "Surface energy and the contact of elastic solids", *Proc. R. Soc. Lond. A* **324**, 301–313 (1971).
[22] Brach, R. M., Li, X. and Dunn, P. F., "An Attachment Theory for Microsphere Adhesion", *J. Adhesion* **69**, 181–200 (1999).
[23] Horn, R. G., Israelachvili, J. N. and Pribac, F., "Measurement of the Deformation and Adhesion of Solids in Contact", *J. Colloid and Interface Sci.* **115**, 492 (1987).
[24] Guttman, I., Wilks, S. and Hunter, J., *Introductory Engineering Statistics*, 3rd edn. (John Wiley, New York, 1982).
[25] Brach, R. M., Li, X. and Dunn, P. F., "Parameter Sensitivity in Microsphere Impact and Capture", accepted for publication, *Aerosol Science and Technology* (1999).

The Adhesion of Irregularly-shaped 8 μm Diameter Particles to Substrates: The Contributions of Electrostatic and van der Waals Interactions

D. S. RIMAI

NexPress Solutions LLC., Rochester, NY 14653-6402, USA

D. J. QUESNEL

University of Rochester, Rochester, NY 14627-0132, USA

R. REIFENBERGER

Purdue University, West Lafayette, IN 47907, USA

The forces needed to remove irregularly-shaped, 8 μm diameter, polyester particles from a polyester substrate were measured using an ultracentrifuge. Measurements were also made on a second set of similar particles where nanometer-size silica clusters had been placed on their surfaces. These silica clusters acted as spacers, reducing direct contact between the particle and the substrate. It was found that the separation forces for the bare particles were consistent with predictions of the JKR theory of adhesion, but were much larger than could be accounted for from simple electrostatic interactions associated with either uniformly-charged particles or particles with localized charged patches. It was found, however, that the forces needed to effect separation decreased with increasing silica concentration. For particles with 2% by weight silica clusters on their surfaces, the separation force was only about 5% of the separation forces of the bare particles. At this concentration of silica, the estimates of the separation forces obtained from JKR theory, from the uniformly-charged model, and from the localized-charged-patch model are all about equal. The numerical estimates are consistent with the experimentally-obtained values.

INTRODUCTION

The nature of the interactions controlling the adhesion of irregularly-shaped, dry particles to substrates has been the subject of interest for many years [1,2]. Early work by Derjaguin [3]; Bradley [4,5] and Hamaker [6] suggest that the force needed to separate a particle from a substrate is determined by surface energy considerations that are, in turn, related to the density of states of the molecules comprising the contacting materials. Subsequently, Lifshitz [7] argued that the surface forces are due to van der Waals interactions, which arise principally from the correlation of instantaneous dipole fluctuations occurring within the contacting materials [8].

Following the work of Krupp [8], numerous theories of particle adhesion were advanced. Among these, the model proposed by Johnson, Kendall and Roberts, hereafter referred to as the JKR theory [9], seems to have gained a predominance in the literature over the others.

According to the JKR theory, for a spherical particle of radius, R, adhering to a substrate by surface forces, the separation of that particle from the substrate occurs upon the application of an externally-applied load, F_S, such that

$$F_S = -\frac{3}{2} w_A \pi R \tag{1}$$

where w_A is the thermodynamic work of adhesion and is related to the surface energies, γ_P, and γ_S, of the particle and substrate, respectively, as well as their interfacial energy, γ_{PS}, by

$$w_A = \gamma_P + \gamma_S - \gamma_{PS}. \tag{2}$$

Assuming a reasonable value of $w_A = 0.05\,\mathrm{J/m^2}$, one finds that Eq. (1) predicts a separation force of the order of 1000 nN for a particle with a radius of 5 μm. This is in reasonable agreement with reported values [10].

While it might appear at this point that particle adhesion is well understood, this is not the case. Indeed, several complicating factors arise. The first is that particles are rarely perfect spheres. Rather, at best, they tend to be spherical, but with surfaces bearing irregularities or asperities. As discussed by numerous researchers including Krupp [8]; Fuller and Tabor [11]; Schaefer *et al.* [12] and Mizes [13], such asperities can significantly decrease the adhesion forces.

The effects of particle geometry are, however, more complicated than simply treating particles as spheres, perturbed by the occurrence of asperities on the surface. Rather, particles are often formed by fracturing larger materials into smaller pieces. This results in their having highly irregular shapes, appearing somewhat similar to lumps of coal. Such shape factors can have a significant impact on the effects of surface forces.

Another complicating factor is that particles often carry electrical charges and the resulting electrostatic forces can interact with neighboring substrates. The relative contributions of the electrostatic and van der Waals forces have been determined for spherical particles in several recent studies. Rimai *et al.* [14] used electrostatic detachment to measure the separation force between spherical polystyrene particles having diameters between 2 μm and 12 μm and a polyester overcoated conducting substrate and concluded that the contributions of the electrostatic force was substantially smaller than those arising from van der Waals interactions. In that study, it was assumed that the charge was uniformly distributed over the surface of the particles, which is reasonable for the highly-regular and spherical particles.

In another series of experiments, Gady *et al.* [15–18] distinguished between the electrostatic and van der Waals contributions to the force of attraction between polystyrene spheres and a variety of substrates. In these studies, they attached the particle to the cantilever of an atomic force microscope (AFM). By measuring the deflection of the cantilever and its resonance frequency as a function of the particle-to-substrate separation, they were able to determine the power-law dependence of the separation on both the force and force-gradient. As the power-laws for the electrostatic and van der Waals forces are different, they were able to ascertain that, for large separations ($>3-10$ nm), the force of attraction was dominated by a localized charge. This charge was, presumably, due to the contact between the particle and substrate that initially occurred during calibration of the displacement when the particle jumped into contact with the substrate. For smaller separations, the force of attraction was dominated by van der Waals forces. These results were confirmed by measuring the separation distance where the particle-to-substrate jump-to-contact occurred and relating that distance to the spring constant of the cantilever. It should be noted that, according to these studies, although the separation force was dominated by van der Waals interactions, electrostatic forces also contributed. This was especially noticed when the particle and substrate were comprised of materials having vastly different triboelectric properties. In that case, the separation force was observed to increase monotonically with the number of times the particle had been allowed to contact the substrate. Finally, there are effects that arise from the combination of the irregular shape of particles with their charge. Specifically, particles become charged through triboelectric interactions. As discussed by Hays [10, 19, 29, 30], it would be quite difficult for particles to become charged in their crevasses or other recessed regions, where their surface cannot be contacted by other materials. The occurrence of the resulting nonuniform charge distribution was demonstrated by Hays during an experiment wherein 99 μm diameter particles were electrostatically detached from one electrode and deposited on another, which was parallel to, but separated from, the first by a gap. Associated with the traversal of the charged particles was a current, as would be expected. However, in addition to that current was a second current associated with the particles flipping during the transit. This was attributed by Hays to the occurrence of a dipole moment on the particle caused by a nonuniform charge distribution.

A commonly-used method of measuring the adhesion force between a particle and a substrate involves the use of an ultracentrifuge because of its ability to provide the high accelerations needed to detach micrometer-size particles from surfaces [8]. Goel and Spencer used ultracentrifugation to determine the force needed to detach toner particles, between 5 and 35 μm in diameter, from photoreceptors used in copiers [20]. Subsequently, Mastrangelo [21] used ultracentrifugation to study the effects of smooth and irregular toner particles from hard and soft photoconductors. They found that smooth particles on soft substrates were held more tightly than were irregularly-shaped particles on a hard substrate.

More recently, Lee and Jaffe [22] measured the force needed to separate irregularly-shaped, 20 μm toner particles from a photoconducting substrate using an ultracentrifuge. They found that the measured forces were in agreement with the values predicted assuming the dominance of van der Waals interactions. However, they argued that an adhesion model *per se* could not possibly be correct for two reasons. First, they claimed that the van der Waals force model overestimates the force of attraction because the irregular shape of the particles resulted in the particles resting on asperities. Second, they also argued that the electrostatically-charged patches actually cause the electrostatic forces to be substantially larger than one would estimate assuming a spherical particle.

However, as discussed by Bowling [23] such a contact may actually increase the effect of van der Waals forces. This is because, if the irregular shape of the particle effectively causes the contact to look planar, rather than the "point" contact of a sphere on a plane, there would be an increase in the size of the van der Waals force adhering the two materials together.

As should be readily apparent, the adhesion of irregularly-shaped particles to a substrate is a complicated problem. This paper examines the roles played by both van der Waals and electrostatic interactions for irregularly-shaped particles, approximately 8 μm in diameter, that had been formed by grinding a larger block of material. This process of formation results in particles that would look somewhat akin to coal particles, rather than highly-irregular particles having sharp asperities. This problem is of both fundamental and practical importance, as the particles closely resemble electrophotographic toner in size, charge, and shape.

EXPERIMENT

The force needed to separate nominal 8.5 μm diameter ground-polyester particles from a polyester substrate was measured using a Beckman LM 70 ultracentrifuge. In addition, similar particles were coated with substantially smaller silica particles, which served as asperities, and the removal force of the coated particles were also measured.

The particles, which were similar to electrophotographic toners, were formed by grinding from a larger block of material and classifying the resulting particles to give the appropriate size. The volume-weighted diameter of the particles was 8.6 μm, as determined using a Coulter Multisizer. Subsequently, the surfaces of the particles were coated with between 0% and 2%, by weight, of particulate silica (Aerosil R972, produced by DeGussa, Inc.). These are shown in Figures 1A, 1B and 1C, respectively. The fundamental size of the silica, as reported by DeGussa, was approximately 16 nm. However, the silica tended to form agglomerates with a diameter of approximately 60 nm, as determined from observation of field emission scanning electron micrographs. For clarity in this paper, unless otherwise mentioned, the term "particle" will be used to refer to the polyester particle, which is the subject of this investigation.

The substrate consisted of a photoconducting polyester coating, approximately 20 μm thick, on a nickelized Estar (polyethylene terphthalate, produced by Eastman Kodak) support. The particles were deposited onto the substrate in a manner analogous to that used in an electrophotographic development process. In short, the particles were mixed with larger, magnetic carrier particles, against which the

(A)

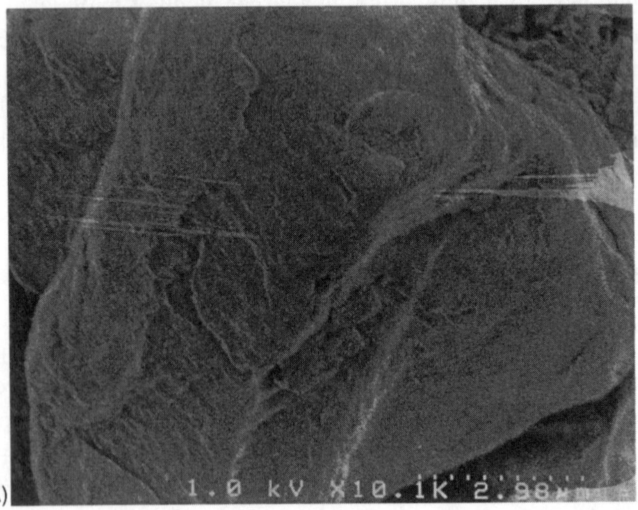

FIGURE 1 Field emission scanning electron micrographs (FESEM) of nominal 8.6 μm, irregularly-shaped, polyester particles overcoated with 0%(A), 1%(B) and 2%(C) silica (nominal cluster size approximately 60 nm).

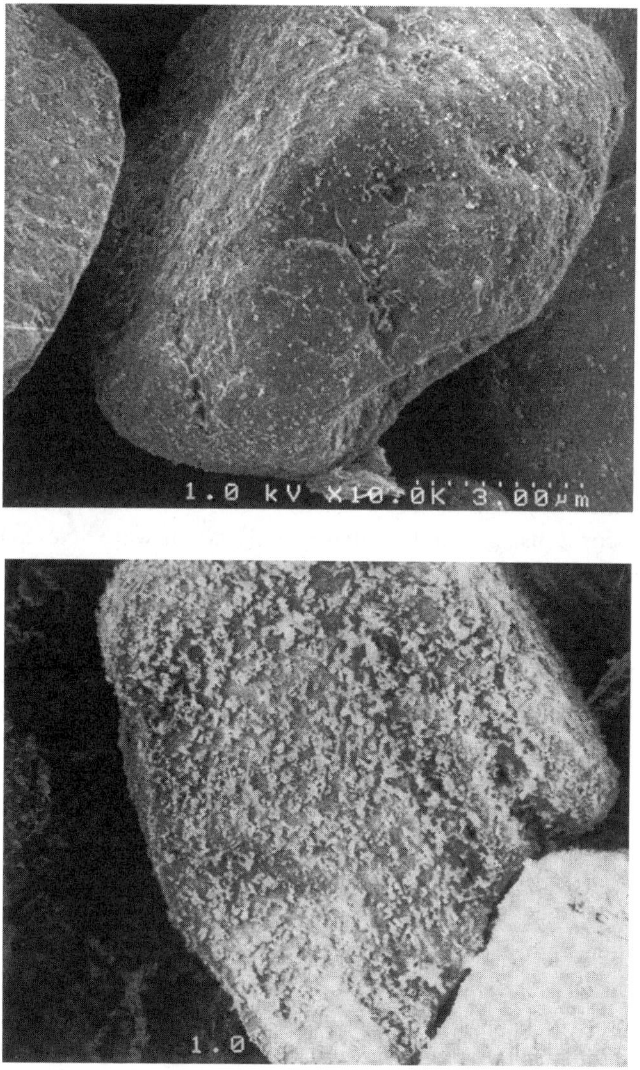

FIGURE 1 (Continued).

particles tribocharged. The charge-to-mass ratio of the particles, measured using standard techniques [24] was approximately $-37 \pm 3\ \mu C/g$ for each of the sets of particles.

Monolayers of particles were deposited onto the substrate by loading the mixture of particles and magnetic carrier particles into a toning station, similar to one used in electrophotographic development, which comprised a rotating stainless steel cylinder and a coaxial cylindrical core comprising a series of magnets. The conductive layer of the substrate was grounded and the shell biased in such a manner so as to result in the appropriate amount of particles being deposited onto the substrate. Typically, coverages of less than a monolayer were used to ensure that particle-substrate, rather than particle–particle, interactions were measured and to facilitate assessing the number of particles present using image analysis methods.

The adhesion of the particles to the substrate was determined by removing the particles in an ultracentrifuge capable of spinning at 70,000 rpm. The initial number of particles on the substrate was determined by counting the particles observed through an optical microscope, using suitable image analysis software. Next, portions of the substrate were placed in the centrifuge and spun at the desired speed. The sample was then removed and the remaining particles counted. This process was repeated for a series of increasing speeds. Centrifugation was performed in a low vacuum of approximately 10^{-2} torr (roughing pump vacuum). Initially, between 50% and 60% of the substrate was covered with particles.

RESULTS

Figure 2 shows the percent of particles removed from the substrate as a function of the mean applied force produced by different centrifuge speeds. Data for three silica concentrations, 0%, 1% and 2%, are shown. The highest force corresponds to a rotational speed of 70,000 rpm on the centrifuge. As can be seen, the general shapes of the curves gradually change for increases in silica concentration. Without silica, the percent removed is nearly linear with the mean applied force over the range investigated. There is no tendency to reach an asymptote. With 2% silica, the curve rapidly rises and then curves to approach asymptotically 100% particle removal as the mean applied force is increased. The result for the particles containing 1% silica is between the other two curves, following the 0% result

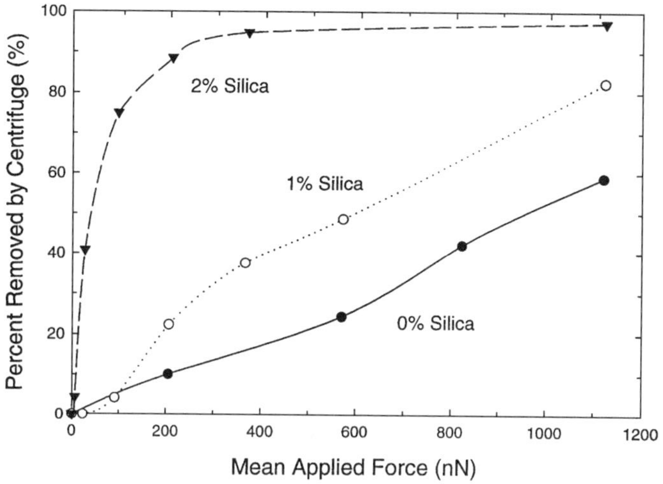

FIGURE 2 The percentage of particles removed by the ultracentrifuge as a function of removal force for three levels of silica: 0%, solid circles; 1%, open circles; and 2%, solid triangles.

TABLE I The measured and estimated forces, for various levels of silica, needed to detach the particles from the substrate

Percent silica	*Measured detachment force (nN)*	*Estimated detachment force from JKR theory (nN)*
0	970	943
1	580	507
2	39	70

initially and then rising as the centrifugation speed and, hence, the mean force is increased. Because there is a distribution in particle sizes, the larger particles would be removed first, as a qualitative view of the samples through an optical microscope suggests.

The mean applied forces reported above were calculated by assuming that the particles were spherical polyester particles with a radius of 4 μm and a mass density of $1.2\,g/cm^3$. The force needed to remove 50% of the particles, P_S, estimated at the 50% removal point, was determined to be 970 nN, 580 nN, and 39 nN for the 0%, 1% and 2% silica-coated particles, respectively. The measured and estimated detachment forces are listed in Table I.

DISCUSSION

As is well known, there is much debate in the literature as to whether the force of adhesion of irregularly-shaped particles to a substrate arises predominantly from surface forces such as those due to van der Waals interactions or from electrostatic forces such as those arising when charged particles polarize the substrate. For the case of a conducting substrate, this polarization would result in the particle generating its so-called image charge. The actual magnitudes of the two types of interactions depend on a number of factors including the conductivity or polarizability of the materials, as well as the size and shape of the particles. Therefore, it is unlikely that this issue will have a single solution for all possible cases. However, in-so-far as this paper addresses the adhesion of commercially-important, toner-like materials and many other commonly-found particles having similar shapes and properties, it is well worthwhile to attempt to resolve this issue for the present case. This is most readily accomplished by estimating the adhesional forces arising from both mechanisms.

Let us first assume that the uncoated particles are spheres with a radius of approximately 4 μm. The particle removal force, F_S, can be calculated from JKR theory using Eq. (1). Assuming a reasonable value of $w_A = 0.05\,\mathrm{J/m^2}$, the particle removal force is estimated to be 943 nN. In light of the approximations made, this value is in reasonable agreement with the experimentally-obtained value of 970 nN.

As previously mentioned, it can be argued that the apparent agreement between the measured force and that calculated from a surface-force model is purely coincidental, due to asperities actually reducing the surface forces. This issue will be discussed more fully later in this paper in relation to the effect of the silica on the separation forces. However, actual measurements of the contact between particles similar to these and planar substrates suggest that the proposed asperities serving as spacers does not actually occur in the present situation. Specifically, for irregularly-shaped particles, independent sets of measurements by Eklund [25] and by Bowen *et al.* [26] both report contact areas being of the order of 10% of the projected cross-sectional area of the particles. In other words, experimental evidence suggests that intimate contact is established between particles similar to the ones used in this study and planar substrates and the particles are not resting on asperities that separate it from the substrate. This would be in direct contradiction to the assumption that the van der Waals forces are weaker than assumed in the spherical-particle approximation due to the spacer-effect of the asperities. Therefore, it would appear that van der Waals interactions by themselves could readily account for the size of the observed separation force.

It is also necessary to estimate the size of the electrostatic contributions to the total force of adhesion. Charged particles generate fields which are approximately radial, falling off with distance. Such fields will polarize neighboring dielectric materials. When the dielectric constant of the intervening medium is less that those of the particle and substrate, the particle and substrate will attract each other [27, 28], with the force of attraction varying with the difference between the dielectric constants of the materials and intervening medium. If the particle and substrate were in intimate contact, thereby xcluding any intervening medium, the force would vary as the difference between the dielectric constants of the two materials. Unfortunately, as discussed by Jones [28], the problem of calculating either the electrostatic forces of attraction or separation between two dielectrics has not been solved. Therefore, the force of attraction between a charged dielectric particle and a conducting substrate will be approximated instead. Because of the highly polarizable nature of an electrically-conducting material, this should actually overestimate the actual force of attraction arising from the electrostatic attraction between two dielectric materials, as will be seen.

Even with the aforementioned assumption, such estimates are not simple to make, owing to effects arising from polarization and charge distributions. Details of this problem are presented elsewhere [10, 29]. However, for the sake of completeness, these issues will be discussed in brief herein.

Let us now assume, for the time being, that the irregularly-shaped particles used in this study can be approximated as a dielectric sphere of radius, R. Let us further assume that each particle has a charge, q, uniformly distributed over its surface. The electrostatic image force of attraction, F_I, between that particle and a conducting substrate is given by

$$F_I = \alpha \frac{q^2}{4\pi\varepsilon_0 (2R)^2}. \quad (3)$$

When $\kappa = 4$, representing a value of the dielectric constant appropriate for a typical polymeric particle, the value of α is 1.9 [30]. In the present situation, however, the particle is not adhered to a conductor. Rather, the substrate comprises a relatively thick polymeric layer whose dielectric constant is similar to that of the particle. Moreover, as previously discussed, experimental evidence argues that the particle-to-substrate contact is intimate and relatively large. In that case, where the dielectric constants of the two contacting materials are equal and there are no air gaps, $\alpha = 1.0$. Presumably, the present case would lie between these extremes. Using the values of charge to mass reported earlier, ($-37 \pm 3\,\mu C/g$, $\rho = 1.2\,g/cm^3$), it is then calculated that F_I would be in the range of 20 to 40 nN, depending on the value chosen for α for the present particles. This value is far less than the measured force needed for detachment shown in Figure 2.

However, as is discussed by Hays [10], the charge on an irregularly-shaped particle may not be uniformly distributed over its surface. Rather, the charge is assumed to reside totally on the high spots of the particle. Accordingly, there would be a finite charge density on these spots and no charge elsewhere on the particle. In that instance, the electrostatic contribution to the force of adhesion, F_E, is related to a surface charge density, σ, and the actual area of contact between the particle and substrate, A_C, by [10]

$$F_E = \frac{\sigma^2 A_C}{2\varepsilon_0}. \tag{4}$$

Using Eq. (4) and assuming that the contact area is approximately 10% of the cross-sectional area [25, 26], one could simply solve for the charge density needed to give the measured removal force. Upon substitution, one finds that $\sigma = 1.85 \times 10^{-3}\,coul/m^2$. Using a parallel-plate-capacitor approximation, one finds that this charge density would result in an electric field of approximately $2.1 \times 10^8\,V/m$.

The problem with a particle having a charge density resulting in this high an electric field is that the air between the particle and substrate, prior to the particle coming into contact with the substrate, cannot support the resulting field. Rather, there is a maximum electric field that air can support, which is generally referred to as the Paschen limit [31, 32]. At fields above the Paschen limit, air spontaneously ionizes and becomes conductive. This limit increases from about

3×10^6 V/m for large air gaps, as would occur when the particle and substrate are widely separated, to approximately 70×10^6 V/m for separations of the order of a micrometer. Exceeding the Paschen limit would result in the electrostatic discharge of the particle. In other words, it would be very difficult, if not impossible, for a particle to maintain this high surface-charge density.

Alternatively, it is worthwhile to estimate F_E within the confines of the Paschen limit. Again, this is not simple to do, as the Paschen limit decreases with increasing air gap. However, one could estimate the field around a charged particle using either a parallel-plate-capacitor approximation when the particle is close to a surface (say a few micrometers) or a spherical particle approximation for larger distances. Moreover, if the field is less than the Paschen limit at a separation distance of approximately 10 μm in air, the particle would be able to approach the surface of the substrate without discharging at other separation distances. In other words, the field associated with the charged particle would increase with decreasing air gap at a slower rate than would the Paschen limit. With a 10 μm gap, air could support a field of approximately 3.5×10^7 V/m [32]. Moreover, the field would fall off sufficiently fast with distance so that the Paschen limit would not be exceeded. The calculated surface-charge density, under this assumption, would be approximately 3×10^{-4} coul/m^2. Therefore, F_E would be of the order of 30 nN, which is consistent with estimates of F_I. This result suggests that the charge on the particle is not highly localized but, rather, uniformly distributed. Therefore, the force of adhesion due to the presence of localized charged patches is much smaller than those contributions that can attributed to van der Waals interactions, and certainly much smaller than the experimentally-determined separation force.

As discussed in the literature [10, 29], if the separation force is determined by electrostatic interactions associated with localized charged patches, it is also necessary to discount the apparent role played by van der Waals forces. This is done by assuming that asperities on the particles separate the particles from the substrate, thereby weakening the van der Waals forces. This proposal can be readily tested by introducing asperities having a controlled size and distribution. This was accomplished in the present study by coating the particles with particulate silica.

Even so, a precise determination of the effect of the silica on the particle-detachment forces will require a detailed knowledge of how the particle and the substrate contact each other and how that contact behaves under the influence of the surface forces. This depends on a number of factors such as the size and distribution of the silica, the shape of the particles, the range of the interactions, and the compliance of the materials. Despite these complications, one may still make some order-of-magnitude estimates of the detachment forces of the silica-treated particles.

The percent of the surface coverage of the particles by the silica can be estimated by assuming both the particles and silica are spherical. For the purpose of this calculation, it was assumed that the weight fraction of the silica is 1%. As previously indicated, the primary particle size of the silica is 16 nm diameter. However, the silica tended to form clusters with an average diameter of about 60 nm. Using $\rho = 1.75\,\mathrm{g/cm^3}$ as the mass density of the silica and $\rho = 1.2\,\mathrm{g/cm^3}$ as the mass density of the particles, and knowing that the particles have a mean diameter of 8.6 μm, the fraction of the surface area of the particles covered by silica clusters is estimated to be approximately 25%. For 2% silica by weight, the area of the particles covered by silica was calculated to be 50%. These estimates are consistent with SEM micrographs of the silica-coated particles.

Again, assuming that the particles are spherical, the contact radius, a_{JKR}, estimated using JKR theory, is given by

$$a_{\mathrm{JKR}} = \left(\frac{6\pi w_A R^2}{\mathrm{E}}\right)^{1/3} \tag{5}$$

where E is the Young's modulus of polyester, approximately 3 GPa [33]. In the absence of silica, $a_{\mathrm{JKR}} = 196\,\mathrm{nm}$. This estimate is of the same magnitude as that reported by Bowen [26] for the contact of an irregularly-shaped particle with a substrate. Assuming a similar contact region exists when silica is present, it was then estimated that approximately 10 silica particles would be in contact with the substrate when the silica concentration is 2%. For the silica-coated particles, the separation force, F'_S, is then approximately given by

$$F'_S = n\frac{3}{2}w_A\pi r \tag{6}$$

where $n = 10$ is the number of contacts and $r = 30\,\text{nm}$ is the radius of the silica particle clusters. Assuming that the work of adhesion for silica to photoconductor remains at $w_A = 0.05\,\text{J/m}^2$, upon substitution it is found that $F'_S \approx 70\,\text{nN}$. The experimentally-obtained value of F'_S was approximately 39 nN. In view of the approximations made, the experimentally-obtained value is in reasonable agreement with that estimated. It is interesting to note that these values are also close to the estimated contributions of the electrostatic image charges to the total force of adhesion. This suggests that, for highly irregular particles with many sharp asperities, electrostatic interactions associated with locally-charged patches can be significant factors in determining the separation forces. Indeed, observations have been made of micrometer-size nickel particles on a silicon substrate [34], where the particles do indeed show the sharp asperities discussed. The electrical conductivity of the nickel, as well as the size and shape of the asperities on the nickel particle, suggest that this particle might be a prime candidate for adhesion that is dominated by electrostatically-charged patches.

The detachment force for the particles containing 1% silica was determined by the centrifuge experiments to be approximately 580 nN, or about an order of magnitude larger than the estimated image-charge contributions. One can conclude from this result that the number of asperities present at 1% was insufficient to decrease significantly the van der Waals attraction of an irregularly-shaped particle to a substrate. Indeed, if it is simply assumed that, with a 1% coating of silica, the particles were not totally separated from the substrate so that there is some, but not total, separation of the particle from the substrate, then the separation force can be estimated by simply taking the mean values obtained from JKR theory. Accordingly, the estimated separation force for the particles with 1% silica was determined to be 507 nN, compared with the experimentally-determined value of 580 nN.

Unfortunately, owing to the rather complex nature of the particle-to-substrate contact and the possibility of statistical variations in particle charge density, it is presently not feasible to estimate quantitatively what the separation force should be under these circumstances. Indeed, with recent advances in finite element modeling software, this would be a suitable topic for future studies.

CONCLUSIONS

The force needed to remove electrically-charged, ground 8.6 μm diameter polyester particles from a polyester substrate was measured using an ultracentrifuge. The size of the measured force was in good agreement with that estimated from van der Waals interactions, but was too large to be attributed to either forces arising from a uniform or localized charge distribution. The size of the force was found to decrease with an increasing concentration of asperities, as introduced by coating the surface of the particles with silica. As the concentration of silica approaches 2%, the separation force decreased to approximately 5% of that observed for the uncoated particles. At this concentration, the estimated contributions of the van der Waals and the electrostatic forces become comparable in magnitude. This suggests that, although the Iocalized-charged-patch model does not appear valid for the present type of particle, it may be appropriate to describe the adhesion of highly-irregular particles with sharp asperities.

Acknowledgement

The authors would like to thank P. Alexandrovich, B. Gady and S. Leone for their technical assistance.

References

[1] Goel, N. S. and Spencer, P. R., *Polym. Sci. Technol.* **9B**, 763 (1975).
[2] Mastrangelo, C. J., *Photo. Sci. Engin.* **22**, 232 (1978).
[3] Derjaguin, B. V., *Kolloid Z.* **69**, 155 (1934).
[4] Bradley, R. S., *Philos. Mag.* **13**, 853 (1932).
[5] Bradley, R. S., *Trans. Faraday Soc.* **32**, 1088 (1936).
[6] Hamaker, H. C., *Physica* **4**, 1058 (1937).
[7] Lifshitz, E. M., *Soviet Phys. JEPT* **2**, 73 (1956).
[8] Krupp, H., *Advan. Colloid Interface Sci.* **1**, 111 (1967).
[9] Johnson, K. L., Kendall, K. and Roberts, A. D., *Proc. R. Soc. London Ser. A* **324**, 301 (1971).
[10] Hays, D. A., In: *Advances in Particle Adhesion*, Rimai, D. S. and Sharpe, L. H., Eds. (Gordon and Breach, Amsterdam, 1996), pp. 41–48.
[11] Fuller, K. N. G. and Tabor, D., *Proc. R. Soc. London, Ser. A* **345**, 327 (1975).
[12] Schaefer, D. M., Carpenter, M., Gady, B., Reifenberger, R., DeMejo, L. P. and Rimai, D. S., *J. Adhesion Sci. Technol.* **9**, 1049 (1995).
[13] Mizes, H. A., In: *Advances in Particle Adhesion*, Rimai, D. S. and Sharpe, L. H., Eds. (Gordon and Breach, Amsterdam, 1996), pp. 155–166.

[14] Rimai, D. S., Quesnel, D. J., DeMejo, L. P. and Regan, M. T., Submitted to *Digital Printing: Science and Technology*.
[15] Gady, B., Reifenberger, R., Rimai, D. S. and DeMejo, L. P., *Langmuir* **13**, 2533 (1997).
[16] Gady, B., Schleef, D., Reifenberger, R., DeMejo, L. P. and Rimai, D. S., *Phys. Rev. B* **53**, 8065 (1996).
[17] Gady, B., Schleef, D., Reifenberger, R. and Rimai, D. S., *J. Adhesion* **67**, 291 (1998).
[18] Gady, B., Reifenberger, R. and Rimai, D. S., *J. Appl. Phys.* **84**, 319 (1998).
[19] Hays, D. A. and Wayman, W. H., *J. Imag. Sci.* **33**, 160 (1989).
[20] Goel, N. S. and Spencer, P. R., *Polym. Sci. Technology* **9B**, 763 (1975).
[21] Mastrangelo, C. J., *Photo. Sci. Eng.* **22**, 232 (1978).
[22] Lee, M. H. and Jaffe, A. B., In: *Particles on Surfaces 1: Detection, Adhesion, and Removal*, Mittal, K. L., Ed. (Plenum, New York City, 1988), pp. 169–178.
[23] Bowling, R. A., In: *Particles on Surfaces 1: Detection, Adhesion, and Removal*, Mittal, K. L., Ed. (Plenum, New York City, 1988), pp. 129–142.
[24] Maher, J. C., *IS&T's Tenth International Congress on Advances in Non-Impact Printing Technologies*, IS and T, Springfield, VA, USA, 1994, pp. 156–159.
[25] Eklund, E. A., Wayman, W. H., Brillson, L. J. and Hays, D. A., In: *IS&T's Tenth International Congress on Advances in Non-Impact Printing Technologies*, IS&T, Springfield, VA, USA, 1994, pp. 142–146.
[26] Bowen, R. C., DeMejo, L. P. and Rimai, D. S., *J. Adhesion* **51**, 191 (1995).
[27] Smythe, W. R., *Static and Dynamic Electricity*, 3rd edn. (Hemisphere Publishing Company, New York, 1989).
[28] Jones, T. B., *Electromechanics of Particles* (Cambridge University Press, Cambridge, 1995).
[29] Hays, D. A., In: *Fundamentals of Adhesion and Interfaces*, Rimai, D. S., DeMejo, L. P. and Mittal, K. L., Eds. (VSP, Utrecht, 1995), pp. 61–72.
[30] Hays, D. A., In: *Particles on Surfaces 1: Detection, Adhesion and Removal*, Mittal, K. L., Ed. (Plenum Press, New York, 1988), pp. 351–360.
[31] Paschen, F., *Wied. Ann.* **37**, 69 (1889).
[32] Cobine, J. D., *Gaseous Conductors* (Dover Publications, New York, 1957).
[33] van Krevelen, D. W., *Properties of Polymers* (Elsevier, Amsterdam, 1976).
[34] DeMejo, L. P., Rimai, D. S. and Bowen, R. C., *J. Adhesion, Sci. Technol.* **2**, 331 (1988).

Electrical Conductivity Through Particles

Copper-based Conductive Polymers: A New Concept in Conductive Resins*

DAVID W. MARSHALL

Textron Systems, 201 Lowell St., Wilmington, MA 01887, USA

The history of making plastic materials conductive to both electric currents and to the transfer of thermal energy has traditionally been accomplished by the addition of metallic particles into a resin matrix. Principally, such metals as aluminum, silver, gold, nickel and copper have been used. Copper has had a limited success due to its tendency to form a non-conductive oxide surface layer and currently such adhesives depend primarily on silver for high conductivity.

An intense research effort to eliminate the problems associated with copper-filled conductive polymers resulted in a treatment and preparation of copper flake that allows the stable formation of a conductive structure within a polymer matrix. Once the activated copper particles are in the resin, the formulation is stable. Most of the resins evaluated have been epoxy resins although certain thermoplastic resins have also been made conductive. Volume resistivities as low as 10^{-5} ohm-cm have been achieved.

INTRODUCTION

Since most plastics are thermal and electrical insulators, it is necessary to add a conductive material to the polymer in sufficient volume that a conductive path can be found through the matrix. The inherent conductivity, the shape of the particle, and the nature of the particle surface are all important but particle-to-particle contact is a

*Presented in part at the 22nd Annual Meeting of The Adhesion Society, Inc., Panama City Beach, Florida, USA, February 21–24, 1999.

necessity [1]. Conductivity is usually determined by measuring volume resistivity which is given by $Vr = RA/L$ where R is the measured resistance between two points, A is the cross sectional area through which the current flows and L is the distance between the two points. Therefore, the lower the volume resistivity, the higher the conductivity. The volume resistivity of some typical metals is given in Table I.

It should be noted that the largest volume of commercially-available conductive polymers is not based on the addition of metals but on the addition of carbon particles. These materials are largely based on thermoplastic resins and are used principally for static shielding. They are relatively inexpensive but have volume resistivities only in the range of 10^1 to 10^3 ohm-cm. This class of conductive materials will not be considered further since their conductivities are low compared with the better metal-filled systems. Also, the discussion below is limited to thermosetting resins with the brief exception of the mention of a copper-filled thermoplastic polyethylene polymer.

The most common metals used include aluminum, silver, gold, nickel and copper, along with some non-conductive materials such as glass fiber or spheres, which have been coated with a conductive metal [2]. Despite the availability of these materials and the large number of powder or flake metals that could be used, virtually all the high-performance conductive resins are based on silver particles. This is primarily due to silver's inherent high conductivity and the fact that silver has an oxide coating which is conductive [3]. From the data in Table I, one would suspect that aluminum, gold and copper would also result in highly conductive polymers. This is true for gold but the cost of gold powder precludes its use in all but the most demanding cases. A layer of aluminum oxide always covers aluminum surfaces.

TABLE I Electrical conductivity of metals

	Specific gravity *gms/cm*^{3}	ρ = *Volume resistivity* *ohm-cm*
Silver	10.5	1.6×10^{-6}
Copper	8.9	1.8×10^{-6}
Gold	19.3	2.3×10^{-6}
Aluminum	2.7	2.9×10^{-6}
Nickel	8.9	10×10^{-6}
Platinum	21.5	21.5×10^{-6}
Eutectic Solders	–	$20-30 \times 10^{-6}$

This oxide is a good insulator and prevents aluminum-to-aluminum particle contact when aluminum particles are used as a filler in a resin formulation. However, aluminum particles are occasionally used to make thermally-conductive polymers because of their low cost.

Copper also forms a non-conductive oxide layer but its oxide layer is not as stable or as tightly attached as that of aluminum. If this oxide could be eliminated or changed to a conductive form and the surface kept in that state, copper-filled compounds would be very effective conductors. One way of eliminating the oxide is to add reducing agents such as certain amines to the resin. These agents partially reduce the oxide to copper metal but have been found to be ineffective in providing long-term protection against oxidation and they often adversely affect the cured properties of the resin system. The overall result of using reducing agents has been a general unreliability with progressive loss of conductivity over time. Since copper is a relatively low cost metal compared with silver or gold, there is a considerable driving force to develop a surface treatment for copper particles that would result in conductivities similar to those of the more-precious metals.

PROCEDURE

An intense research effort to eliminate the problems associated with copper conductives produced a novel treatment for the copper which results in the ability to use copper particles in a resin matrix to form an electrically-stable system. Much of the technical work was done some years ago but little has been published since the process was originally patented [4] and the techniques used were considered proprietary. The treatment consists of a four-step chemical process to activate the copper surface. These steps include using a solvent to remove any coating on the particles, then chemically removing the oxide from the surface, drying the particles under vacuum and storing them under nitrogen or vacuum until ready for use.

For example, to treat copper flake for incorporation into a typical two-component epoxy resin, the following steps were taken. A one-pound blend (454 grams) of flakes consisting of a 50–50 mixture of US Bronze Powder's USB 6500 and C-100 copper flakes was made.

These flakes have a thickness of approximately 2 microns and an average diameter of 20 microns and 60 microns, respectively. Purity of the flake was 99.99%. Lower purity could result in poor conductivity. The flake was placed in a glass vessel and covered with trichloroethylene equal in volume to 8 times the flake volume. The mixture was stirred for one-half hour and then filtered under vacuum through a Büchner funnel. The flake was rinsed in the funnel three or four times with denatured alcohol. This removed any processing aid from the flake mixture.

The flake was then removed from the funnel and mixed with one quart (0.95 l) of 1 molar citric acid and stirred overnight. The mixture was then filtered again through the Büchner funnel and rinsed with distilled water until the filtrate was clear. The flake was then rinsed with denatured alcohol. The damp flake was then placed in a vacuum oven and dried under vacuum at 100°F(38°C). One can tell when the copper is dry by the vacuum held. When it is stable around 77 mm the flake should be dry. It can be stored under vacuum or under nitrogen. The copper flake should not be removed from the oven before it cools to room temperature or it may ignite.

Depending on the particle size of the cleaned flake, it can be left in air for several minutes to several hours before incorporating into a resin. The flake can be added to any epoxy resin by simply mixing into the liquid using a low-shear mixer such as those made by Ross Machine Co. Once the particles have been added to a typical epoxy resin the uncured system is stable for a year or more. Loading of the copper flake in the resin can vary from 40% by weight to 60%. Higher loading is difficult to achieve because of the rapid build-up of viscosity. Lower loadings have insufficient copper to provide particle-to-article contact.

Mechanical, thermal and electrical properties were determined for several formulations containing copper treated in the above manner. The following ASTM methods were used: Tensile strength and modulus-ASTM D638, Lap shear-ASTM D1002, Density-ASTM D1622, and Volume resistivity-ASTM D2739. Thermal conductivity measurements were made using a calorimeter. Viscosity measurements were performed using a Brookfield viscometer with a Heliopath stand. Speed and spindle were chosen to give a reading near the middle of the viscometer's scale. Test results are discussed in the following sections.

DISCUSSION OF RESULTS

Although copper powders could have been used, copper flake is easier to treat and most formulations studied used particles of that shape. The flake nature of the copper filler enhances the electrical conductivity and its stability in the resin matrix. One reason for this is discernable under microscopic examination where a "House of Cards" structure can be seen [4]. The edges of flakes rest against the flat portion of other flakes and form a structure similar to a "House of Cards". In most cases, the interface between particles or flakes can not be differentiated since the contact is so intimate. This structure makes it difficult to screen-print formulations made with the treated copper since the "particle association" tends to clog the screen.

Unlike prior copper-filled epoxy systems, a simple amine cured bisphenol A epoxy resin with copper flake treated in the above manner to form an activated copper surface will retain its conductivity indefinitely.

Copper flake fillers with an average diameter of 30 microns and a thickness of 2 microns are the easiest to treat. Varying the ratio of the two flake sizes used can vary the viscosity of a resin to a certain extent using this blend. Viscosities tend to be high, a few million centipoise, but typical formulations are thixotropic and, if necessary, can be thinned with ethyl alcohol. An activated copper flake loading of 56% by weight in a typical bisphenol A resin cured with an aliphatic amine will have a volume resistivity of around 10^{-3} ohm-cm. In some systems, volume resistivities as low as 10^{-5} ohm-cm have been achieved. It should be noted that the "percolation theory" [5] which describes the mechanism of electrical conduction through metal particles in a polymeric matrix does not apply in the case of these flake-filled composites nor does it apply for metallic fibers such as stainless steel [6]. Metal fibers or fibrils typically are made from stainless steel and have a diameter around 8 microns with an aspect ratio of about 750. These fibers can be effective at loadings as low as 3% by weight.

Assuming the values for silver can be used for copper, since they are similar in conductivity, the percolation theory gives a critical value of 30 percent by volume [7]. The volume of copper in the above formulations is around 15%. Since photomicrographs of the filled resin exhibit the "House of Cards" structure, it is evident that the structure requires less copper for conductivity. Because of the flake alignment and the activation treatment, it is theorized that a charge is present on the flake surface with the edges being either positively or negatively charged and the flat surface oppositely charged. This would account for the structure and the relatively small amount of copper necessary for electrical conductivity.

Particles can also be treated but their stability in air is limited and much higher loadings (around 80% by weight) are required to obtain high electrical conductivity. The volume percent copper necessary for conduction, in this case, more nearly matches that predicted by the percolation theory. Such compositions tend to be more sensitive to the ingredients used in the resin formulation. The resistivity is affected principally by the cure temperature but also by the chemical composition of the resin itself. Lovinger [8] has verified that there can be an effect of chemical composition on the conductivity of metal-filled composites. Acetate and hydroxyl groups were found to be beneficial. Stearates and other processing aids were detrimental. In the case of copper-filled epoxies, amines, anhydrides, and certain diamides and imidazoles have been found by the author to be conducive to conductivity but not sufficient for long-term conductivity. High cure temperatures, particularly in the case of particle-filled resins, promoted the formation of copper oxide and reduced conductivity. Activated copper was found to be quite sensitive to this and storage life outside the resin is limited, even at room temperature [4].

Some compositions, particularly single-component epoxies, cured at elevated temperatures, exhibit a current inrush effect. This occurs when resistance is initially high but drops precipitously when a sufficiently high voltage (usually around 12 volts) is momentarily applied [9]. Once the resistance has dropped, it remains at the lower level indefinitely. It is likely that either a thin layer of epoxy, or copper oxide, formed on the surface of the particle. This necessitated a sufficiently high voltage to break through the insulating layer. Properties of a typical conductive two-component epoxy resin, consisting of a bisphenol-A resin such as Dow's DER 332 and triethylenetetramine (TETA) curing agent, in the ratio of 13 parts TETA to 100 parts DER 332, and containing 56% of the copper blend mentioned in the previous section, are given in Table II. A conductive single-component

TABLE II Typical copper-filled epoxies

Type	*2-Component rigid*	*1-Component rigid*
Copper Wt. %	56	56
Cure	24 hrs @ 75°F (24°C) or 1(1/2) hrs. @ 150°F (66°C)	1 hr @ 350°F (177°C)
Density	2.29 g/cc	2.29 g/cc
Tensile Strength	34,500 kPa	31,000 kPa
Elongation	1%	–
Lap Shear, Al–Al	8,300 kPa	8,300 kPa
Modulus	7×10^5	–
Volume Resistivities	0.001 ohm-cm	0.01 ohm-cm
Thermal Conductivity W/m/°C	~3	~3
Viscosity	6×10^6 cps	12×10^6 cps
Shelf Life at 70°F (21°C)	12 months	6 months
Comments	Good all around properties	Best high temperature properties

TABLE III 2-Component flexible adhesive

• Designed to bond materials with different thermal expansions	
• Tensile Strength	22,000 kPa
• Lap shear (Al–Al)	14,500 kPa
• Elongation	4%
• Modulus	1.4×10^6 kPa
• Volume Resistivity	.001 ohm-cm
• Viscosity	11×10^6 cps.

epoxy resin containing the same amount of copper in a DER 332 dicyandiamide blend cured at 350°F(177°C) is also shown in Table II.

A flexible version of the two-component system shown in Table II using a polyglycol modified epoxy is shown in Table III. For comparison purposes, typical values for other metal-filled polymers is shown in Table IV along with values using activated copper. Note that treated copper, the earlier antioxidant method for making conductive copper-filled epoxies, has poor thermal stability.

Other resin systems have been evaluated, including polyolefins, some of which have been moderately successful and provide conductivities superior to those of carbon-filled polyolefins. However, copper-filled epoxy polymers have exhibited the highest conductivities and most of the data collected have been on those systems.

TABLE IV Comparison table: conductive metal-filled resin

Metal filler (in epoxy resin)	*Volume resistivity range in ohm-cm (initial)*	*Cured stability at 100°C at normal atmosphere*
Aluminum	1.0×10^{10}	Good
Silver	$.1-1.0 \times 10^{-3}$	Good
Gold	$.1-1.0 \times 10^{-4}$	Good
Treated Copper	$1.0-5.0 \times 10^{-3}$	Poor
Activated Copper	$.5-1.0 \times 10^{-3}$	Good

TYPICAL MATERIAL PROPERTIES AND APPLICATIONS

Electrical

One of the early products made using a copper-filled polymer was a heating panel. This panel was made with a polyamide-cured epoxy resin. The intended application was an economical home-heating panel that could be operated at low voltage. The panel was made by spreading the resin on a strip of gypsum board 2.4 meters long and 0.6 meters wide. Copper foil strips were embedded in the resin at each end as contact points. The panel was operated at 12 volts. This resulted in a surface temperature of about 38°C. Because of the low voltage used, the panel exhibited a minimal shock hazard making it suitable for walls as well as ceilings. If a nail were driven into it or a cut made, it would continue to operate without presenting a shock hazard to the worker. Figure 1 illustrates the panel's stability. As the graph shows, there was essentially no change in either surface temperature or resistance over approximately 900 days. Because of the costs involved in stepping down normal household voltage at current electrical rates this particular application was not pursued.

Other products have shown various copper-filled adhesive formulations to have excellent long-term stability. In one case, a "Positive Temperature Coefficient" (PTC) heating device was bonded to an aluminum plate in a food-warming tray [9]. This tray consisted of two aluminum inserts set into a conventional plastic cafeteria tray. The PTCs (based on barium titanate) were bonded to the back of the inserts and connected to terminals located on the bottom of

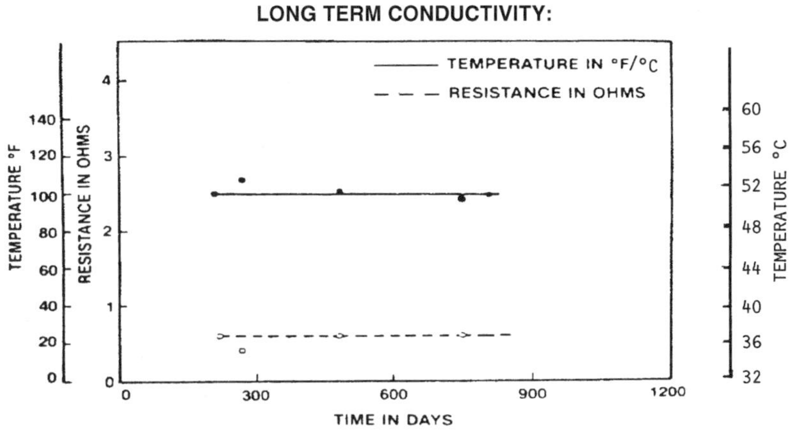

Stability of a 50% copperflake filled epoxy resin.

FIGURE 1 Stability of the conductivity of copper-flake-filled epoxy resin.

FIGURE 2 Heated cafeteria tray.

the tray. Figure 2 shows the tray and Figure 3 the larger insert with three PTC disks bonded to the aluminum insert.

Several trays fit into a cart that contains a 12-volt battery as a power supply. A connection is made with the tray causing the PTC

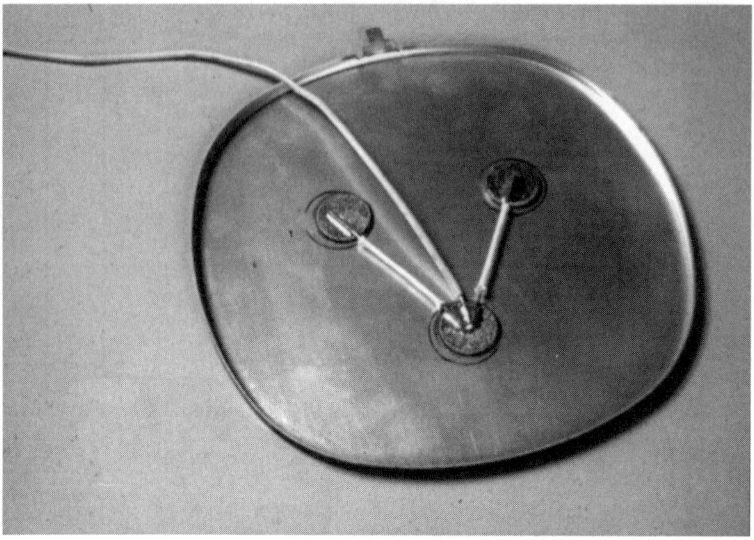

FIGURE 3 Cafeteria tray insert showing PTC heating devices.

devices to heat and keep the food placed on the inserts warm. The trays were designed for use in hospitals and similar institutions where food had to be distributed from a central kitchen. The trays originally used a silver-filled adhesive to bond the PTC devices but, during use, severe corrosion problems occurred resulting in debonding of the PTC disks. When copper was substituted for the silver and the trays run through several commercial dishwasher cycles, as well as thermal stability tests, the trays outlasted silver-filled epoxy bonded trays and did it at a lower cost. The copper-based adhesive used in this application is a single-component resin. Its properties are given in Table V. Some of the testing involved bonding a PTC device to an aluminum plate and measuring the resistance through the device over time at 177°C to determine the electrical stability of the bond. The copper adhesive performed very well as can be seen from Table VI. It should be noted that the operating temperature of the tray is only around 66°C.

Tables VII and VIII show the superiority of the copper adhesive in both cycling and drop tests. Note that after three years bond failure is rare.

TABLE V Copper based tray adhesive

Properties	
Copper wt%	56
Density	2.29 g/cc
Tensile strength	31,000 kPa
Thermal conductivity	~3 W/m/°C
Volume resistivity	0.01 ohm-cm
Lap shear strength	10,300 kPa
Cure	1 hour @ 177°C
Viscosity	6×10^6 cps

TABLE VI Stability of one-component copper based PTC adhesive

177°C Aging test cure time 1 hour @ 177°C	
Test time (hours)	*Resistance (ohms)*
0	5.2
150	5.2
800	5.8
1500	5.8
1600	5.3
2300	5.7

TABLE VII Cycle tests Test I

Heater heater bond	*Heater resistance (ohms)*	*Temp. °C*	*% Change*
Silver Epoxy			
Initial	4.6	96	–
40 cycles	4.8	89	4.3
100 cycles	7.1	79	54.3
Copper Epoxy			
Initial	4.1	84	–
40 cycles	4.0	82	2.4
100 cycles	4.9	82	19.5

- On/off cycle for 5000 cycles. No difference between copper and silver adhesives.
- On/off cycle in water 6 min. on -6 min. off.

A single-component, copper-filled epoxy adhesive has been used for a number of years to bond the base and provide electrical conductivity with the filaments in 6 and 12-volt T2 type lamps (see Fig. 4). These lamps were used in push-button telephones at one time but are now largely used as indicator lights in various types of equipment. These bases were originally bonded with silver-filled

TABLE VIII Completed tray Test-II

- 1000 dishwasher cycles plus
  686 cycles of 6 min. on – 6 min. off
- Drop-tested onto a concrete floor from a six foot (2 meter) height-Twice.
- Copper adhesive passed
- Silver adhesive, after approximately 400 cycles, had one of three disks unbonded and the rest so corroded test was stopped
- After 3 years of copper adhesive use, bond failure is rare.
  ≫ When failures occur it is usually a solder joint that failed.

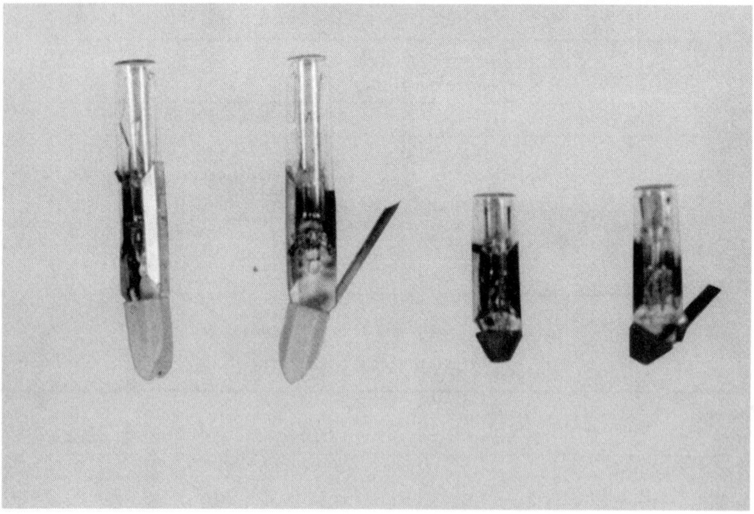

FIGURE 4 Minature lamps showing the bonding of the lamp base. (See Color Plate III).

epoxy resins. The copper system was much less expensive and a single system replaced two grades of silver epoxy with one high-conductivity copper adhesive. This application requires dispensing at a high rate of speed on a production line.

Thermal

The inherent high thermal conductivity of copper can also be utilized. Heat transfer properties of copper-filled *vs.* silver-filled resins are within the same range, with copper providing an equivalent thermal conductivity at a fraction of the cost.

Table IX illustrates thermal conductivity of various filled epoxy resins.

A room-temperature-curing, flexible, copper-filled epoxy resin has been used to replace solder in solar heating panels in the "do-it-yourself" home market. In this case, it is much easier to bond the tubing to the collector plate rather than to try and solder the large sections of tubing.

EMI Shielding

The electrical properties of the copper-filled epoxy systems also result in their providing excellent shielding against EMI radiation [10]. Again, this protection is similar to that of silver-filled resins at a fraction of the cost. The following table, Table X, shows the shielding effectiveness of the room-temperature-curing, copper-filled systems and other high conductivity shielding materials. Table XI gives the properties of a copper-based coating under both a magnetic and plane wave fields. As a result of FCC regulations, allowable electromagnetic emissions of commercial and industrial equipment are limited. Commercial or industrial equipment such as computers or high voltage connectors are designated as Class A by the FCC and require a

TABLE IX Thermal conductivity of epoxy resin system *vs.* type filler

Filler type	*Wt. % filler in resin*	*Thermal conductivity* $Wm^{-1}\ deg^{-1}$
Activated Copper	56	2.8–4.3
Silica	65	0.8–1.1
Aluminum	50	1.1–1.4
Calcium Carbonate	50	0.7–1
Silver	50	2.8–6.9
Alumina	50	0.8–1.1

TABLE X EMI shielding effectiveness

	10 *Mhz to* 1 *Ghz Attenuation* (*db*)
Silver Epoxy	59–87
Copper Epoxy	67–73
Graphite filled	27–40
Nickel filled	35–50
Zinc Metal	70–90

TABLE XI EMI shielding effectiveness of activated copper systems

	Magnetic field		*Plane wave*		
Frequency	150 KHz	200 KHz	400 MHz	1 GHz	10 GHz
Avg. Attenuation	70 db	67 db	70 db	73 db	75 db

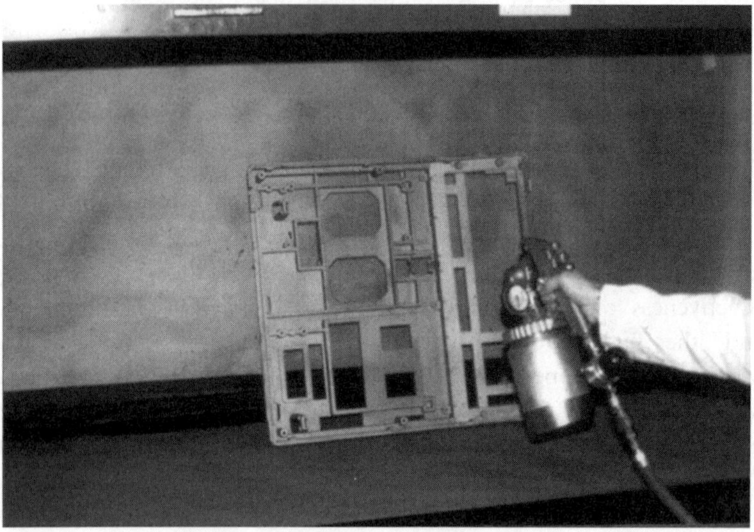

FIGURE 5 The spraying of a computer housing.

minimum attenuation of around 60 db. Most military EMI requirements are for attenuations of 70 db or higher.

Figure 5 shows a computer housing being sprayed with an activated copper coating to shield it against EMI radiation. The coating resin was made by thinning a 2-component, activated copper-filled epoxy resin with ethanol to make a sprayable solution. It can then be applied with conventional spray equipment.

CONCLUSIONS

A treatment for copper particles has been developed that, when combined with the proper resin systems, results in a stable conductive material. Although preliminary work has been successful with certain thermoplastics such as polyethylene and polycarbonate, most of the development has been concentrated on epoxy resin systems. These systems have been shown, in most cases, to be comparable with, or superior to, silver-filled epoxies at a significantly lower cost. They offer superior environmental resistance over silver epoxies where corrosive conditions exist but are not yet suitable for screen-printing or for applications where intermittent surface contact is needed, as in a membrane switch.

References

[1] Bolger, J. *et al.*, "Conductive Adhesives: How and Where they Work", *Adhesives Age*, pp. 17–20, June, 1984.

[2] Bolger, J., Astle, R. and Morano, S., "Conductive Inks and Coatings", *Adhesives for Industry Conference*, El Segundo, CA, June, 1986.

[3] Pandini, S., "The Behavior of Silver Flakes in Conductive Epoxy Adhesives", *Adhesives Age*, pp. 31–35, October, 1987.

[4] Marshall, D. W. *et al.*, "Copper Filled Conductive Epoxy", *US Patent No. 3,983,075*, 28 September, 1976.

[5] Kirkpatrick, S., *Reviews of Modern Physics* **45**, 574–588 (1973).

[6] Borgmans, C. and Gerteisen, S., "Shielding Electronic Components with Conductive Plastics", *Machine Design*, pp. 59–62, November, 1998.

[7] Gurland, J., *Trans. Metallurgical Society*, *AIME* **236**, 642–646 (1966).

[8] Lovinger, A., "Development of Electrical Conduction in Silver–Filled Epoxy Adhesives", *J. Adhesion* **10**, 1–15 (1979).

[9] Marshall, D. W. and Mack, A., "An Electrically Conductive Adhesive used in a Food Warming Tray", ANTEC 1982, *The 40th Annual Technical Conference and Exhibition of the Society of Plastics Engineers*, Vol. XXVIII, May, 1982.

[10] Marshall, D. W., "Copper Based Conductive Polymers, A New Concept in Conductive Resins", *Proceedings 1983 NEPCON Central Meeting*, 16 November, 1983.

Exploring Particle Adhesion with Single Particle Experiments

Interactions Between Micron-sized Glass Particles and Poly(dimethyl siloxane) in the Absence and Presence of Applied Load*

GARY TOIKKA, GEOFFREY M. SPINKS and HUGH R. BROWN

Institute for Steel Processing and Products, University of Wollongong, NSW 2522, Australia

A technique using a scanning electron microscope to view a fine particle in contact with a flat substrate whilst under load and during its removal is described. The particle is attached to an atomic force microscope cantilever so that the magnitude of the load can be estimated directly from the imaged deflection. Interactions between 5 to 60 μm spherical glass particles and cross-linked poly(dimethyl siloxane) were studied in the presence and absence of load. W_A was estimated to be 74 mJ/m^2 from the size of the contact area in the absence of load. Using highly flexible cantilevers to apply load resulted in large shear displacements and forces, which distorted the contact area and assisted in particle removal. These shear effects were eliminated by using a more rigid cantilever to measure a normal pull-off force for which the interface toughness, G_c, exceeded 950 mJ/m^2. The large adhesion hysteresis indicated the presence of chemical bonding, presumed to occur between silanol and siloxane groups. The mode of particle detachment varied significantly with the choice of cantilever, showing evidence of both cohesive failure and interfacial crack propagation. The relevance of these results to the interpretation of AFM data is discussed.

*Presented in part at the 22nd Annual Meeting of The Adhesion Society, Inc., Panama City Beach, Florida, USA, February 21–24, 1999.

INTRODUCTION

Significant effort has been expended on the study of fine particle interactions, largely due to their importance in technological processes ranging from mineral flotation to photocopying. In most of these processes the particles make *contact* with dissimilar surfaces, hence, the magnitude of the adhesion or, most importantly, the ease of their removal becomes of considerable practical interest. As the adhesion is controlled by processes that occur within the contact area, both the contact mechanics and the interfacial properties must be given consideration. Given the small size of the particles and the magnitude of the forces involved, highly specific techniques are required to study the adhesion directly.

The development of the atomic force microscope (AFM) has permitted forces less than 10^{-11} N to be directly measured between imaging tips [1, 2] and flat surfaces. Unfortunately, since the tips are only nm in size and difficult to characterise [3] the data obtained are not directly useful for understanding fine particle adhesion. The problem has been resolved by attaching micron [4] and sub-micron [5] sized particles to AFM imaging tips to measure forces against both flat [6, 7] and spherical substrates [8, 9]. Studies focussed on the measurement of adhesion [10 – 12] have given values distinctly less than that expected on a theoretical basis. The findings were rationalised in terms of surface roughness where asperities increased the effective separation to reduce the intimate contact area. The explanation coincides well with increases in the adhesion using deformable surfaces and stiff cantilevers [13], as load more readily places the particles in intimate contact. However, a number of effects inherent in the normal operation of the AFM limit its use as a dedicated adhesion-measuring device.

AFM's are unable to hold a single contact between a particle and substrate over an extended time period. Since adhesion often increases with time, the measured values are likely to be poor indicators in most real applications. Several contacts are also made over relatively short time periods, which has the potential to deform the substrates and to make the overall interaction geometry ill-defined. The measured forces are assumed to be normal to the plane of the interaction, following the use of a laser and split photodiodes to detect cantilever deflections. Whilst the assumption is reasonable as the particle approaches the substrate, the very geometry of the technique makes shear forces at contact inevitable. As the extent of shear forces in AFM experiments is difficult to know, the interpretation of contact data is somewhat unclear.

Scanning electron microscopes (SEM) have also been used to study the adhesion between micron-sized particles and flat substrates [14–19]. The technique has enabled particles in extended contact with flat substrates to be viewed directly. It was found that significant deformation occurred in the contact area, which, in instances, allowed the adhesion to be indirectly determined. The values were normally an order or more greater in magnitude than those measured using an AFM. It is reasonable that longer contact times lead to increases in the measured adhesion as more intimate contact forms. Further increases in the measured adhesion can also be expected when removing particles from SEM experiments as separation processes include hysteresis mechanisms [20].

Whilst the detachment of a particle from a surface is ultimately governed by the magnitude of the adhesion, the efficiency of a removal processes may also depend on shear forces. For instance, the use of hydrodynamic flow to clean silicon wafers is based on drag and lift forces, for which the latter may be negligible inside the boundary layer. Smaller particles must then be dislodged at an angle to the interaction plane until the adhesion is overcome [21]. Quantitative differences in the magnitude of the normal and shear force required to remove micron-sized particles have been directly measured using an atomic force microscope [22]. It was found that removal was more easily facilitated by shear forces than lift (pull-off) forces for interactions between rigid surfaces. The relationship between the two forces may change if deformable substrates are used, as the particles can be displaced into the surface. One of the aims of this work is to examine explicitly the effects of shear forces on adhesion measurements and particle removal.

JKR [23] adhesion theory predicts a relationship between the thermodynamic work of adhesion W_A [24] and the elastic deformation of macroscopic bodies in contact. The features between an ideal rigid spherical particle and a flat deformable surface are illustrated in Figure 1. As the model is based on an energy balance, it neglects any surface forces outside the contact zone to predict, unrealistically,

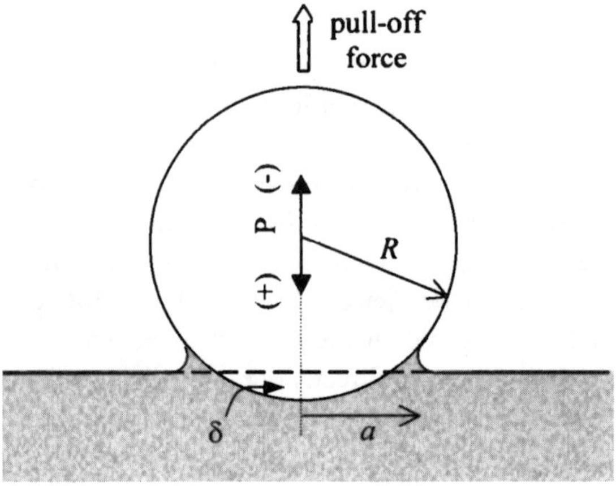

FIGURE 1 Contact size (radius, a) and shape between an elastically deformed flat surface and a particle (radius R). The central axis (dotted line) is indicated, along which displacement (δ), normal load (P) and pull-off force are determined.

infinite tensile stresses at the contact edge. The presence of the tensile forces in the outer contact region result in the distinctive non-Hertzian contact shape described as a "neck". The size of the contact radius, a, is given by

$$a^3 = \frac{R}{K}\left\{P + 3W_A\pi R + \left[6W_A\pi RP + (3W_A\pi R)^2\right]^{1/2}\right\} \tag{1}$$

where R is the particle radius, K is related to the elastic modulus, E, and Poisson ratio, v, of the materials and P is the normal applied load. It should be noted that, by convention, a compressive (tensile) load is positive (negative) and expected to increase (decrease) the contact radius. For the contact radius to decrease in size, a crack must propagate along the interface between the two materials and Eq. (1) can be considered in terms of fracture mechanics. Whilst crack growth is typically not an equilibrium process, the equation remains valid as long as the strain energy release rate, G, is substituted for W_A. Sufficient reduction in the load causes crack propagation to occur when G reaches its critical value, G_c, which itself may depend on crack propagation rate. In this paper we shall consider crack propagation in terms of G_c and the reverse, contact growth, as an approximate measure of W_A. The difference between the values of G_c and W_A obtained in this way is often referred to as "adhesion hysteresis".

The adhesion also places the central contact region under compressive stress, even in the absence of external load, displacing the particle into deformable surfaces along the central axis by δ [25]

$$\delta = \frac{a^2}{3R} + \frac{2P}{3aK} \tag{2}$$

Hence the displacement, as in the case of the contact radius, can be expected to increase with compressive load. It is possible to displace the particle in a negative direction above the interaction plane to cause stable crack propagation, which reduces the contact patch until a finite size of the crack becomes unstable and separation occurs. The negative normal load at this instant is known as the pull-off force and is directly related to G_c *via*

$$F_p = \frac{3}{2} G_c \pi R \tag{3}$$

In the absence of any load, Eq. (1) is reduced so that a linear relationship between the contact radius and the particle radius to the power 2/3 can be expected. The relationship has been independently verified for mm-sized rubber hemi-spheres [23] using optics to view the contact directly. It has also been observed between micron-sized glass particles and polyurethane [15, 26] using a SEM. The findings confirmed the extension of current elastic deformation theory to particles $> 5\,\mu m$ in the absence of load. An extremely useful feature of the approach is that values for W_A may be experimentally determined. These values are able to include many of the non-idealities, such as surface contamination or roughness, which are typically encountered in real systems but are not amenable to theoretical analysis.

This paper describes a novel technique to view and directly measure fine particle adhesion, based on the principles of both an AFM and a SEM. The technique has been used to study interactions between spherical micron-sized glass particles and flat polydimethylsiloxane (PDMS) surfaces. This model system was chosen because (a) PDMS is known to deform mainly in an elastic manner with little viscoelastic effects, (b) it is possible to measure glass-PDMS adhesion on a much larger size scale using a JKR apparatus, and (c) the adhesion between PDMS and silicon oxide shows interesting effects that have been ascribed to hydrogen bonding between silanol groups and the siloxane groups of the PDMS [27]. These effects have been seen in studies of the adhesion of PDMS to the oxide surface of a silicon wafer and the adhesion of a plasma-oxidised PDMS surface to an unoxidised surface. It would not be unreasonable to expect similar adhesion mechanisms between oxidised glass surfaces and PDMS. If contact is formed over longer time period, significant increases in the adhesion may also occur due to surface reconstruction reactions, as has been seen with hydrolysed PDMS surfaces [27–29].

EXPERIMENTAL

Spherical 5–60 μm glass particles were obtained from Duke Scientific (California, USA) and used without further treatment. PDMS sheets were prepared by using 0.0542 g of methylhydrosiloxane to cross-link 1.5087 g of vinyl-terminated, 9600 molecular weight poly (dimethylsiloxane) in the presence of ~8 ppm platinum-divinyl-1-tetramethyldisiloxane complex used as catalyst (Gelest, Inc). A 68% stoichiometric excess of cross-linker was used to ensure that a fully-reacted elastomer network was produced. Both flat and (~1 mm) hemispherical substrates were made and cured in air at 75°C for 2 hours. The elastic modulus (E) of the PDMS, 0.91 MPa, was determined using a JKR set-up to measure symmetrical interactions (self-adhesion) between the two geometrically different substrates. The flat PDMS samples were also cut into approximately $1 \times 1\,\mathrm{cm}^2$ pieces and placed on SEM stubs and the force rig (see below) sample holder. The glass particles were either sprinkled onto the PDMS from heights less than 1 cm, or placed directly into contact using a micro-manipulator or AFM cantilevers. All interactions were kept at 20°C for 7 days prior to imaging to allow (near) equilibrium conditions to be reached. The particle surfaces were also imaged using an atomic force microscope (Digital Instruments, Inc.) to reveal an average RMS surface roughness of 2.7 ± 0.4 nm over select $1 \times 1\,\mu\mathrm{m}^2$ areas. Significantly rougher regions, which were more readily observed using light microscopy and SEM, could also be detected. Overall, the glass was considered rough on both microscopic and macroscopic scales. The PDMS appeared smooth in SEM observations and was not explored further as significant deformation of the surfaces were expected.

All SEM images were measured using a secondary-electron emission electron microscope (Leica 440 Stereoscan). The tungsten filament was operated at 20 kV whilst varying the probe current between 100 – 300 pA to obtain best resolution. Due to the low conductivity of PDMS and glass all samples were sputter coated (Dynavac Magnetron, SC100MS) with platinum immediately prior to being imaged, to avoid artefacts due to either charging or Joule heating. To ensure the platinum thickness was insignificant in the measurement of contact radii, it was kept to approximately 6 nm by sputtering for 30 seconds at 50 mA (~ 2 Å/sec), as determined using a quartz microbalance. All images were measured at $< 2°$ off the interaction plane to best observe the contact regions. At such small angles, care was required to avoid any foreground roughness which may have prevented the contact areas from being directly viewed.

The force rig (Fig. 2a) was designed to allow an AFM cantilever to be manipulated with sub-micron resolution whilst inside a SEM chamber. A high-precision sliding rail (Del-Tron) was used to move

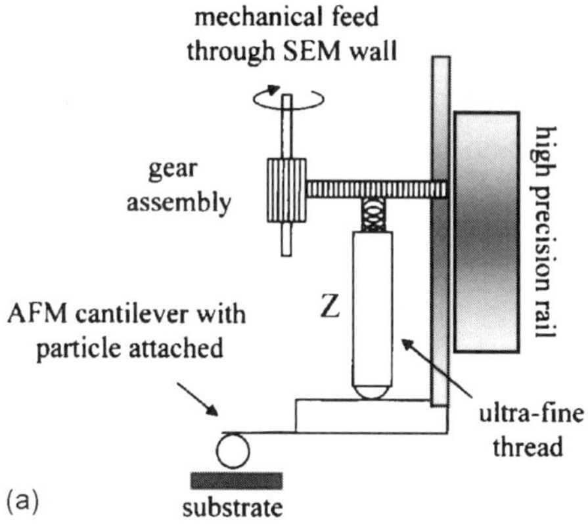

FIGURE 2 (a) Schematic of the force rig designed to apply loads to micron-sized particles whilst inside a SEM, described in text. (b) SEM micrograph of interactions between glass particles, in the size range of 3.3 – 20 μm (radius), and cross-linked PDMS. The particle on the right can be seen under the load of an AFM cantilever and the other five in the absence of applied load.

(b) 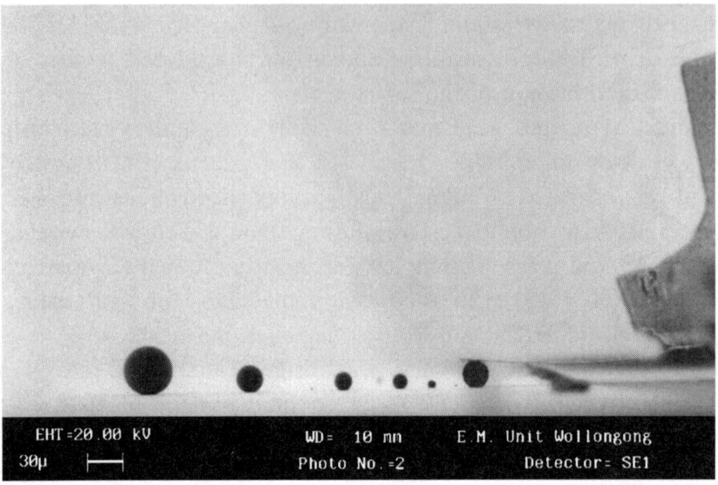

FIGURE 2 (Continued).

the cantilever in a z-direction (normal) to the sample. Their separation was controlled using an ultra-fine thread (Newport) and mechanical gearing. A locking screw was used to ensure that no changes occurred in the z-displacement whilst fitting the force rig into the SEM chamber. It also enabled the contact made between the glass particles and the PDMS under a light microscope to be maintained for 7 days before imaging in the SEM. The integrity of the rail and the complete force rig was tested as follows for run-out or "glitch". A polished tin/lead alloy flat was indented using a stiff (tapping mode) cantilever and its relative position imaged as the load and separation was altered. Out of contact, less than 0.2 µm lateral movement in the image plane, perpendicular to the z-direction, could be detected. To operate the force rig externally to the SEM a mechanical feed-through was installed in the microscope vacuum chamber door. An infrared viewing camera (Robinson Chamber View) was fitted inside the vacuum chamber to ensure that no contact was made between the (approximately 100 mm long) force rig and the SEM interior.

The glass particles were attached to highly-flexible, triangular-shaped (contact mode) and rigid-beam-shaped (tapping mode) AFM cantilevers (Digital Instruments, California), using a technique described elsewhere [11]. The stiffness of the cantilevers was determined from measurements of their resonance frequencies, in the absence and presence of a known mass. Hooke's law was used to evaluate the load on the particles once the deflection of the cantilever, away from its zero-force position, was estimated from the SEM images. Interactions between several glass particles and PDMS can be seen on the sample holder of the SEM force rig in Figure 2b. The particle on the right is under the load of an AFM cantilever, whilst the other (five) particles were kept in the absence of applied load. Their inclusion enabled confidence to be gained that there was no significant variation in the conditions between experiments. Note that the individual legs of the triangular-shaped cantilever can barely be discerned to confirm that the interactions were imaged close to the interaction plane.

RESULTS AND DISCUSSION

Interactions between glass particles and PDMS were studied after seven days in contact in (a) the absence of any load, (b) in the presence of small loads applied using highly flexible cantilevers and (c) large loads applied using stiff cantilevers. In the absence of load the contact between a 10.9 μm glass particle and PDMS (Fig. 3) was found to be significantly larger than one would expect between two totally rigid surfaces. The contact shape was clearly non-Hertzian, providing evidence of tensile forces in the outer region of the contact. As the interaction occurred in the absence of any load the deformation of the PDMS was solely attributed to surface forces. Note that the elastic modulus of the PDMS is almost five orders of magnitude less than that of glass, hence any deformation of the latter has not been considered in this study. The contact radius between several particles and PDMS were measured and plotted (Fig. 4), in log form, against the particle radius to find if the deformation was indeed elastic. A linear regression revealed a 0.66 power law dependence between the respective radii, which is in excellent agreement with the JKR elastic deformation theory of Eq. (1). The measurement of contact between a wide range of rigid small particles and deformable surfaces (or *vice versa*) has revealed similar power laws to that observed here and also at 0.42 [17] and 0.75 [30]. It was, of course, expected that a cross-linked polymer, such as the PDMS used here, would behave in an elastic manner well above its glass transition temperature (−123°C). The large

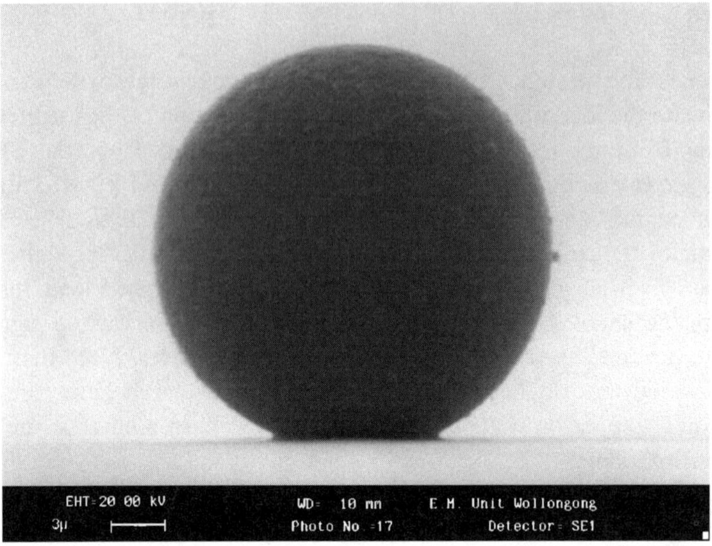

FIGURE 3 A 10.9 µm radius glass particle kept in contact with PDMS for 7 days, in the absence of applied load. The formation of a "neck" is clearly visible and has been predicted by JKR [23] theory.

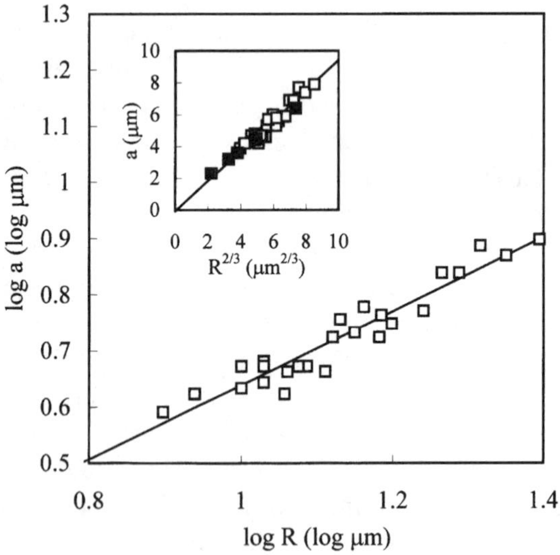

FIGURE 4 Log plot of contact radius, a, as a function particle radius, R, between glass particles and PDMS in the absence of external load. A linear line of best fit, $a = 0.66R - 0.02$ (Rsq, 0.91), is indicated. The plot of a as a function of $R^{2/3}$ (inset) results in a slope of 0.0095 and an intercept of 0.0773 (µm). Data (black squares) from the particles in Figure 2b have been included.

experimental scatter (Rsq. 0.91), also observed elsewhere, was most likely due to the excessive macroscopic scale roughness of the glass particles.

Having established the nature of the deformation, the value of G, $74\,\mathrm{mJ/m^2}$, was readily obtained from the slope of the contact radius as a function of the particle radius to the power 2/3 (Fig. 4, inset). Interactions between larger PDMS elastomers in (~ 30 minutes) contact with glass slides were also measured to yield significantly lower values, between $44-47\,\mathrm{mJ/m^2}$, for G. The reason for the increased adhesion in the smaller interactions is not known but may simply have arisen from the longer contact times (both systems are currently the subject of further study). Under the experimental conditions, the value of G (for contact growth) is assumed to have reached equilibrium and has, for the purpose of this study, been equated to W_A. The zero (within experimental error) intercept confirmed the assumption of no external forces having contributed to the deformation. In the event of the adhesion having occurred solely as a result of dispersion forces, the interfacial energy of each respective surface can be related through the approximation [31] $W_{A(\mathrm{glass/PDMS})} \approx 2(\gamma_{\mathrm{glass}}\gamma_{\mathrm{PDMS}})^{1/2}$. The interfacial energy of the PDMS surface, $\gamma_{\mathrm{PDMS}} = 23.4\,\mathrm{mJ/m^2}$, was measured in self-adhesion measurements and implies a value of $57.8\,\mathrm{mJ/m^2}$ for the glass surface. Whilst glass is a high energy surface [31] the low value obtained here is reasonable as the particles were most likely covered with adventitious hydrocarbons prior to making their contact with the PDMS. Neither the measured nor derived value confirms (or discounts) the origin of the adhesion between the glass and the PDMS. The W_A reported here lies well within the broad boundaries of 44 [30] and 120 [26] $\mathrm{mJ/m^2}$ previously determined between polyurethane elastomers and micron-sized glass particles.

The contact between the (10.5 μm) glass particle and PDMS under compressive load in Figure 2b can be seen at a higher magnification in Figure 5. A highly flexible cantilever (stiffness, 0.029 N/m) was used to apply the small load, estimated to be $0.80 \pm 0.03\,\mu\mathrm{N}$ in magnitude. The 10.9 μm glass particle in the absence of load shown in Figure 3 was also from the same experiment (second from the left in Fig. 2b). Given that both particles were kept under identical experimental conditions and are similar in size, the effect of load on the contact could be determined unequivocally. On closer inspection, it is evident that

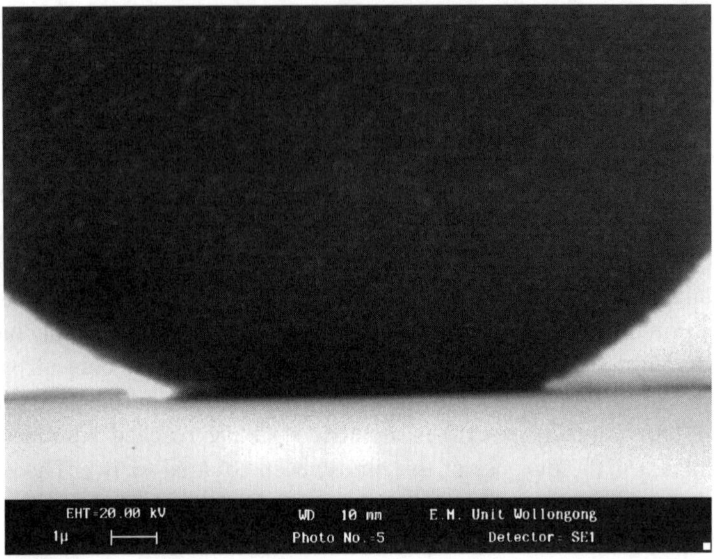

FIGURE 5 Contact area between the 10.5 µm glass particle observed under load in Figure 2 and PDMS. Its contact patch can be unequivocally compared with the particle in Figure 3 as both particles were kept in contact with the same PDMS and are virtually identical in size.

the particle under load is displaced further into the PDMS and the surrounding contact region is somewhat displaced. Under the conditions, the central displacement was expected to increase from 0.67 µm to 0.78 µm with the load, and whilst an increase could be confirmed, the respective measured values of 0.45 ± 0.05 µm and 0.70 ± 0.05 µm were both smaller than predicted. It was also anticipated that the contact radius would increase from the measured 4.67 ± 0.1 µm to the predicted 4.85 µm with the load; surprisingly, it was found to have been reduced to 3.95 ± 0.1 µm. The observation can be rationalised in terms of shear forces, which arise from using the highly flexible cantilever to apply the load. The initially straight AFM cantilever can be seen, in Figure 2, to bend into an arc as contact is made between the particle and the PDMS. Considering that the cantilever approached the interaction plane at approximately 12°, in a similar set-up to that used in an AFM, shear forces from both slip and rotation of the particle are likely to arise. Barquins [32] observed similar reductions in the contact between mm-sized glass hemi-spheres and flat rubber surfaces whilst studying the effect of shear forces. He discovered that the contact shape became asymmetric as it was reduced in size. Once the glass began to slip, the rubber immediately surrounding the contact was also displaced in an asymmetric manner. There is evidence of both these effects in the contact and immediate surroundings of the PDMS and the glass particle in Figure 5. Unfortunately, the contrast resolution of the SEM and the physical impossibility of imaging a profile at 0° to the interaction plane prevented a more quantitative comparison from being made. It was also not possible to unequivocally determine the nature of the contact shape under load as only one (left) side can be discerned as non-Hertzian.

The load on the glass particle in Figure 5 was gradually decreased until a tensile force was applied (Fig. 6a). It can be seen that the cantilever bends away from the PDMS surface near its fixed end, and towards the PDMS surface closer to the particle contact area. The observed S-shape shows that the mechanical action of the AFM

(a)

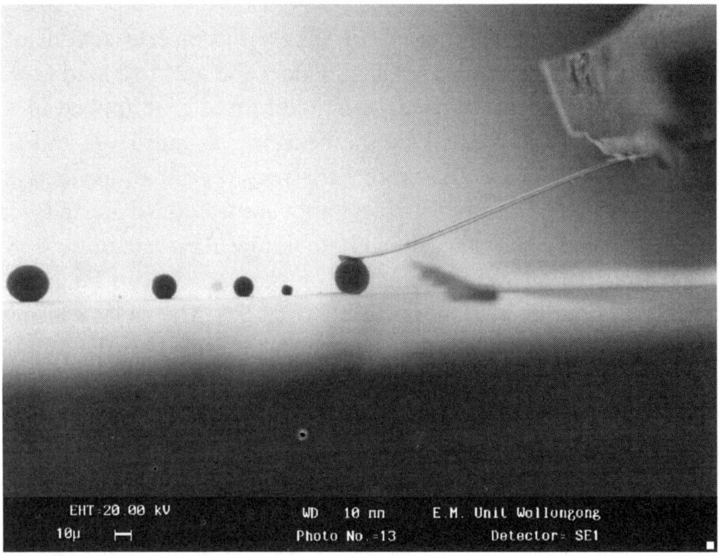

FIGURE 6 (a) Tensile load applied to the glass particle observed in Figure 5 can be seen to result in an S-shape bend of the highly flexible (stiffness, 0.029 N/m) AFM cantilever, which (b), distorts the PDMS in a manner clearly not normal to the plane of interaction.

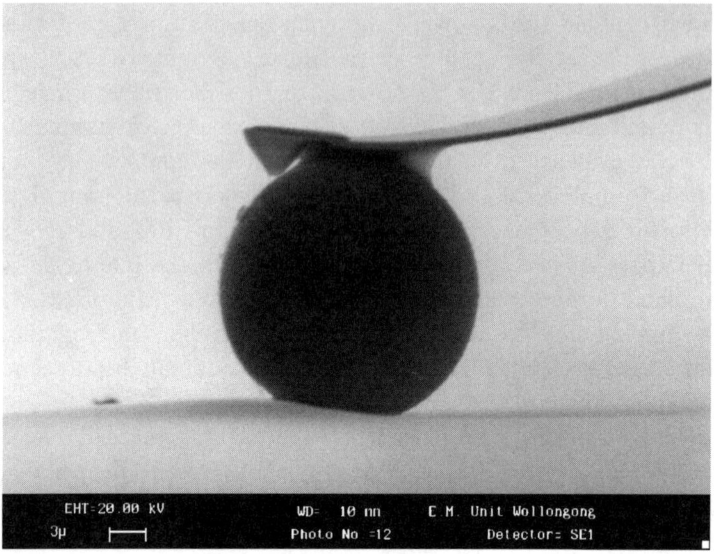

(b)

FIGURE 6 (Continued).

cantilever during unloading is not simply the reverse action to that when a compressive load is applied. Under the negative load there are clearly two forces on the particle, in addition to that applied in the z-direction: a torsion created by the bend in the cantilever and a pull towards the force rig as the cantilever struggles to span the distance to the contact area. Both effects inevitably contribute to shear forces in the contact area but it is difficult to know their magnitude. Some insight into the overall force acting on the particle can be gained by viewing the contact area more closely (Fig. 6b). The PDMS is severely distorted in an asymmetric manner which confirms that the total force applied on the glass particle is not normal to the interaction plane. Evidence of the torsion can also be seen in the relative position of the glass particle to the AFM imaging tip in Figures 2b and 6. As load is applied to the cantilever, the two-part epoxy glue deforms and allows the particle to move closer to the imaging tip. This observation confirms the need for careful selection of the adhesives used to attach particles to AFM cantilevers. Incidentally, if one monitored the deflection of the cantilever in Figure 6 using only a laser reflected off the tip, it would be difficult to avoid the conclusion that the particle was, in fact, under a positively applied load. Hence, in an AFM experiment it would be hard to distinguish the current situation from that observed in Figure 2. It is interesting to note that in the presence of shear, the size of the particle contact radius was larger under the tensile load ($4.77 \pm 0.1\ \mu m$) than under the compressive load ($3.95 \pm 0.1\ \mu m$).

To determine the value of G_c required to propagate a crack which leads to the removal of a glass particle from the PDMS, a pull-off force normal to the plane of interaction must be applied until separation occurs (Eq. (3)). This condition is often assumed in the use of cantilever-based techniques, such as the AFM. From the previous discussion it is clear that a flexible cantilever, such as the one used here, cannot directly measure G_c. There is, however, the prospect that the mechanics of the cantilever could be analysed, prior to detachment, in order to gain an understanding of the forces acting on the particle. This approach could prove useful in the analysis of real processes where both shear and normal forces are inevitable during detachment. Figure 7 shows the glass particle viewed in Figure 6,

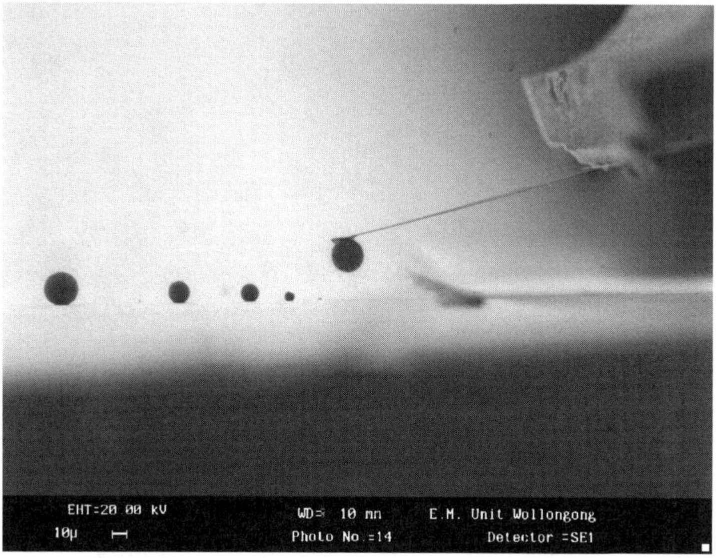

FIGURE 7 Displacement between the glass particle in Figure 6 and PDMS immediately following detachment. The straightness of the cantilever confirms that no permanent bends resulted from it being placed under load.

immediately following its detachment from the PDMS surface. Unfortunately, it was not possible to obtain images of the actual detachment process, as it occurred over a very short (less than 1 second) time period. The linearity of the cantilever in the absence of load confirms that it was not permanently deformed and validates the assumption of the zero-force position used in the estimate of load. The (z) displacement between the particle and the PDMS corresponds to a "normal" pull-off force equivalent to a G_c value of 10.6 mJ/m^2.

Figure 8 shows the detached glass particle and the PDMS imaged at 10° above the interaction plane. Substantial deformation of the PDMS can be seen from this angle and can be readily attributed to cohesive failure since a large piece of the elastomer is also visible on the glass particle. Had the experimental conditions been such that the particle could not be viewed after it was removed, the deformation may well have (in less elastic systems) been mistaken as being plastic in nature. Since the PDMS piece on the glass particle is smaller in size ($\sim$4 μm) than the contact area was prior to detachment, the mechanism for particle removal in the presence of shear forces

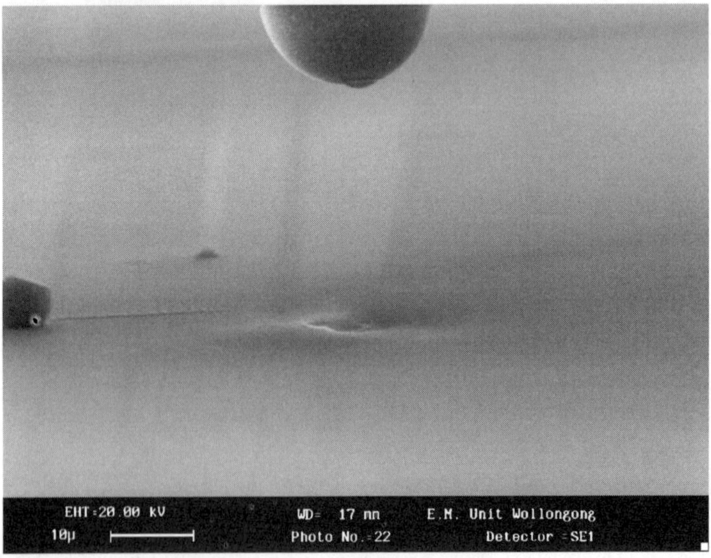

FIGURE 8 A close-up of the glass particle and PDMS in Figure 7 imaged at 10° to the interaction plane. Cohesive failure of the PDMS is evident as a large piece of the elastomer remains attached to the glass particle.

was through a reduction in the contact radius until cohesive failure occurred. Several more experiments were conducted using highly-flexible cantilevers to determine if the cohesive failure could be eliminated. It was found that by using small increments of load, the particles could be detached without any visible signs of PDMS on them. In these instances, an equivalent normal G_c up to $43\,\mathrm{mJ/m^2}$ could be measured; however, in none of the experiments were we able to obtain a value for $G_c > W_A$. Even under the most ideal conditions, the value of G_c must exceed its equilibrium adhesion value, W_A, otherwise detachment could not occur. The result obtained here clearly demonstrates the importance of shear forces in particle removal. It also explicitly shows why an AFM cannot measure strong adhesion using a highly flexible cantilever.

To minimise the effect of shear, a cantilever with a stiffness of 42.7 N/m was used to place a 12.8 μm glass particle (Fig. 9) under positive load against PDMS. The inclusion of a zero-force line (arrow) reveals that the deflection could easily be resolved at a load of $57 \pm 11\,\mu\mathrm{N}$, although the precision was significantly less than when

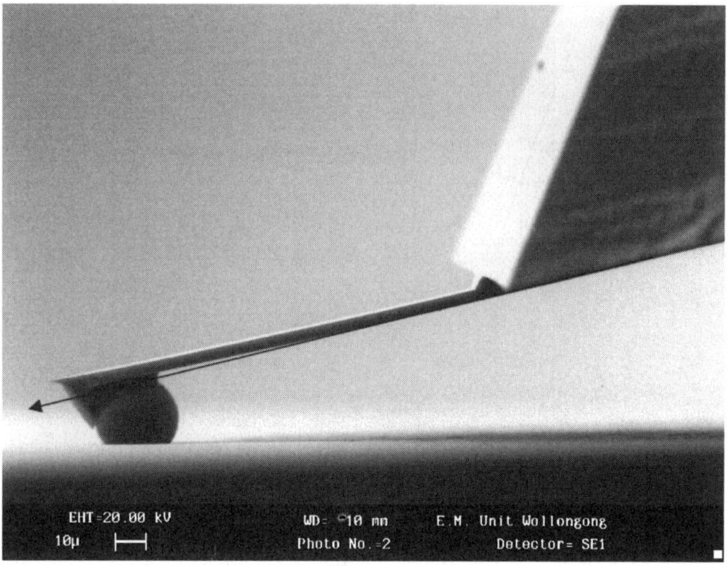

FIGURE 9 Compressive load applied to a 12.8 μm glass particle in contact with PDMS using a stiff (42.7 N/m) AFM cantilever. The arrow illustrates the small deflection required to apply a load which significantly deforms the PDMS.

using the more flexible cantilevers. The actual contact between the particle and the PDMS cannot be directly viewed as the particle is displaced well into the PDMS surface. Its expected [33] profile is illustrated in Figure 10 which predicts the contact radius to be 9.2 μm and the central displacement 4.8 μm, under the estimated load. The measured displacement, 4.8 ± 0.2 μm, coincides well with theory to suggest that the deformation had remained elastic in nature. In order to inspect the contact, the load on the particle was reduced to approximately zero and imaged (Fig. 11a) immediately afterwards. The size of the contact radius, 9.0 ± 0.2, was (within experimental error) the same as that predicted prior to having reduced the load. This suggested that little, if any, change had occurred in the contact region during the process. Hence, the clearly visible "neck" was assumed to have formed under compressive load and has only previously been observed between larger (mm) sized substrates. The interaction profile in Figure 10 does not explicitly accommodate spherical particles (as it is based on a parabola) and lacks in continuity immediately next to the

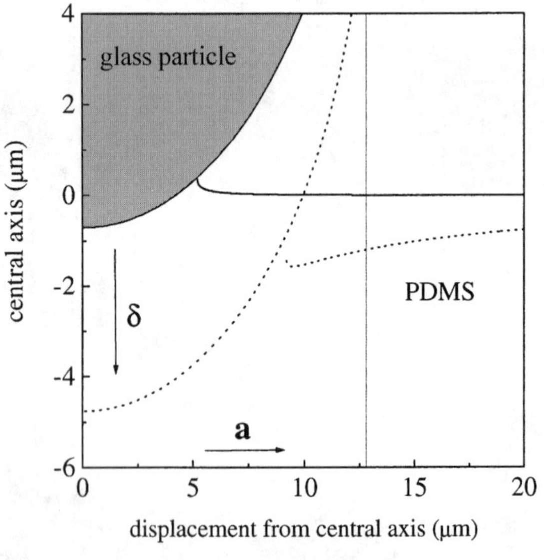

FIGURE 10 Predicted [32] interaction profile of PDMS under zero (solid line) and 57 μN load (dashed line) of a 12.8 μm glass particle. The values used in the calculations were W_A, 74 mJ/m^2, E, 0.91 MPa; ν, 0.5 and E, 70 GPa; ν, 0.3 for the respective surfaces. Increases in the contact radius, a, and the central displacement, δ, are expected with load.

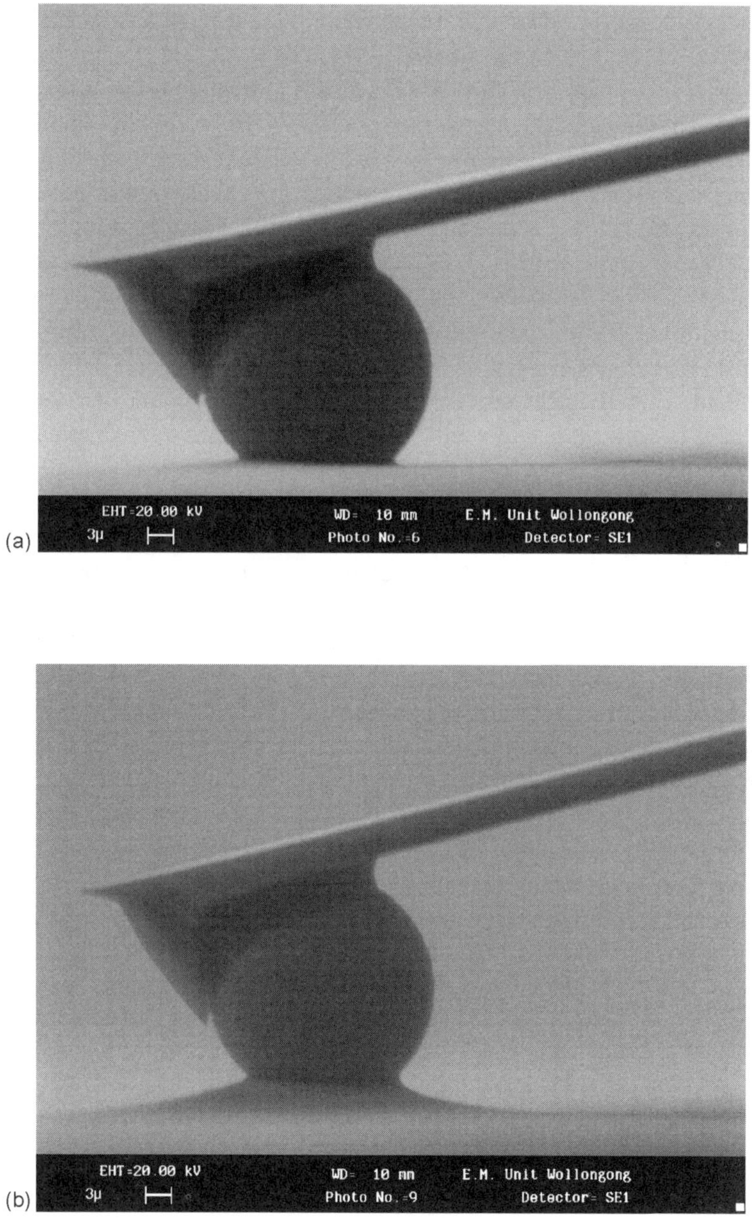

FIGURE 11 (a) Contact between the glass particle and PDMS observed in Figure 9 immediately after the applied load was reduced to (approximately) zero and (b) later under a load of (−) $57 \pm 11\,\mu N$.

particle; however, it can still be seen that the non-Hertzian "neck" was expected.

On the basis of the (near) equilibrium W_A derived in the absence of load, a radius of 5.20 μm is predicted under the zero load conditions in Figure 11a. However, to reach this value from the 9.0 μm radius, a crack must propagate between the materials. Crack growth is *not* an equilibrium process and can only be expected to occur once the critical strain energy release rate, G_c, is reached. Since there was no reduction in the contact radius at zero load, even after several minutes, it was soon evident that G_c was greater in magnitude than W_A for the system. The load was reduced to a tensile force of (−) 57 ± 11 μN (Fig. 11b) where a small reduction in the contact radius, 8.7 ± 0.2 μm, could be detected. Under the large tensile force the particle can be seen to have remained strongly attached to the PDMS, which deforms well above the interaction plane. The particle is, in fact, negatively displaced at (−) 1.2 ± 0.13 μm above the (flat) PDMS surface. At this point, which represents the final load increment prior to detaching the particle, a conservative value was estimated for G_c (from the load and the displacement) at 950 ± 180 mJ/m^2. It should be noted that under these conditions the value of the contact radius was expected to be slightly smaller (8.1 μm) than that measured. The massive difference in the magnitudes of W_A and G_c indicates a large adhesion hysteresis which makes the removal of the particle difficult. The origin of the increased adhesion cannot simply be rationalised by an increase in intimate contact with load, as it would require the interfacial energy of the glass from dispersion forces alone to be 9.6 J/m^2. The value is unrealistically large and indicates the presence of chemical bonds, most likely between silanol and siloxane groups of the glass and PDMS surfaces [27].

The interaction between a glass particle and PDMS under tensile load using a stiff cantilever in Figure 11b is remarkably different from that using a highly-flexible cantilever (*cf.* Fig. 6b). It can be seen that whilst the particle remains in contact the interaction is close to symmetric, providing evidence that the shear forces have been minimised. There remains, however, a small element of asymmetry as the bend in the cantilever causes the particle to "lean" slightly to the right. On closer inspection, small differences between the left and right hand side profiles can also be discerned, which, in essence, must have different values of G_c. It would, therefore, be unreasonable to expect detachment to occur *via* a symmetric reduction in the contact radius until instability is reached at a given load. Figure 12 shows an instant in the actual detachment process of the glass particle from the PDMS to confirm the supposition. A crack can be seen to propagate from the left side of the particle as the right side remains firmly in contact, the process occurring over several 10's of seconds. The mechanism for particle removal under minimal shear force was through a small reduction in the contact radius until a more rapid crack propagated from one side. There was no visible evidence of cohesive failure of the PDMS, unlike when removal occurred in the presence of strong shear forces. Once the glass particle was detached, the PDMS returned to its flat geometry within a few seconds with a minimal amount of residual build-up (of unknown origin) surrounding the contact patch. Whilst the work presented here has been concerned with the study of fine particle interactions, its significance to AFM use warrants a short summary. It has been illustrated that both the adhesion and the applied load, between a probe and a sample, result in cantilever distortions which give rise to non-normal forces. Whilst higher applied loads and higher adhesion forces than those usually encountered in

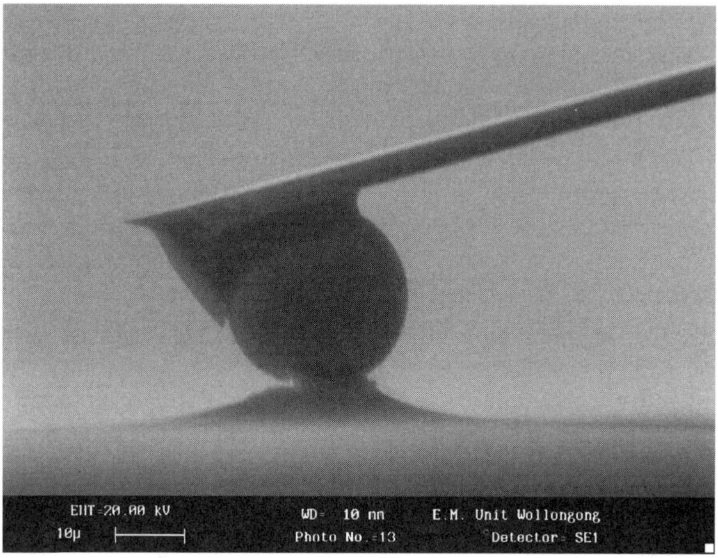

FIGURE 12 Detachment of the glass particle observed in Figure 11 from the PDMS surface can be seen to occur *via* the propagation of a crack on the (left) side farthest away from the fixed end of the cantilever.

AFM measurements have been studied here, there is little reason to suspect that the relative contribution of non-normal forces under small loads or weak adhesion would be less significant. Attard [34] has calculated that applying loads as small as 5 nN on an imaging tip in contact with a flat sample causes hysteresis in the constant compliance region due to shear caused by friction. Since cantilever distortions are difficult to characterise or even to detect in an AFM, the contribution of shear forces to measured data cannot be unequivocally determined. It has also been shown here that the role of compliant materials, used for particle attachment and surface modifications, must be given serious consideration.

CONCLUSIONS

A new technique has been used to study interactions between micron-sized particles and flat substrates under applied load. Its advantages include the ability to examine contact mechanics in the smaller particle size domain, to measure directly and to observe particle removal under variable shear forces and to provide insight into the interpretation of AFM contact data.

Contact between micron-sized glass particles and PDMS led to elastic deformation of the latter, both in the presence and absence of load. The (near) equilibrium work of adhesion, W_A, was readily derived from the size of the contact patch under zero load. It was, however, a poor indication of G_c and, hence, the normal force required to remove the particles. Whilst the large disagreement between the two values was easily rationalised, it highlights the need for direct removal of particles if normal detachment forces are to be determined.

Shear forces deformed the contact between PDMS and glass particles under normal load to change the interaction geometry significantly. Understanding such changes are important to applications based on the use of fine particles as abrasives or lubricants. The effect of shear was also distinct in the removal of glass particles as it was able to reduce the measure of G_c and even to change the mode of particle detachment. Both these effects illustrated the relationship between shear and normal forces in the removal of rigid particles from deformable surfaces. The direct measure of fine particle removal forces whilst observing their detachment should prove useful to the study of many applied processes.

Acknowledgements

This work was jointly sponsored by the Australian Research Council (ARC), the University of Wollongong and BHP Australia. The PDMS and the JKR measurements were made by Robert Oslanec, whose help is greatly appreciated.

References

[1] Eastman, T. and Zhu, D. M., *Langmuir* **12**(11), 2859 (1996).
[2] Lin, X. Y., Creuzet, F. and Arribart, H., *J. Phys. Chem.* **97**(28), 7272 (1993).
[3] Drummond, C. J. and Senden, T. J., *J. Colloids Surf.* **A87**, 217 (1994).
[4] Ducker, W. A., Senden, T. J. and Pashley, R. M., *Nature* **353**, 239 (1991).
[5] Toikka, G. and Hayes, R. A., *J. Colloid Interface Sci.* **191**, 102 (1997).
[6] Rabinovich, Y. I. and Yoon, R. H., *Langmuir* **10**(6), 1903 (1993).
[7] Toikka, G., Hayes, R. A. and Ralston, J., *J. Chem. Soc., Faraday Trans.* **93**(19), 3523 (1997).
[8] Larson, I., Drummond, C. J., Chan, D. Y. C. and Grieser, F., *J. Phys. Chem.* **99**(7), 2114 (1995).
[9] Toikka, G., Hayes, R. A. and Ralston, J., *Langmuir* **12**(16), 3783 (1996).
[10] Schaefer, D. M., Carpenter, M., Gady, B., Reifenberger, R., DeMejo, L. P. and Rimai, D. S., *J. Adhesion Sci. Technol.* **9**(8), 1049 (1995).
[11] Toikka, G., Hayes, R. A. and Ralston, J., *J. Colloid Interface Sci.* **180**, 329 (1996).
[12] Mizes, H. A., *J. Adhesion* **51**, 155 (1995).
[13] Biggs, S. and Spinks, G., *J. Adhesion Sci. Technol.* **12**(5), 461 (1998).
[14] DeMejo, L. P., Rimai, D. S. and Bowen, R. C., *Particles on Surfaces*, Mittal, K. L., Ed. (Plenum, New York, 1989), pp. 49–58.
[15] DeMejo, L. P., Rimai, D. S. and Bowen, R. C., *J. Adhesion Sci. Technol.* **5**(11), 959 (1991).
[16] DeMejo, L. P., Rimai, D. S., Chen, J. H. and Bowen, R. C., *Particles on Surfaces*, Mittal, K. L., Ed. (Marcel Dekker, Inc, New York, 1995), pp. 33–45.
[17] Bowen, R. C., DeMejo, L. P. and Rimai, D. S., *J. Adhesion* **51**, 201 (1995).
[18] Rimai, D. S., Moore, R. S., Bowen, R. C., Smith, V. K. and Woodgate, P. E., *J. Mater. Res.* **8**(3), 662 (1993).
[19] Rimai, D. S., DeMejo, L. P. and Bowen, R. C., *J. Adhesion* **51**, 139 (1995).
[20] Israelachvili, J. N. In: *Fundamentals of Friction: Microscopic and Macroscopic Processes*, Singer, I. L. and Pollock, H. M. Eds. (Kluwer, Dordrecht, 1992), pp. 351–385.
[21] Soltani, M. and Ahmadi, G., *J. Adhesion Sci. Technol.* **8**(7), 763 (1994).
[22] Toikka, G., Hayes, R. A. and Ralston, J., *J. Adhesion Sci. Technol.* **11**(12), 1479, (1997).
[23] Johnson, K. L., Kendall, K. and Roberts, A. D., *Proc. R. Soc. London, Ser.* **A324**, 301 (1971).
[24] Dupre, A., *Theorie Mechanique de La Chaleur* (Gauthier-Villars, Paris, 1869).
[25] Barquins, M., *J. Adhesion* **14**, 63 (1982).

[26] Rimai, D. S., DeMejo, L. P., Vreeland, W., Bowen, R. C., Gaboury, S. R. and Urban, M. W., *J. App Phys.* **71**(5), 2253 (1992).
[27] Kim, S., Choi, G. Y., Ulman, A. and Fleischer, C., *Langmuir* **13**, 6850 (1997).
[28] Perutz, S., Kramer, E. J., Baney, J. M. and Hui, C.-Y., *Macromolecules* **30**(25), 7964 (1997).
[29] Perutz, S., Kramer, E. J., Baney, J. M., Hui, C.-Y. and Cohen, C., *J. Polym. Sci. Part A: Polym. Chem.* **36**(12), 2129 (1998).
[30] Rimai, D. S., DeMejo, L. P., Vreeland, W. B. and Bowen, R. C., *Langmuir* **10**(11), 4361 (1994).
[31] Israelachvili, J. N., *Intermolecular and Surface Forces*, 2nd edn. (Academic Press, London, 1992).
[32] Barquins, M., *M. Mater. Sci. Eng.* **73**, 45 (1985).
[33] Maugis, D., *J. Colloid Interface Sci.* **150**(1), 243 (1992).
[34] Attard, P., *Langmuir* **15**(2), 553 (1999).

Atomic Force Microscope Techniques for Adhesion Measurements

D. M. SCHAEFER

Department of Physics, Astronomy and Geosciences, Towson University, Towson, MD 21252, USA

J. GOMEZ

Dept. Fisica de la Materia Condensada, Universidad Autonoma de Madrid, 28049-Madrid, Spain

The Atomic Force Microscope (AFM) has become a powerful apparatus for performing real-time, quantitative force measurements between materials. Recently the AFM has been used to measure adhesive interactions between probes placed on the AFM cantilever and sample surfaces. This article reviews progress in this area of adhesion measurement, and describes a new technique (Jump Mode) for obtaining adhesion maps of surfaces. Jump mode has the advantage of producing fast, quantitative adhesion maps with minimal memory usage.

I. INTRODUCTION

The ability to measure interaction forces between an Atomic Force Microscope [1] probe and a surface was exploited shortly after the invention of the AFM. Examples of these forces include electrostatic [2], magnetic [3], double layer [4], van der Waals [5], and frictional forces [6]. Of particular interest to this discussion is the ability of the AFM to measure the force of adhesion between the probe and substrate [7, 8]. Such measurements have been performed on a wide range of conducting and insulating materials [9] in a variety of media, including UHV [10], ambient [11], and liquid [12] environments. Initially, the high force resolution was exploited to produce three-dimensional topographical images of a sample surface. To optimize the lateral resolution during such imaging, sharp probes were used. Recently, however, researchers have modified the AFM probes in order to study specific probe/substrate interactions. By placing a well-characterized micrometer – size sphere on the end of the AFM cantilever, surface force interactions between the sphere and atomically-flat substrates have been reported [13]. Parameters such as lift-off force and work of adhesion have been investigated as a function of applied load. These studies have allowed researchers to investigate the effects of surface roughness and environmental conditions on the measured adhesive force [14]. Additionally, molecular interactions have been studied using the AFM [15]. The force needed to separate individual molecules between functionalized AFM probes and substrates demonstrate the possibility of sensors with single-molecule recognition capabilities.

The following article describes an application of the AFM that combines the high-resolution imaging ability with force measurement. Performing force measurements at various points on a surface provide images representing surface topography, stiffness, and adhesion.

II. FORCE MEASUREMENTS WITH AFM

All quantitative force data derive from a force measurement that results in a force *vs.* displacement curve. Figures 1 and 2 illustrate how force measurements are performed between an AFM probe and a surface. The probe and the surface are brought into close proximity using a coarse approach mechanism. For most force measurements, the probe position is initially several hundred nanometers above the surface. The substrate is then moved toward the probe in a controlled way using a piezoelectric tube. The interaction between the probe and the substrate is monitored by sensing the deflection of the cantilever. Knowledge of the spring constant allows a conversion from cantilever displacement into force. A representation of the force exerted on the probe as a function of substrate displacement is shown in Figure 2. As the probe is brought close to the substrate, any forces acting between the probe and

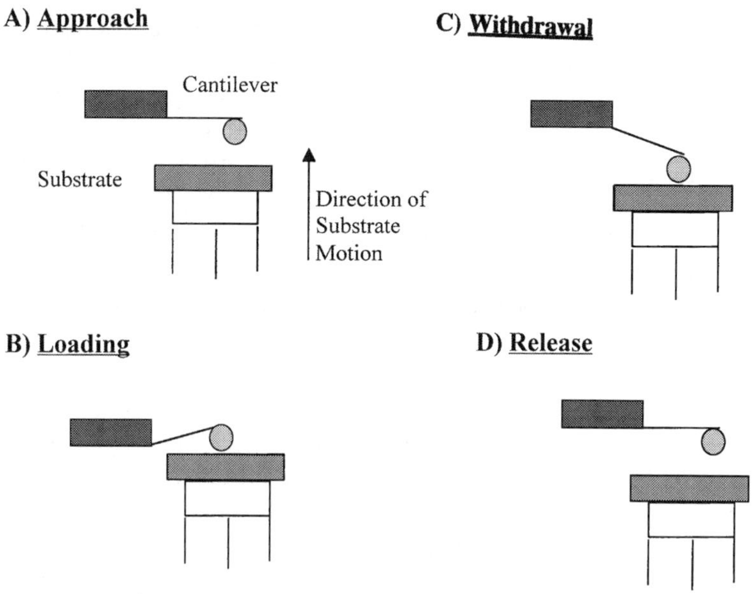

FIGURE 1 Schematic illustrating the process of force measurement with the AFM.

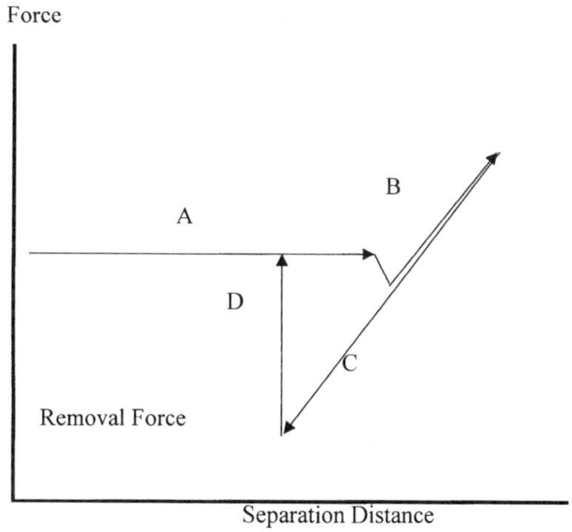

FIGURE 2 Typical shape of the force curve obtained using the AFM.

substrate causes a deflection of the cantilever. The distance-dependence of this deflection is useful for identifying the origin of the interaction force. When the tip is only a few tens of nanometers from the substrate, an instability occurs. The cantilever flexes and the probe jumps into contact with the substrate. The sample is continually moved forward until a specified positive load is applied by the probe onto the substrate (Fig. 1B). The sample is then withdrawn (Fig. 1C) at the same rate as the approach rate. During unloading, the cantilever motion initially retraces the loading curve (for elastic materials). However, due to binding forces along the contact region, the probe may not separate from the substrate at the same point where contact was originally established. Rather, a substantially larger negative load is required before an abrupt separation occurs (Fig. 1D). This force is the removal force, or adhesion force.

Schaefer *et al.* [16] studied the adhesive forces by systematically measuring the force required to remove different particles from a variety of substrates. In this study, the relative lift-off forces were found to scale consistently with the relative works of adhesion in a manner qualitatively consistent with the predictions of Johnson, Kendall and Roberts (JKR) theory of adhesion [17]. The absolute magnitude of the lift-off force was found to be smaller than expected. This effect was attributed to surface roughness of the particle. Although this study was confined to well-characterized, spherical particles and atomically-flat substrates, the results obtained indicate that the AFM techniques developed are capable of providing useful, quantitative information about the particle adhesion that is difficult to obtain using more conventional techniques.

III. ADHESION MAPS

A natural extension of the previous results is to measure the lift-off force as a function of sample location between the AFM cantilever and a substrate. Figure 3 illustrates the method of producing these adhesion maps. The probe, placed on the AFM cantilever, is raster-scanned from point to point in an $n \times n$ array across an area of the surface. At each point in the array, the surface topography is determined and plotted using the usual false color techniques. A force curve is also taken at each point, and the lift-off force is determined from the minimum of the retrace curve. This force is then plotted using the same false coloring techniques as employed for the topography scans.

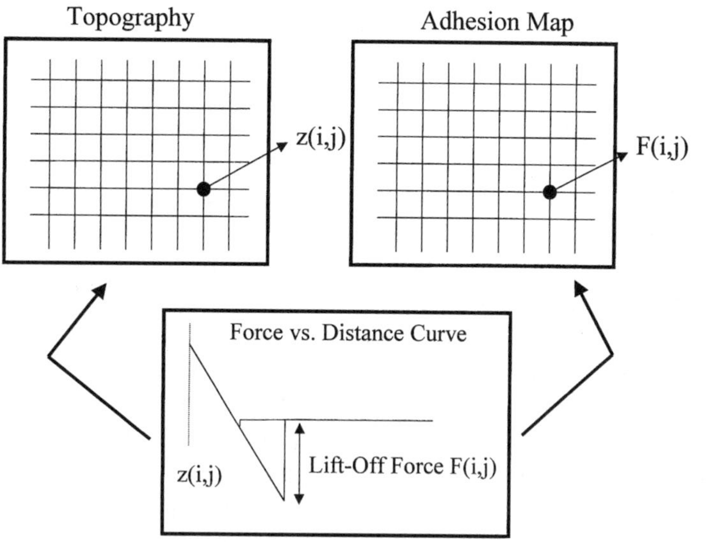

FIGURE 3 Schematic of an adhesion map showing how the topography and adhesion of the tip to the substrate are measured simultaneously.

Different measuring modes have been developed to study the spatial dependence of the adhesion force. Mizes *et al.* [18] first used a commercial AFM to spatially map the adhesion force between a bare Si_3N_4 tip and doped and undoped polycarbonate films. In this study, individual force curves were taken at each imaging point on the sample. A computer algorithm processed each force curve in real time and stored only the minimum of the retrace curve. Adhesion measurements were demonstrated to be repeatable to within 2% when taken over the same region of the sample. It was noted that this result conflicts with macroscopic measurements where repeated contacts cause charge transfer and an associated increase in adhesion due to Coulombic attraction. Mizes argued that the enhanced adhesion is not seen at the microscopic scale because (1) triboelectric charging does not contribute significantly to the adhesive force at this scale, (2) all the tribocharging occurs with first contact, or (3) the sample is slightly conducting over the whole surface between contacts. Experiments by Gady [12, 19, 20] have shown that the charge on a particle attached to an AFM cantilever will increase with increasing contact. This would suggest that (2) is not responsible for the repeatability.

Additionally, experiments by Schaefer [21] have demonstrated that charging can play a significant role in adhesion force for large charge densities. The amount of charge accumulated on the AFM tip during Mizes' experiments, however, would probably be sufficiently small such that charging would not significantly contribute to the adhesive force. For larger scan sizes, the adhesion force was shown to vary from 0.76 to 3.63 mdyn. This variation was explained by proposing a changing contact area between tip and surface due to topographic variations.

Koleske *et al.* [22] utilized a similar technique to obtain adhesion maps using a home-built system. In this technique the entire force curve was taken at each point and stored in computer memory. Adhesion forces were determined in post-acquisition processing. The major disadvantages of performing force mapping in this way are the extensive data storage space necessary and the time for imaging. In the studies by Mizes, 2×2 micron scans are reported to take 40 minutes, and Koleske reports that a 256×256 pixel image (1500 nm $\times$ 1500 nm) taken on a 2400 line/mm grating spanned 4 hours. As observed by Koleske, the effects of thermal and piezoelectric drift can become significant over these time scales.

The effects due to long imaging times were eliminated by van der Werf *et al.* [23] by using analog electronics to control the tip motion and record topography and adhesion information. Detection electronics were used to determine the sample topography, the minimum value of the withdraw curve (removal point), the width of the removal trace, and the area of the adhesion trace. Adhesion maps were obtained in air and liquid, with force curves taken at a rate of 550 Hz (air) and 70 Hz (liquid). Thus, in air, the time for the adhesion map to be completed would be on the order of a few minutes. Van der Werf was able to use this apparatus to show the effects of humidity on the adhesive force in air. By comparing the adhesive force between a Langmuir-Blodgett monolayer film (DPDA) placed on a glass slide with the adhesive force of the bare glass substrate, it was shown that the adhesive force on the DPDA (270 nN) was less than that on the glass (300 nN). Because the DPDA is less hydrophilic than glass, the glass surface will accumulate a larger water content. This water will, therefore, produce an increase in the net adhesive forces due to capillary force contributions. Experiments were also done in liquid, demonstrating a reduction of adhesion due to the absence of a capillary contribution. In these liquid studies, van der Werf performed adhesion maps on gold films covered by a self-assembling monolayer. As the liquid medium was varied, the adhesion force was observed to change. These results were in agreement with observations found by Hoh *et al.* [24] who showed that adhesion forces should be diminished by adding ions to the solution. A similar approximation can also be seen in the work of Hans-Ulrich Krotil *et al.* [25].

The experiments described above can be categorized as "force-volume" and "pulsed-force microscopy" (as is defined by de Pablo *et al.* [26]). In force-volume mode (Koleske), the entire force curves are stored during imaging and the relevant parameters are determined during post-acquisition processing. In pulsed-force microscopy (Mizes, van der Werf) the physically-interesting data are extracted from the force curves during the imaging process. In the remainder of this paper, a new technique for performing quantitative, real-time force mapping is discussed in detail. This mode of adhesion mapping is called Jump Mode (JM) Scanning Force Microscopy [27].

IV. JUMP MODE SCANNING FORCE MICROSCOPY

IV.1. Jump Mode Description

Simultaneous, real-time images of the sample topography and the lift-off force are performed in Jump Mode as illustrated in Figure 4, where the motion of the tip as a function of time along (a) the x-axis and (b) the z-axis is plotted. In addition, (c) the cantilever deflection recorded as the tip is moved along the z-axis is also illustrated. In (d), a schematic of the resulting force curve is shown.

The AFM tip is first brought into contact with the sample during the coarse approach, under feedback control, until the setpoint cantilever displacement is achieved. The feedback is then disabled, and a ramp to the z-piezo for the sample is applied, withdrawing the sample by a prescribed amount. With the tip in this position, the topography and adhesion maps can be performed. At each point on the image, the control unit ramps the z-piezo voltage to move the sample through a given distance, δ, *toward* the tip. This brings the tip into contact with

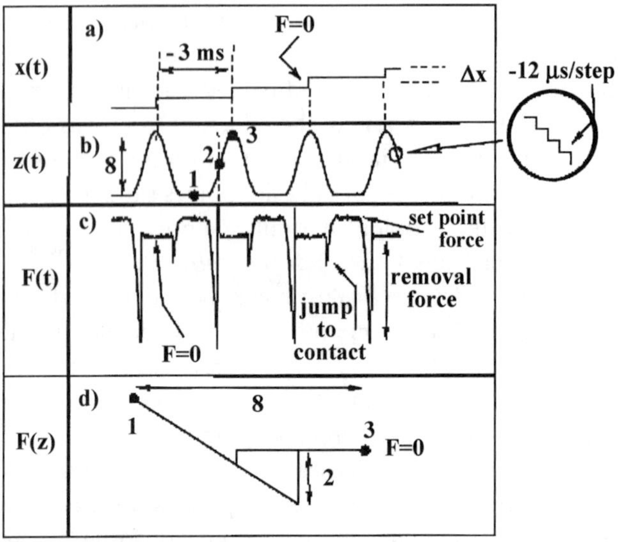

FIGURE 4 Oscilloscope traces showing (a) the position of the sample along the x-axis as a function of time, (b) the position of the sample along the z-axis as a function of time, (c) the normal force applied to the substrate by the tip as a function of time and (d) a schematic of the resulting force curve obtained during one cycle of the z-piezo.

the sample. The cantilever displacement is monitored during this process, providing the approach segment of the force curve. When the ramp is complete, the feedback is then enabled for a short time (typically 1 ms) and its output is stored as the topographic height. The feedback is then disabled, and an inverse ramp is applied to the z-piezo, withdrawing the tip from the sample. Again, the cantilever is monitored, providing the "withdrawal" portion of the force curve.

When the largest z value is reached (*i.e.*, the maximum tip – sample distance), a voltage step is applied to the x-piezo and the tip is quickly moved parallel to the surface. This allows motion without lateral force, and is similar to a "tapping-mode" scan. This process is repeated for all points in the image, providing topographic and force curve data at each point. The entire force curve at each point is, however, not stored in memory. Storing such data would significantly slow the scanning speed and require large amounts of memory. After each force curve has been completed, the relevant local parameters (such as lift-off force or elastic modulus) are extracted from the force curve, and only these values are stored in memory. Following this approach, an entire adhesion map consisting of 256 × 256 data points takes about 10 minutes to complete.

In order to perform quantitative adhesion maps in real time, it is important to optimize the system's performance. Since the modulation of the *z*-position during force curves in Jump Mode is typically performed by ramping at a frequency of 1 kHz, resonant frequencies can easily be excited in the system. In order to prevent these excitations, a sine wave ramp is applied to the *z*-piezoelectric driver. The frequency of the sinusoidal ramp is chosen away from the resonance frequencies of the system.

IV.2. Comparison of Jump Mode with Other Adhesion Mapping Modes

Jump Mode combines the features of lift mode, of intermittent contact (tapping) mode scanning, of pulsed-force mode, as well as force-volume. As in the last three modes, in JM the lateral displacement of the tip and sample is done when they are not in contact, thus avoiding shear forces and the corresponding damage to tip and sample. As in force-volume mode, tip-sample interaction is measured at each point. However, in JM the interaction is evaluated and characterized usually by one parameter, which is displayed as an image in real time. The advantages of JM compared with force-volume are, on the one hand, a dramatic reduction of stored data and, on the other hand, its direct visualization without the need for post-acquisition processing. JM is similar to lift mode, since in both cases topography and tip-sample interaction are measured in different phases of an acquisition cycle. In JM, one cycle is performed at every image point, whereas, in lift mode, a whole scan line is acquired in a cycle. This implies that tip-sample distance is much better controlled in JM, since it is "refreshed" at every point after feedback is performed. In lift mode, this happens only after each scan line and, in fact, it has been observed that this distance varies along one scan line, presumably due to piezo creep, hysteresis, and/or cross talk between different piezotube axes. Finally, pulsed-force microscopy can be considered a special application of JM to measure adhesion, even though data processing is performed by analog electronics, as opposed to digital processing in the case of JM.

IV.3. Jump Mode Applications

Two different AFM heads were used to perform the following experiments presented here. The first AFM is a commercially-available instrument from Nano Tec^{TM} which features a modular design, easy adjustment of the optical system, top view of the AFM scanning region which makes it easy for optical microscope inspection, low thermal drift and a humidity-controlled system. The scanning range is up to $50\,\mu m \times 50\,\mu m$. The cantilever deflection was monitored by a laser beam deflection system. The system uses a large piezotube to minimize hysteresis effects. SiN_3 pyramidal cantilevers from Olympus ($k \sim 0.4\,N/m$) were used in the system.

The second system has been described previously [28, 29] and is mounted in a small stainless steel vacuum chamber, allowing for control of environmental conditions. To avoid problems associated with absorbed water, the system was repeatedly pumped out to pressures of 20 mTorr, followed by a backfill with dry nitrogen gas. To study the effects of ambient conditions on adhesion, the system could be vented to the atmosphere. Si ultracantilevers from Park Scientific ($k \sim 2.0\,N/m$) were used in the system.

Several procedures have been followed during these adhesion studies. Briefly, detection of the AFM cantilever displacement is performed using laser deflection techniques. Calibration of the cantilever's spring constant is performed by measuring the resonance frequency and using manufacturer-specified parameters such as length, width, elastic modulus, and density to calculate the spring constant [30]. In the second AFM system, a second laser deflection system is used to monitor the motion of the sample piezoelectric tube. This is critical for eliminating nonlinearities and creep when performing force measurements.

Both AFMs were controlled by a PC-based Nano Tec^{TM} control unit. This system of hardware and software controls all aspects of data acquisition, processing, and feedback, and is now a standard feature of the Nano Tec^{TN} software package. The core of the system is a Digital Signal Processor (DSP) with 4 simultaneous ADC/DAC channels, each with 16-bit accuracy. The DAC outputs drive a high-voltage amplifier unit which provides the scanning signals. Control and data signals are input through the ADC channels. Scanning and force measurements are performed in real time under the execution of a C program which resides in the DSP memory.

V. RESULTS

Adhesion maps have been performed on a wide variety of different materials to explore fully the capabilities of this new technique. The following are a few selected examples.

V.1. Single-walled Carbon Nanotubes on SiO_2

As an example of adhesion mapping using Jump Mode, topographic and adhesion images were obtained on single-walled carbon nanotubes placed on a flat SiO_2 substrate. A sample of the results is shown in Figure 5. Simultaneous images of topography and adhesion are shown for a 250 nm × 250 nm area. The adhesion map shows three distinct regions with different adhesive characteristics. On the SiO_2 substrate, the adhesion is observed to be relatively constant, with slight changes in removal force appearing due to surface roughness. The removal force on the nanotubes is observed to be significantly less than that of the SiO_2 substrate (2 nN difference). A third region observed in the adhesion image is an area of high adhesion on either side of the carbon nanotubes. These results can be understood by applying the Derjaguin approximation [31, 32] which describes the interaction force $F(D)$ between two spherical bodies in close proximity (contact) as a function of separation distance, D.

$$F(D) = W\left(\frac{1}{C_t + C_s}\right) \tag{1}$$

Here, we see that the adhesion force will be a function of two factors: (a) the local geometry of the tip and surface (represented by the curvature of the tip, C_t, and the curvature of the sample, C_s), and (b) all other factors including chemical composition (represented in the form of W, the energy per unit area of two flat surfaces at separation D). From a purely geometric perspective, if the tip is located on a perfectly flat surface, $C_s = 0$. If the tip is located on top of a region

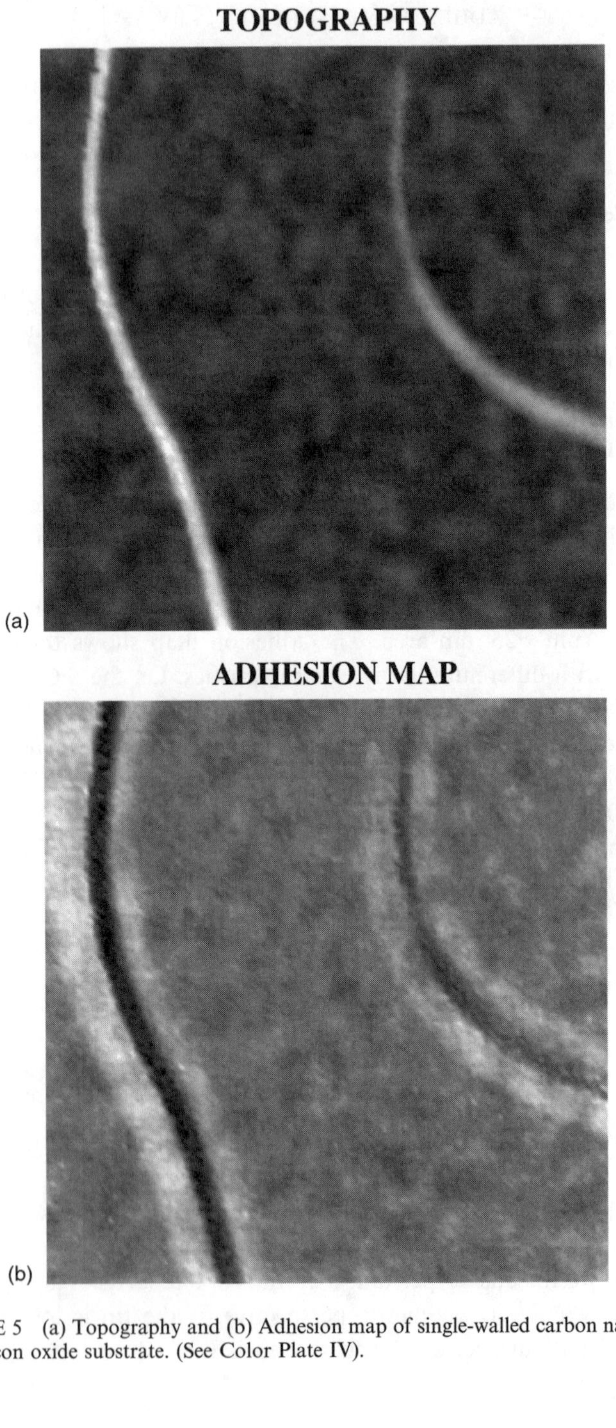

FIGURE 5 (a) Topography and (b) Adhesion map of single-walled carbon nanotubes on a silicon oxide substrate. (See Color Plate IV).

with positive surface curvature (such as on top of the nanotube), then Eq. (1) predicts that the adhesion force will be smaller than over the flat area. Conversely, if the curvature of the surface is negative (when the tip is on the side of the nanotube), the adhesion force will be greater. Therefore, the observations in the adhesion map can be qualitatively understood from these geometric considerations. Additionally, the chemical composition of the surfaces will also be a factor. The surface free energy for graphite is lower than that of silicon oxide, which, again, will act to produce a lower adhesion on the nanotube.

V.2. Au Bridge on Glass

In order to investigate further the effects of surface roughness and composition on adhesion maps, two thin gold electrodes were thermally evaporated onto a clean glass slide in such a way as to form two separate electrical contact pads, each 60 nm thick and separated by $\sim$10 μm. A nominal 30 nm thick Au film was then evaporated through a mask positioned at right angles to these electrodes to form a thin electrical connection (*i.e.*, a bridge) between the two contact pads.

Both topography and adhesion maps were obtained in the region of the Au bridge as shown in Figure 6. It is clear from a careful examination of the topographic images that the roughness of the Au film increases in proportion to the thickness. It is interesting to see how this surface roughness influences the adhesion. An analysis of the relevant adhesion maps permits this study to be performed. Adhesion maps show that the adhesion (*i.e.*, lift-off force) increases as the gold surface becomes rougher. In addition, clear evidence for a local decrease in adhesion due to particulate contamination of the glass substrate is also evident in Figure 6.

The variation in adhesion was quantitatively determined by calculating the RMS roughness and average adhesion force in the three relevant areas. These results are shown in Table I. As was observed qualitatively, a clear correlation is observed between the surface roughness and the adhesion force.

It is also instructive to examine the distribution of heights and adhesion forces over the entire images from Figure 6. The topography and adhesion histograms are shown in Figure 7. An examination of

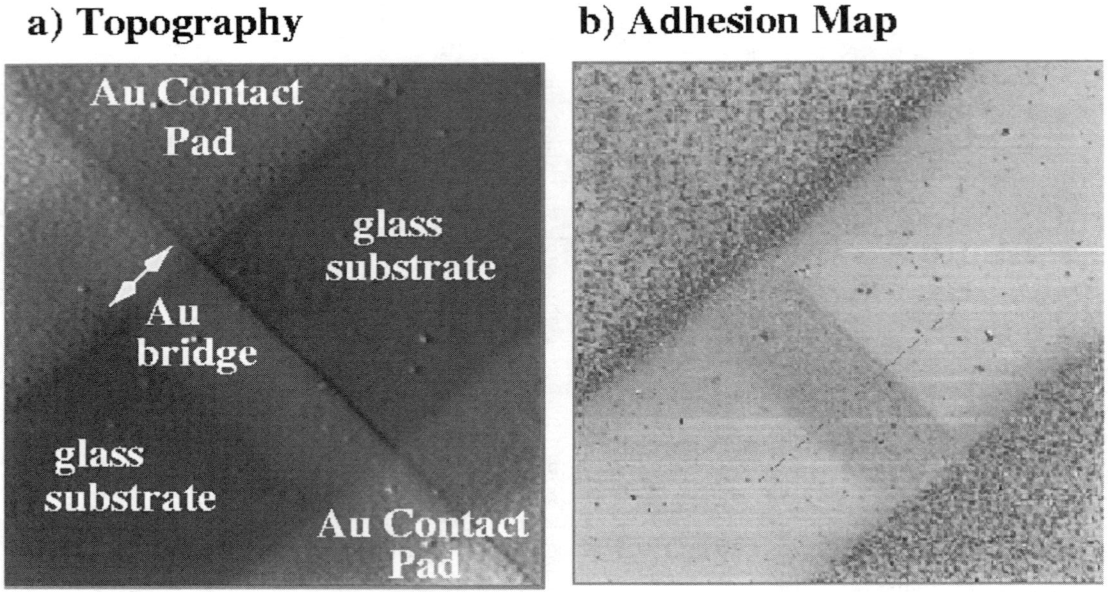

FIGURE 6 (a) A 10 μm × 10 μm AFM topographic scan (contact mode) of two Au contact pads spanned by a thin Au bridge supported on a glass substrate. In (b), an adhesion map of the same region. A mapping algorithm was used such that a lighter color (grey-scale in B & W) implies a larger adhesion force. (See Color Plate V).

TABLE I Quantitative comparison of topographic and adhesion force variations

RMS roughness	*Contact pad*	*Gold bridge*	*Glass substrate*
Topography (nm)	7.1	3.5	2.7
Adhesion force (a.u.)	619	278	202

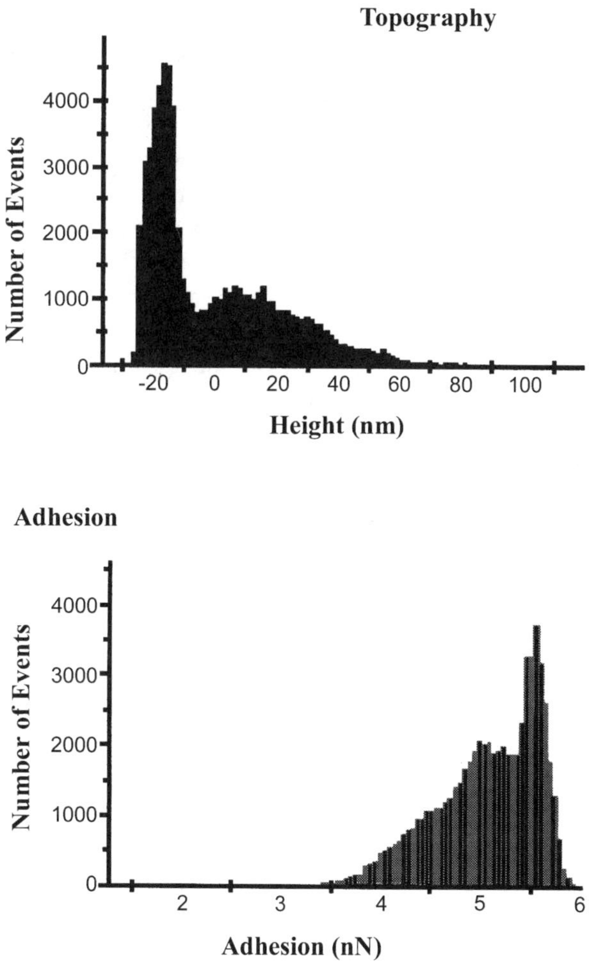

FIGURE 7 A histogram from the images shown in Figure 8 displaying the distribution of heights and lift-off forces.

the topographic histogram shows two features which represent (1) the glass surface (centered at $-20\,\text{nm}$) and (2) the Au bridge plus the Au pads (centered around $\sim$30–40 nm). The spread in height measured from the Au contact pad is evident and indicates a greater roughness characterizing this thicker Au film. By contrast, the glass substrate is smoother, as can be seen by the narrower peak in the topographic histogram.

The adhesion histogram also reveals two features that can be identified with the same topographic features. The adhesion between the glass substrate and AFM tip was found to require on average a 5.5 nN lift-off force. Because of the smoothness of the glass substrate, we believe the adhesion is dominated by the inter-molecular forces acting between the tip and glass. The region of the histogram with adhesion forces ranging from approximately 5 to 5.3 nN corresponds to forces found on the gold bridge. Forces below 5 nm correspond to the region of the gold contact pad. The variation in force over the contact pad region is due to the large variation in surface roughness. From Figure 7, the adhesion force from the Au bridge is found to be more uniform than from the Au contact pads. This correlated with the smoother topography associated with the Au bridge.

V.3. Adhesion from an Argon – Ion Sputtered HOPG Substrate

It is clear that both topography and chemistry play an important role in determining the shape of an adhesion map. For this reason, it is useful to investigate a surface that minimizes the chemical effects to adhesion. Highly-ordered pyrolytic graphite (HOPG) is ideal for this purpose because it is atomically flat over large regions. Furthermore, due to the bonding of the carbon atoms to form graphite sheets, the surface of flat HOPG is known to be unreactive. However, if HOPG is roughened, the adhesion properties should change.

A roughened sample of HOPG was prepared to study the alteration of adhesion due to surface roughness. An HOPG sample was masked in such a way so as to expose approximately one-half of the surface to an Argon – ion discharge for approximately 10 minutes at an Argon pressure of $\sim 10^{-3}$ Torr with the sample biased at a potential of 1000 V with respect to ground. The boundary region demarcating the part of the surface exposed to the discharge and that covered by the mask was then investigated both in the topography and adhesion map modes. The results are shown in Figure 8 which shows three maps of both the topography and adhesion near the interface region. A histogram analysis allows the quantitative determination of the roughness by measuring the standard deviation in topography and the average lift-off force from the adhesion map. These values are listed under the three images. A clear correlation between increasing roughness and adhesion can be established in this way.

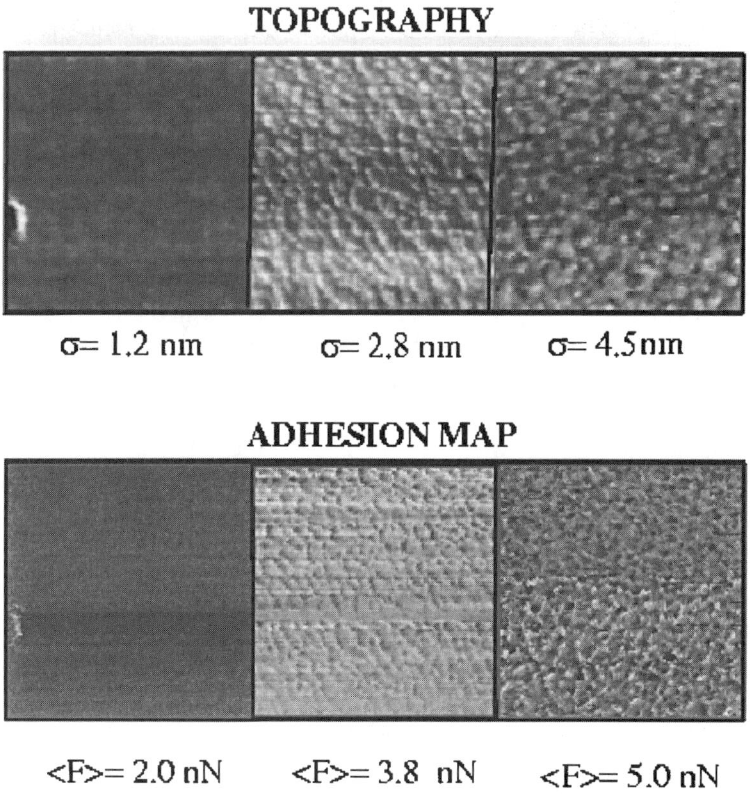

FIGURE 8 Three images, each spanning 1 μm × 1 μm, of an HOPG surface exposed to an Argon – ion discharge for 10 minutes. Values for the roughness (calculated from the standard deviation of the topographic image (σ)) and the average lift-off force ($\langle F \rangle$) (acquired from the adhesion map) are listed. A quantitative correlation between substrate roughness and adhesion can be estimated in this way. (See Color Plate VI).

VI. CONCLUSION

The Atomic Force Microscope has developed into a quantitative tool for the measurement of adhesion forces between materials with nanometer-scale lateral resolution. Exploiting this ability, adhesion maps have been produced using a new technique called Jump Mode to study the adhesion force as a function of position on a material. This technique utilizes a DSP to acquire, process, and store topographical and adhesion data during AFM operation. Jump mode accommodates fast data acquisition rates, does not require large amounts of memory for data storage, and can be modified through software for other applications. This imaging method requires only the electronics required to perform contact-mode imaging. Additionally, lateral forces are virtually eliminated, so that mapping of delicate samples with high resolution in air and fluids is easily possible. Experiments were performed on a variety of materials to illustrate the effects of surface roughness and material composition on adhesion.

Acknowledgements

The authors would like to thank S. Howell and R. Reifenberger (Purdue University) and D. Rimai (Nexpress) for many helpful comments and suggestions. Additionally, the authors would like to thank P. J. de Pablo, J. Colchero, and A. M. Baro for their contributions toward this work. The work in Madrid was supported by project CICYT PB95-0169. The work at Purdue was partially funded by the Office of Imaging Division of Eastman Kodak.

References

[1] Binnig, G., Quate, C. F. and Gerber, Ch., *Phys. Rev. Lett.* **56**, 930 (1986).
[2] Hao, H. W., Baro, A. M. and Saenz, J. J., *J. Vac. Sci. Technol.* **B9**(2), 1323 (1991).
[3] Abraham, D. W., Williams, C. C. and Wickramasinghe, H. K., *Microscopy* **152**, 863 (1988).
[4] Ducker, W. A., Senden, T. J. and Pashley, R. M., *Nature* (London) **353**, 239 (1991).
[5] Gady, B., Schleef, D., Reifenberger, R., DeMejo, L. P. and Rimai, D. S., *Phys. Rev. B* **53**, 8065 (1996).
[6] Meyer, E., Overney, R., Howard, L., Brodbeck, D., Luthi, R. and Guntherodt, H. J., In: *Fundamentals of Friction: Macroscopic and Microscopic Processes*, Singer, I. L. and Pollock, H. M. (Eds.) (Kluwer Academic Publishers, Netherlands, 1992), pp. 427–436.

[7] Schaefer, D. M., Carpenter, M., Reifenberger, R., DeMejo, L. P. and Rimai, D. S., *J. Adhesion Sci. Technol.* **8**, 197 (1994).
[8] Mizes, H. A., *J. Adhes. Sci. Technol.* **8**, 937 (1994).
[9] Rugar, D. and Hansma, P., *Physics Today*, p. 23 Oct. (1990).
[10] Koshiba, K., Tanaka, I., Nakamura, Y., Nobe, H. and Sakaki, H., *Appl. Phys. Lett.* **70**, 883 (1997).
[11] Binnig, G. K., *Physica Scripta* **T19**, 53 (1987).
[12] Drake, B., Prater, C. B., Weisenhorn, A. L., Gould, S. A., Albrecht, T. R., Quate, C. F., Cannell, D. S., Hansma, H. G. and Hansma, P. K., *Science* **243**, 1586 (1989).
[13] Schaefer, D. M., Carpenter, M., Reifenberger, R., DeMejo, L. P. and Rimai, D. S., *J. Adhesion Sci. Technol.* **8**, 197 (1994).
[14] Schaefer, D. M., Carpenter, M., Gady, B., Reifenberger, R., DeMejo, L. P. and Rimai, D. S., *J. Adhesion Sci. Technol.* (1995).
[15] Lee, G. U., Kidwell, D. A. and Colton, R. J., *Langmuir* **10**(2), 354 (1994).
[16] Schaefer, D. M., Carpenter, M., Gady, B., Reifenberger, R., DeMejo, L. P. and Rimai, D. S., *J. Adhesion Sci. Technol.* **9**, 1049 (1995).
[17] Johnson, K. L., Kendall, K. and Roberts, A. D., *Proc. R. Soc. London* **A324**, 301 (1971).
[18] Mizes, H. A., Loh, K. G. R., Miller, J. D., Ahuja, S. K. and Grabowski, E. F., *Appl. Phys. Lett.* **59**, 22, 2901 (1991).
[19] Gady, B., Reifenberger, R., Rimai, D. S. and DeMejo, L. P., *Langmuir* **13**, 2533 (1997).
[20] Gady, B., Reifenberger, R. and Rimai, D. S., *J. Appl. Phys.* **84**, 319 (1998).
[21] Schaefer, D. M., Unpublished.
[22] Koleske, D. D., Lee, G. U., Gans, B. I., Lee, K. P., Dilella, D. P., Wahl, K. J., Barger, W. R., Withman, L. J. and Colton, R. J., *Rev. Sci. Instrum.* **66**, 9, 4566 (1995).
[23] van der Werf, K. O., Putman, C. A., de Grooth, B. G. and Greve, J., *Appl. Phys. Lett.* **65**, 9, 1194 (1994).
[24] Hoh, J. H., Cleveland, J. P., Prater, C. B., Revel, J. P. and Hansma, P. K., *J. Amer. Chem. Soc.* **114**, 4917 (1992).
[25] Hans-Ulrich Krotil, J., Stifer, T., Waschipky, H., Weishaupt, K., Hild, S. and Marti, O., *Surf. Interface Anal.* **27**, 336–340 (1999).
[26] de Pablo, P. J., Colchero, J., Gomez-Herrero, J. and Baro, A. M., *Appl. Phys. Lett.* **73**, 22, 3300 (1998).
[27] de Pablo, P. J., Colchero, J., Gomez, J., Baro, A. M., Schaefer, D. M., Howell, S., Walsh, B. and Reifenberger, R., *J. Adhesion* to be published.
[28] Mahoney, W., Schaefer, D. M., Patil, A., Andres, R. P. and Reifenberger, R., *Surf. Sci.* **316**, 383 (1994).
[29] Schaefer, D. M., Andres, R. P. and Reifenberger, R., *Phys. Rev. B* **51**, 5322 (1995).
[30] Sarid, D., *Scanning Force Microscopy* (Oxford University press) (1991).
[31] Mizes, H. A., *J. Adhesion* **51**, 155 (1995).
[32] Israelachvili, J. N., *Intermolecular and Surface Forces* (Academic Press, London, 1985), pp. 130–133.

Limitation of the Young-Dupré Equation in the Analysis of Adhesion Forces Involving Surfactant Solutions

J. DRELICH, E. BEACH and A. GOSIEWSKA

Department of Metallurgical and Materials Engineering, Michigan Technological University, Houghton, MI 49931, USA

J. D. MILLER

Department of Metallurgical Engineering, University of Utah, Salt Lake City, UT 84112, USA

The atomic force microscope was used to measure adhesion forces between polyethylene particles, serving as model oil droplets, and mineral substrates (fluorite and quartz) in aqueous solutions of ethoxylated alcohols. Also, contact angles were measured in the kerosene–ethoxylated alcohol solution–mineral systems. Correlations obtained between adhesion and surfactant concentration for the polyethylene–aqueous solution–quartz system differs significantly from those predicted by the Young-Dupré equation for the kerosene–aqueous solution–quartz system. Interactions, characteristic for such aqueous systems, which contribute to the pull-off forces measured by atomic force microscopy are not included in the Young-Dupré equation, and are primarily responsible for the inconsistency in the adhesion *versus* surfactant concentration relationship obtained from contact angle measurements.

INTRODUCTION

The *ex-situ* remediation of oil-contaminated soil by flotation and/or washing represents a possible method for the removal of a wide range of petroleum products from soil, sand and gravel. The efficiency of such a technology is rooted in the nature of the contaminants, the type of soil and the selection of separation reagents. Both flotation and washing technologies utilize interfacial properties to remove contaminants selectively from soil particles suspended in an aqueous solution of surfactant(s). The dual, hydrophobic and hydrophilic, nature of surfactants causes them to adsorb at interfaces, thereby reducing interfacial energies (note that in many systems with a specific adsorption of surfactants to a solid, a value of the solid-water interfacial tension can increase).

The reduction and control of interfacial energies of particulates dispersed in aqueous surfactant solutions are often the keys to optimizing the effectiveness of oil removal from a soil matrix. Controlling the micro-processes such as: (i) the oil roll-up or oil drop necking and release from a mineral, and (ii) the attachment of released oil droplets to air bubbles, are important in the design of oily soil remediation processes. Simply stated, the use of surfactants minimizes the amount of energy required to remove oil and purify the contaminated soil. Ethoxylated alcohol surfactants have been largely overlooked for this application [1, 2].

The adhesion forces for oil attachment at mineral particle surfaces are poorly understood at both the microscopic and molecular scale of observation. A common approach in describing the affinity of oil for a solid surface in an aqueous environment is through the measurement of oil–water–solid contact angles (θ) and the measurement of oil–water interfacial tensions (γ_{ow}) [3–5]. The work of adhesion (W_A) is then calculated using the Young-Dupré equation [3–5]:

$$W_A = \gamma_{ow}(1 + \cos\theta) \qquad (1)$$

It follows directly from the above equation that adding surfactant to the aqueous phase can reduce the work of adhesion between an oil drop and a solid surface by changing the interfacial tensions, decreasing the oil–water interfacial tension in particular. Indeed, this approach is typically used in practice.

Although the thermodynamic basis for Eq. (1) is well established, experimental validation, particularly for oil–water–solid systems, has never been accomplished. The lack of appropriate instrumentation is probably the major reason the equation has not been validated for aqueous systems. Recent invention of atomic force microscopy (AFM) [6] and its application to adhesion force measurements certainly has the potential to help in this validation. The principles of the AFM are well described in the literature [7, 8] and will not be repeated here.

In this note, we present the results of AFM pull-off (adhesion) force measurements between polyethylene particles and mineral (fluorite and quartz) surfaces in aqueous solutions of ethoxylated alcohols. The particle–mineral adhesion forces were calculated based on the AFM pull-off results and then compared with the adhesion forces calculated from the measurements of kerosene–aqueous phase–mineral contact angles and kerosene–aqueous phase interfacial tensions. Significant discrepancies in the adhesion forces determined from the AFM study and contact angle/interfacial tension measurements are reported and discussed.

EXPERIMENTAL PROCEDURE

Materials and Reagents

Polished synthetic quartz and fluorite disks were purchased from Harrick Scientific Corporation. The RMS roughness of the mineral substrates was determined with the AFM, being 20 ± 10 nm for quartz and 10 ± 5 nm for fluorite. The disks of quartz were cleaned with a chromic solution and washed with deionized water before use. Fluorite was water-wet repolished on Chemomet cloth (Buehler) and washed with deionized water before use.

Low-density polyethylene powder (MW 1800 and melting point of 117° C) was purchased from Scientific Polymer Products, Inc., and used to prepare spherical particles according to the procedure presented in previous work [9]. Such prepared PE particles have surface irregularities in the height range from 10 to 50 nm [9]. PE particles with a size from about 10 to 14 μm in diameter were used in the AFM study.

A commercial mixture of ethoxylated alcohols, $(CH_2CH_2O)_6$ $CH_3(CH_2)_{7-9}CH_2OH$, known as Rhodasurf 91-6, with a purity of 99.5 wt%, was received from Rhodia Inc. Freshly prepared solutions of Rhodasurf 91-6 in 0.001M KCl were used in all experiments. The pH value of the solutions was pH 6.0 ± 0.2.

Kerosene was purchased from a local gas station. Kerosene was selected for this study because it is non-polar liquid with a value of kerosene–water interfacial tension equal to about 46 mN/m. This is very close to the polyethylene–water interfacial tension that is 45 ± 5 mN/m as discussed in the next part.

Atomic Force Microscopy Study

The interfacial force measurements were performed using a Nanoscope III AFM, with a Nanoscope E scanner, Digital Instruments, Inc. Polyethylene particles were glued to the AFM cantilevers by means of a micromanipulator. V-frame Si_3N_4 tipless cantilevers from Digital Instruments with a spring constant 0.58 N/m were used in this study. The adhesion force measurements were done in a fluid cell, offered by Digital Instruments, using freshly prepared solutions of Rhodasurf 91-6. Loading of the PE particle on the mineral substrate before a pull-off force measurement was nearly constant in all experiments, from 10 to 30 nN for quartz and from 20 to 50 nN for fluorite. Forty to one hundred measurements were carried out for each experimental condition examined.

The deflection in the cantilever was calculated as the horizontal distance from the point where the particle comes into contact with the surfaces (the point of attachment) to the point where the particle is removed from the surface and the cantilever springs back to the undeflected position (point of detachment). The pull-off (adhesion) force was calculated by multiplying the distance (Δz) of cantilever deflection by the force constant (k) of the cantilever.

$$F_A = k \times \Delta z \tag{2}$$

The diameter of the PE particle (R) was determined from scanning electron microscopy (SEM) micrographs after completion of adhesion force measurements.

Contact Angle Measurements

The mineral disk was located in a rectangular polystyrene cell and next covered with kerosene. A 20–40 μl drop of water or surfactant solution was placed on the surface of the mineral disk. A Krüss Drop Shape Analysis System (G10) was used to measure the contact angle for a sessile water drop situated on the mineral surface in kerosene. This contact angle corresponds to the receding contact angle when measured for the oil phase (reported in the next part of the paper), and advancing contact angle when measured for the aqueous phase. Measurements were done in 2–3 minutes from the moment of water drop deposition to the moment the image of the drop was taken by the image analysis system. No detectable change in contact angle was noted after 2–3 minutes.

RESULTS AND DISCUSSION

Figure 1 shows the work of adhesion, calculated from Eq. (1) based on measured oil contact angles and interfacial tensions, for kerosene at quartz and fluorite surfaces for varying concentration of the ethoxylated alcohol surfactant mixture, Rhodasurf 91-6. The decrease in the work of adhesion, for both minerals, was mainly a result of decreasing kerosene–water interfacial tension with increasing concentration of surfactant. The interfacial tension decreased from about 46 mN/m for a kerosene–water system to 29 mN/m for a kerosene –0.01 g/L Rhodasurf solution. On the other hand, the kerosene–water–mineral contact angles remained relatively constant over the

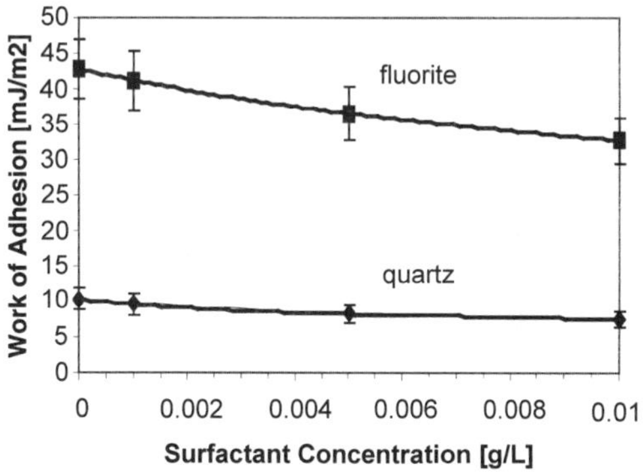

FIGURE 1 Work of adhesion for kerosene at fluorite and quartz surfaces in aqueous solutions of Rhodasurf 91-6 at pH 6 and 20–22° C.

range of surfactant concentration from 0 to 0.01 g/L. In the system with fluorite, the contact angle dropped from 94 degrees to 82 degrees when water was replaced with a 0.01 g/L Rhodasurf solution. In similar experiments, the contact angle dropped from 141 degrees to 138 degrees when measurements were performed on a quartz surface.

Next, the AFM technique was adapted in our laboratory for the examination of oil–water–mineral systems. In this work, we used solid polyethylene particles instead of oil droplets, as we cannot yet overcome the problem of interfacial force measurements for systems involving deformable interfaces. Nevertheless, the selection of polyethylene as a model "oil" particle in the AFM studies should not be questioned. First, in many real-world situations, separated contaminants occur in a solid or semi-solid state. Second, polyethylene has surface properties very similar to those measured for non-polar oils, such as kerosene. In particular, the surface tension of polyethylene is 31–35 mN/m [10]. The polyethylene–water interfacial tension is expected to be 45 ± 5 mN/m, based on available polyethylene surface tension and water-on-polyethylene contact angle data available in the literature [10]. The same properties for kerosene were determined by a pendant drop technique to be 28 mN/m and 46 mN/m, respectively.

Figure 2 shows the values of normalized pull-off forces (force per radius (R) of particle) measured for polyethylene particles on quartz and fluorite substrates in aqueous solutions of ethoxylated alcohols. Both the mineral substrates and Rhodasurf 91-6 solutions were the same as used in surfactant solution–kerosene–mineral contact angle measurements, the results of which are shown in Figure 1.

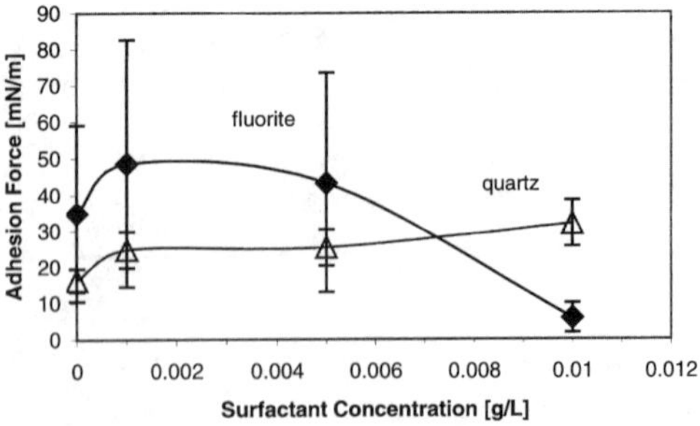

FIGURE 2 Adhesion force measured for a polyethylene particle at the surface of fluorite and quartz in aqueous solutions of Rhodasurf 91-6 at pH 6 and 20–22° C.

Figure 3 shows the comparison between results from Figures 1 and 2. The potential energy of interaction (W_T), per unit area (called the work of adhesion in the next part of the paper), between polyethylene on quartz in Rhodasurf 91-6 solutions was calculated from the adhesion force (F_A) using the following equation [11, 12]:

$$W_T = F_A/(2\pi R) \tag{3}$$

If W_T is assumed to be equal to W_A, an assumption often made in the literature, the relationship between measured adhesion force and concentration of surfactant should be similar to that observed in Figure 1. Because both kerosene and polyethylene demonstrate similar surface properties (see text above), the use of ethoxylated alcohol surfactant solutions should reduce the strength of polyethylene

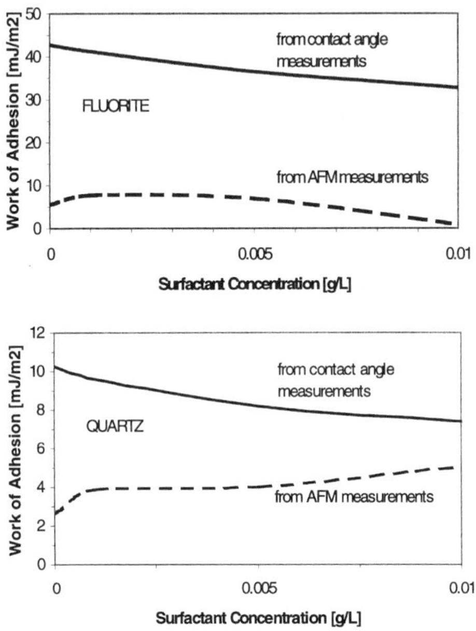

FIGURE 3 The correlation between the work of adhesion and Rhodasurf 91-6 concentration for a kerosene drop (solid line) and a polyethylene particle (broken line) on the fluorite and quartz surfaces. The solid line is the representation of results shown in Figure 1; calculated values from the Young-Dupré Eq. (1) using the results of contact angle and interfacial tension measurements. The broken line is the representation of the experimental results presented in Figure 2; the values are calculated from Eq. (3).

attachment to the mineral surface. As shown in Figures 2 and 3, this tendency was observed for a polyethylene particle at the fluorite substrate surface, although the correlation is slightly different from that for kerosene as shown in Figure 1. A small increase in the polyethylene adhesion force observed at low surfactant concentration is probably associated with wash-off impurities that could slightly contaminate the fluorite surface. Additionally, although we observed a similar effect of surfactant concentration on adhesion of kerosene and polyethylene to a fluorite surface, the relative values of the work of adhesion were significantly different. Specifically, the values of the work of adhesion between polyethylene and fluorite calculated from AFM measurements were about $\Delta W = 32 - 33\,\text{mJ/m}^2$ smaller that those calculated for kerosene on fluorite from contact angles and interfacial tensions. Surface nano-roughness of the polyethylene particle and the fluorite is probably the major factor causing the presence of this systematic difference in the ΔW value. Surface asperities could reduce the contact area between polyethylene and fluorite surface in our adhesion force measurements and normalized data are probably not true representation of adhesion.

Further, as shown also in Figures 2 and 3, the adhesion force and work of adhesion values follow a different pattern for the polyethylene – solution – quartz system. The force required to pull off the polyethylene particle from the quartz surface increased with increasing surfactant concentration – opposite to what was observed for fluorite. This cannot be attributed to experimental error. We observed the same effect of ethoxylated alcohol on adhesion force between polyethylene and quartz in other experiments [13].

The major problem, not always recognized, is that the condition of $W_T = W_A$ is only applicable to the systems in vacuum and inert gas. In the process of oily soil remediation or detergency, whenever separation takes place in a liquid the value of W_T might differ significantly from the W_A value. This is because the Young-Dupré equation was derived based on consideration of the interfacial free energies of the three-phase system taking into account two states of the system: an initial state corresponding to the contact between bodies, and a final state when interfaces are infinitely separated. The analysis completely neglects the intermediate states of the system at different distances of one interface from another. Specifically, in the system under consideration such as the polyethylene–water–mineral system, the interactions associated with overlapping of electrical double layers around polyethylene (oil) and mineral [14] as well as steric/molecular interactions of ethoxylated alcohols adsorbed on polyethylene and mineral should be taken into account (details of interfacial and molecular forces can be found in [12]). Both the polyethylene and quartz are negatively charged in water and ethoxylated alcohol solutions at pH 6. Consequently, repulsive forces contribute to the work of adhesion for a polyethylene particle at a mineral surface. This effect should be included in the results of AFM measurements but certainly is not included in the Young-Dupré equation. However, the electrostatic interactions cannot be the only factor responsible for the discrepancy of the work of adhesion values shown in Figure 3 for the systems with quartz. Ethoxylated alcohols are non-ionic surfactants and their adsorption at interacting interfaces has no major effect on surface charge. In other words, the electrostatic interactions will approximately equally contribute to the interaction between polyethylene or kerosene and quartz over the entire concentration of Rhodasurf 91-6. As we estimated, a reduction in the work of adhesion between kerosene (polyethyelene) and quartz caused by electrostatic repulsion should be at a level of $0.5-1.5\,mJ/m^2$ at the conditions of experiments used in this study, over the entire surfactant concentration.

We expect that in the polyethylene–solution–quartz system, molecular interactions between adsorbed surfactants might be important for an interpretation of the experimental data in Figure 2 and 3. The contact angle measurements performed in our laboratory indicate adsorption of ethoxylated alcohols on hydrophobic polyethylene with the hydrophobic part of the molecule attached to the polyethylene surface and the polar group oriented into the aqueous phase (results to be published). It is possible that OH-groups on the quartz surface interact with polar groups of ethoxylated alcohol adsorbed on the polyethylene surface. Increasing adsorption of ethoxylated alcohols on polyethylene increases a number of possible interacting polar groups. As the result, the adhesion force increased with increased concentration of surfactant in the solution (Fig. 2). Obviously, this is only speculation at the moment. The literature [15] indicates, for example, that ethoxylated alcohols also have a tendency to adsorb on silica surfaces. In this regard, other mechanisms for interacting molecular layers cannot be ignored.

In conclusion, we expect that the following equation should be valid in general:

$$W_T = W_A + \Sigma W_i \tag{4}$$

where W_i represents the contribution of forces characteristic for systems in a liquid such as electrostatic interactions, specific molecular and/or steric interactions (see [12, 14] for review of interfacial and molecular forces). Some researchers have already recognized [14, 16, 17] that electrostatic interactions must be included in the analysis of adhesion between solids and liquids suspended in a liquid. The results presented here indicate that adsorption of surface-active compounds at these interfaces would require consideration of molecular interactions in the analysis of adhesion forces.

CONCLUSIONS

We found that the work of adhesion of kerosene at mineral surfaces, calculated using kerosene–water–mineral contact angles and kerosene–water interfacial tensions, decreases in the presence of ethoxylated alcohol. Direct AFM measurements of the pull-off force (adhesion force) in the model oil (polyethylene)–water–quartz system showed the reverse correlation, *i.e.*, the ethoxylated alcohol might improve the strength of attachment of oil at the quartz surface. This unexpected result shed important light on the mechanisms of oil release from mineral surfaces. In view of this discovery it is clear that the Young-Dupré equation should be used with care in the analysis of three-phase systems involving liquid(s). The Young-Dupré equation has limited application to systems that involve suspended matter in liquids. Simply, the effects associated with overlapping of interfacial regions have not been included in the original Young-Dupré equation which limits its application to only a few systems of practical importance, and probably not to the removal of oil from mineral surfaces in the presence of natural or artificial surface-active compounds.

Acknowledgments

The authors would like to express their appreciation to Dr. Jakub Nalaskowski for his technical support in AFM measurements and discussion of AFM results. JD appreciates comments from Prof. B. Chibowski (M. Curie-Sklodowska University, Lublin, Poland) and Prof. B. Summ (Moscow State University, Moscow, Russia). Financial support from the U.S. Environmental Protection Agency–National Center for Environmental Research and Quality through the University of Utah is gratefully recognized. Although the research described in this article has been funded by the United States Environmental Protection Agency through grant number R825396-01-0 to the University of Utah, it has not been subjected to the Agency's required peer and policy review and therefore does not necessarily reflect the views of the Agency and no official endorsement should be inferred.

References

[1] Abdul, A. S. and Gibson, T. I., *Environ. Sci. Technol.* **25**, 665 (1991).
[2] Verma, S. and Kumar, V. V., *J. Colloid Interface Sci.* **207**, 1 (1998).
[3] Cutler, W. G. and Davis, R. C. Eds., *Detergency: Theory and Test Methods* (Marcel Dekker, Inc., New York, 1972).
[4] Drelich, J. and Miller, J. D., *Fuel* **73**, 1504 (1994).
[5] Carroll, B. J., *Colloids Surf. A: Physicochem. Eng. Aspects* **74**, 131 (1993).
[6] Binnig, G., Quate, C. F. and Gerber, C., *Phys. Rev. Lett.* **56**, 930 (1986).
[7] Sarid, D., *Scanning Force Microscopy* (Oxford University Press, New York, 1991).
[8] Wiesendanger, R., *Scanning Probe Microscopy and Spectroscopy: Methods and Applications* (Cambridge University Press, Cambridge, 1994).
[9] Nalaskowski, J., Drelich, J., Hupka, J. and Miller, J. D., *J. Adhesion Sci. Technol.* **13**, 1 (1999).
[10] Wu, S., *Polymer Interface and Adhesion* (Marcel Dekker, Inc., New York, 1982).
[11] Cappella, B. and Dietler, G., *Surface Sci. Rep.* **34**, 1 (1999).
[12] Israelachvili, J., *Intermolecular and Surface Forces* (Academic Press, San Diego, 1992).
[13] Beach, E., Gosiewska, A., Fang, Ch., Dudeck, K. and Drelich, J., In: "*The Proceedings of US United Engineering Foundation Conference on Environmental Technology for Oil Pollution*" (Hupka, J. and Miller, J. D. Eds), pp. 89–93, Technical University of Gdansk, Gdansk, Poland, 1999.
[14] Van Oss, C. J., *Interfacial Forces in Aqueous Media*, (Marcel Dekker, Inc., New York, 1994).
[15] Tiberg, F., Jonsson, B. and Lindman, B., *Langmuir* **10**, 3714 (1994).
[16] Van Oss, C. J., *Colloids Surf. A: Physicochem. Eng. Aspects* **78**, 1 (1993).
[17] Ontiveros, A., Duran, J. D. G., Gonzalez-Caballero, F. and Chibowski, E., *J. Adhesion Sci. Technol.* **10**, 999 (1996).

Mechanical Detachment of Nanometer Particles Strongly Adhering to a Substrate: An Application of Corrosive Tribology

J. T. DICKINSON, R. F. HARIADI and S. C. LANGFORD

Department of Physics and Materials Science Program, Washington State University, Pullman, WA 99164-2814, USA

The tip of a scanning probe microscope was used to detach nanometer-scale, single crystal NaCl particles grown on soda lime glass substrates. After imaging a particle at low contact forces, a single line scan at high contact force was used to detach the particle from the substrate. The peak lateral force at detachment is a strong function of particle contact area and humidity. As the relative humidity is raised from low to high values, the strength of the particle-substrate bond decreases dramatically. We interpret these results in terms of detachment by chemically-assisted crack growth along the NaCl-glass interface. Numerical estimates of the electrostatic and dispersive contributions to the work of adhesion are also discussed.

1. INTRODUCTION

Particles smaller than about a micron in diameter can be extremely difficult to remove from many surfaces. The technologies of chemical-mechanical polishing [1] and laser-assisted particle removal [2, 3], among others, have been developed in part to ensure that such particles are removed prior to critical surface operations. Particle removal is often facilitated by the presence of a liquid phase, which reduces the adhesive forces binding the particle to the substrate. The nature of the interfacial binding forces, the character and role of interfacial defects, and the effect of the liquid phase are not presently well understood.

In this work, we study the effect of humidity on the force required to detach submicron salt crystals from soda lime glass substrates. The chemical activity of water vapor is readily varied over a wide range by changing the pressure of the vapor. Particles were detached with the tip of a scanning force microscope (SFM). The SFM tip is analogous to a well-characterized asperity which "rubs" adhering particles from the substrate. Sodium chloride crystals of the appropriate size are readily grown by evaporation from a dilute aqueous NaCl solution. The resulting crystals adhere strongly to soda lime glass, forming symmetric protrusions on a reasonably flat substrate. Both NaCl and soda lime glass are hydrophilic and, thus, interact strongly with water. The effect of water vapor on crack growth in soda lime glass has been well studied [2].

Previous studies of particle adhesion using scanning probe microscopy include that of Meyer *et al.* [3] (adhesion and motion of C_{60} molecules on NaCl) and Junno *et al.* [4] (silver particles on semiconductor surfaces). Lebreton *et al.* [5] found that atom removal with scanning tunneling microscopy (STM) was facilitated by humidities above a critical value. Significantly, a large fraction of the interfacial bonds binding very small particles to the substrate lie along their perimeter; this strongly amplifies the response of the particle to chemically active agents in the surrounding atmosphere. In this work, we show that raising the humidity from low values dramatically lowers the lateral force required to fracture the NaCl-glass bond as the SFM tip is drawn across the particle.

2. EXPERIMENT

Submicron NaCl crystals were deposited on soda lime glass substrates by dissolving 1 – 3 grains ($\sim 1\,\text{mm}^3$) of commercial salt in a drop of de-ionized water on a clean microscope glass slide. The solution was gently spread across the slide with a cotton swab and allowed to evaporate to dryness at room humidity and temperature. Both evaporation and sample storage were under ambient laboratory atmosphere conditions – typically 20 – 40% relative humidity (RH).

A Digital Instruments (Santa Barbara, CA) Nanoscope III scanning force microscope mounted in a controlled environment chamber was used to image and manipulate the particles. This work employed commercial Digital Instruments Si_3N_4 cantilevers. A representative sample of tips was characterized by imaging niobium "artifacts" from Electron Microscopy Sciences, Fort Washington, PA [6], yielding an average tip radius of 37 ± 4 nm. This value is consistent with previous measurements in our laboratory, as well as radii measurements inferred from image deconvolution. Due to the small size of the NaCl crystals employed in this work, deconvolution significantly improved our measurements of particle dimensions. Deconvolution was performed with the DECONVO program from Silicon-MDT of Moscow, Russia [7, 8]. As the particle bases are not imaged, even in the deconvoluted images, the area of the top of the particle was taken as the particle area. This is equivalent to assuming that the particles are rectangular parallelepipeds, like their macroscopic counterparts in table salt.

Considerable effort was devoted to tip characterization and calibration. The spring constant for normal deflections, k_{normal}, of a sample of typical SFM cantilevers was determined by analyzing the power spectra of thermal fluctuations in the cantilever deflection [9 – 11], yielding $k_{normal} = 0.31$ N/m. The spring constant for lateral deflections, $k_{lateral}$, was estimated by multiplying k_{normal} by the expected $k_{normal} : k_{lateral}$ (0.0032, from finite element calculations [12]), yielding $k_{lateral} = 96$ N/m [13]. This k_{normal} to $k_{lateral}$ ratio is within 5% of an experimental measurement using the method outlined by Bhushan *et al.* [14]. The detector sensitivity in the normal direction was determined to be 9.22 nm/V by performing the force calibration procedure on a stiff substrate (aluminum oxide), where the deformation of the substrate and tip can be neglected [3].

Particle detachment experiments were initiated by setting the humidity to the desired value by adjusting the flow of dry air and humidified air into the chamber. The RH in the chamber was continuously measured with a BioForce Laboratory humidity sensor.

The glass surface was then scanned in contact mode to locate attached particles of the appropriate size. High tip velocities (typically 42 μm/s) and low applied normal forces (less than 20 nN) were employed during imaging to minimize the effect of scanning on particle adhesion. Scanning was continued until the tip was positioned to cross the center of the chosen particle (the white line in Fig. 1(b)); then the contact force raised to a high value ($\leq$ 320 nN).

When an appropriate particle was found, the tip was positioned to cross the center of the chosen particle and the tip speed reduced to 0.20 μm/s. As the tip approached the particle, the normal force was raised to a high value (180 – 320 nN). Given a particle of the appropriate size, the particle would then be detached from the substrate upon the tip's first encounter with the particle. After each detachment event, the signal corresponding to zero lateral force was determined by averaging the signals during trace and retrace portions of the linear scan.

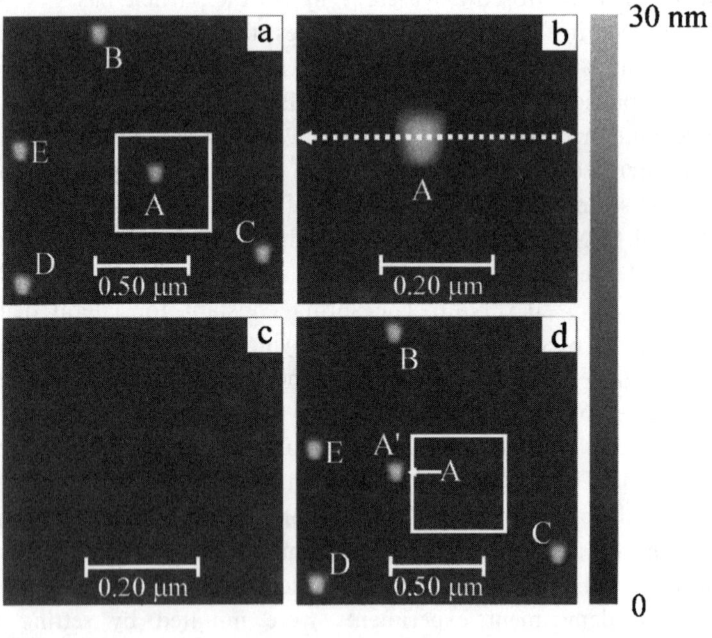

FIGURE 1 Low contact force images of NaCl particles (a) before particle removal, (b) a close-up immediately prior to the linear scan that removed the particle, (c) a close-up immediately after the line scan that removed the particle, and (d) an additional scan of region imaged in (a), showing the removed particle at the end of the 500 nm linear scan.

Subsequently, low contact force, high scan rate images were acquired to locate the detached particle. Room temperature during data acquisition was typically between 21 and 24°C.

3. RESULTS

Images taken before and after particle detachment at low RH are shown in Figure 1. Particle A in Figure 1(a) was selected for removal, and imaged at higher magnification in Figure 1(b). Note that the rectangular form of the particles in these images reflects the pyramidal tip geometry and not the actual particle shape. After aligning the SFM along the dotted line in Figure 1(b), the contact force was raised to a high value and the tip was drawn once across the position of the particle. Low contact force imaging was then resumed, confirming that the particle had indeed been removed (Fig. 1(c)). Imaging at lower magnification (Fig. 1(d)) subsequently located the detached particle at the end of the high contact force scan. Subsequent images do not always show the detached particle, especially at high relative humidities. Although it is possible that these particles are forcibly ejected some distance after detachment (a "slap shot"), it appears that these particles often adhere to the SFM tip and are moved some distance without being imaged.

Typical height and lateral force signals during scanning across a NaCl particle at 10% RH are shown in Figure 2. The signals in Figures 2(a) and 2(c) were acquired at low applied normal forces (10 nN) and high tip velocities (42 μm/s) prior to particle detachment. The height signal in Figure 2(a) shows the particle profile, while the lateral force signal in Figure 2(c) shows characteristic upward and downward spikes as the tip slides up onto and down off of the sample, respectively. Interestingly, the lateral force as the tip passes over the center of the NaCl particle is measurably lower than the lateral force as the tip passes over the glass substrate. This difference cannot be attributed to elastic deformation effects, as the glass is actually stiffer than the NaCl. Young's modulus in soda lime glass is typically about 73 GPa; in NaCl the modulus (directionally averaged) is 37 GPa [15]. The salt-to-glass lateral force ratio ranges from about 0.8 at 3% RH to 0.3 at 68% RH. The changes in lateral force with humidity suggest

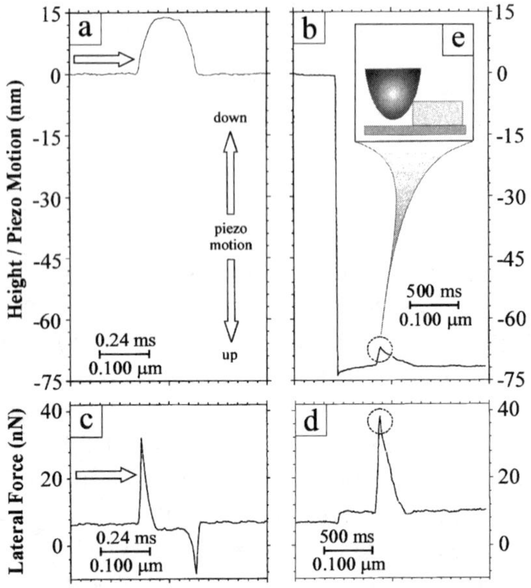

FIGURE 2 (a) and (b) Height, (c) and (d) lateral force signals during the low contact force scan used to align the tip on the particle, and during the slow, high contact force scan used to remove the particle from the substrate. Inset (e) shows the tip and NaCl position just before fracture.

that the adsorbed water layer on these hydrophilic surfaces strongly affects the apparent coefficient of friction, perhaps due to capillary effects.

The height and lateral force signals during detachment are shown in Figures 2(b) and 2(d). The tip velocity was reduced from 42 μm/s to 0.20 μm/s before beginning this scan. Approximately a quarter of the way through the scan, the nominal contact force was raised to 57 nN, producing a dramatic drop in the height signal of Figure 2(b). (To increase the normal force, the piezoelectric transducer must move downward, which changes the height signal in much the same way as a change in topography.) Increasing the normal force also produced a small, stepwise increase in the lateral force signal (Fig. 2(d)) due to the accompanying increase in friction. Midway through the scan, the tip encountered the particle, lifting slightly ($\sim$4 nm) onto the edge of the particle before detachment. Because the tip is not in contact with the glass at the moment of detachment (Fig. 2(b)), the friction between the tip and the glass does not contribute to the measured lateral force at detachment. Thus, the entire lateral force at detachment is applied to the particle, and it is this force that induces detachment.

We identify the peak in lateral force and height signals with the detachment event itself. An upper bound on the energy per unit area required to fracture the interface is provided by the area under the rising portion of the lateral force *vs*. displacement plot prior to failure. In Figure 2(d), this amounts to about $50\,mJ/m^2$. In the fracture of more-macroscopic, highly-brittle samples (including soda lime glass and NaCl), it is not uncommon for two-thirds of this energy to be dissipated *via* plastic deformation and similar processes. Since plastic deformation is strongly hindered in nanometer-scale systems [16], its role in the removal of nanometer-scale particles is an open question.

Both the height and lateral force signals show distinct tails after detachment. Their duration depends strongly on RH and can extend much further than the physical extent of the particle. Thus, these tails cannot be attributed to incomplete detachment. We attribute these tails to the gradual drop in lateral force as the particle slips out from under the SFM tip. At low RH, the particle slides more freely along the glass, and the lateral force drops rapidly. However, at high RH, the particle encounters wet glass and slides more slowly.

The peak lateral force (at detachment) is a strong function of tip velocity. Figure 3 displays the lateral force signal during two successive

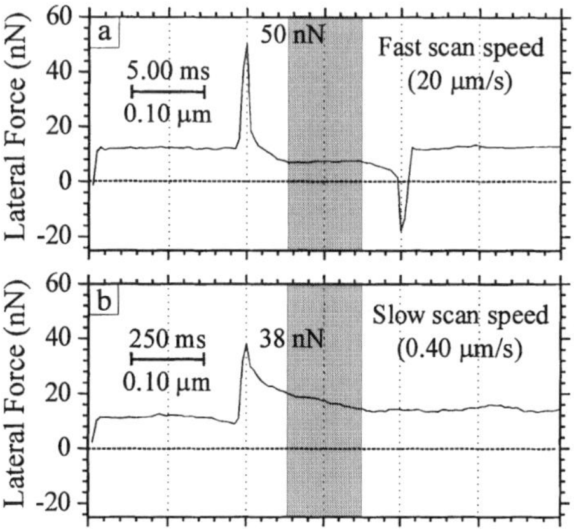

FIGURE 3 Lateral force signal during linear scans across the same NaCl particle at tip velocities of (a) 20 μm/s and (b) 0.40 μm/s. The normal force in each case was 17 nN.

linear scans across the same NaCl particle at an applied normal force of 17 nN and a high relative humidity (51%). After the scan of Figure 3(a) and before the scan of Figure 3(b), the tip velocity was reduced from 20 μm/s to 0.40 μm/s. During the first scan (at 20 μm/s), the SFM tip passes cleanly over the particle. However, during the slow scan (at 0.4 μm/s) the particle is detached. Furthermore, the peak lateral force during the slow scan (38 nN) is significantly lower than the peak lateral force during the fast scan (50 nN), which did not remove the particle. Experience indicates that a single scan at the faster rate of 20 μm/s does not significantly weaken the NaCl/glass bond. (A large number of scans at 20 μm/s do weaken the interface.) The dramatic effect of scan rate suggests that the duration of the applied force is critical, and that particle detachment is not merely due to the application of some critical stress associated with the intrinsic strength of the interface.

These results suggest that particle detachment involves the relatively slow growth of an interfacial crack. At low scan rates, relatively low crack velocities can detach the particle before the SFM tip passes onto the particle. Lower stresses are required to produce these lower crack velocities, so that failure occurs at lower applied lateral forces. At high scan rates, particle detachment during tip-particle contact requires higher crack velocities and higher lateral forces. The failure stresses reported below were all determined with a minimum of prior, low contact force scanning at relatively high tip velocities (>42 μm/s) and low contact forces (<20 nN), followed by a single, high contact force scan (typically 150 nN) at tip velocity of 0.20 μm/s. This procedure yielded consistent failure stress measurements. The crack velocities required for particle detachment at these scan rates ($1-10$ μm/s) are very low – consistent with crack velocities measured in soda lime glass and similar materials at low stresses in chemically-active environments [2].

For comparison purposes, we define a nominal shear strength, σ_c, of the interface equal to the peak lateral force divided by interfacial area, A. Values of σ_c from a large number of particle detachment events at several relative humidities are plotted in Figure 4 as a function of particle size. We have scaled the values of σ_c for 3% RH by a factor of ten for presentation purposes; the smallest particle at 3% RH failed at an nominal shear stress of 55 MPa. At lower humidities,

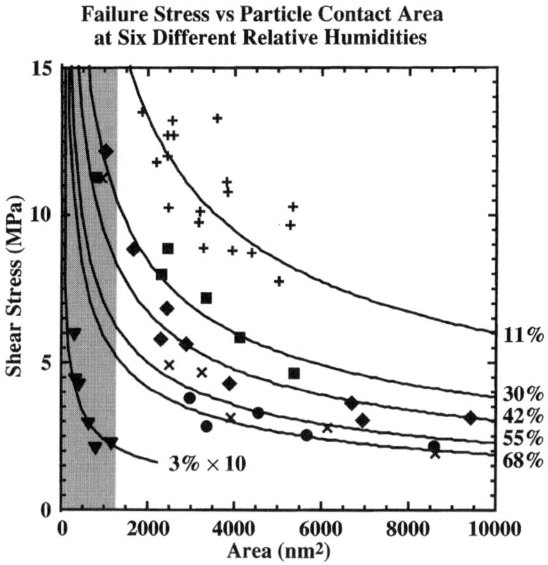

FIGURE 4 Nominal shear stress at failure *vs*. NaCl/glass contact area for NaCl cubes of various sizes at relative humidities ranging from 3% to 68%. Note that the failure stresses at 3% RH have been scaled (reduced) by a factor of ten for convenient display; the smallest particle tested at 3% RH failed at 55 MPa. The dark lines represent one-parameter, least-squares fits to the data, assuming that the nominal shear stress is proportional to $A^{-0.5}$, where A is the area of the NaCl cube in contact with the glass. The shaded region of the graph marks the size range in which the cubes are smaller than the radius of curvature of the SFM tip.

raising the humidity dramatically decreases σ_c. Particle size (here parameterized by the contact area, A) also strongly affects the shear strength, especially for the smaller particles. Both effects impose experimental limits on the size of particles amenable to study at the lowest humidities, where the SFM tip may break before a large particle will be removed. This is consistent with anecdotal reports of the difficulty of particle removal from surfaces under dry conditions. Significantly, increasing the humidity beyond 50% has little additional effect on the shear stress at failure. Similarly, the particle-size dependence becomes weak for particles with contact areas larger than 3000 – 4000 nm^2.

The effect of humidity on shear strength is quantified in Figure 5, where the failure stress for particles with nearly equal contact areas (5000 nm^2) is plotted as a function of relative humidity. (The value for 3% RH has been extrapolated due to our inability to remove

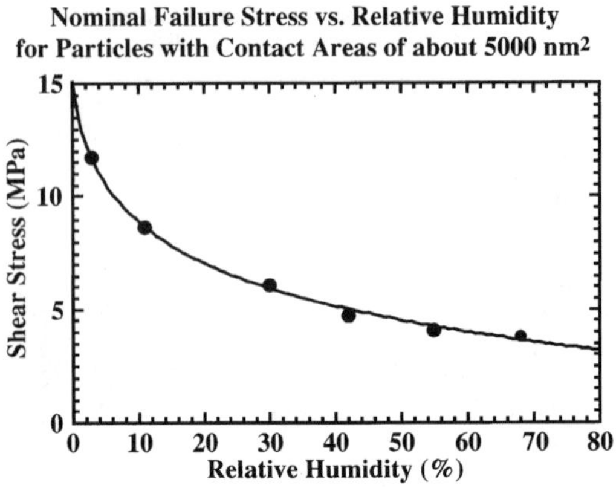

FIGURE 5 A plot of the peak lateral forces observed for cubes of nominal contact area 5000 nm^2 (~70 nm × 70 nm) as a function of relative humidity. The point for the lowest RH (= 3%) is extrapolated from the data, due to the difficulty of removing particles of this size at RH = 3%.

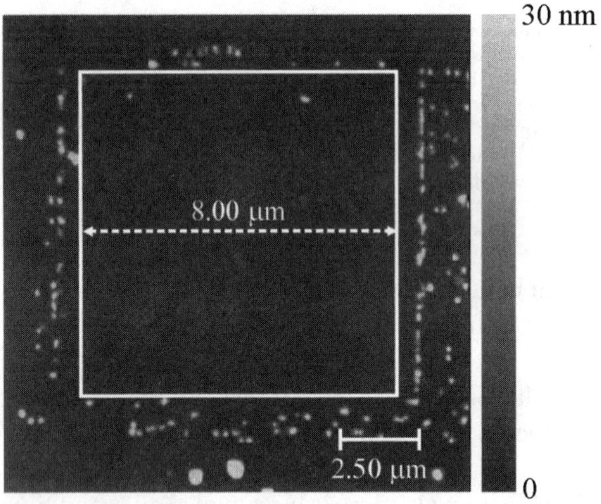

FIGURE 6 A SFM image of a region of the substrate cleared of adhering NaCl particles at 55% RH and a contact force of 51 nN.

particles of this size at low RH.) The failure stress drops rapidly with increasing RH in the lower humidity range, and falls more gradually at the higher humidities. The dark line represents a least squares fit of a model to the data, described below.

At high RH, relatively large particles can be removed from the substrate at modest contact forces. Under these conditions, the SFM tip can be used to remove virtually all the particles from large areas. The results of such a sweeping procedure at 55% RH and 51 nN contact force are shown in Figure 6. An $8 \times 8\,\mu m^2$ area has been swept clean of NaCl particles, and the resulting debris is piled up around the edges, exhibiting a fair amount of order due to their alignment. This illustrates the effectiveness of *combined* chemical and mechanical stimuli in particle manipulation.

4. DISCUSSION

4.1. Chemically-assisted Crack Growth

Considerable effort has been devoted to the understanding of chemically-assisted crack growth, often designated environmental crack growth. Nevertheless, a universal, detailed understanding of it has not yet been achieved. For the purposes of discussion, we adopt a description outlined by Lawn [2] who treats crack growth as a stress-activated chemical reaction. This picture grew out of studies by Wiederhorn, Michalske, Freiman, Bunker, and Lawn. (See Ref. [2] and references therein.) In this approach, the crack velocity, V, is represented as the product of a classical "attempt frequency", f_0 (a typical vibrational frequency), a "crack propagation distance per successful attempt", a_0 (a typical intermolecular spacing), and a probability factor equal to the fraction of attempts yielding a broken bond. Ignoring the reverse reaction (crack healing), classical kinetic theory yields:

$$V = a_0 f_0 \exp(-\Delta F/\mathrm{k}T) \qquad (1)$$

where k is Boltzmann constant, T is the temperature, and ΔF is the free energy change associated with fracture in a chemically-active environment.

The effect of chemical environment and stress on ΔF is conveniently described in terms of the relevant energy terms. The chemical environment affects the energy per unit area required to form new fracture surfaces, R. Classically, increasing the stress increases the available elastic strain energy per unit area to drive crack growth, G (the strain energy release rate). In classical fracture (Griffith) theory, crack growth occurs when $G > R$. An alternate approach, adopted here, is to replace G with the appropriate stress intensity factor, K, where $K = (GE)^{1/2}$ and E is the relevant Young's modulus. K reflects the magnitude of the stress singularity at the tip of an infinitely sharp crack. Replacing G with K is not a trivial change; in chemical terms the difference corresponds roughly to the difference between a stress-activated process (with G) and a volume-activated process (with K). We adopt the stress intensity description because it provides a more consistent description of our data. In general, the question of which description is to be preferred in chemically-assisted crack growth has not been settled. In previous studies of stress-enhanced dissolution of atomic steps on single crystal surfaces, a similar comparison of stress- *vs.* volume-activated mechanisms also favored a volume-activated process [17, 18].

In our case, it is convenient to express the effect of stress on crack velocity in terms of the excess available mechanical energy (proportional to K) over the energy required to form the fracture surfaces (proportional to R). Expanding ΔF in terms of this difference yields

$$\Delta F = \Delta F_0 + (\alpha K - \beta R) + \cdots \tag{2}$$

where $\Delta F_0 = \Delta F$ when $\alpha K = \beta R'$, and α and β are constants of proportionality. Equation (1) then becomes

$$V = a_0 f_0 \exp(-\Delta F_0/\mathrm{k}T) \exp[(\alpha K - \beta R)/\mathrm{k}T] \tag{3}$$

In this formalism, the effect of mechanical stress and chemical attack on crack growth reduces to finding appropriate expressions for K and R. Although we do not measure the crack velocity *per se*, the exponential dependence of crack velocity on K and R assures that velocities which yield fracture on a given experimental time scale will be associated with similar values of $(K - R)$. In the limit of low crack speeds (typically $< 10\,\mu\text{m/s}$), R is a simple function of RH. At sufficiently high crack speeds, R becomes less dependent on humidity due to the inability of diffusing vapor to keep up with the advancing crack tip. The strong effect of humidity in this work indicates that the time-to-failure is dominated by crack growth in the low velocity limit. In all cases, failure on a given experimental time scale requires that the crack velocity [and, thus, $(K-R)$] be large enough to produce failure while the tip is applying a significant lateral force to the particle. We treat the minimum K value for failure on a given time scale (here determined by the tip velocity) as a critical value for failure and denote it as K_c. In general, raising the humidity will lower R and, thus, lower K_c.

The applied stress intensity, K, is a function of sample geometry, including crack geometry, and is generally treated numerically. For a cube bound to a planar substrate with a small, sharp interfacial crack, K is expected to scale as

$$K \sim a\sigma_{xy}c^{1/2} \tag{4}$$

where σ_{xy} is the nominal shear stress applied to the particle (lateral force divided by particle/substrate contact area), c is the crack length, and a is a constant of order 3 [19]. A complication in our loading scheme is the effect of the normal (compressive) stress exerted by the SFM tip on the adhering particles. This compressive stress tends to close any existing crack and effectively reduces K. For the purpose of analysis, we assume this effect is independent of particle size. (Proportionally higher compressive stresses are expected for particles shorter than the radius of curvature of the SFM tip. The shaded region in Figure 4 identifies data points where this may be a factor.) Assuming that all particles of a given size have (at least statistically) similar interfacial flaws, we can replace K_c with σ_c, where $\sigma_c = \sigma_{xy}$ at detachment.

In Figure 4 we presented these critical values of shear stress as a function of the nominal particle/substrate contact area, A, for several values of relative humidity. The monotonically decreasing shear stress with area led us first to a two-parameter curve fit of $\sigma_c \sim A^{-n}$ for each value of RH. Each fit yielded $n=0.5$ to within the uncertainty of the curve-fitting procedure (typically ± 0.1). The dark lines in the figure represent one-parameter least-squares fits of a function proportional to $A^{-1/2}$. This dependence is quite reasonable if we assume that the size of the initial interfacial flaws at the perimeter of the particles scale with particle area. [Assuming failure occurs at $K=K_c$ in Eq. (4), constant for a given RH, $\sigma_{xy} \sim K_c\, c^{-1/2} \sim A^{-1/2}$ requires that c is proportional to A.] This model suggests that flaw size is responsible for the area dependence of the failure stress.

Humidity influences σ_c through its effect on R. Raising the partial pressure of water vapor lowers R, which in turn lowers K_c and σ_c. For interfacial failure,

$$R = \gamma_{\text{glass}} + \gamma_{\text{salt}} - \gamma_{\text{salt-glass}} - \Delta U_{\text{glass}} - \Delta U_{\text{salt}}, \tag{5}$$

where γ_{glass} and γ_{salt} are the two surface energies after detachment ($\gamma_{\text{glass}} \neq \gamma_{\text{salt}}$), $\gamma_{\text{salt-glass}}$ is the interfacial energy before detachment, and ΔU_{glass} and ΔU_{salt} are the *changes* in the two surface energies due to the reactions with water vapor.

The adsorption energy is a complex function of humidity and surface coverage, especially at high relative humidities where more than one monolayer of water may cover the surface. For submonolayer coverages, the Langmuir isotherm describes the adsorption in many adsorbate/substrate systems. It also provides a convenient empirical description of adsorption in the present case. For the Langmuir isotherm, $\Delta U = 2\Gamma_m \ln(p/p_0)$ [2], where Γ_m is the adsorption energy per unit area for full (monolayer) coverage, p is the partial pressure of the adsorbate in the surrounding atmosphere (here proportional to the relative humidity), and p_0 is the partial pressure at which the coverage equals half a monolayer. In the present context, full coverage will amount to several monolayers of water on both glass and NaCl, so the interpretation of these values will be somewhat strained. This problem is mitigated somewhat by the hydrophilic nature of glass and salt surfaces, which ensures that water–glass and water–salt bonds are much stronger than water–water bonds. Thus, the total energy of water sorption will be dominated by the energy of the first monolayer. With these assumptions, the crack velocity as a function of σ_{xy} and partial pressure, p, become:

$$V = a_0 f_0 (1 + p/p_0)^{2\alpha\Gamma_m} \exp[-(2\gamma + \Delta F_0)/\mathrm{k}T] \exp[\alpha' \sigma c^{1/2}/\mathrm{k}T]. \tag{6}$$

Again assuming that interfacial fracture occurs on experimental time scales for cracks that reach a critical speed, σ_c for NaCl cubes of a given contact area will scale as:

$$\sigma_c = m \ln[n/(1 + p/p_0)] \tag{7}$$

where m, n, and p_0 are parameters. A fit of Eq. (7) to experimental data for particles with contact areas close to 5000 nm^2 appears in Figure 5. The model describes the data quite adequately.

The kinetics approach to environmental crack growth (Eq. (6)) predicts that temperature changes alter the crack velocity. We plan to explore the effect of temperature on particle removal in future work.

4.2. Work of Adhesion Estimate

Particle detachment requires that work be done against interfacial forces whose origins are electronic in nature. In many systems, dispersion (van der Waals) forces dominate adhesion between dissimilar materials. However, the highly ionic nature of NaCl suggests that electrostatic forces (*via* image charges) may contribute. An understanding of these effects is necessary if the effect of water and other solvents is to be understood on the molecular level. As a first step in this direction, we have estimated the contribution of electrostatic and dispersion forces to adhesion in the salt – glass system.

The electrostatic contribution to the work of adhesion was estimated by performing a Madelung-like sum of the electrostatic potential energy of an array of plus and minus charges (the NaCl crystal) in the presence of a similar array of reduced strength (image charges in the glass). The NaCl crystal was treated as a lattice of charges of magnitude $\pm e$ (where e is the electron charge) with the rocksalt structure and nearest neighbor distance of 2.82 Å (as in NaCl). The glass was treated as a continuous dielectric positioned parallel to one edge of the NaCl lattice, with the interface arbitrarily positioned one-half the nearest neighbor distance from the nearest NaCl plane. (This choice of interface is analogous to treating the glass as a continuous extension of the NaCl crystal.) The magnitude of the dielectric constant, κ, is a key uncertainty in this estimate, due to the contribution of ionic diffusion (which increases in the low frequency limit). For this work, a typical value of κ for soda lime glass at 100 Hz was used ($\kappa = 8.3$) [1]. Reasonable convergence was obtained with NaCl lattices containing 2000 ion pairs, with a corresponding work of adhesion of $20\,\mathrm{mJ/m^2}$.

The contribution of dispersion forces was estimated by computing an appropriate Hamaker constant and assuming an interfacial separation equal to one nearest-neighbor distance in NaCl. The Hamaker constant for the system was taken as the geometric mean of the Hamaker constants for soda lime glass and NaCl, which, in turn, were estimated from the Clausius-Mosotti equation for the

Hamaker constant in terms of the bandgap and high-frequency dielectric constant [20]. This procedure yielded Hamaker constants of 6.2×10^{-20} J and 3.1×10^{-20} J for the salt and the glass, respectively. The resulting contribution to the interfacial work of adhesion amounts to $15\,mJ/m^2$.

Our estimate of the total work of adhesion, W_{ads}, is equal to the sum of the electrostatic and dispersion contributions, or $35\,mJ/m^2$. In the near future, we hope to extend these calculations to include the effect of water vapor.

The work of adhesion is related to the energy per unit area required for crack growth, R, by the relation $W_{ads} = \gamma_{glass} + \gamma_{salt} - \gamma_{salt-glass}$ [20]. Equation (5) then yields $R = W_{ads} - \Delta U_{glass} - \Delta U_{salt}$, with $R = W_{ads}$ in the limit of zero humidity. The area under the lateral force curve at 11% RH provides an upper bound on R and, thus, W_{ads} of $50\,mJ/m^2$. Our estimate of $W_{ads} \sim 35\,mJ/m^2$ lies comfortably below this upper bounds. Our estimate is also well below the experimental fracture surface energies for NaCl ($\sim 0.3\,J/m^2$) and soda lime glass ($4-5\,J/m^2$) [1], consistent with failure along the interface. Although the precise nature of the interface is not known, the surface exposed by particle removal was indistinguishable from the surrounding surface within the resolution of the SFM. If failure is not ideally interfacial, the deviation cannot be much more than a monolayer.

5. CONCLUSIONS

The lateral forces required to remove particles from a soda lime glass substrate with the tip of a scanning force microscope are strong functions of particle size and relative humidity. At 3% RH, only the smallest particles could be removed by the SFM tip at any accessible contact force. Increasing the RH to 30% dramatically reduces the stress required for particle removal and promotes the removal of much larger particles. Increasing the relative humidity further yields smaller decreases in the failure stress. Clearly, a *combined* chemical and mechanical attack is most effective for particle removal.

The humidity dependence of the failure stress is well described in terms of the energy required to form the final fracture surfaces, where these surfaces are in equilibrium with a given partial pressure of water vapor. This reduces the energy and, thus, the stress, required for slow crack growth. Although somewhat empirical, the Langmuir isotherm adequately describes the humidity dependence of the stresses required for particle removal. The small size of these particles is important for clear observation of this effect, due to the large perimeter-to-area ratio of the interface. For these particles, the interaction of water vapor with the perimeter of the interface has a significant effect on the strength of the interface as a whole.

Particle removal is an important factor in a number of technologies, including integrated circuit manufacture and optical component manufacture. An improved understanding of these processes should facilitate intelligent redesign of particle removal processes as new particle removal challenges present themselves.

Acknowledgments

This work was supported by the National Science Foundation Surface and Tribology Program under Grant CMS-98-00230 and a National Science Foundation Instrumentation Grant DMR-9201767. We wish to thank Louis Scudiero, WSU, for his assistance in this work and preliminary SFM studies of the NaCl particles on glass, and John Hutchinson, Harvard University, for helpful discussions on fracture mechanics.

References

[1] Kingery, W. D., Bowen, H. K. and Uhlmann, D. R., *Introduction to Ceramics*, 2nd edn. (John Wiley, New York, 1976), pp. 797.
[2] Lawn, B., *Fracture of Brittle Solids*, 2nd edn. (Cambridge University Press, Cambridge, 1993), pp. 378.
[3] Meyer, E., Luthi, R., Howald, L. and Bammerlin, M., In: *Micro/Nanotribology and its Applications*, Bhusan, B., Ed. (Kluwer Academic, Dordrecht, 1997), pp. 193–215.
[4] Junno, T., Deppert, K., Montelius, L. and Samuelson, L., *Appl. Phys. Lett.* **66**, 3627–3629 (1995).
[5] Lebreton, C. and Wang, Z. Z., *J. Vac. Sci. Technol. B* **14**, 1356–1359 (1996).
[6] Westra, K. L. and Thomson, D. J., *J. Vac. Sci. Technol. B* **12**, 3176–3181 (1994).
[7] Williams, P. M., Shakesheff, K. M., Davies, M. C., Jackson, D. E., Roberts, C. J. and Tendler, S. J. B., *Langmuir* **12**, 3468–3471 (1996).
[8] Villarrubia, J. S., *J. Res. Natl. Inst. Stand. Technol.* **102**, 425–454 (1997).
[9] Hutter, J. L. and Bechhoefer, J., *Rev. Sci. Instrum.* **64**, 1868–1873 (1993).
[10] Drexler, K. E., *Nanosystems: Molecular Machinery, Manufacturing, and Computation* (John Wiley & Sons, 1992), pp. 576.

[11] Walters, D. A., Cleveland, J. P., Thomson, N. H., Hansma, P. K., Wendman, M. A., Gurley, G. and Elings, V., *Rev. Sci. Instrum.* **67**, 3583–3590 (1996).
[12] Neumeister, J. M. and Ducker, W. A., *Rev. Sci. Instrum.* **65**, 2527–2531 (1994).
[13] Labardi, M., Allegrini, M., Salermo, M., Frediani, C. and Ascoli, C., *Appl. Phys. A* **74**, 3–10 (1994).
[14] Bhushan, B., *Tribology and Mechanics of Magnetic Storage Devices*, 2nd edn. (Springer, New York, 1996), pp. 499–540.
[15] Chung, D. H. and Buessem, W. R., In: *Anisotropy in Single-Crystal Refractory Compounds*, Vahldiek, F. W. and Mersol, S. A., Eds. (Plenum, New York, 1968), pp. 217–245.
[16] Bull, S. J., Page, T. F. and Yoffe, E. H., *Philos. Mag. Lett.* **59**, 281–288 (1989).
[17] Park, N.-S., Kim, M.-W., Langford, S. C. and Dickinson, J. T., *J. Appl. Phys.* **80**, 2680–2686 (1996).
[18] Scudiero, L., Langford, S. C. and Dickinson, J. T., *Tribology Lett.* **6**, 41–55 (1999).
[19] Hutchinson, J. W., *Personal communication* (1998).
[20] Adamson, A. W., *Physical Chemistry of Surfaces*, 5th edn. (John Wiley, New York, 1990).

Advances in Controlling the Attachment and Removal of Groups of Particles

The Effect of Relative Humidity on Particle Adhesion and Removal

AHMED A. BUSNAINA and TAMER ELSAWY

Microcontamination Research Laboratory, Clarkson University, Potsdam, NY 13699-5725, USA

The removal of small particles is vital for contamination-free manufacturing. In humid environments liquid can condense between the particle and substrate and give rise to a very large capillary force, which increases the total force of adhesion. The removal and adhesion forces of polystyrene latex (PSL) particles and pigmented coating chips were measured on silicon, polyethylene terephthalate, metallized and polyester coating substrates as a function of humidity. The results indicate that the capillary force is significant at a relative humidity above 50% and dominates at a relative humidity above 70%. At relative humidity below 45%, the electrostatic force becomes significant. The adhesion forces varied depending on the particles and substrates used, but the trend of high adhesion at high and low relative humidity was observed for all PSL particles/substrate systems. The pigmented coating chips/substrate system however, exhibited high adhesion at high relative humidity and low adhesion at low relative humidity.

1. INTRODUCTION

Adhesion between small particles and solid surfaces is a concern to many technologies. Understanding of particle adhesion and removal is vital for contamination-free manufacturing in many industries such as semiconductor manufacturing, xerography, pharmaceuticals, aerospace, *etc*. Efficient removal of particles from surfaces requires knowledge of the magnitude of the forces involved. Experimental studies on the removal of particles and a detailed review of particle-surface interactions have been widely reported [1 – 12]. Kurz [13] used a rotating silicon wafer to generate the hydrodynamic force required to remove 1 μm and larger particles. Removal rates above 90% were reported for particles larger than 2 μm. Taylor, Busnaina and coworkers [13 – 15] measured the removal force for sub-micrometer particles on silicon substrates and correlated it with the theoretical van der Waals force. That same technique is used in this study to determine the total adhesion force including the van der Waals force.

The primary adhesion forces for a dry uncharged particle on a dry uncharged surface are the van der Waals and electrostatic forces. The van der Waals forces can increase due to particle and/or surface deformation that increases the particle contact area. Electrostatic forces, although they predominate for particles larger than 50 μm, play a significant role in bringing particles to surfaces. In humid environments, liquid can condense between the particle and substrate, giving rise to a very large capillary force, which increases the total force of adhesion. Zimon [7] concluded that when the relative humidity is above 70%, the capillary force dominates and should be the only adhesion force considered. The capillary force consists of two main components: the surface tension at the perimeter of the meniscus and the pressure difference between the liquid and vapor phases [18]. The effect of relative humidity on the removal force has also been studied by Bowden and Throssel [17]. They suggested that the force required to remove 98% of 50-micron glass particles from a glass substrate increases with increasing relative humidity. The study, however, did not provide the adhesion force as a function of humidity. Many studies have been conducted to investigate particle adhesion and removal. Very few studies, however, considered the effect of relative humidity. The overall objective of this study is to develop an understanding of the effects of the relative humidity on particle adhesion for different particle/substrate systems. The specific objective of this study is to determine the effect of relative humidity on the adhesion force between the following particle/substrate systems:

1. Polystyrene latex (PSL) particles/silicon substrate system.
2. Polystyrene latex (PSL) particles/polyethylene terephthalate (PET), metallized, and polyester coating layer (PCL) substrates.
3. Pigmented coating chips (selenium)/polyethylene terephthalate (PET), metallized, and polyester coating layer (PCL) substrates.
4. Skin flakes/polyethylene terephthalate (PET), metallized, and polyester coating layer (PCL) substrates.

They were obtained from CP Films, Fieldale, Virginia, USA. The metallized (Ti/Zr) and polyester coating layers are coatings on PET. All experiments were conducted in a class 10 clean room. Relative humidity between 10% – 90% was generated in a Plexiglas humidity chamber built for the purpose of conducting these experiments.

2. BACKGROUND

2.1. Adhesion Force

The van der Waals force is the dominant adhesion force for small particles (less than 50 μm). It arises from the high frequency movement of the electrons in the atoms or molecules giving rise to momentary areas of charge concentrations called dipoles [9]. The van der Waals force for a spherical particle attached to a planar substrate is given as:

$$F_{\text{vdw}} = \frac{AR}{6z^2} \tag{1}$$

where A is the Hamaker constant, R is the radius of the spherical particle, and z is the separation distance between the particle and the substrate. The average separation distance between the two surfaces is taken as 4 Å (for smooth surfaces). Electrostatic force constitutes the main force of attraction for particles larger than 50 μm in diameter. For dry particle-substrate systems the electrostatic force becomes important. The presence of electrostatic charge can drastically alter the total adhesion force. Zimon [7] reported that the force of adhesion could be increased by a factor of two when the net number of unit charges per particle on 40 – 60 μm particles increases from 700 to 2500. Most particles carry some electric charge, and some may be highly charged. Particles at low humidity were found to retain their charge and are held to surfaces by an attractive electrostatic force [9]. A charged particle experiences an electrostatic force in the vicinity of a conductive or insulating surface. The electrostatic image force (F_{el}) is given as [20]:

$$F_{\mathrm{el}} = \frac{Q^2}{6(D + z_0)^2} \tag{2}$$

where Q is the particle charge, D is the particle diameter and z_0 is the separation distance. This electrostatic force will deteriorate with time due to the dissipation of the charge.

When moisture is present in the air medium, condensation can take place between the particle and substrate. The capillary condensation gives rise to a capillary force, F_C, which is given by [20]:

$$F_C = 4\pi r \gamma_{\mathrm{LV}} \tag{3}$$

where γ_{LV} is the surface tension (for the liquid – air interface), and r is the radius of the spherical particle. Equation (3) is applicable to smooth surfaces and represents the maximum force that could be experienced as a result of capillary condensation. However, the force of adhesion approaches the values predicted by Eq. (3) only at relative humidity near 100% [19]. The capillary force consists of two components, as mentioned previously: the surface tension at the perimeter of the meniscus and the difference in pressure between the liquid and vapor phases. The existence of surface tension in a liquid – gas interface causes a difference in hydrodynamic pressure across the interface if the interface is curved. The quantitative expression for this pressure difference, as a function of surface tension and curvature, was derived by Laplace in 1806. He postulated an intermolecular force of attraction that diminishes rapidly with increasing distance [18]. Laplace's equation is given by:

$$P_C = P_1 - P_2 = \gamma\left(\frac{1}{r_2} + \frac{1}{r_2}\right) \tag{4}$$

where P_C is called the capillary pressure, P_1 is the pressure within the liquid phase, P_2 is the pressure within the gas phase, and r_1 and r_2 are the principal radii of curvature of the surface at the point of interest. The resulting capillary force depends on several parameters such as particle size, surface tension of the condensed fluid and the wettability (contact angle) of the substrate surface. Capillary forces are proportional to particle size, and according to Corn [19] the adhesion of large particles increases with relative humidity.

2.1.1. The Effect of Capillary Force on Adhesion

Equation (3) applies only to smooth spheres in a saturated atmosphere. Zimon [7] found that in air at a relative humidity near 100%, the majority of particles are held with forces less than that predicated by Eq. (3). Kordecki and Orr [22] observed that capillary condensation begins to appear at relative humidity above 50%. Luzhnov [24] reported that adhesion due to capillary force occurs when the relative humidity exceeds 70%. The same effect was also reported by Zimon [7], who concluded that at relative humidity of 50%, and particularly at humidity below 50%, capillary forces have no effect on the adhesion force. But all agreed that at relative humidity between 50% – 65%, the capillary force starts to have an effect on the total adhesion force. Zimon went on to conclude that between 70% – 100% relative humidity, the capillary force dominates the other adhesion forces and should be the only adhesion force considered.

2.2. Removal Forces

The removal forces required to dislodge particles off the substrate were generated by spinning the substrate about its axis and using air as the fluid medium, as shown in Figure 2.1.

A stationary particle in a moving fluid stream will experience a drag force, F_D, due to the difference in pressure across the surface of the particle. A gradient in the shear flow of the fluid will result in the Saffman or hydrodynamic lift force, F_L. Centrifugal force, F_{Cent}, is another contributing factor to the total removal force and is due to the angular velocity of the substrate about its axis. When the resultant removal force (drag, lift, and centrifugal force) exceeds the force

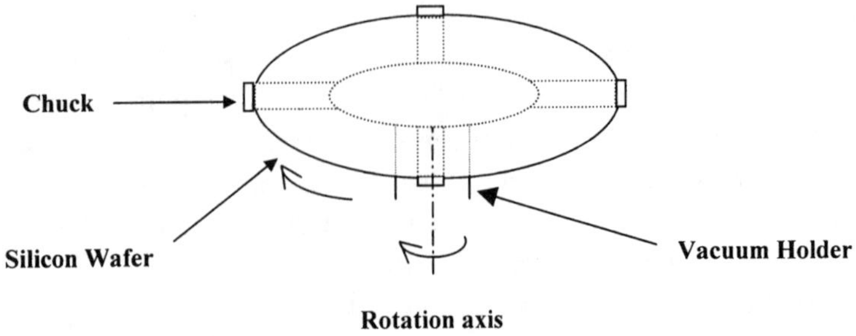

FIGURE 2.1 Schematic of the experimental setup for the particle removal experiment.

of adhesion, the particles are detached from the surface [15]. The Reynolds number based on the velocity at any radial distance, R, from the axis of rotation is given by [15, 16]:

$$\Re = \frac{R^2\omega}{\nu} \tag{5}$$

where R is the wafer radius, ω is the rotational speed of the substrate, and ν is the kinematic viscosity of the fluid. In the case of laminar flow, the following formulas are used for the drag and lift forces [15]. For the Stokes drag:

$$F_D = 3\pi\mu dV \tag{6}$$

For the Saffman lift:

$$F_L = 1.62\,\mu\,d^2\left(\frac{1}{\nu}\frac{du}{dx}\right)^{1/2} V \tag{7}$$

where μ is the viscosity of the fluid, d is the particle diameter, V is the relative velocity between the particle and the fluid, ν is the kinematic viscosity of the fluid and u is the fluid velocity. If the Reynolds number ($\Re^* = (dU^*/\nu)$ based on the particle diameter) exceeds 3×10^5, the flow is presumed turbulent. The two forces acting on the particle in turbulent flow are the drag force which is given as [15]:

$$F_D = 8\rho\nu^2\left(\frac{dU^*}{\nu}\right)^2 \tag{8}$$

and the Saffman lift force that is given as [15]:

$$F_L = 10.1\rho\nu^2\left(\frac{dU^*}{\nu}\right)^3 \tag{9}$$

where d is the particle diameter, ρ is the fluid density, and ν is the kinematic viscosity of the fluid. U^* is the friction velocity. The friction velocity has the dimensions of velocity and is defined as:

$$U^* = \left(\frac{\tau_0}{\rho}\right)^{1/2} \tag{10}$$

where τ_0 is the shear stress.

The centrifugal force in both flow regimes (laminar and turbulent) depends on the mass of the particle as well as the speed of rotation and is given as [20]:

$$F_{\text{Cent}} = mR\omega^2 \tag{11}$$

where m is the mass of the particle, ω is the rotational speed of the substrate, and R is the radial distance from the axis of rotation to the particle position on the substrate. The Reynolds number is calculated at any radial distances, R, from the center of the substrate. Equations (6) and (7) are used to calculate the drag and lift forces for laminar flow, and Eqs. (8) and (9) are used to calculate the drag and lift forces for turbulent flow. The centrifugal force is calculated using Eq. (11) and is vectorially added to the drag and lift forces to calculate the total removal force.

2.2.1. Adhesion Force Measurements

Busnaina, Taylor and Kashkoush [15] utilized the method outlined in Section 2.2 to determine the magnitude of the removal forces. They then correlated the hydrodynamic removal force with the removal efficiency and adhesion force, and introduced the following empirical correlation:

$$RE = 37 + 86.3\,FR - 39\,FR^2 + 5.7\,FR^3 \tag{12}$$

where FR is the force ratio ($F_{\text{removal}}/F_{\text{adhesion}}$). By using this correlation and calculating the hydrodynamic removal forces analytically and measuring the removal efficiency for any particle/substrate system experimentally, the particle adhesion force can be estimated [16]. The above equation is utilized to determine the total adhesion force of the particles considered in this study. Experimentally, the removal efficiency is defined as [20]:

$$\text{Removal efficiency} = \frac{n_{\text{before}} - n_{\text{after}}}{n_{\text{before}}} \tag{13}$$

where n_{before} and n_{after} are the numbers of particles before and after cleaning in the area to be cleaned. Equation (12) is not valid at a removal force equal to zero.

3. EXPERIMENTAL PROCEDURE

All experiments were conducted in the Microcontamination Research Laboratory's class 10 cleanroom at Clarkson University. Polished silicon wafers, 5 inches (125 mm) in diameter, were used as the deposition substrates in the experiments. The PET, metallized, and PCL substrates were glued on the silicon wafer. Polystyrene latex spheres (22 μm), pigmented coating chips and skin flakes were used to simulate particulate contaminants. The PSL spheres were manufactured by Duke Scientific. A standard pharmaceutical nebulizer consisting of a pump, reservoir and nozzle was used. To eliminate contamination from the pump, a 1 μm Millipore air filter was used between the pump and reservoir. A Headway Research photoresist spinner was used to spin the wafer up to a maximum speed of 10,000 rpm. A laser surface scanner, Particle Measuring System (PMS), provides a detailed color-coded map of particle sizes, positions, and exact count of particles. It then sorts them into different sizes ranging from 0.1 μm to 10 μm. Particles larger than 10 μm in diameter are shown as 10 μm. An Olympus BH-2-UMA optical microscope is used to count the particles on substrates other than silicon wafers. The microscope is connected to a CCD camera, a video monitor, and video microscaler. A calcium sulfate desiccant was used to reduce humidity below 40%. To study the effect of levels between 10% and 90% relative humidity, a humidity chamber was designed and built (using Plexiglas) to provide the required humidity.

The chamber is equipped with two independent humidity sensors, and an opening through which particles can be deposited, a platform to place the wafer during deposition, and an inlet for the moisture being supplied through a PVC valve from a humidifier. When the chamber sensor is set, the moisture input to the chamber is controlled to maintain the desired relative humidity. The following experimental procedures were used for each of the particle/substrate systems:

3.1. The PSL Particles/Silicon Substrate

1. To generate relative humidity between 10% and 50%, a half-pound (228 grams) of calcium sulfate desiccant is placed in the humidity chamber. For relative humidity between 50% and 90%, a humidifier is turned on, and in both cases the system is monitored until the desired humidity level is reached.
2. The silicon substrate is placed in the humidity chamber for two minutes.
3. Using a nebulizer, the 22 μm PSL particles are deposited and left for two minutes; 12 – 16 seconds are required to deposit more than 300 particles.
4. The laser surface scanner is used to obtain the exact number (n_{before}) of the 22 μm PSL particles deposited on the substrate.
5. Using air as the medium, the spin coater generates the removal forces that consist of drag, lift and centrifugal forces. The silicon substrate is rotated at 8500 rpm for 120 seconds.
6. The laser surface scanner is used to determine the exact number (n_{after}) of the 22 μm PSL particles remaining on the silicon substrate.
7. The above procedure is repeated at various humidity levels, 10% to 90% relative humidity.

3.2. The PSL Particles/(PET, Metallized and PCL) Substrates

1. The substrates (PET, metallized, and PCL) are cut into five-inch (12.7 cm) disks and glued onto the surface of a silicon wafer. A small (1.5 cm × 1.5 cm) square is marked on the substrate surface, at 4 cm (in the radial direction) from the center of the circular substrate.

2. Depending on the desired relative humidity, the substrate is left for two minutes in that humidity and then the 22 μm PSL particles are deposited using a nebulizer and left for two minutes.
3. The substrate is then taken out of the humidity chamber and a microscope is used to count the number (n_{before}) of PSL particles in the marked square.
4. Using air as the medium, the spin coater generates the removal force that consists of lift, drag and centrifugal forces. The substrate is rotated at 8500 rpm for 120 seconds.
5. The microscope is again used to count the number of PSL particles (n_{after}) still remaining in the marked square.
6. The above procedure is repeated for the PET, metallized surface and PCL substrates and at different humidity levels.

3.3. The Pigmented Coating Chips/ (PET, Metallized and PCL) Substrates

The procedure was the same as in Section 3.2 with the following exceptions:

1. The pigmented coating chips are placed on a glass plate and cut into smaller pieces. A tweezer was used to deposit the pigmented coating chips.

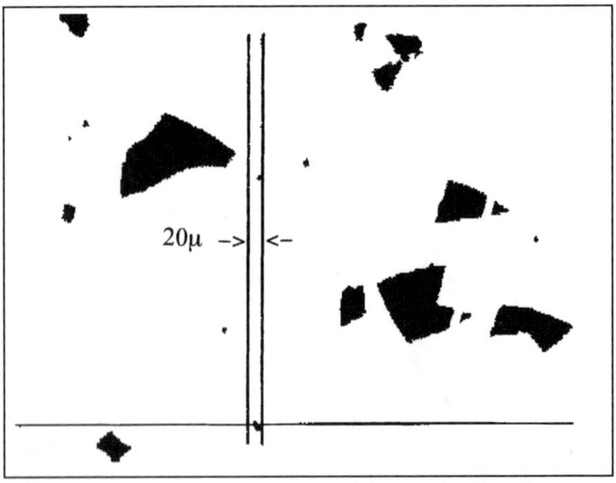

FIGURE 1 Pigmented coating chips.

2. The number of pigmented coating chips deposited (n_{before}) ranged from 65 – 120 in the marked square.
3. The size of the deposited pigmented coating chips ranged from 20 μm – 120 μm. There was a significant variation in the shapes of the pigmented coating chips. Figure 1 shows the different shapes and sizes observed under the microscope.

3.4. Skin Flakes/(PET, Metallized and PCL) Substrates

The same experimental procedure was followed as in Section 3.2, with the following exceptions:

1. Skin flakes were generated by rubbing dry skin over the substrate.
2. The number of skin flakes deposited (n_{before}) was from 10 – 12 in the marked square.
3. Different sizes and shapes were observed under the microscope. The size ranged from 30 μm – 100 μm.

4. RESULTS AND DISCUSSION

4.1. The Effect of Relative Humidity

A wafer surface scanner is used to measure the removal of PSL particles on silicon wafers which gives an accurate count of the number of particles on the five-inch (12.7 cm) wafer. This is the only part of the study which uses a scanner to count the particles. The rest of the experiments involved counting the particles visually using an optical microscope in pre-marked areas only. This produced a larger data scatter compared with the measurements using the surface scanner. For example, a typical standard deviation for the PSL/silicon substrate experiments is about 4 – 5%. For the rest of the data the standard deviation is 14 – 18%. Therefore, these experiments are at best qualitative since only the particles on that pre-marked area of the substrate are counted. However, the data show clear trends for the effect of relative humidity on particle adhesion and removal.

Figures 2 and 3 show the effect of relative humidity on the removal efficiency and adhesion force for 22 μm PSL particles on a silicon substrate. The relative humidity was between 15% and 85% in the humidity chamber. The highest removal efficiency is achieved at 45%

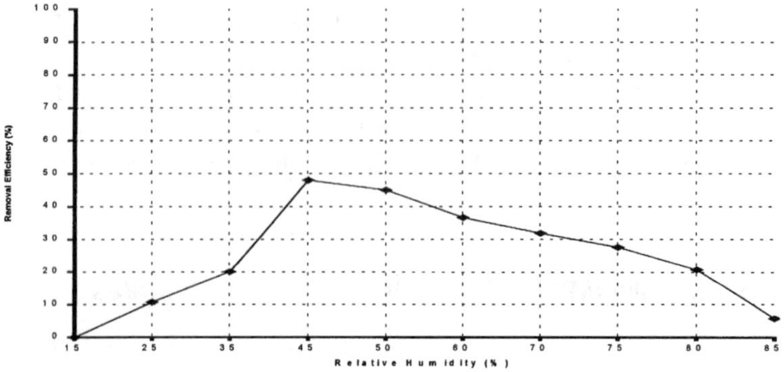

FIGURE 2 The effect of relative humidity on the removal efficiency (PSL particles/silicon wafer).

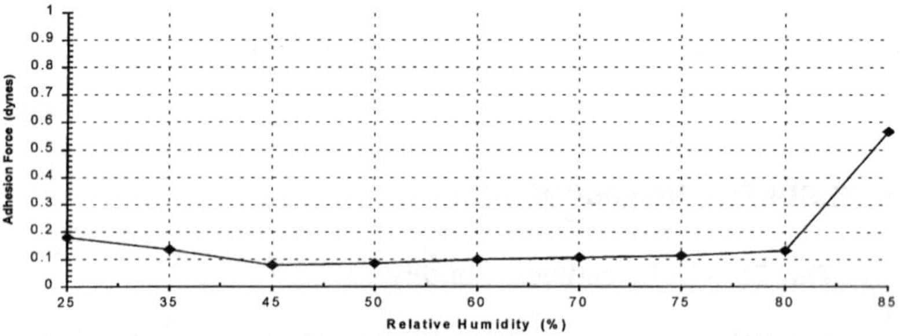

FIGURE 3 The effect of relative humidity on the adhesion force (PSL particles/silicon wafer).

relative humidity. As the relative humidity is reduced (below 45%), the adhesion force increases, and the removal efficiency decreases to zero at 15% relative humidity. The increase in the adhesion force is due to the charge build-up in low relative humidity since both the PSL particles and the silicon wafer (with native oxide film) are insulating.

At relative humidities between 45% and 80% the removal efficiency decreases, and the adhesion force gradually increases. Above 80% relative humidity, the adhesion force increases linearly due to capillary condensation which gives rise to capillary forces. This is due to the fact that both surfaces are hydrophillic, which gives rise to a large capillary force. Also, above 70% relative humidity the capillary force becomes the dominant force of adhesion. The theoretical capillary force value obtained using Eq. (3) is 1.0 dynes while the capillary force obtained from Figure 3 is 1.2 dynes. The difference is about 16%, which could be attributed to the van der Waals force. Using Eq. (1), the van der Waals force was calculated to be 0.4 dynes. Figures 4 and 5 show the effect of relative humidity on the removal efficiency and adhesion force for 22 μm PSL particles and three substrates (PET, metallized and PCL). The bell-shaped curve is again observed with the highest removal efficiency, with lowest adhesion force for all three substrates occurring at 45% relative humidity. As the relative humidity increases above 45%, the capillary force increases the total adhesion force. The PSL particles and the substrates (PET, metallized and PCL) are hydrophilic which gives rise to a large capillary force resulting in a low

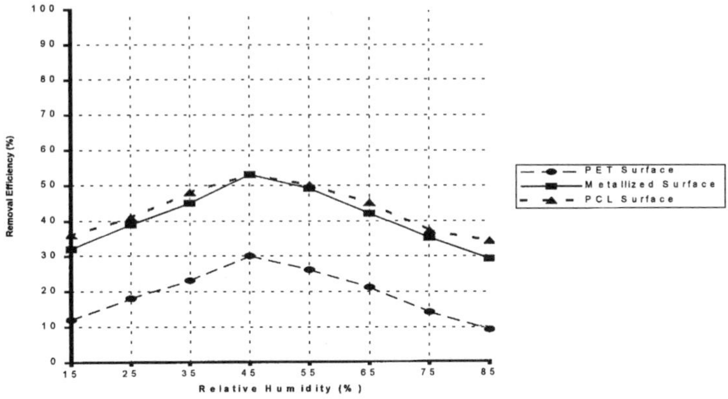

FIGURE 4 The effect of relative humidity on the removal efficiency (PSL particles/(PET, metallized, and PCL) surfaces).

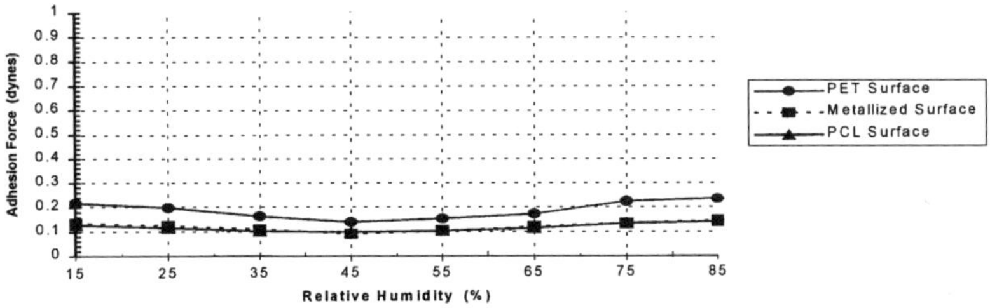

FIGURE 5 The effect of relative humidity on the adhesion force (PSL particles/(PET, metallized, and PCL) surfaces).

removal efficiency at high relative humidity. As the relative humidity decreases (below 45%) the electrostatic force increases the total adhesion force and decreases the removal efficiency. Figure 4 shows that the PSL particles on a PET substrate were the hardest to remove, compared with the metallized and PCL substrates that had similar removal efficiencies. This is due to the fact that the PET substrate and the PSL particle are both insulating (nonconductive) materials. At high relative humidity, the PET surface is more hydrophilic (a contact angle of 35°) as compared with the metallized (contact angle of 50°) and PCL substrates (a contact angle of 60°).

Figures 6 and 7 show the effect of relative humidity on the removal efficiency and adhesion force for pigmented coating chips on the three considered substrates (PET, metallized and PCL). The shape of the curve is different from that observed in the previous two figures where the removal efficiency is very high at low relative humidity below 25% (a removal efficiency above 90% can be achieved at 15% relative humidity). As the relative humidity is increased (above 25%), the adhesion force steadily increases. The high removal efficiency at low humidity could be attributed to the composition of the pigmented coating chips, which are made from selenium (a *p*-type semiconductor). The selenium will not build an electrostatic charge (when exposed to light) as compared with the PSL particles and skin flakes, which are not conductive (insulating). The conductivity of selenium is sensitive to light. The conductivity increases when the

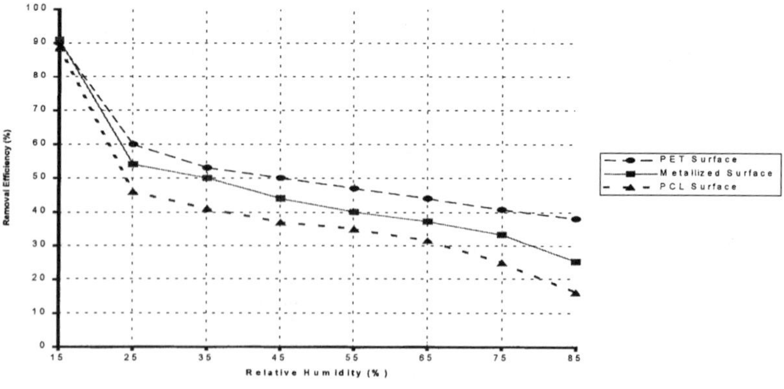

FIGURE 6 The effect of relative humidity on the removal efficiency (pigmented coating chips/(PET, metallized, and PCL) substrates).

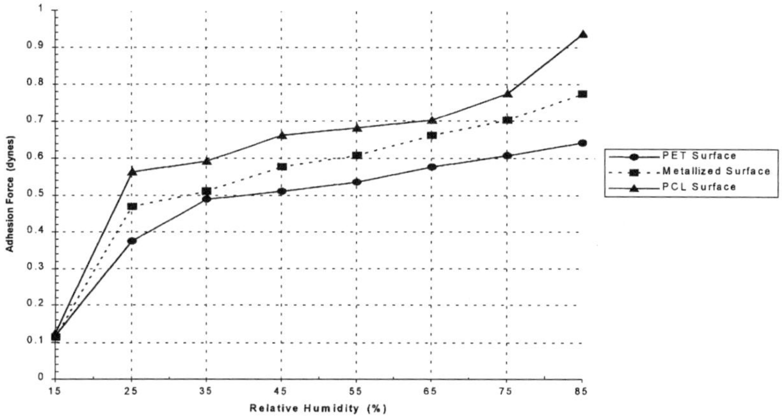

FIGURE 7 The effect of relative humidity on the adhesion force (pigmented coating chips/(PET, metallized, and PCL) substrates).

selenium is illuminated [25] which occurred throughout the experiments. The removal experiment leads to electrostatic charge build up by the high-speed rotation of the substrates and the particles.

Figures 8 and 9 show the effect of relative humidity on the removal efficiency and adhesion force for skin flakes on the three considered substrates (PET, metallized and PCL). As the relative humidity decreases (below 45%) the adhesion force also increases. Between 45% and 75% relative humidity the removal efficiency increases and then

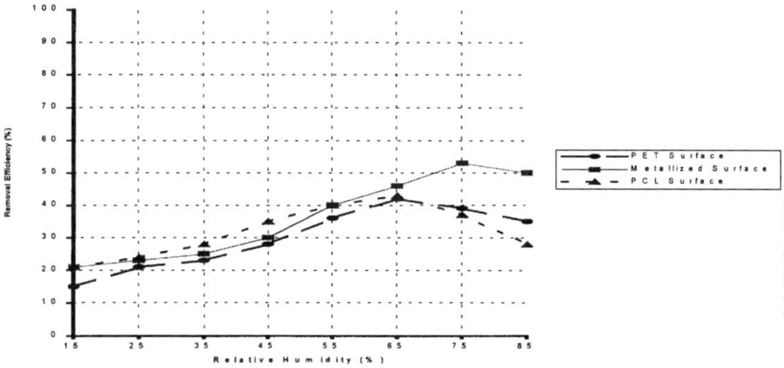

FIGURE 8 The effect of the relative humidity on the removal efficiency (skin flakes/(PET, metallized, PCL) substrates).

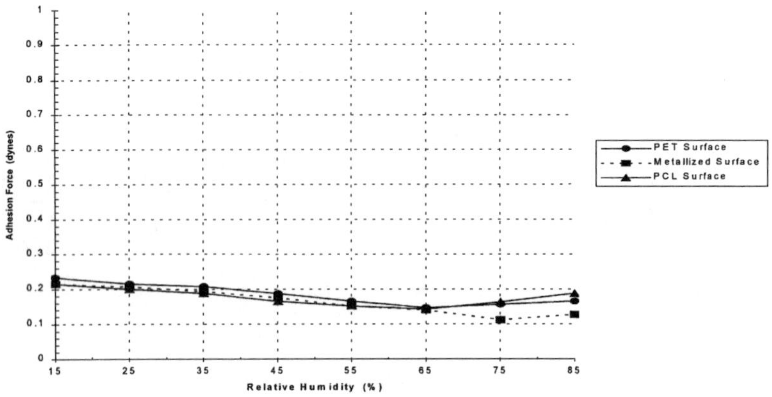

FIGURE 9 The effect of the relative humidity on the adhesion force (skin flakes/(PET, metallized, PCL) substrates).

starts to decrease. For the PET and PCL surfaces the adhesion force starts to increase at 65%, and for the metallized surface at 75%. The reason the removal efficiency is higher at higher humidity (65% – 75%) than for the other particles may be due to the fact that skin flakes absorb water and, consequently, may slightly increase in size and mass.

4.2. The Effect of Particle Size

The size of a particle has a profound effect on the adhesion force. To evaluate this effect, two different experiments were conducted using the three considered surfaces (PET, metallized and PCL) and the pigmented coating chips as contaminants. Figure 10 shows the effect of the particle size of the pigmented coating chips on the removal efficiency. At first, the pigmented coating chips (sizes between 100 and 500 μm) were deposited directly at different values of relative humidity. This is represented by the dotted line in Figure 10. The same experiments were conducted under the same conditions, but using smaller pieces of pigmented coating chips (that are chopped into sizes between 20 and 60 μm). The removal efficiency is lower for the small particle size (dotted lines). This is because the hydrodynamic removal force is a function of the particle size (cross sectional area of the particle). This results in higher removal force for larger particles [9].

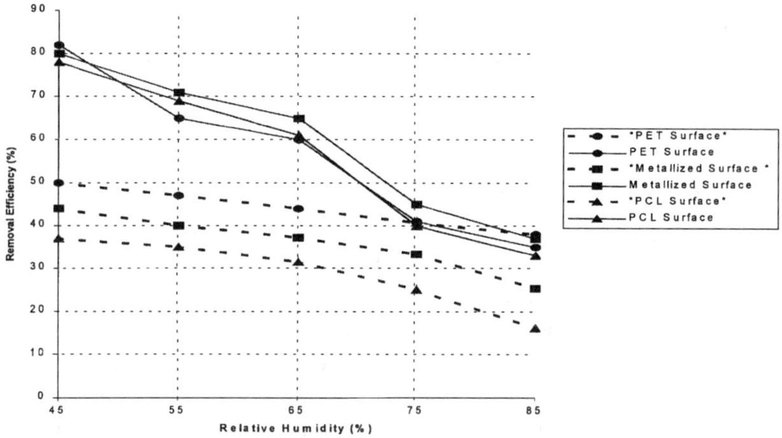

FIGURE 10 The effect of particle size on the removal efficiency (pigmented coating chips/(PET, metallized, PCL) surfaces). The smaller particles are indicated by the dotted line.

5. CONCLUSIONS

The results indicate that the capillary force is significant at a relative humidity above 50% and dominates at a relative humidity above 70%. At relative humidity below 45%, the electrostatic force became significant. Removal and adhesion results varied depending on the particles and substrates used but the trend of high adhesion at high and low humidity was observed for all particle/substrate systems. The results also indicate that the larger particles removal efficiency is higher overall.

1. The effect of relative humidity on the adhesion and removal for the 22 μm PSL particles on silicon substrates was very significant. The removal of PSL particles was very low at high and low relative humidity. The lowest adhesion force (highest removal efficiency, 49%) occurs at 45% relative humidity.
2. The effect of relative humidity on the adhesion and removal for the 22 μm PSL particles on the PET, metallized, and PCL substrates is very similar to PSL on a silicon substrate, showing the same bell-shaped curve. The removal of PSL particles was very low at high and low relative humidity. The highest removal efficiency (lowest

adhesion force) for the all three substrates occurred at 45% relative humidity. The PSL particles on a PET substrate surface were the hardest to remove, compared with the metallized and PCL substrates that had similar removal efficiencies.

3. The effect of relative humidity on the removal efficiency for pigmented coating chips on the three considered substrates (PET, metallized and PCL) was different from the PSL particles on the same substrates. The removal efficiency is very high at 90% (low adhesion force) at low relative humidity (below 25%). The removal efficiency is low (between 16 and 40%) at high relative humidity.
4. The effect of relative humidity on the removal efficiency of skin flakes from the three considered substrates (PET, metallized and PCL), was somewhat similar to the PSL behavior. At low relative humidity (below 45%) the removal efficiency was low. The maximum removal efficiency (lowest adhesion force) occurs between 45% and 75% relative humidity. For the PET and PCL surfaces the removal efficiency starts to decrease at 65%, and for the metallized surface at 75%.
5. The effect of particle size on removal efficiency is significant. The data indicate that when the size of particles decreases, the removal efficiency decreases.

References

[1] Rimai, D. S. and Busnaina, A. A., *J. Particulate Science and Technology* **13**, 249 (1995).
[2] Mittal, K. L., Ed., *Particles on Surfaces 1: Detection, Adhesion, and Removal* (Plenum Press, New York, 1988).
[3] Mittal, K. L., Ed., *Particles on Surfaces 2: Detection, Adhesion, and Removal* (Plenum Press, New York, 1989).
[4] Mittal, K. L., Ed., *Particles on Surfaces 3: Detection, Adhesion, and Removal* (Plenum Press, New York, 1991).
[5] Hubbe, M., *Ph.D. Dissertation* Clarkson University (1983).
[6] Tabor, D., *J. Colloid Interface Sci.* **58**, 2 (1977).
[7] Zimon, A. D., *Adhesion of Dust and Powder* (Plenum Press, New York, 1969).
[8] Ranade, M. B., *Aerosol Sci. Technol.* **7**, 161 (1987).
[9] Hinds, W. C., *Aerosol Technology* (John Wiley, New York, 1982).
[10] Musselman, R. P. and Yarbrough, T. W., *J. Environmental Sci.* **51**, 51–56 (1987).
[11] Saffman, P. G., *J. Fluid Mech.* **22**, 385 (1965).
[12] Visser, J., In: *Surface and Colloid Science*, Matijevic, E., Ed., Vol. 8 (John Wiley and Sons, New York, 1976).
[13] Kurz, M., Busnaina, A. A. and Kern, F. W., *Proceedings, IES 35th Meeting* Anaheim, CA, May 1–5, pp. 340–347 (1989).

[14] Taylor, J., Busnaina, A. A., Kern, F. W. and Kunesh, R., *Proceedings, IES 36th Meeting* New Orleans, LA, April 23–27, pp. 422–426 (1990).
[15] Busnaina, A. A., Taylor, J. and Kashkoush, I., *J. Adhesion Sci. Technol.*, **7**(5), 441 (1993).
[16] Krishnan, S., Busnaina, A. A., Rimai, D. S. and DeMejo, L. P., *J. Adhesion. Sci. Technol.* **8**, 1357 (1994).
[17] Bowden, F. P. and Throssel, W. R., *Proc. Roy. Soc.* **A209**, 297 (1951).
[18] Schwartz, A. M., In: *Chemistry and Physics of Interfaces-II*, Ross, S., Ed. (Am. Chem. Socy. Publications, Washington, DC, 1971), pp. 1–13.
[19] Davies, C. N., *Aerosol Science* (Academic Press, New York, 1966).
[20] Taylor, J., *M.S. Thesis* Clarkson University (1992).
[21] Kurz, M. R., *M.S. Thesis* Clarkson University (1988).
[22] Kordecki, M. C. and Orr, C. Jr., *Archives of Enviro. Health* **1**, 7 (1960).
[23] Hamaker, H. C., *Physica* **4**, 1058 (1937).
[24] Luzhnov, Yu. M., *Research in Surface Forces* (Consultants Bureau, New York, 1971).

The Effect of Time and Humidity on Particle Adhesion and Removal

JING TANG and AHMED A. BUSNAINA

Microcontamination Research Laboratory, Center for Advanced Materials Processing, Clarkson University, Potsdam, NY 13699-5725, USA

Time and humidity greatly influence particle adhesion and removal in many particle-substrate systems. The effect of time (aging) and humidity on the adhesion and removal of 22 μm PSL (Polystyrene Latex) particles on polished silicon wafers is investigated. The results show that the effect of time on the adhesion and removal of the 22 μm PSL particles on silicon substrates in high humidity environment is very significant. The removal efficiency of PSL particles significantly decreased after the samples were aged for more than one day in high humidity environment. The combined effect of the van der Waals force and the capillary force tend to accelerate the adhesion-induced deformation process. When capillary force occurs at the particle substrate interface, the removal efficiency decreases quickly by more than 50% within 24 hours. Without the capillary force, the adhesion-induced deformation is negligible within the first 24 hours.

1. INTRODUCTION

Particle adhesion and removal is of scientific and technological interest in many industries such as semiconductor manufacturing, xerography and pharmacology. The mitigation of defects caused by particulate contamination depends on the understanding of the mechanisms and factors that influence particle adhesion and removal. Many models have been introduced to predict particle adhesion and deformation [1–3]. The effect of time and humidity on adhesion and removal on different particle/substrate systems has also been studied [4–8]. Krishnan, Busnaina *et al.*, used scanning electron microscopy to observe adhesion-induced deformation of sub-micrometer PSL (Polystyrene Latex) particles on silicon substrates as a function of time only. The contact area between the particle and the substrate was found to increase with time for a period of approximately 72 hours before reaching a constant value. They also related this result to the study of particle removal conducted using hydrodynamic and centrifugal removal forces for different time periods. The removal efficiency correlates well with the increase of particle adhesion force with time as observed from the SEM measurement [4]. Busnaina and Elsawy [5, 7, 8] measured the removal and adhesion force as a function of humidity. The trend of high adhesion at high and low humidity was observed. At high humidity, the capillary force constitutes a large adhesion force in addition to the van der Waals force and as a result the removal efficiency significantly drops. At low humidity, the low removal efficiency is attributable to the electrostatic force because of the higher charge that can build up during airflow over the silicon substrate (an insulator) [5].

2. THEORETICAL

For small, uncharged particles on uncharged surfaces, the primary force is usually attributed to van der Waals (vdW) interaction. For the ideal case in which both the spherical particle and the surface are not deformed, the vdW force is proportional to the radius of the sphere. When deformation occurs, the magnitude of the adhesion force will also depend on the contact area between the particle and the surface. According to Bowling [9], when a sphere and a flat substrate come into contact with one another, the attractive force, $F_{(\mathrm{vdW})}$, deforms the interface and a circular adhesion area is formed. The total adhesion force consists of two additive components, the force acting between the adherents before deformation, $F_{(\mathrm{vdW})}$, and the force acting on the contact area due to the deformation, $F_{(\mathrm{vdW\_deform})}$:

$$F_{(\mathrm{vdW\_total})} = F_{(\mathrm{vdW})} + F_{(\mathrm{vdW\_deform})} \tag{1}$$

Bowling gave the total van der Waals force including the component due to the deformation as:

$$F_a = F_0\left(1 + \frac{a^2}{Rz_0}\right) \tag{2}$$

where $F_0 = (\mathrm{A}R/6z_0^2)$ is the van der Waals force for the spherical particle, A is the Hamaker–van der Waals constant, R is the radius of the spherical particle, z_0 is the separation distance between the particle and the substrate (For smooth surfaces, it is taken as 4 Å) and a is the contact radius between the deformed particle and surface.

When moisture is present in the air medium, condensation can take place between the particle and substrate as shown in Figure 1. This capillary condensation gives rise to a very large capillary force, which increases the total force of adhesion. The capillary force is made up of two components: surface tension at the perimeter of the meniscus and the difference in pressure between the liquid and vapor phases. Zimon [10] concluded that when the relative humidity is above 70%, the capillary force dominates and should be the only adhesion force considered. Bowden and Throssel [11] found that the force required for removing 98% of 50-micron glass particles from a glass substrate increases with increasing relative humidity. At relative humidity near 100%, the capillary force can be predicted by [12]:

$$F_c = 4\pi R\gamma_{LV} \tag{3}$$

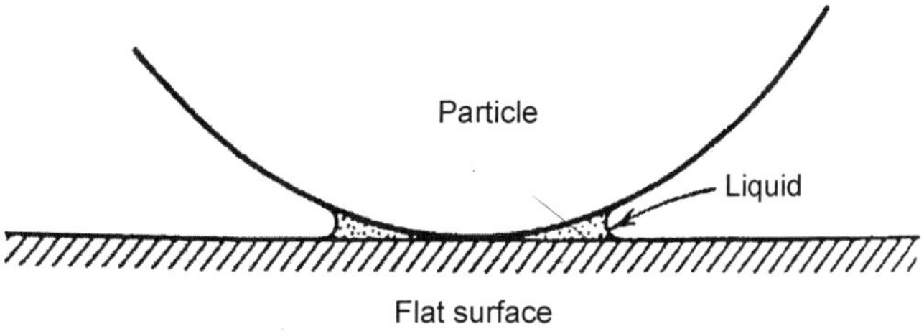

FIGURE 1 Condensed film between particle and substrate.

3. EXPERIMENTAL

The objective of this study is to determine the combined effect of time and relative humidity on the removal efficiency of PSL particles from silicon substrates.

3.1. Experimental Facilities

Polished silicon wafers, 5 inches (125 mm) in diameter, were used as the deposition substrates in the experiments. A dry powder of PSL spheres (22 μm) manufactured by Duke Scientific Inc. was used to simulate particulate contaminants. A humidity chamber system that consists of a humidifier with a controller that allows settings at different humidity was used to provide the environment in which particles were deposited. A dessicator was used to keep the samples during the aging period. A Headway research photo-resist spinner was used to generate the removal force required to detach particles. The spinner has a maximum rotational speed of 10,000 rpm. An Olympus BH-2 UMA optical microscope equipped with a CCD camera and a computer-controlled auto-focus micro-stage was used to count the particles on the substrate.

3.2. Experimental Procedure

The substrate was placed on the platform in the humidity chamber with the desired relative humidity for two minutes. Dry powder of 22 μm PSL particles was deposited very gently on the substrate in a 0.5×0.5 cm square, which is about 2 cm from the center of the wafer. Usually, there were about 100 to 200 particles in the square. The sample was kept in the dessicator with the desired relative humidity for the desired time. The microscope was used to obtain the exact number (n_{before}) of the PSL particles deposited on the substrate before the sample was cleaned. Using air as the medium, the spin coater generated the removal force that consists of drag, lift and centrifugal forces. The silicon substrate was rotated at 3000 rpm or 6000 rpm, successively, each for 120 seconds. The microscope was used again to determine the number (n_{after}) of the particles still present on the silicon substrate. The above procedure was repeated for different values of relative humidity.

4. RESULTS AND DISCUSSION

Figure 2 illustrates the theoretical calculation results that show the contact area between particle and substrate with and without the capillary force. The MP (Maugis–Pollock) model [3] is used to calculate the contact area:

$$P = \pi a^2 H \tag{4}$$

where P is the total force causing the deformation, a is the radius of contact and H is the hardness of particle material. Figure 2a depicts that the contact area between the particle and the substrate due to adhesion-induced deformation with the capillary force is more than eight-fold that without the capillary force. When the capillary force occurs (at high humidity), calculations in Figure 2b show that before deformation commences, the capillary force is the dominant force of adhesion while after deformation occurs the van der Waals force

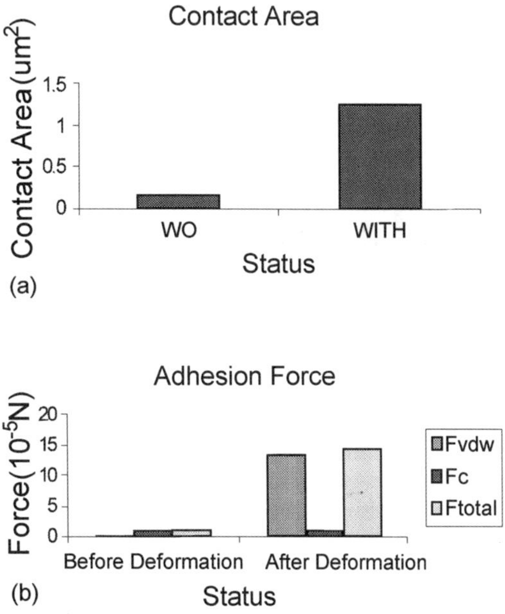

FIGURE 2 a: Comparison of Contact Areas with and without Capillary Force. b: Comparison of Adhesion Forces Before and After Deformation (with capillary force).

becomes the dominant force because of the adhesion-induced deformation. The increase in the contact area due to the adhesion-induced deformation caused by the combined van der Waals and capillary forces significantly increases the final van der Waals force between the particle and the substrate.

Figures 3 and 4 show the particle removal efficiency results ($1-(n_{after}/n_{before})$) corresponding to intermediate and high humidity environments with low and high rotational speeds at 3000 rpm and 6000 rpm. Figure 3 illustrates particle removal efficiency results at 45% relative humidity using low and high rotational speeds (corresponding

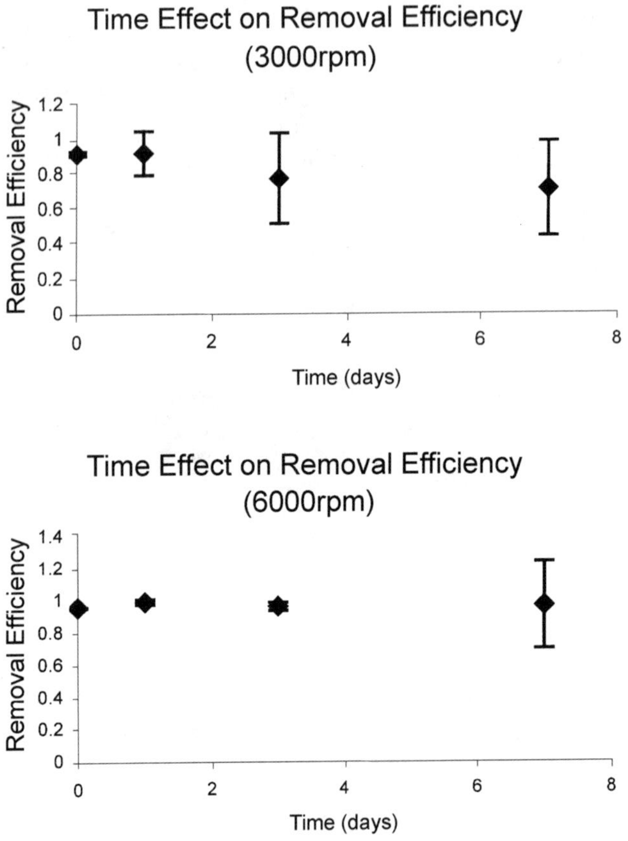

FIGURE 3 Time Effect on Removal Efficiency ($1-(n_{after}/n_{before})$) at 45% Relative Humidity.

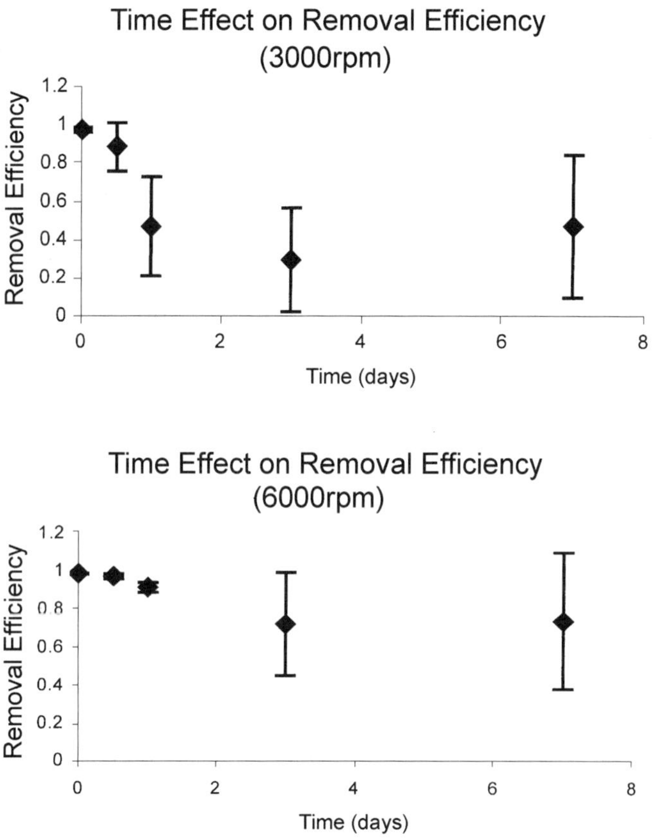

FIGURE 4 Time Effect on Removal Efficiency ($1-(n_{\text{after}}/n_{\text{before}})$) at 100% Relative Humidity.

to low and high removal forces). The figure shows that the removal efficiency of PSL particles on silicon substrates decreases with time. The decrease of removal efficiency is caused by the increase of the adhesion force due to the adhesion-induced deformation of the PSL particles due to the van der Waals force only and without the capillary force effect. When capillary condensation occurs the adhesion-induced deformation also increases due to the additional capillary force. Figure 4 shows the combined effect of the van der Waals force and the capillary force on the removal efficiency. The combined effect of the van der Waals force and the capillary force tends to accelerate the deformation process. In Figure 4, the removal efficiency decreases by 50% after one day at 3000 rpm and 10% at 6000 rpm at 100% relative humidity. By comparison, at 45% relative humidity the decrease in the removal efficiency after one day is negligible at 3000 rpm and 6000 rpm. The decrease in removal efficiency over a week is 50% at high humidity (100%) and low rotational speed (3000 rpm) where at 45% relative humidity and the same speed the decrease is about 20%. At higher speed (6000 rpm) the decrease in the removal efficiency at high humidity (100%) and long aging time of one week is 25–30%. However, at lower humidity (45%) the decrease in the removal efficiency over a week was negligible.

5. CONCLUSIONS

The effect of time on the adhesion and removal of 22 μm PSL particles on silicon substrates in high humidity environment is very significant. The removal efficiency of PSL particles significantly decreased after the samples were aged for more than one day in high humidity environment. The combined effect of the van der Waals force and the capillary force tends to accelerate the adhesion-induced deformation process. Both time and humidity have a remarkable effect on the adhesion force. When the capillary force occurs, the removal efficiency decreases quickly by more than 50% within 24 hours. Without the capillary force, the adhesion-induced deformation is negligible within the first 24 hours.

References

[1] Johnson, K. L., Kendall, K. and Roberts, A. D., *Proc. Roy. Soc. A* **324**, 301–313 (1972).
[2] Deryaguin, B. V., Muller, V. M. and Toporov, Yu. P., *J. Colloid Interface Sci.* **53**, 314–326 (1984).
[3] Maugis, D. and Pollock, H. M., *Acta Metall.* **32**, 1323–1334 (1984).
[4] Krishnan, S., Busnaina, A. A., Rimai, D. S. and Demejo, L. P., *J. Adhesion Sci. Tech.* **8**(11), 1357–1370 (1994).
[5] Busnaina, A. A. and Elsawy, T., "The Effect of Relative Humidity on Particle Adhesion and Removal", *J. Adhesion*, December, 1999.
[6] Busnaina, A. A., Taylor, J. and Kashkoush, I., *J. Adhesion Sci. and Tech.* **7**(5), 441–455 (1993).

[7] Busnaina, A. A. and Elsawy, T., "The Measurement of Particle Adhesion Forces in Humid and Dry Environments", *The Adhesion Society Proceedings, 21st Annual Meeting*, Savannah, GA, USA, Feb. 22–25, 1998, pp. 394–397.

[8] Busnaina, A. A. and Elsawy, T., "The Effect of Relative Humidity on Particle Adhesion and Removal", *The Adhesion Society Proceedings, 21st Annual Meeting*, Savannah, GA, USA, Feb. 22–25, 1998, pp. 315–317.

[9] Bowling, R. A., In: *Particles on Surface 1: Detection, Adhesion and Removal*, Mittal, K. L. Ed. (Plenum Press, New York, 1988), pp. 129–142.

[10] Zimon, A. D., *Adhesion of Dust and Powder* (Plenum Press, New York, 1969), pp. 108–119.

[11] Bowden, F. P. and Throssel, W. R., *Proc., Roy. Soc.* **A209**, 297 (1951).

[12] Ranade, M. B., "Adhesion and Removal of Fine Particles on Surfaces", *Aerosol Science and Technology* **7**, 161–176 (1987).

Recent Theoretical Results for Nonequilibrium Deposition of Submicron Particles

VLADIMIR PRIVMAN

Department of Physics, Clarkson University, Potsdam, New York 13699-5820, USA

Selected theoretical developments in modeling of deposition of sub-micrometer size (submicron) particles on solid surfaces, with and without surface diffusion, of interest in colloid, polymer, and certain biological systems, are surveyed. We review deposition processes involving extended objects, with jamming and its interplay with in-surface diffusion yielding interesting dynamics of approach to the large-time state. Mean-field and low-density approximation schemes can be used in many instances for short and intermediate times, in large enough dimensions, and for particle sizes larger than few lattice units. Random sequential adsorption models are appropriate for higher particle densities (larger times). Added diffusion allows formation of denser deposits and leads to power-law large-time behavior which, in one dimension (linear substrate, such as DNA), was related to diffusion-limited reactions, while in two dimensions (planar substrate), was associated with evolution of the domain-wall and defect network, reminiscent of equilibrium ordering processes.

1. INTRODUCTION

1.1. Surface Deposition of Submicron Particles

Surface deposition of submicron particles is of immense practical importance [1–4]. Typically, particles of this size, colloid, protein or other biological objects, are suspended in solution, without sedimentation due to gravity. In order to maintain the suspension stable, one has to prevent aggregation (coagulation) that results in larger flocks for which gravity pull is more profound. Stabilization by particle–particle electrostatic repulsion or by steric effects, *etc.*, is usually effective for a sufficiently dilute suspension. But this means that even if a well-defined suspension of well-characterized particles is available, it cannot be always easily observed experimentally in the bulk for a wide range of particle interactions. For those interaction parameters for which the system is unstable with respect to coagulation, the time of observation will be limited by the coagulation process which can be quite fast.

One can form a dense deposit slowly, if desired, *on a surface*. Indeed, particles can be deposited by diffusion, or more realistically by convective diffusion [5] from a flowing suspension, on collector surfaces. The suspension itself need not be dense even though the on-surface deposit might be quite dense, depending of the particle–particle and particle–surface interactions. Dilution of suspension generally prolongs an experiment aimed at reaching a certain surface coverage. Thus, surface deposition has been well established as an important tool to probe interactions of matter objects on the submicron scale [1–4].

1.2. Particle Jamming and Screening at Surfaces

Figure 1 illustrates possible configurations of particles at a surface. From left to right, we show particles deposited on the surface of a collector, then particles deposited on top of other particles. The latter is possible only in the absence of significant particle–particle repulsion. The two situations are termed monolayer and multilayer deposition even though the notion of a layer beyond the one exactly at the surface is only approximate. We next show two effects that play important roles in surface growth. The first is jamming: a particle marked by an open circle cannot fit in the lowest layer at the surface. A more realistic two-dimensional (2*D*) configuration is shown in the inset.

The second effect is screening: surface position marked by the open circle is not reachable. Typically, in colloid deposition monolayer or few-layer deposits are formed and the dominant effect is jamming, as

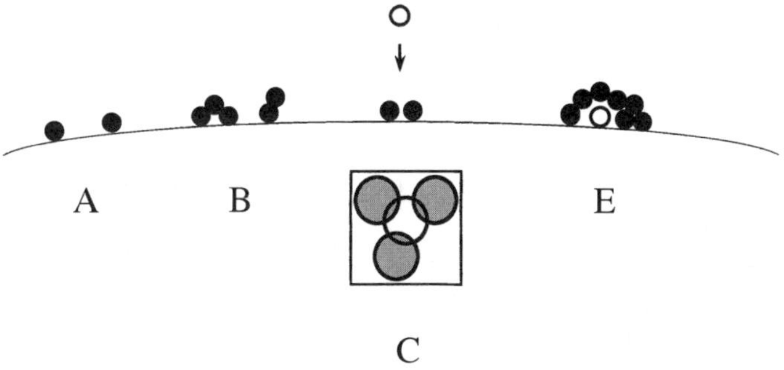

FIGURE 1 Possible configurations of particles at surfaces. From left to right, A – particles deposited directly on the collector; B – particles deposited on top of other particles. We next show an example of jamming, C – a particle marked by an open circle cannot fit in the lowest layer at the surface. A top view of a more realistic two-dimensional (2*D*) surface configuration is shown in the inset. The rightmost example, E, illustrates screening: surface position marked by the open circle is not reachable.

will be discussed later. Screening plays the dominant role in deposition of multiple layers and, together with the transport mechanism, determines the morphology of the growing surface. In addition, the configuration on the surface depends on the transport mechanism of the particles to it and on the particle motion on the surface, as well as possible detachment. Particle motion is typically negligible for colloidal particles but may be significant for proteins.

1.3. Role of Dimensionality and Relation to Other Systems

An important feature of surface deposition is that for all practical purposes it is essentially a 2*D* problem. As a result, any mean-field, rate-equation, effective-field, *etc.*, approaches which are usually all related in that they ignore long-range correlations and fluctuation effects, may not be applicable. Indeed, it is known that as the dimensionality of a many-body interacting system decreases, fluctuations play a larger role. Dynamics of important physical, chemical, and biological processes [6–7] provide examples of strongly fluctuating systems in low dimensions, $D=1$ or 2. These processes include surface adsorption on planar substrates or on large collectors. The surface of the latter is semi-two-dimensional owing to their large size as compared with the size of the deposited particles.

The classical chemical reaction–diffusion kinetics corresponds to $D=3$. However, heterogeneous catalysis generated interest in $D=2$. For both deposition and reactions, some experimental results exist even in $D=1$ (see later). Finally, kinetics of ordering and phase separation, largely amenable to experimental probing in $D=3$ and 2, attracted much recent theoretical effort in $D=1$, 2.

Models in $D=1$, and sometimes in $D=2$, allow derivation of analytical results. Furthermore, it turns out that all three types of model: deposition–relaxation, reaction–diffusion, phase separation, are interrelated in many, but not all, of their properties. This observation is by no means obvious. It is model-dependent and can be firmly established [6, 7] only in low dimensions, mostly in $D=1$.

Such low-dimensional nonequilibrium models pose several interesting challenges theoretically and numerically. While many exact, asymptotic, and numerical results are already available in the literature [6, 7], this field presently provides examples of properties which lack theoretical explanation even in $1D$. Numerical simulations are challenging and require large scale computational effort already for $1D$ models. For more experimentally relevant $2D$ cases, where analytical results are scarce, difficulty in numerical simulations has been the limiting factor in the understanding of many open problems.

1.4. Outline of this Review

The purpose of this article is to provide an introduction to the field of nonequilibrium surface deposition models of extended particles. By "extended" we mean that the main particle–particle interaction effect will be jamming, *i.e.*, mutual exclusion. No comprehensive survey of the literature is attempted. The relation of deposition to other low-dimensional models mentioned earlier will be referred to in detail only in few cases. The specific models and examples selected for a more detailed exposition, *i.e.*, models of deposition with diffusional relaxation, were biased by the author's own work.

The outline of the review is as follows. The rest of this introductory section is devoted to defining the specific topics of surface deposition to be surveyed. Section 2 describes the simplest models of random sequential adsorption. Section 3 is devoted to deposition with relaxation, with general remarks followed by definition of the simplest, $1D$ models of diffusional relaxation for which we present a more detailed description of various theoretical results. Multilayer deposition is also commented on in Section 3. More numerically-based $2D$ results for deposition with diffusional relaxation are surveyed in Section 4. Section 5 presents brief concluding remarks.

Surface deposition is a vast field of study. Our emphasis here will be on those deposition processes where the particles are "large" as compared with the underlying atomic and morphological structure of the substrate and as compared with the range of the particle – particle and particle – substrate interactions. Thus, colloids, for instance, involve particles of submicron to several micron size. We note that 1 μm = 10000 Å, whereas atomic dimensions are of order 1 Å, while the range over which particle – surface and particle – particle interactions are significant, as compared with kT, is typically of order 100 Å or less. Extensive theoretical study of such systems is relatively recent and it has been motivated by experiments where submicron-size colloid, polymer, and protein "particles" were the deposited objects [1 – 4, 8 – 18].

Perhaps the simplest and the most studied model with particle exclusion is Random Sequential Adsorption (RSA). The RSA model, to be described in detail in Section 2, assumes that particle transport (incoming flux) onto the surface results in a uniform deposition attempt rate, R, per unit time and area. In the simplest formulation, one assumes that only monolayer deposition is allowed. Within this monolayer deposit, each new arriving particle must either fit in an empty area allowed by the hard-core exclusion interaction with the particles deposited earlier, or the deposition attempt is rejected.

The basic RSA model will be described shortly, in Section 2. Recent work has been focused on its extensions to allow for particle relaxation by diffusion, see Sections 3 and 4, to include detachment processes, and to allow multilayer formation. The latter two extensions will be briefly surveyed in Section 3. Several other extensions will not be discussed [1 – 4].

2. RANDOM SEQUENTIAL ADSORPTION

2.1. The RSA Model

The irreversible Random Sequential Adsorption (RSA) process [19, 20] models experiments of submicron particle deposition by assuming a planar $2D$ substrate and, in the simplest case, continuum (off-lattice) deposition of spherical particles. However, other RSA models have also received attention. In $2D$, noncircular cross-section shapes as well as various lattice-deposition models were considered [19, 20]. Several experiments on polymers and attachment of fluorescent units on DNA molecules [18] (the latter is usually accompanied by motion of these units on the DNA and detachment) suggest consideration of the lattice-substrate RSA processes in $1D$. RSA processes have also found applications in traffic problems and certain other fields. Our presentation in this section aims at defining some RSA models and outlining characteristic features of their dynamics.

Figure 2 illustrates the simplest possible monolayer lattice RSA model: irreversible deposition of dimers on the linear lattice. An arriving dimer will be deposited if the underlying pair of lattice sites are both empty. Otherwise, it is discarded, which is shown schematically by the two dimers above the surface layer. Their deposition on the surface is not possible unless detachment and/or motion of monomers or whole dimers clear the appropriate landing sites.

Let us consider the irreversible RSA without detachment or diffusion. The substrate is usually assumed to be empty initially, at $t=0$. In the course of time t, the coverage, $\rho(t)$, increases and builds up to order 1 on the time scales of order $(RV)^{-1}$, where R was defined earlier as the deposition attempt rate per unit time and area of the surface, while V is the particle D-dimensional "volume". For deposition of spheres on a planar surface, V is actually the cross-sectional area.

At large times the coverage approaches the jammed-state value where only gaps smaller than the particle size were left in the monolayer. The resulting state is less dense than the fully-ordered, close-packed coverage. For the $D=1$ deposition shown in Figure 2 the fully-ordered state would have $\rho=1$. The variation of the RSA coverage is illustrated by the lower curve in Figure 3.

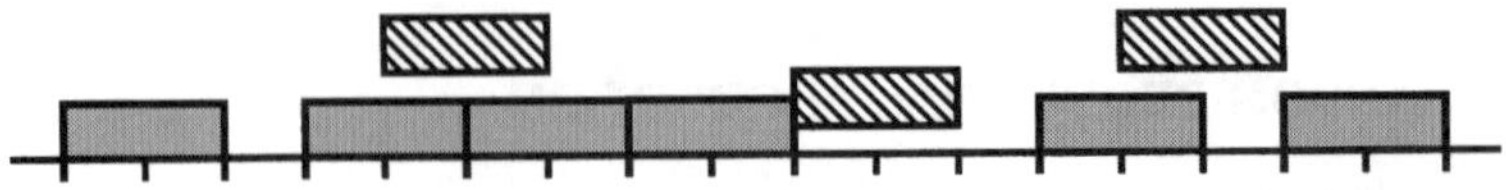

FIGURE 2 Deposition of dimers on the $1D$ lattice. Only one of the three hatched dimers can deposit on the surface, which then becomes fully jammed in the interval shown.

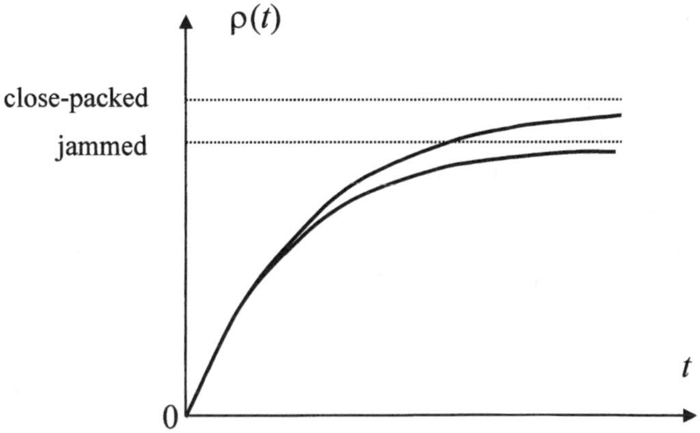

FIGURE 3 Schematic variation of the coverage, $\rho(t)$, with time for deposition without (lower curve) and with (upper curve) diffusional or other relaxation. The "ordered" density corresponds to close packing.

At early times the monolayer deposit is not dense and the deposition events are largely uncorrelated. In this regime, mean-field-like, low-density approximation schemes are useful [21–23]. Deposition of k-mer particles on the linear lattice in $1D$ was in fact solved exactly for all times [24]. In $D=2$, extensive numerical studies were reported [23, 25–36] of the variation of coverage with time and large-time asymptotic behavior which will be discussed shortly. Some exact results [24] for correlation properties are available in $1D$. Numerical results [27] for correlation properties have been obtained in $2D$.

2.2. The Large-time Behavior in RSA

The large-time deposit has several characteristic properties. For lattice models, the approach to the jammed-state coverage is exponential [36–38]. This was shown to follow from the property that the final stages of deposition are in few sparse, well separated surviving landing sites. Estimates of decrease in their density at late stages suggest that

$$\rho(\infty) - \rho(t) \sim \exp(-R\ell^D t), \tag{1}$$

where ℓ is the lattice spacing and D is the dimensionality of the substrate. The coefficient in Eq. (1) is of order ℓ^D/V if the coverage is defined as the fraction of lattice units covered, *i.e.*, the dimensionless fraction of area covered, also termed the coverage fraction, so that coverage as density of particles per unit volume would be $V^{-1}\rho$. The detailed behavior depends of the size and shape of the depositing particles as compared with the underlying lattice unit cells.

However, for continuum off-lattice deposition, formally obtained as the limit $\ell \to 0$, the approach to the jamming coverage is power-law. This interesting behavior [37, 38] is due to the fact that for large times the remaining voids accessible to particle deposition can be of sizes arbitrarily close to those of the depositing particles. Such voids are, thus, reached with very low probability by the depositing particles, the flux of which is uniformly distributed. The resulting power-law behavior depends on the dimensionality and particle shape. For instance, for D-dimensional cubes of volume V,

$$\rho(\infty) - \rho(t) \sim \frac{[\ln(RVt)]^{D-1}}{RVt}, \qquad (2)$$

while for spherical particles,

$$\rho(\infty) - \rho(t) \sim (RVt)^{-1/D}. \qquad (3)$$

For $D > 1$, the expressions Eqs. (2, 3), and similar relations for other particle shapes, are actually empirical asymptotic laws which have been verified, mostly for $D = 2$, by extensive numerical simulations [4, 25–36]. The most studied $2D$ geometries are circles (corresponding to the deposition of spheres on a planar substrate) and squares. The jamming coverages are

$$\rho_{\rm squares}(\infty) \simeq 0.5620 \quad \text{and} \quad \rho_{\rm circles}(\infty) \simeq 0.544 \text{ to } 0.550, \qquad (4)$$

much lower than the close-packing values, 1 and $(\pi/2\sqrt{3}) \simeq 0.907$, respectively. For square particles, the crossover to continuum in the limit $k \to \infty$ and $\ell \to 0$, with fixed $V^{1/D} = k\ell$ in deposition of $k \times k \times \cdots \times k$ lattice squares, has been investigated in some detail [36], both analytically (in any D) and numerically (in $2D$).

The correlations in the large-time jammed state are different from those of the equilibrium random gas of particles with density near

$\rho(\infty)$. In fact, the two-particle correlations in continuum deposition develop a weak singularity at contact, and correlations generally reflect the full irreversibility of the RSA process [24, 27, 38].

3. DEPOSITION WITH RELAXATION

3.1. Detachment and Diffusional Relaxation

Monolayer deposits may relax, *i.e.*, explore more configurations, by particle motion on the surface, by their detachment, as well as by motion and detachment of the constituent monomers or recombined units. In fact, detachment has been experimentally observed in deposition of colloid particles which were otherwise quite immobile on the surface [39]. Theoretical interpretation of colloid particle detachment data has proved difficult, however, because binding to the substrate, once the particle is deposited, can be different for different particles, whereas the transport to the substrate, *i.e.*, the flux of the arriving particles in the deposition part of the process, typically by convective diffusion, is more uniform. Detachment also plays a role in deposition on DNA molecules [18].

Recently, more theoretically motivated studies of the detachment relaxation processes, in some instances with surface diffusion allowed as well, have led to interesting model studies [40–46]. These investigations did not always assume detachment of the original units. Models involving monomer recombination prior to detachment, of k-mers in $D=1$, have been mapped onto certain spin models and symmetry relations were identified which allowed derivation of several exact and asymptotic results on the correlations and other properties [40–46]. We note that deposition and detachment combine to drive the dynamics into a steady state, rather than a jammed state as in ordinary RSA. These studies have been largely limited thus far to $1D$ models.

We now turn to particle motion on the surface, in a monolayer deposit, which was experimentally observed in deposition of proteins [17] and also in deposition on DNA molecules [18, 47]. From now on, we consider diffusional relaxation, *i.e.*, random hopping on the surface in the lattice case. The dimer deposition in $1D$, for instance, is shown in Figure 2. Hopping of dimer particles one site to the left or to the right is allowed only if the target site is not occupied. Such hopping can open a two-site gap to allow additional deposition. Thus, diffusional relaxation lets the deposition process reach denser, in fact close-packed, configurations. Initially, for short times, when the empty area is plentiful, the effect of the in-surface particle motion will be small. However, for large times, the density will exceed that of the RSA process, as illustrated by the upper curve in Figure 3.

It is important to emphasize that deposition and diffusion are two independent processes going on at the same time. External particles arrive at the surface with a fixed rate per unit area. Those finding open landing sites are deposited; others are discarded. At the same time, internal particles, those already on the surface, attempt, with some rate, to hop to a nearby site. They actually move only if the target site is available.

3.2. One-dimensional Models

Further investigation of this effect is much simpler in $1D$ than in $2D$. Let us, therefore, consider the $1D$ case first, postponing the discussion of $2D$ models to the next section. Specifically, consider deposition of k-mers of fixed length, V. By keeping the length fixed, we can also naturally consider the continuum limit of no lattice by having the lattice spacing vanish as $k \to \infty$. This limit corresponds to continuum deposition if we take the underlying lattice spacing $\ell = V/k$. Since the deposition attempt rate, R, was defined per unit area (unit length here), it has no significant k-dependence. However, the added diffusional hopping of k-mers on the $1D$ lattice, with the attempt rate to be denoted by H, and hard-core or similar particle interaction, must be k-dependent. Indeed, we consider each deposited k-mer particle as randomly and independently attempting to move one lattice spacing to the left or to the right with the rate $H/2$ per unit time. Particles cannot run over each other so some sort of hard-core interaction must be assumed, *i.e.*, in a dense state most hopping attempts will fail. However, if left alone, each particle would move diffusively for large time scales. In order to have the resulting diffusion constant, $\mathcal{D}$, finite in the continuum limit $k \to \infty$, we must assume that

$$H \propto \mathcal{D}/\ell^2 = \mathcal{D}k^2/V^2 \tag{5}$$

which is only valid in $1D$.

Each successful hopping of a particle results in motion of one empty lattice site. It is useful to reconsider the dynamics of particle hopping in terms of the dynamics of this rearrangement of empty area fragments [48–50]. Indeed, if several of these empty sites are combined to form large enough voids, deposition attempts can succeed in regions of particle density which would be jammed in the ordinary RSA. In terms of these new "diffuser particles", which are the empty lattice sites of the deposition problem, the process is in fact that of reaction–diffusion. Indeed, k reactants (empty sites) must be brought together by diffusional hopping in order to have finite probability of their annihilation, *i.e.*, disappearance of a group of consecutive nearest-neighbor empty sites due to successful deposition. Of course, the k-group can also be broken apart due to diffusion. Therefore, the k-reactant annihilation is not instantaneous in the reaction nomenclature. Such k-particle reactions are of interest on their own [51–57].

3.3. Beyond the Mean-field Approximation

The simplest mean-field rate equation for annihilation of k reactants describes the time dependence of the coverage, $\rho(t)$, in terms of the reactant density $1-\rho$, *i.e.*, the density of the empty spaces,

$$\frac{d\rho}{dt} = \Gamma(1-\rho)^k, \tag{6}$$

where Γ is the effective rate constant. Note that we assume that the close-packing coverage is 1 in $1D$. There are two problems with this approximation. Firstly, it turns out that for $k=2$ the mean-field approach breaks down. Diffusive-fluctuation arguments for non-mean-field behavior have been advanced for several chemical reactions [51, 53, 58, 59]. In $1D$, several exact calculations support this conclusion [60–66]. The asymptotic large-time behavior turns out to be

$$1-\rho \sim 1/\sqrt{t} \quad (k=2,\ D=1), \tag{7}$$

rather than the mean-field prediction $\sim 1/t$. The coefficient in Eq. (7) is expected to be universal, when expressed in an appropriate dimensionless form by introducing the single-reactant diffusion constant.

The power law, Eq. (7), was confirmed by extensive numerical simulations of dimer deposition [67] and by exact solution [68] for one particular value of H for a model with dimer dissociation. The latter work also yielded some exact results for correlations. Specifically, while the connected particle–particle correlations spread diffusively in space, their decay time is nondiffusive [68]. Series expansion studies of models of dimer deposition with diffusional hopping of the whole dimers or their dissociation into hopping monomers, has confirmed the expected asymptotic behavior and also provided estimates of the coverage as a function of time [69].

The case $k=3$ is marginal with the mean-field power law modified by logarithmic terms. The latter were not observed in Monte Carlo studies of deposition [49]. However, extensive results are available directly for three-body reactions [53–56], including verification of the logarithmic corrections to the mean-field behavior [54–56].

3.4. Continuum Limit of Off-lattice Deposition

The second problem with the mean-field rate equation is identified when one attempts to use it in the continuum limit corresponding to off-lattice deposition, *i.e.*, for $k \to \infty$. Note that Eq. (6) has no regular limit as $k \to \infty$. The mean-field approach is essentially the fast diffusion approximation assuming that diffusional relaxation is efficient enough to equilibrate nonuniform density fluctuations on time scales which are short as compared with the time scales of the deposition events. Thus, the mean-field results are formulated in terms of the uniform properties, such as the density. It turns out, however, that the simplest, kth-power of the reactant density form Eq. (6) is only appropriate for times $t \gg e^{k-1}/(RV)$.

This conclusion was reached [48] by assuming the fast-diffusion, randomized hard-core reactant system form of the inter-reactant distribution function in $1D$. This approach, not detailed here, allows estimation of the limits of validity of the mean-field results and it correctly suggests mean-field validity for $k=4, 5, \ldots$, with logarithmic corrections for $k=3$ and complete breakdown of the mean-field assumptions for $k=2$. This detailed analysis yields the modified mean-field relation

$$\frac{d\rho}{dt} = \frac{\gamma RV(1-\rho)^k}{(1-\rho+k^{-1}\rho)} \quad (D=1), \tag{8}$$

where γ is some effective dimensionless rate constant. This new expression applies uniformly as $k \to \infty$. Thus, the continuum deposition is also asymptotically mean-field, with the essentially-singular rate equation

$$\frac{d\rho}{dt} = \gamma(1-\rho)\exp[-\rho/(1-\rho)] \quad (k=\infty,\ D=1). \tag{9}$$

The approach to the full, saturation coverage for large times is extremely slow,

$$1-\rho(t) \approx \frac{1}{\ln(t\ln t)} \quad (k=\infty,\ D=1). \tag{10}$$

Similar predictions were also derived for k-particle chemical reactions [53].

3.5. Comments on Multilayer Deposition

When particles are allowed to attach also on top of each other, with possibly some rearrangement processes allowed as well, multilayer deposits will be formed. It is important to note that the large-layer structure of the deposit and fluctuation properties of the growing surface will be determined by the transport mechanism of particles to the surface and by the allowed relaxations (rearrangements). Indeed, these two characteristics determine the screening properties of the multilayer formation process which in turn shape the deposit morphology, which can range from fractal to dense, and the roughening of the growing deposit surface. There is a large body of research studying such growth, with recent emphasis on the growing surface fluctuation properties.

However, the feature characteristic of the RSA process, *i.e.*, the exclusion due to particle size, plays no role in determining the universal, large-scale properties of thick deposits and their surfaces. Indeed, the RSA-like jamming will only be important for detailed morphology of the first few layers in a multilayer deposit. However, it turns out that RSA-like approaches (with relaxation) can be useful in modeling granular compaction [70].

In view of the above remarks, multilayer deposition models involving jamming effects were relatively less studied. They can be divided into two groups. Firstly, structure of the deposit in the first few layers is of interest [71–73] because they retain memory of the surface. Variation of density and other correlation properties away from the wall has structure on the length scale of particle size. These typically oscillatory features decay away with the distance from the wall. Numerical Monte Carlo simulation aspects of continuum multilayer deposition (ballistic deposition of $3D$ balls) were reviewed in Ref. [73]. Secondly, few-layer deposition processes have been of interest in some experimental systems. Mean-field theories of multilayer deposition with particle size and interactions accounted for were formulated [74] and used to fit such data [15, 16, 75, 76].

4. TWO-DIMENSIONAL DEPOSITION WITH DIFFUSIONAL RELAXATION

4.1. Combined Effects of Jamming and Diffusion

We now turn to the $2D$ case of deposition of extended objects on planar surfaces, accompanied by diffusional relaxation, assuming monolayer deposits. We note that the available theoretical results are limited to few studies [34, 77–79]. They indicate a rich pattern of new effects as compared with $1D$. In fact, there exists extensive literature [81] on deposition with diffusional relaxation in other models, in particular those where the jamming effect is not present or plays no significant role. These include deposition of monomer particles, usually of atomic dimensions, which align with the underlying lattice without jamming, as well as models where many layers are formed (mentioned in the preceding section).

The $2D$ deposition with relaxation of extended objects is of interest in certain experimental systems where the depositing objects are proteins [17]. Here we focus on the combined effect of jamming and diffusion, and emphasize dynamics at large times. For early stages of the deposition process, low-density approximation schemes can be used. One such application was reported [34] for continuum deposition of circles on a plane.

In order to identify new features characteristic of $2D$, let us consider deposition of 2×2 squares on the square lattice. The particles are exactly aligned with the 2×2 lattice sites as shown in Figure 4. Furthermore, we assume that the diffusional hopping is along the lattice directions $\pm x$ and $\pm y$, one lattice spacing at a time. In this model dense configurations involve domains of four phases as shown in Figure 4. As a result, immobile fragments of empty area can exist. Each such single-site vacancy (Fig. 4) serves as a meeting point of four domain walls.

Here by "immobile" we mean that the vacancy cannot move due to local motion of the surrounding particles. For it to move, a larger empty-area fragment must first arrive, along one of the domain walls. One such larger empty void is shown in Figure 4. Note that it serves as a kink in the domain wall. Existence of locally immobile ("frozen") vacancies suggests possible frozen glassy behavior with extremely slow relaxation, at least locally. The full characterization of the dynamics of this model requires further study. The first numerical results [77] do provide some answers which will be reviewed shortly.

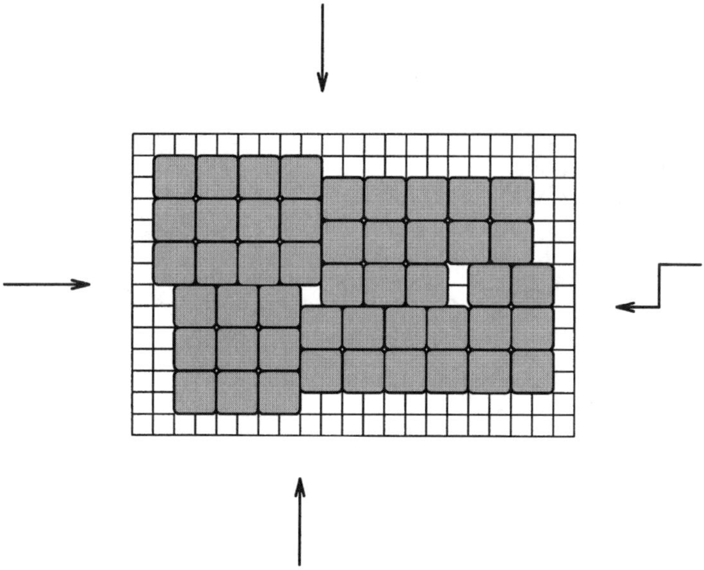

FIGURE 4 Fragment of a deposit configuration in the deposition of 2×2 squares. Illustrated are one single-site frozen vacancy at which four domain walls meet (indicated by arrows), and one dimer vacancy which causes a kink in one of the domain walls.

4.2. Ordering by Shortening of Domain Walls

We first consider a simpler model depicted in Figure 5. In this model [78, 79] the extended particles are squares of size $\sqrt{2} \times \sqrt{2}$. They are rotated 45° with respect to the underlying square lattice. Their diffusion, however, is along the vertical and horizontal lattice axes, by hopping one lattice spacing at a time. The equilibrium variant of this model (without deposition, with fixed particle density) is the well-studied hard-square model [82] which, at large densities, phase separates into two distinct phases. These two phases also play role in the late stages of RSA with diffusion. Indeed, at large densities the empty area is stored in domain walls separating ordered regions. One such domain wall is shown in Figure 5. Snapshots of actual Monte Carlo simulation results can be found in Refs. [78, 79].

Figure 5 illustrates the process of ordering which essentially amounts to shortening of domain walls. In Figure 5, the domain wall gets shorter after the shaded particles diffusively rearrange to open up a deposition slot which can be covered by an arriving particle.

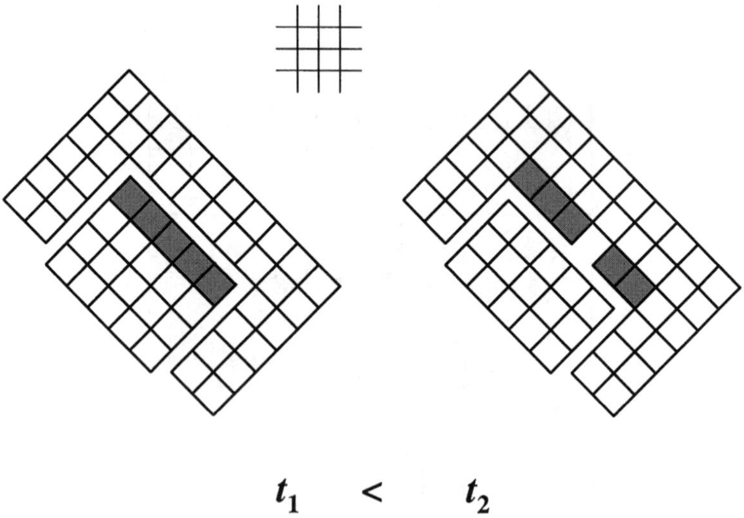

FIGURE 5 Illustration of deposition of $\sqrt{2} \times \sqrt{2}$ particles on the square lattice. Diffusional motion during time interval from t_1 to t_2 can rearrange the empty area "stored" in the domain wall to open up a new landing site for deposition. This is illustrated by the shaded particles.

Numerical simulations [78, 79] find behavior reminiscent of the low-temperature equilibrium ordering processes [83–85] driven by diffusive evolution of the domain-wall structure. For instance, the remaining uncovered area vanishes according to

$$1 - \rho(t) \sim \frac{1}{\sqrt{t}}. \tag{11}$$

This quantity, however, also measures the length of domain walls in the system (at large times). Thus, disregarding finite-size effects and assuming that the domain walls are not too convoluted (as confirmed by numerical simulations), we conclude that the power law, Eq. (11), corresponds to typical domain sizes growing as $\sim \sqrt{t}$, reminiscent of the equilibrium ordering processes of systems with nonconserved order parameter dynamics [83–85].

4.3. Numerical Results for Models with Frozen Vacancies

We now turn again to the 2×2 model of Figure 4. The equilibrium variant of this model corresponds to hard-squares with both nearest and next-nearest neighbor exclusion [82, 86, 87]. It has been studied in lesser detail than the two-phase hard-square model described in the preceding paragraphs. In fact, the equilibrium phase transition has not been fully classified (while it was Ising for the simpler model). The ordering at low temperatures and high densities was studied [86]. However, many features noted, for instance large entropy of the ordered arrangements, require further investigation. The dynamical variant (RSA with diffusion) of this model was studied numerically [77]. The configuration of the single-site frozen (locally immobile) vacancies and the associated network of domain walls turn out to be boundary-condition sensitive. For periodic boundary conditions the density freezes at values $1 - \rho \sim L^{-1}$, where L is the linear system size.

Preliminary indications were found [77] that the domain size and shape distributions in such a frozen state are nontrivial. Extrapolation $L \to \infty$ indicates that the power law behavior similar to Eq. (11) is nondiffusive: the exponent 1/2 is replaced by ~ 0.57. However, the density of the smallest mobile vacancies, *i.e.*, dimer kinks in domain walls, one of which is illustrated in Figure 4, does decrease diffusively. Further studies are needed to clarify fully the ordering process associated with the approach to the full coverage as $t \to \infty$ and $L \to \infty$ in this model.

Even more complicated behaviors are possible when the depositing objects are not symmetric and can have several orientations as they reach the substrate. In addition to translational diffusion (hopping), one has to consider possible rotational motion. The square-lattice deposition of dimers, with hopping processes including one-lattice-spacing motion along the dimer axis and 90° rotations about a constituent monomer, was studied [80]. The dimers were allowed to deposit vertically and horizontally. In this case, the full close-packed coverage is not achieved at all because the frozen vacancy sites can be embedded in, and move by diffusion in, extended structures of different topologies. These structures are probably less efficiently demolished by the motion of mobile vacancies than the elimination of localized frozen vacancies in the model of Figure 4.

5. CONCLUSION

In summary, we reviewed theoretical developments in the description of deposition processes of extended objects, with jamming and diffusional relaxation. While significant progress has been achieved in $1D$, the $2D$ systems require further study. Most of these investigations will involve large-scale numerical simulations.

Other research directions that require further work include multilayer deposition and particle detachment, especially the theoretical description of the latter, including the description of the distribution of values/shapes of the primary minimum in the particle–surface interaction potential. This would allow one to advance beyond the present theoretical trend of studying deposition as mainly the process of particle transport to the surface, with little or no role played by the details of the actual particle–surface and particle–particle double-layer and other interactions. Ultimately, we would like to interrelate the present deposition studies and approaches in the study of adhesion [4], of typically larger particles of sizes up to several microns, at surfaces.

References

[1] Particle Deposition at the Solid-liquid Interface, Tardos, Th. F. and Gregory, J. Eds., *Colloids Surf.* **39**(1/3), 30 August, 1989.

[2] *Advances in Particle Adhesion*, Rimai, D. S. and Sharpe, L. H. Eds. (Gordon and Breach Publishers, Amsterdam, 1996).

[3] Particle Deposition & Aggregation. *Measurement, Modeling and Simulation*, Elimelech, M., Gregory, J., Jia, X. and Williams, R. A. (Butterworth-Heinemann Woburn, MA, 1995).

[4] *Adhesion of Submicron Particles on Solid Surfaces*, Privman, V. Ed., *Colloids Surf. A* (in print, 2000).

[5] Levich, V. G., *Physiochemical Hydrodynamics* (Prentice-Hall, London, 1962).

[6] Privman, V., *Trends in Statistical Physics* **1**, 89 (1994).

[7] *Nonequilibrium Statistical Mechanics in One Dimension*, Privman, V. Ed. (Cambridge University Press, 1997).

[8] Feder, J. and Giaever, I., *J. Colloid Interface Sci.* **78**, 144 (1980).

[9] Schmitt, A., Varoqui, R., Uniyal, S., Brash, J. L. and Pusiner, C., *J. Colloid Interface Sci.* **92**, 25 (1983).

[10] Onoda, G. Y. and Liniger, E. G., *Phys. Rev.* **A33**, 715 (1986).

[11] Kallay, N., Tomić, M., Biškup, B., Kunjašić, I. and Matijević, E., *Colloids Surf.* **28**, 185 (1987).

[12] Aptel, J. D., Voegel, J. C. and Schmitt, A., *Colloids Surf.* **29**, 359 (1988).

[13] Adamczyk, Z., *Colloids Surf.* **39**, 1 (1989).

[14] Adamczyk, Z., Zembala, M., Siwek, B. and Warszyński, P., *J. Colloid Interface Sci.* **140**, 123 (1990).

[15] Ryde, N., Kihira, H. and Matijević, E., *J. Colloid Interface Sci.* **151**, 421 (1992).

[16] Song, L. and Elimelech, M., *Colloids Surf.* **A73**, 49 (1993).

[17] Ramsden, J. J., *J. Statist. Phys.* **73**, 853 (1993).

[18] Murphy, C. J., Arkin, M. R., Jenkins, Y., Ghatlia, N. D., Bossmann, S. H., Turro, N. J. and Barton, J. K., *Science* **262**, 1025 (1993).

[19] Bartelt, M. C. and Privman, V., *Internat. J. Mod. Phys.* **B5**, 2883 (1991).

[20] Evans, J. W., *Rev. Mod. Phys.* **65**, 1281 (1993).

[21] Widom, B., *J. Chem. Phys.* **58**, 4043 (1973).

[22] Schaaf, P. and Talbot, J., *Phys. Rev. Lett.* **62**, 175 (1989).

[23] Dickman, R., Wang, J.-S. and Jensen, I., *J. Chem. Phys.* **94**, 8252 (1991).

[24] Gonzalez, J. J., Hemmer, P. C. and Høye, J. S., *Chem. Phys.* **3**, 228 (1974).

[25] Feder, J., *J. Theor. Biology* **87**, 237 (1980).

[26] Tory, E. M., Jodrey, W. S. and Pickard, D. K., *J. Theor. Biology* **102**, 439 (1983).

[27] Hinrichsen, E. L., Feder, J. and Jøssang, T., *J. Statist. Phys.* **44**, 793 (1986).

[28] Burgos, E. and Bonadeo, H., *J. Phys.* **A20**, 1193 (1987).

[29] Barker, G. C. and Grimson, M. J., *J. Phys.* **A20**, 2225 (1987).

[30] Vigil, R. D. and Ziff, R. M., *J. Chem. Phys.* **91**, 2599 (1989).

[31] Talbot, J., Tarjus, G. and Schaaf, P., *Phys. Rev.* **A40**, 4808 (1989).

[32] Vigil, R. D. and Ziff, R. M., *J. Chem. Phys.* **93**, 8270 (1990).

[33] Sherwood, J. D., *J. Phys.* **A23**, 2827 (1990).

[34] Tarjus, G., Schaaf, P. and Talbot, J., *J. Chem. Phys.* **93**, 8352 (1990).

[35] Brosilow, B. J., Ziff, R. M. and Vigil, R. D., *Phys. Rev.* **A43**, 631 (1991).

[36] Privman, V., Wang, J.-S. and Nielaba, P., *Phys. Rev.* **B43**, 3366 (1991).

[37] Pomeau, Y., *J. Phys.* **A13**, L193 (1980).

[38] Swendsen, R. H., *Phys. Rev.* **A24**, 504 (1981).

[39] Kallay, N., Biškup, B., Tomić, M. and Matijević, E., *J. Colloid Interface Sci.* **114**, 357 (1986).

[40] Barma, M., Grynberg, M. D. and Stinchcombe, R. B., *Phys. Rev. Lett.* **70**, 1033 (1993).

[41] Stinchcombe, R. B., Grynberg, M. D. and Barma, M., *Phys. Rev.* **E47**, 4018 (1993).
[42] Grynberg, M. D., Newman, T. J. and Stinchcombe, R. B., *Phys. Rev.* **E50**, 957 (1994).
[43] Grynberg, M. D. and Stinchcombe, R. B., *Phys. Rev.* **E49**, R23 (1994).
[44] Schütz, G. M., *J. Statist. Phys.* **79**, 243 (1995).
[45] Krapivsky, P. L. and Ben-Naim, E., *J. Chem. Phys.* **100**, 6778 (1994).
[46] Barma, M. and Dhar, D., *Phys. Rev. Lett.* **73**, 2135 (1994).
[47] Bossmann, S. H. and Schulman, L. S., In: *Nonequilibrium Statistical Mechanics in One Dimension*, Privman, V. Ed. (Cambridge University Press, 1997), p. 443.
[48] Privman, V. and Barma, M., *J. Chem. Phys.* **97**, 6714 (1992).
[49] Nielaba, P. and Privman, V., *Mod. Phys. Lett.* **B6**, 533 (1992).
[50] Bonnier, B. and McCabe, J., *Europhys. Lett.* **25**, 399 (1994).
[51] Kang, K., Meakin, P., Oh, J. H. and Redner, S., *J. Phys.* **A17**, L665 (1984).
[52] Cornell, S., Droz, M. and Chopard, B., *Phys. Rev.* **A44**, 4826 (1991).
[53] Privman, V. and Grynberg, M. D., *J. Phys.* **A25**, 6575 (1992).
[54] ben-Avraham, D., *Phys. Rev. Lett.* **71**, 3733 (1993).
[55] Krapivsky, P. L., *Phys. Rev.* **E49**, 3223 (1994).
[56] Lee, B. P., *J. Phys.* **A27**, 2533 (1994).
[57] Grynberg, M. D., *Phys. Rev.* **E57**, 74 (1998).
[58] Kang, K. and Redner, S., *Phys. Rev. Lett.* **52**, 955 (1984).
[59] Kang, K. and Redner, S., *Phys. Rev.* **A32**, 435 (1985).
[60] Racz, Z., *Phys. Rev. Lett.* **55**, 1707 (1985).
[61] Bramson, M. and Lebowitz, J. L., *Phys. Rev. Lett.* **61**, 2397 (1988).
[62] Balding, D. J. and Green, N. J. B., *Phys. Rev.* **A40**, 4585 (1989).
[63] Amar, J. G. and Family, F., *Phys. Rev.* **A41**, 3258 (1990).
[64] ben-Avraham, D., Burschka, M. A. and Doering, C. R., *J. Statist. Phys.* **60**, 695 (1990).
[65] Bramson, M. and Lebowitz, J. L., *J. Statist. Phys.* **62**, 297 (1991).
[66] Privman, V., *J. Statist. Phys.* **69**, 629 (1992).
[67] Privman, V. and Nielaba, P., *Europhys. Lett.* **18**, 673 (1992).
[68] Grynberg, M. D. and Stinchcombe, R. B., *Phys. Rev. Lett.* **74**, 1242 (1995).
[69] Gan, C. K. and Wang, J.-S., *Phys. Rev.* **E55**, 107 (1997).
[70] de Oliveira, M. J. and Petri, A., *J. Phys.* **A31**, L425 (1998).
[71] Xiao, R.-F., Alexander, J. I. D. and Rosenberger, F., *Phys. Rev.* **A45**, R571 (1992).
[72] Lubachevsky, B. D., Privman, V. and Roy, S. C., *Phys. Rev.* **E47**, 48 (1993).
[73] Lubachevsky, B. D., Privman, V. and Roy, S. C., *J. Comp. Phys.* **126**, 152 (1996).
[74] Privman, V., Frisch, H. L., Ryde, N. and Matijević, E., *J. Chem. Soc. Farad. Tran.* **87**, 1371 (1991).
[75] Ryde, N., Kallay, N. and Matijević, E., *J. Chem. Soc. Farad. Tran.* **87**, 1377 (1991).
[76] Zelenev, A., Privman, V. and Matijević, E., *Colloids Surf.* **A135**, 1 (1998).
[77] Wang, J.-S., Nielaba, P. and Privman, V., *Physica* **A199**, 527 (1993).
[78] Wang, J.-S., Nielaba, P. and Privman, V., *Mod. Phys. Lett.* **B7**, 189 (1993).
[79] James, E. W., Liu, D.-J. and Evans, J. W., in Ref. [4].
[80] Grigera, S. A., Grigera, T. S. and Grigera, J. R., *Phys. Lett.* **A226**, 124 (1997).
[81] Vernables, J. A., Spiller, G. D. T. and Hanbücken, M., *Rept. Prog. Phys.* **47**, 399 (1984).
[82] Runnels, L. K., In: *Phase Transitions and Critical Phenomena*, Vol. 2, Domb, C. and Green, M. S. Eds. (Academic, London, 1972), p. 305.
[83] Gunton, J. D., San Miguel, M. and Sahni, P. S., In: *Phase Transitions and Critical Phenomena*, Vol. 8, Domb, C. and Lebowitz, J. L. Eds. (Academic, London, 1983), p. 267.
[84] Mouritsen, O. G., In: *Kinetics of Ordering and Growth at Surfaces*, Lagally, M. G. Ed. (Plenum, NY, 1990), p. 1.
[85] Sadiq, A. and Binder, K., *J. Statist. Phys.* **35**, 517 (1984).
[86] Binder, K. and Landau, D. P., *Phys. Rev.* **B21**, 1941 (1980).
[87] Kinzel, W. and Schick, M., *Phys. Rev.* **B24**, 324 (1981).

Aerosol Particle Removal and Re-entrainment in Turbulent Channel Flows – A Direct Numerical Simulation Approach

HAIFENG ZHANG and GOODARZ AHMADI

Department of Mechanical and Aeronautical Engineering, Clarkson University, Potsdam, NY 13699-5727, USA

Aerosol particle removal and re-entrainment in turbulent channel flows are studied. The instantaneous fluid velocity field is generated by the direct numerical simulation (DNS) of the Navier–Stokes equation *via* a pseudospectral method. Particle removal mechanisms in turbulent channel flows are examined and the effects of hydrodynamic forces, torques and the near-wall coherent vorticity are discussed. The particle resuspension rates are evaluated, and the results are compared with the model of Reeks. The particle equation of motion used includes the hydrodynamic, the Brownian, the shear-induced lift and the gravitational forces. An ensemble of 8192 particles is used for particle resuspension and the subsequent trajectory analyses. It is found that large-size particles move away roughly perpendicular to the wall due to the action of the lift force. Small particles, however, follow the upward flows formed by the near-wall eddies in the low-speed streak regions. Thus, turbulent near-wall vortical structures play an important role in small particle resuspension, while the lift is an important factor for re-entrainment of large particles. The simulation results suggests that small particles (with $\tau_p^+ \leq 0.023$) primarily move away from the wall in the low-speed streaks, while larger particles (with $\tau_p^+ \geq 780$) are mostly removed in the high-speed streaks.

INTRODUCTION

Particle detachment from surfaces has increasingly become the subect of considerable attraction because of its importance in the semiconductor and imaging industries. Numerous studies concerning the particle detachment mechanisms from various surfaces have been reported by Mittal [1]. Extensive reviews of the particle adhesion mechanism have been provided by Corn [2], Krupp [3], Visser [4], Tabor [5], Bowling [6], and Ranade [7]. Accordingly, the van der Waals force makes the major contribution to the particle adhesion force on a surface under dry conditions.

The effect of contact deformation on adhesion was first considered by Derjaguin [8]. More recently, Johnson *et al.* [9] used the surface energy and surface deformation effects to develop an improved contact model called the JKR theory. According to this model, at the moment of separation, the contact area does not disappear entirely; instead, a finite contact area exists.

Derjaguin *et al.* [10] developed a new theory based on the Hertzian profile assumption. In this model (the so-called DMT theory) the force required to detach the particle from the surface is 4/3 as large as in the JKR theory. Further progress was reported by Muller [11, 12]. Accordingly, for a system that has a high Young's modulus, low surface energy, and small-diameter particles, the DMT theory applies. In contrast, for a system that has a low Young's modulus, high surface energy, and large particle size, the JKR theory is more suitable. Recently, Maugis [13] analyzed the adhesion of spheres to plane surfaces based on the assumption that the adhesion force is constant in the region near the contact boundary. His analytical results clearly show the transition between the JKR and DMT theories. The JKR theory is further generalized by Maugis and Pollock [14] by allowing for plastic deformation.

Tsai *et al.* [15] studied the elastic flattening and particle adhesion and argued that the JKR theory is not correct for hard systems, and also that there is a violation of the static equilibrium in the DMT theory. They proposed a new (TPL) model which considers the effect of material properties in the deformation and adhesion force of particle-surface systems. Rimai *et al.* [16] performed a series of experimental studies and reported significant effects of the Young's modulus and material properties on the surface-force-induced contact radii of spherical particles. Soltani and Ahmadi [17, 18] studied the particle removal mechanisms from smooth and rough walls subject to substrate accelerations.

Numerous experimental and computational studies related to particle transport in turbulent flows were reported in the literature (Hinze [19], Ahmadi [20]). Extensive reviews on particle removal process from surfaces were provided by Healy [21], Sehmel [22], Nicholson [23], and Smith *et al.* [24]. Braaten *et al.* [25] performed an experimental study of particle re-entrainment in turbulent flow. They concluded that ejection-sweep events and macrosweep flow patterns near a wall strongly affect the particle resuspension process. However, based on their flow visualization experiments, Yung *et al.* [26] reported that the bursting phenomenon has a small effect on entrainment of particles within the viscous sublayer.

A sublayer model for particle resuspension and deposition in turbulent flows was proposed by Cleaver and Yates [27–29]. In particular, they suggested that the particle entrainment most likely results from the wall ejection events, while their deposition occurs by the inrush process. A dynamic model for the long-term resuspension of small particles from smooth and rough surfaces in turbulent flow was developed by Reeks *et al.* [30] and Reeks and Hall [31]. A kinetic model for particle resuspension was proposed by Wen and Kasper [32] and compared with the data from industrial high-purity gas systems and with controlled experiments using Latex particles of 0.4–1 μm. Wang [33] studied the effect of inceptive motion on particle detachment from surfaces and concluded that the removal of spherical particles is more easily achieved by the rolling motion, rather than sliding or lifting. This result is consistent with the experimental observation of Masironi and Fish [34]. A flow-structure-based model for turbulent resuspension was developed by Soltani and Ahmadi [35, 36].

A series of direct numerical simulations (DNS) of particle deposition in wall-bounded turbulent flows were performed by McLaughlin [37] and Ounis *et al.* [38, 39]. These studies were concerned with providing insight into the particle deposition mechanisms in turbulent flows. Brooke *et al.* [40] performed detailed DNS studies of vortical structures in the viscous sublayer. Recently, Pedinotti *et al.* [41] used the DNS to investigate the particle behavior in the wall region of turbulent flows. They reported that an initially uniform distribution of particles tends to segregate into low speed streaks and resuspension occurs by particles being ejected from the wall. The DNS simulation was used by Soltani and Ahmadi [42] to study the particle entrainment process in a turbulent channel flow. They found that the wall coherent structure plays a dominant role in the particle entrainment process.

Squires and Eaton [43] simulated a homogeneous, isotropic, non-decaying turbulent flow field by imposing an excitation at low wave numbers, and studied the effects of inertia on particle dispersion. They also used the DNS procedure to study the preferential micro-concentration structure of particles as a function of the Stokes number in turbulent, near-wall flows [44]. Rashidi *et al.* [45] performed an experiment to study the particle-turbulence interactions near a wall. They reported that the particle transport is mainly controlled by the turbulence burst phenomena.

In this work, the particle removal mechanism from the smooth surface in turbulent channel flows is studied. The theories of rolling and sliding detachments are used, and the critical removal condition is analyzed. Effects of various forces and turbulent near-wall coherent eddies on the turbulent resuspension process are studied. An ensemble of 8192 particles is used in these simulations, and it is shown that the turbulent near-wall vortical structure and the lift force, respectively, play important roles on small and large particle re-entrainment and resuspension processes.

TURBULENT FLOW FIELD VELOCITY

The instantaneous fluid velocity field in the channel is evaluated by the direct numerical simulation (DNS) of the Navier–Stokes equation. It is assumed that the fluid is incompressible, and a constant mean pressure gradient in the x-direction is imposed. The corresponding governing equations of motion are:

$$\nabla \cdot \mathbf{u} = 0 \tag{1}$$

$$\frac{\partial \mathbf{u}}{\partial t} + \mathbf{u} \cdot \nabla \mathbf{u} = \nu \nabla^2 \mathbf{u} - \frac{1}{\rho^f} \nabla P \tag{2}$$

where $\mathbf{u} = (u_x, u_y, u_z)$ is the fluid velocity vector, P is the pressure, ρ^f is the density, and ν is the kinematic viscosity. The fluid velocity is assumed to satisfy the no-slip boundary conditions at the channel walls. In the simulations, a channel that has a width of 250 wall units, and a 630×630 periodic segment in x- and z-directions is used. The schematics of the flow domain and the periodic cell is shown in Figure 1a. A $16 \times 64 \times 64$ computational grid in the x-, y- and z-directions is also employed. The grid spacing in the x- and z-directions is constant, while the variation of grid points in the y-direction is represented by the Chebyshev series. The distance of the ith grid point in the y-direction from the centerline is given as

$$y_i = \frac{h}{2}\cos(\pi i/M), \quad 0 \leq i \leq M \tag{3}$$

Here h is the channel height, $M = 64$, and there are 65 grid points in the y-direction.

The channel flow code used in this study is the one developed by McLaughlin [37]. To solve for the velocity components by pseudospectral methods, the fluid velocity is expanded in a three-dimensional Fourier–Chebyshev series. The fluid velocity field in the x- and z-direction is expanded by Fourier series, while in the y-direction the

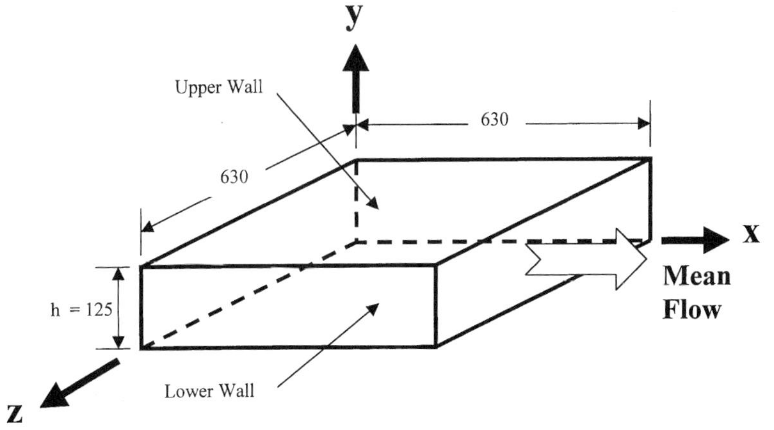

FIGURE 1a Schematics of the channel flow and the computational periodic cell used.

Chebyshev series is used. The code uses an Adams–Bashforth–Crank–Nickolson (ABCN) scheme to compute the nonlinear and viscous terms in the Navier–Stokes equation and performs three fractional time steps to forward the fluid velocity from time step (N) to time step (N+1). The details of the numerical techniques were described by McLaughlin [37]. In these computer simulations, wall units are used; and all variables are nondimensionalized in terms of shear velocity, u^*, and kinematic viscosity, ν.

MacLaughlin [37] showed that the near wall root-mean-square fluctuation velocities as predicted by the present DNS code are in good agreement with the high resolution DNS code of Kim *et al.* [46]. Zhang and Ahmadi [47] showed that the present DNS with a grid size of $16 \times 64 \times 64$ can produce first-order and second-order turbulence statistics that are reasonably accurate when compared with the results of high resolution grids of $32 \times 64 \times 64$ and $32 \times 128 \times 128$. In this paper, for the sake of computational economy, the coarser grid is used.

Figure 1b shows the geometry of the flow and a sample instantaneous velocity field at $t^+ = 100$ in different planes. While the velocity field in the $y-z$ plane (at $x^+ = 157.5$) shown in Figure 1b has a random pattern, near-wall coherent eddies and flow streams towards and away from the wall can be observed from this figure. The random deviations from the expected mean velocity profile are clearly seen from Figures 1c, and d shows that the flow is predominantly in the x-direction. The near wall low- and high-speed streaks are also noticeable from this figure.

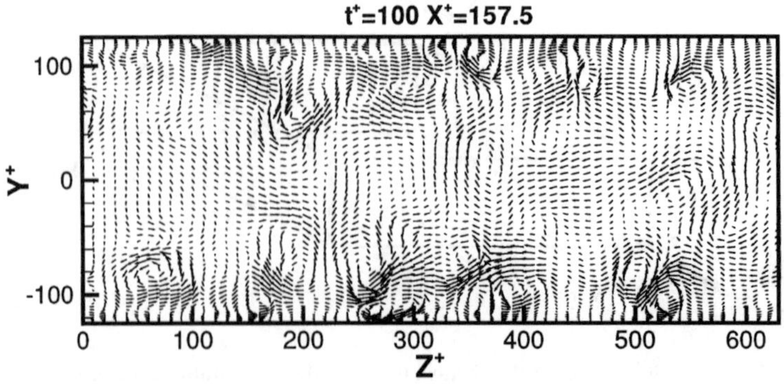

FIGURE 1b Sample velocity vector plot in the $y-z$ plane.

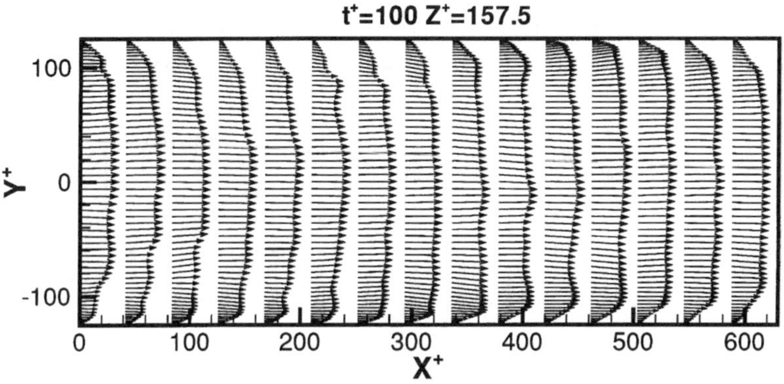

FIGURE 1c Sample velocity vector plot in the $x-y$ plane.

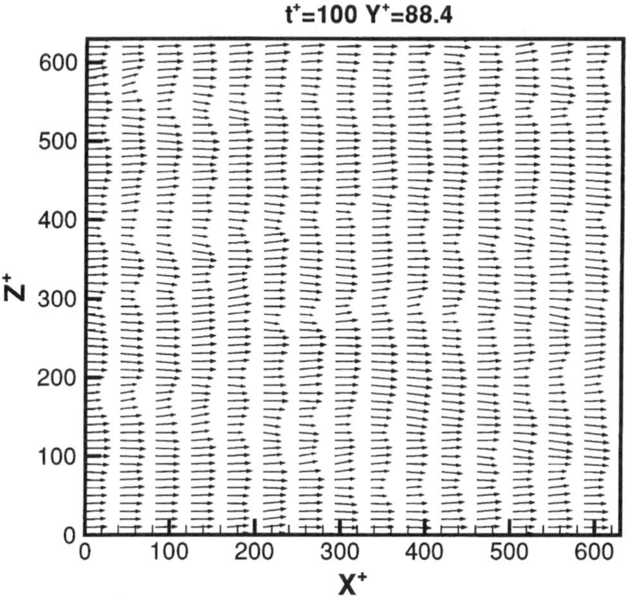

FIGURE 1d Sample velocity vector plot in the $x-z$ plane.

ADHESION MODELS

JKR Model

The Hertz contact theory is modified in this model by taking into account the surface energy effects and by allowing for the deformation of the particle and substrate surfaces. Accordingly, a finite contact area forms and the radius of the contact circle, a, is given as

$$a^3 = \frac{d}{2K}\left[P + \frac{3W_A\pi d}{2} + \sqrt{3\pi W_A dP + \left(\frac{3\pi W_A d}{2}\right)^2}\,\right] \tag{4}$$

where

$$K = \frac{4}{3}\left[\frac{(1-\nu_1^2)}{E_1} + \frac{(1-\nu_2^2)}{E_2}\right]^{-1} \tag{5}$$

is the composite Young's modulus. Here, d is the diameter of the spherical particle, W_A is the thermodynamic work of adhesion, P is the applied normal load, and ν_i and E_i are, respectively, the Poisson's ratio and the Young's modulus of material i ($i=1$, or 2).

According to the JKR model, to detach a sphere from a plane surface, the required pull-off force, F_{P_o}, is given by

$$F_{P_o}^{\text{JKR}} = \frac{3}{4}\pi W_A d \tag{6}$$

At the moment of separation, the contact radius is finite and is given by

$$a = \frac{a_0}{4^{1/3}} \tag{7}$$

where a_0 is the contact radius at zero applied load given as

$$a_0 = \left(3\pi W_A d^2/2K\right)^{1/3} \tag{8}$$

For a number of common interfaces, the corresponding material properties are listed in Table I. In this table, A is the Hamaker constant, and k is the friction coefficient.

TPL Model

Based on a detailed molecular interaction analysis, Tsai *et al.* [15] found the following equation for the force required to detach a particle

TABLE I Material properties

Material	E ($10^{10}\,N/m^2$)	A ($10^{-20}\,J$)	W_A ($10^{-3}\,J/m^2$)	ρ ($10^3\,kg/m^3$)	ν_i	k
Silicon – Silicon	17.90	23.50	38.9	2.3	0.27	0.9
Graphite – Graphite	67.50	46.90	77.75	2.2	0.16	0.1
Copper – Copper	13.00	28.30	46.91	8.89	0.34	1.6
Glass – Glass (dry air)	6.9	8.5	14.1	2.18	0.2	0.9
Glass – Glass (moist air)	6.9	320	530	2.18	0.2	0.9
Steel – Steel	21.5	21.2	35	7.84	0.28	0.58
Glass – Steel	–	–	150	–	–	0.6
Polyst – Polyst	0.28	6.37	10.56	1.05	0.33	0.5
Polyst – Nickel	–	14.27	23.65	–	–	–
Rubber – Rubber	2.4e-4	20.5	34	1.13	0.5	0.8

from a plane surface:

$$F_{P_o}^{TPL} = F_0\{0.5\exp[0.124(\coprod -0.01)^{0.439}] + 0.2\coprod\} \tag{9}$$

where the adhesion parameter, $\coprod$, is defined as

$$\coprod = \left[\frac{25A^2d}{288z_0^7K^2}\right]^{1/3} \tag{10}$$

and

$$F_0 = \pi W_A d \tag{11}$$

Here, z_0 is the minimum separation distance and A is the Hamaker constant. The corresponding contact radius at the moment of separation is given as

$$\frac{a}{d} = \sqrt{\frac{K_{20}z_0}{2d}} \tag{12}$$

where K_{20} is the deformation parameter at the equilibrium condition and, according to Tsai *et al.* [15], is given by

$$K_{20} = 0.885[\exp(0.8\coprod^{0.5}) - 1.0], \quad \coprod \leq 1.6 \tag{13}$$

$$K_{20} = 0.735\coprod^{0.178} + 0.52\coprod, \quad \coprod > 1.6 \tag{14}$$

The adhesion parameter, ∐, for particle diameters between 0.01 and 100 μm varies from 0.01 to 5 for metals and oxides, and from 5 to 200 for polymers.

DETACHMENT MODEL

A particle may be detached from a surface when the applied forces overcome the adhesion forces. A particle may lift-off from the surface, slide over it, or roll on the surface. These detachment mechanisms have been discussed by Wang [33]. The moment and sliding detachment mechanisms which are important for particle removal by fluid flows are briefly described here.

Moment Detachment

The critical moment model for the detachment of particles from a surface was studied by Tsai *et al.* [15] and Soltani and Ahmadi [35, 36]. Figure 2 shows the geometric features of a spherical particle attached to a plane surface. A particle will be detached when the external force moment about point "*o*", which is located at the rear perimeter of the contact circle, overcomes the resisting moment due to the adhesion force. That is

$$M_t + F_t\left(\frac{d}{2} - \alpha_0\right) + F_L a \geq F_{P_o} a \tag{15}$$

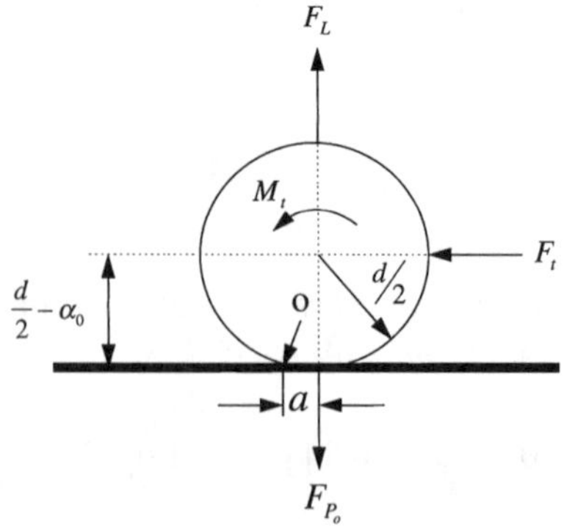

FIGURE 2 Geometric features of a spherical particle attached to a smooth surface.

where F_t is the tangential external force acting on the particle (*e.g.*, the fluid drag force), α_0 is the relative approach between the particle and surface (at equilibrium conditions), M_t is the external moment of the surface stresses about the center of the particle, F_L is the lift force, and F_{P_o} is the particle adhesion force.

Sliding Detachment

Wang [33] studied the effect of inceptive motion on particle detachment. Accordingly, the particle will be removed by sliding if

$$F_t \geq kF_{P_o} \tag{16}$$

Here F_t is external force (*i.e.*, the fluid drag force) acting on the particle parallel to the surface, and k is the coefficient of static friction.

PARTICLE EQUATIONS OF MOTION

The equations of motion for a spherical particle moving in a channel flow are given as:

$$\frac{d\mathbf{v}^+}{dt^+} = \mathbf{g}^+ + \mathbf{F}_d^+ + \mathbf{F}_L^+ + \mathbf{n}^+(t^+) \tag{17}$$

and

$$\frac{d\mathbf{x}^+}{dt} = \mathbf{v}^+ \tag{18}$$

where $\mathbf{g}^+$ is the gravity, $\mathbf{F}_d^+$ is the drag force, $\mathbf{F}_L^+$ is the lift force, and $\mathbf{n}^+(t^+)$ is the Brownian random force. (Note that only the y-component of the lift force is considered in this study.) All variables are nondimensionlized by the fluid viscosity, ν, and shear velocity, u^*. That is

$$\mathbf{x}^+ = \frac{\mathbf{x}u^*}{\nu}, \quad \mathbf{v}^+ = \frac{\mathbf{v}}{u^*}, \quad t^+ = \frac{tu^{*2}}{\nu}, \quad \mathbf{g}^+ = \frac{\nu}{u^{*3}}\mathbf{g},$$
$$\mathbf{F}_d^+ = \frac{\nu}{u^{*3}}\mathbf{F}_d, \quad \mathbf{F}_L^+ = \frac{\nu}{u^{*3}}\mathbf{F}_L \tag{19}$$

Drag Force

The combined effect of the translational motion, rotational motion and fluid shear due to the presence of the wall was studied by Goldman *et al.* [48]. For a sphere with no externally-applied torque moving in a wall-bounded channel flow, the drag force can be expressed as

$$\mathbf{F}_d = \frac{6\pi\mu a C_N}{C_c}\begin{pmatrix} C_x^w(u_x - v_x) \\ C_y^w(u_y - v_y) \\ C_z^w(u_z - v_z) \end{pmatrix} \tag{20}$$

where $\mathbf{u} = (u_x, u_y, u_z)$ is the fluid velocity, $\mathbf{v} = (v_x, v_y, v_z)$ is the particle velocity, $a = d/2$ is the particle radius, μ is the coefficient of viscosity, C_i^w is the wall correction factor, and C_c is the Cunningham correction factor given by

$$C_c = 1 + \frac{2\lambda}{d}[1.257 + 0.4\exp(-1.1d/2\lambda)] \tag{21}$$

Here, λ is the mean free path of the gas.

When the Reynolds number of the particle based on particle-fluid slip velocity is not small, the drag force deviates from the Stokes expression. The nonlinear correction coefficient to the Stokes drag is given as (Hinds [49])

$$C_N = 1 + 0.15\,\mathrm{Re}_p^{0.687} \tag{22}$$

Equation (22) agrees with experiments in the range of $1 < \mathrm{Re}_p < 200$, where Re_p is the Reynolds number of the particle defined as

$$\mathrm{Re}_p = \frac{|\mathbf{v} - \mathbf{u}|d}{\nu} \tag{23}$$

Based on a synthesis of available experimental results, Clift *et al.* [50] recommended the following nonlinear drag correction factors:

$$C_N = 1 + 0.1875\mathrm{Re}_p \quad \text{for } \mathrm{Re}_p \leq 0.01 \tag{24}$$

and

$$C_N = 1 + 0.1315\mathrm{Re}_p^{0.82 - 0.0217\ln(\mathrm{Re}_p)} \quad \text{for } 0.01 < \mathrm{Re}_p < 20 \tag{25}$$

Combining the results of Goldman *et al.* [48, 51] and Brenner [52], Li [53] suggested for the following wall correction factors:

$$C_x^w = \frac{v_x f_x^{t*} - (f_x^{r*}/t_z^{r*})\left(v_x t_z^{t*} + (aG_x/2)t_z^{s*}\right) + yG_x f_x^{s*}}{u_x - v_x} \tag{26}$$

$$C_y^w = f_y^{t*} \tag{27}$$

$$C_z^w = \frac{v_z f_z^{t*} - (f_z^{r*}/t_x^{r*})\left(v_z t_x^{t*} + (aG_z/2)t_x^{s*}\right) + yG_z f_z^{s*}}{u_x - v_x} \tag{28}$$

where y is the distance from the particle center to the wall, G_x, G_z, are the shear rates in the streamwise and spanwise directions, $f_i^{t*}, f_i^{r*}, f_i^{s*}$, t_i^{t*}, t_i^{r*} and t_i^{s*} are the nondimensional coefficients that depend only on a/y (or δ/a) where $\delta = y - a$.

In the lubrication limit ($\delta/a \leq 0.01$) according to the asymptotic solutions of Goldman *et al.* [48], the nondimensional coefficients are given as

$$f_x^{t*} = f_z^{t*} = 0.5333\ln(\delta/a) - 0.9588 \tag{29}$$

$$f_x^{r*} = f_z^{r*} = -0.1333\ln(\delta/a) - 0.2526 \tag{30}$$

$$t_x^{r*} = t_z^{r*} = 0.4\ln(\delta/a) - 0.3817 \tag{31}$$

$$f_x^{s*} = f_z^{s*} = 1.701 \tag{32}$$

$$t_x^{s*} = t_z^{s*} = 0.944 \tag{33}$$

In the region $0.01 \leq \delta/a \leq 10$, the following formula can be used to fit the exact solutions tabulated by Goldman *et al.* [51, 48]:

$$\begin{aligned} f_x^{t*} = f_z^{t*} = &-1.388 + 0.2739\ln\delta/a - 5.216\times10^{-2}(\ln\delta/a)^2 \\ &- 2.526\times10^{-3}(\ln\delta/a)^3 + 1.709\times10^{-4}(\ln\delta/a)^4 \end{aligned} \tag{34}$$

$$\begin{aligned} f_x^{r*} = f_z^{r*} = &\; 6.837\times10^{-3} - 1.638\times10^{-2}\ln\delta/a \\ &- 1.123\times10^{-2}(\ln\delta/a)^2 \\ &- 1.741\times10^{-3}(\ln\delta/a)^3 - 2.662\times10^{-4}(\ln\delta/a)^4 \end{aligned} \tag{35}$$

$$t_x^{r*} = t_z^{r*} = -1.045 + 7.832 \times 10^{-2} \ln \delta/a - 3.805 \times 10^{-2} (\ln \delta/a)^2 - 3.603 \times 10^{-3} (\ln \delta/a)^3 + 6.976 \times 10^{-4} (\ln \delta/a)^4 \tag{36}$$

$$f_x^{s*} = f_z^{s*} = 1.701 - \frac{1}{1.423 + 1.287(\delta/a)^{-1.021}} \tag{37}$$

$$t_x^{s*} = t_z^{s*} = 0.994 - \frac{1}{17.45 + 7.012(\delta/a)^{-1.282}} \tag{38}$$

In the region very near the wall for $\delta/a \leq 0.2$, Cox and Brenner [54] suggested that

$$f_y^{t*} = \frac{a}{\delta}\left(1 - \frac{\delta}{5a}\ln\frac{\delta}{a} + 0.9713\frac{\delta}{a}\right) \tag{39}$$

For $0.2 \leq \delta/a \leq 10$, the following fit can be used to the exact solution tabulated by Cox and Brenner [54]:

$$f_y^{t*} = 0.9871 + 1.138\left(\frac{a}{\delta}\right)^{0.9634} \tag{40}$$

Table II summarizes the range of applications of the various expressions for the nonlinear correction and the wall coefficient for the drag force. In the present simulation, the appropriate expressions for the drag force are selected according to this table.

The hydrodynamic drag force acting on a spherical particle attached to a smooth plane surface is given as:

$$F_d = \frac{3\pi f \mu d C_N}{C_c}\sqrt{u_x^2 + u_z^2} \tag{41}$$

TABLE II Regions of validity for various drag nonlinear and wall correction coefficients

Region for C_N	*C_N*	*Region for wall correction coefficients*	*Wall correction coefficients*
$Re_p \leq 0.01$	Eq. (24)	$\delta/a \leq 0.01$	Eqs. (29–33), (39)
$0.01 < Re_p < 20$	Eq. (25)	$0.01 < \delta/a \leq 0.2$	Eqs. (34–38), (39)
$20 \leq Re_p < 200$	Eq. (22)	$0.2 < \delta/a \leq 10$	Eqs. (34–38), (40)
$Re_p \geq 200$	Eq. (22)	$\delta/a > 10$	$C_i^w = 1$

where u_x, u_z are, respectively, the streamwise and spanwise components of fluid velocity at the center of sphere, $f(=1.7009)$ is a dimensionless correction factor for the wall effect given by O'Neil [55], and C_N is the nonlinear drag correction given by Eq. (22).

Lift Force

Saffman [56, 57] obtained an expression of lift force for a spherical particle moving in an unbounded shear flow field which is given as

$$\mathbf{F}_L = -\text{sgn}(G)6.46\mu a^2 U_s \left[\frac{|G|}{\nu}\right]^{1/2} \hat{\mathbf{y}} \tag{42}$$

where sgn denotes the signum function, $\hat{\mathbf{y}}$ denotes the unit vector in the direction perpendicular to the wall, $U_s = v_x - u_x$ is the particle-fluid slip velocity, and G is the shear rate defined as

$$G = \frac{du_x}{dy} \tag{43}$$

In his derivation, Saffman assumed that the Reynolds numbers defined in terms of slip velocity, Re_s, and velocity gradient, Re_G, respectively, given by

$$\text{Re}_s = \frac{|U_s|d}{\nu} \tag{44}$$

and

$$\text{Re}_G = \frac{|G|d^2}{\nu} \tag{45}$$

were small compared with unity and satisfy:

$$\text{Re}_s \ll \text{Re}_G^{1/2} \tag{46}$$

However, in a DNS study of aerosol motion in a turbulent channel flow at a moderate Reynolds number, McLaughlin [58] reported that the value of Re_G is typically of the order of 0.04, whereas Re_s is of the order of unity, indicating that Re_s is not small compared with $\text{Re}_G^{1/2}$.

Furthermore, Saffman's formula also is not accurate for predicting the lift force for particles near the wall.

Vasseur and Cox [59] derived the expression for the lift force acting on a sphere moving parallel to a rigid wall through a motionless fluid. Accordingly,

$$\mathbf{F}_L^+ = -\frac{27U_s^+ G}{8\pi y^{+2} a^+ S |U_s^+ G|} I_u \hat{\mathbf{y}} \tag{47}$$

where superscript '+' denotes nondimensional quantities as defined by Eq. (19), y^+ is the distance from the particle center to the wall, S is the particle-to-fluid density ratio, I_u is an integral, the values of which are tabulated as a function of Re_l in Table III, and Re_l is the Reynolds number based on the distance from the wall defined as

$$\mathrm{Re}_l = |U_s^+| y^+ \tag{48}$$

Asmolov [60] and McLaughlin [58, 61] extended Saffman's work by removing the limitation imposed by Eq. (46). However, the small Reynolds number limitations ($\mathrm{Re}_G, \mathrm{Re}_s \ll 1$) are still required in their analysis. The dimensionless lift force then is given by

$$\mathbf{F}_L^+ = \frac{27|G^+|^{1/2} U_s^+}{2\pi^2 d^+ S} \frac{G^+}{|G^+|} J \hat{\mathbf{y}} \tag{49}$$

where J is an integral depending on the distance from the wall and the nondimensional parameter given by

$$\varepsilon = \frac{\sqrt{|G^+|}}{U_S^+} \tag{50}$$

TABLE III Values of I_u for various Re_l

Re_l	0.100	0.200	0.300	0.400	0.500	0.600	0.700	0.800	0.900
I_u	0.004	0.016	0.034	0.060	0.092	0.130	0.173	0.220	0.271
Re_l	1.000	1.100	1.200	1.300	1.400	1.500	1.600	1.700	1.800
I_u	0.326	0.383	0.443	0.504	0.566	0.630	0.693	0.757	0.821
Re_l	1.900	2.000	2.500	3.000	3.500	4.000	4.500	5.000	6.000
I_u	0.885	0.948	1.248	1.514	1.741	1.929	2.081	2.203	2.374
Re_l	7.000	8.000	9.000	10.00	20.00	50.00	100.0	200.0	500.0
I_u	2.475	2.532	2.561	2.572	2.423	2.050	1.832	1.705	1.617

McLaughlin [61] tabulated the values of J as a function of y^+ and ε which are reproduced in the Tables IV and V.

Cox and Hsu [62] gave an expression for the lift force when the wall lies in the inner region of the particle disturbance flow, and this equation is applied in the region $a \ll y \ll \min(L_G, L_S)$. Here, L_G and L_S are called the Saffman length and the Stokes length, respectively, defined as

$$L_G = \left(\frac{\nu}{|G|}\right)^{1/2} \tag{51}$$

TABLE IV Values of J for positive ε

y^+	$\varepsilon=0.2$	0.4	0.6	0.8	1.0	1.5	2.0	∞
0.1	3.07	1.65	1.14	0.881	0.720	0.505	0.409	0.143
0.2	2.82	1.69	1.23	0.982	0.826	0.615	0.521	0.255
0.4	2.06	1.56	1.25	1.07	0.943	0.766	0.686	0.455
0.6	1.52	1.42	1.25	1.12	1.03	0.891	0.827	0.631
0.8	1.16	1.30	1.23	1.15	1.09	0.983	0.934	0.771
1.0	0.903	1.19	1.20	1.17	1.13	1.05	1.01	0.886
1.2	0.727	1.08	1.17	1.18	1.16	1.12	1.10	1.01
1.4	0.580	0.977	1.12	1.17	1.18	1.17	1.17	1.12
1.6	0.475	0.889	1.08	1.16	1.19	1.21	1.23	1.22
1.8	0.398	0.816	1.04	1.15	1.20	1.25	1.52	1.30
2.0	0.342	0.766	1.01	1.14	1.21	1.28	1.69	1.37
3.0	0.192	0.572	0.908	1.13	1.27	1.44	1.52	1.69
4.0	0.126	0.463	0.857	1.15	1.34	1.58	1.69	1.89
5.0	0.090	0.396	0.848	1.19	1.42	1.70	1.82	2.02
∞	−0.0125	0.408	1.024	1.436	1.686	1.979	2.094	2.255

TABLE V Values of J for negative ε

y^+	$\varepsilon=-0.2$	−0.4	−0.6	−0.8	−1.0	−1.5	−2.0	$-\infty$
0.1	−2.90	−1.46	−0.952	−0.695	−0.542	−0.338	−0.223	0.143
0.2	−2.55	−1.34	−0.844	−0.589	−0.435	−0.230	−0.114	0.255
0.4	−1.68	−0.980	−0.566	−0.334	−0.191	0.001	0.110	0.455
0.6	−1.11	−0.704	−0.340	−0.119	0.018	0.204	0.308	0.631
0.8	−0.745	−0.492	−0.157	0.057	0.191	0.371	0.471	0.771
1.0	−0.504	−0.317	0.0015	0.211	0.342	0.515	0.610	0.886
1.2	−0.358	−0.178	0.146	0.362	0.495	0.666	0.757	1.01
1.4	−0.239	−0.048	0.287	0.508	0.642	0.809	0.896	1.12
1.6	−0.162	0.051	0.404	0.634	0.771	0.937	1.02	1.22
1.8	−0.111	0.126	0.501	0.743	0.884	1.05	1.13	1.30
2.0	−0.076	0.182	0.576	0.827	0.972	1.14	1.22	1.37
3.0	−0.016	0.314	0.805	1.12	1.30	1.51	1.59	1.69
4.0	−0.003	0.354	0.898	1.25	1.46	1.70	1.80	1.89
5.0	−0.0007	0.370	0.939	1.32	1.54	1.81	1.91	2.02
∞	−0.0125	0.408	1.024	1.436	1.686	1.979	2.094	2.255

$$L_S = \frac{\nu}{|U_s|} \tag{52}$$

Cherukat and McLaughlin [63] extended Cox and Hsu's work by removing the limitation, $y \gg a$. The corresponding dimensionless lift force then is given as:

$$\mathbf{F}_L^+ = \frac{3U_s^2 G}{4\pi a^+ S|G|} I\hat{\mathbf{y}} \tag{53}$$

where I is a dimensionless factor depending on the distance of the particle from the wall, the shear rate and the particle size. For a freely-rotating sphere in the near-wall region, I is given as:

$$\begin{aligned} I = {} & (1.7631 + 0.3561\kappa - 1.1837\kappa^2 + 0.8452\kappa^3) \\ & - (3.2414/\kappa + 2.676 + 0.8248\kappa - 0.4616\kappa^2)\Lambda_G \quad \text{for } U_S^+ \neq 0 \\ & + (1.8081 + 0.8796\kappa - 1.9009\kappa^2 + 0.9815\kappa^3)\Lambda_G^2 \\ I = {} & 336\pi/576 \quad \text{for } U_S^+ = 0 \end{aligned} \tag{54}$$

In this equation, the nondimensional parameters, κ and Λ_G, are defined as

$$\kappa = \frac{a}{y} \tag{55}$$

$$\Lambda_G = -\frac{Ga}{U_s} \tag{56}$$

Table VI summarizes the validity regions of various expressions for the lift force. In the present simulation, appropriate expressions for the lift force are selected according to this table.

TABLE VI Regions of validity for various lift force expressions

	$\lvert\varepsilon\rvert < 0.2$	$\lvert\varepsilon\rvert \geq 0.2$
$y^+ < 0.1$	Eq. (53)	Eq. (53)
$0.1 \leq y^+ \leq 5$	Eq. (47)	Eq. (49)
$y^+ > 5$	Eq. (42)	Eq. (49)

Leighton *et al.* [64] derived an expression for the lift force for a sphere attached to a wall in a linear shear flow, given as:

$$\mathbf{F}_L^+ = \frac{6.915 U_s^2 G}{\pi a^+ S|G|}\hat{\mathbf{y}} \tag{57}$$

Cherukat and McLaughlin [63] showed that the integral given by Eq. (54) tends to $I = 9.22$ as $y/a \rightarrow 1$ which is compatible with Eq. (57) of Leighton *et al.* [64]. Therefore, Eq. (57) is applied for evaluating the lift force for a spherical particle attached to a surface.

Hydrodynamic Torque

For a sphere in contact with a plane surface, the hydrodynamic torque acting on the particle is given by

$$M_t = 2\pi\mu C_M d^2 \sqrt{u_x^2 + u_z^2} \tag{58}$$

where $C_M = 0.943993$ is the correction factor for the wall effects as derived by O'Neill [55].

Brownian Force

The nondimensional Brownian force is given as (Ounis *et al.* [38, 39]);

$$n_i^+(t^+) = \frac{6\nu}{\pi\rho^p d^3 u^{*3}} N_i(t^+) \tag{59}$$

The spectral intensity of n_i^+ was given by Ounis *et al.* [38, 39], *i.e.*,

$$S_{n_i^+ n_j^+}(\omega^+) = \frac{648}{\pi C_c^2} \frac{\delta_{ij}}{S_c S^2 d^{+4}} = \frac{2}{\pi S_c \tau_p^{+2}} \delta_{ij} \tag{60}$$

where

$$S_c = \frac{\nu}{D} = \frac{3\pi\nu d\mu}{C_c kT} \tag{61}$$

is the Schmidt number, ω^+ is the frequency in wall units, T is the air temperature, μ is the air viscosity, $k = 1.38 \times 10^{-23}$ J/K is the

Boltzmann constant, and D is the particle Brownian diffusivity. In Eq. (60) τ_p^+ is the nondimensional particle relaxation time. The dimensional and nondimensional particle relaxation time are given as:

$$\tau_p = \frac{C_c S d^2}{18\nu}, \quad \tau_p^+ = \frac{C_c S d^{+2}}{18} \tag{62}$$

The Stokes–Cunningham slip correction factor, C_c, is given by Eq. (21). At every time step in a simulation, the dimensionless Brownian force is given as:

$$n_i^+(t^+) = G_i \sqrt{\frac{\pi S_0}{\Delta t^+}} \tag{63}$$

where G_i is a zero-mean independent Gaussian random number, S_0 is the spectral intensity, and Δt^+ is the time increment.

RESULTS AND DISCUSSION

In this section, particle removal, resuspension and re-entrainment in the turbulent channel flows are studied. A temperature of $T = 298$ K, a kinematic viscosity of $\nu = 1.5 \times 10^{-5}\,\mathrm{m^2/s}$, and a density of $\rho^f = 1.12\,\mathrm{kg/m^3}$ for air, a density ratio of $S = 1964$ for graphite particle, and a shear velocity of $u^* = 1.0$ m/s are assumed. In this case, the Reynolds number based on the shear velocity, u^*, and the half-channel width is 125, while the flow Reynolds number based on the hydraulic diameter and the centerline velocity is about 8000. This condition corresponds to a channel width of 3.75 mm. To keep the computational effort within an acceptable limit and to reduce the statistical error, ensembles of 8192 (2^{13}) particles for each diameter are used in these simulations. The gravity is assumed to be in the direction perpendicular to the lower wall.

REMOVAL AND ADHESION

In this section, ensembles of different size particles are initially randomly distributed on the lower wall, and all forces acting on each particle are computed at every time step. When the detachment condition (Eqs. (15) or (16)) is satisfied, the particle is assumed to be resuspended in the turbulent flow, and its subsequent motion is simulated using Eqs. (17) and (18).

Figure 3a shows variations of the number of removed 40 and 50 μm graphite particles *versus* time as predicted by the JKR and the TPL adhesion models. It is observed that the number of particles removed as predicted by the JKR model is larger than that obtained from the TPL model. This trend is consistent with the earlier results reported by Soltani and Ahmadi [17]. Since the JKR model has been more accepted in the literature, it is used in the following simulations for particle removal from surfaces.

Variations of the number of particles that are resuspended and those that remain attached or redeposit on the lower wall as a function of time are shown in Figure 3b. This figure shows that the number of resuspended particles increases up to about $t^+ = 400$ and then

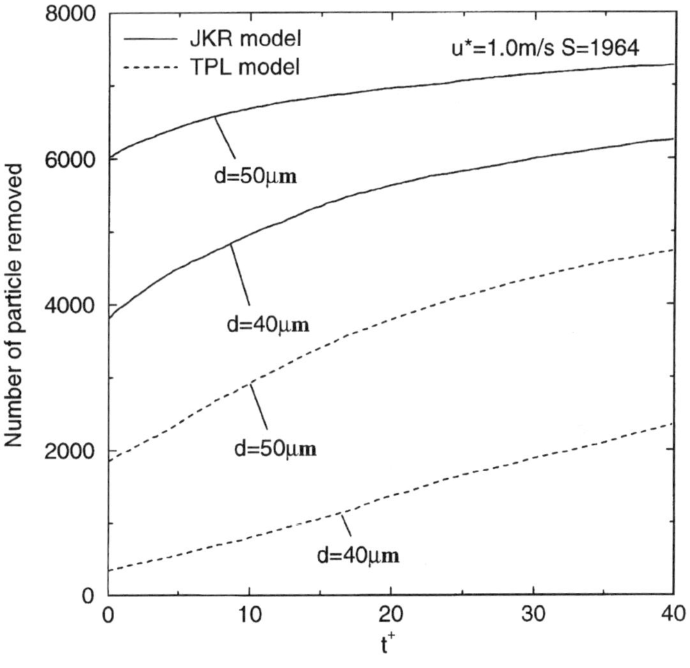

FIGURE 3a Variations of the number of removed particles *versus* time for different adhesion models.

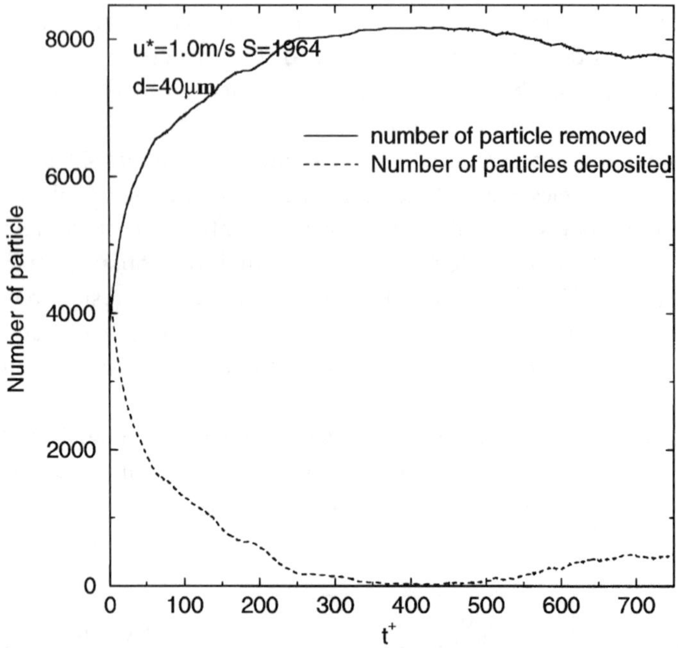

FIGURE 3b Variations of the number of removed and deposited particles *versus* time.

equilibrates at about 7700. At the equilibrium condition, about 95% of particles are resuspended. For $t^+ < 400$, the particle resuspension process is dominant; but, for $t^+ > 400$, the resuspension and deposition processes tend to come to equilibrium. The simulation was repeated for 40 μm glass, polystyrene, silicon and rubber particles. Figure 3c shows the number of removed particles *versus* time for different materials. It is observed that graphite, glass, polystyrene and silicon particles are more easily removed when compared with rubber particles. This is because the softer rubber particles form a larger contact area and have a higher adhesion force.

Particle Detachment Mechanism

In this section, the statistical properties of various nondimensional forces and moments acting on the particles attached to the lower wall of the channel are evaluated at every time step, and their variations are discussed. The results for an ensemble of 8192 graphite particles with a

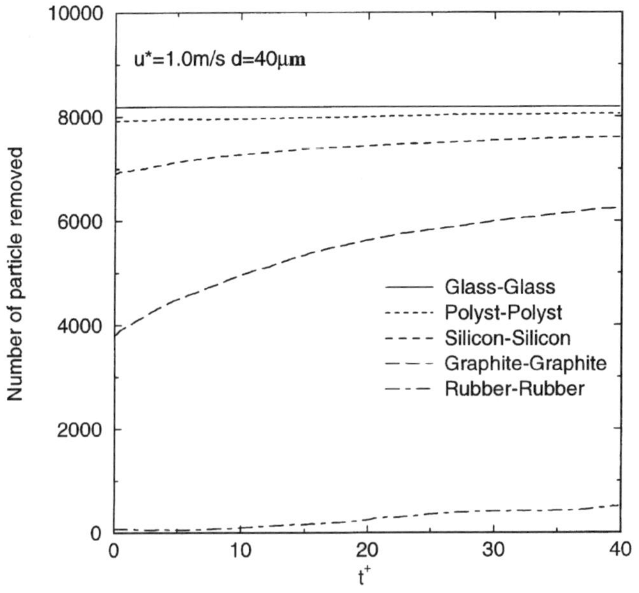

FIGURE 3c Variations of the number of removed particles *versus* time for different materials.

diameter of $d = 40\,\mu m$ ($\tau_p^+ = 780$) are described. It is found that all particles detach from the wall by the rolling detachment mechanism. This is consistent with the suggestion of Soltani and Ahmadi [17] that the rolling detachment is the dominant resuspension mechanism of spherical particles in turbulent flows.

Figure 4a shows the probability density function of dimensionless drag and hydrodynamic torque for particles that are removed and the particles that remain attached to the wall. The statistical results are obtained in the time period (0, 40) in wall units. The density function is evaluated using $f(\xi) = N_\xi/N$, where N_ξ is the number of particles with the nondimensional drag force in the region $[\xi, \xi + \Delta\xi]$, and N is the total number in the sample. Here, $\Delta\xi = 1.0 \times 10^{-4}$ is assumed, and the density function satisfies the normalization condition, $\sum_{i=1}^{200} f(\xi_i) = 1$, where 200 is the number of bins used in the analysis. It is observed that there exists a threshold value of about 0.003 for the drag force for the removed particles. Particles will detach from the wall when the drag force is beyond the threshold value. Figure 4a also shows that the dimensionless drag force for particles that remain attached to the wall

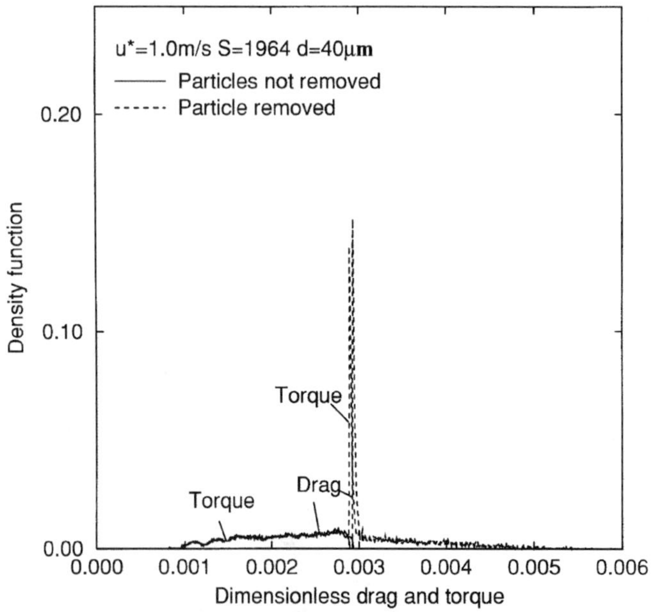

FIGURE 4a Probability density function of dimensionless drag and torque for removed and deposited particles.

is distributed roughly uniformly between 0.001 and 0.003, while the density function of drag force for removed particles has a high peak which is slightly larger than the threshold value. This indicates that most detachment occurs when the drag force begins to exceed the threshold value. Figure 4a also shows that the density function of torque is similar to that of the drag force and the threshold value of torque is slightly smaller than that for the drag force. Similar probability density functions for the lift force are shown in Figure 4b. While there is an approximate critical value of 0.0004 for the nondimensional lift force, there is a noticeable spread in the distributions, and the density functions of the lift force for particles removed and attached overlap. This indicates that the lift force is not the critical factor for particle detachment.

The mean values of nondimensional forces and hydrodynamic torque acting on the removed and attached 40 μm particles are shown in Table VII. The mean value of adhesion force is at least three orders of magnitude larger than those of the gravitational and lift forces.

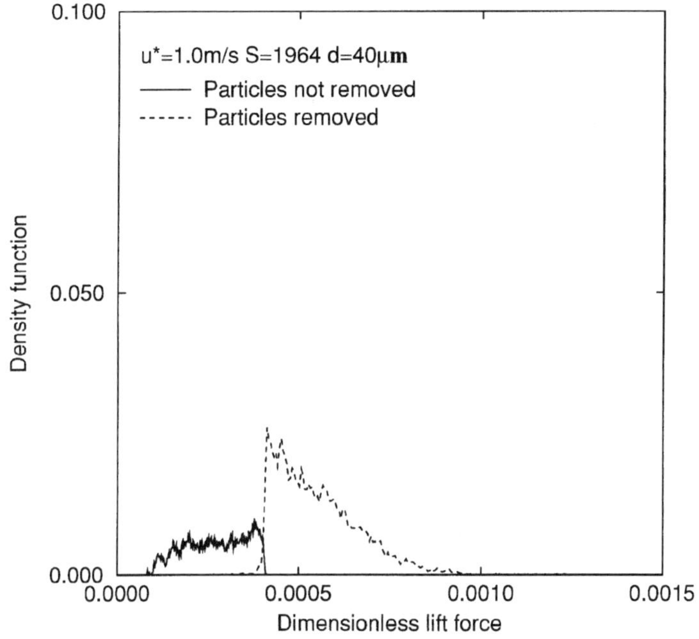

FIGURE 4b Probability density function of dimensionless lift force for removed and deposited particles.

TABLE VII Mean values of various forces and moments for $d = 40\,\mu m$ graphite particles

	Lift force	*Gravity*	*Adhesion force*	*Drag*	*Moment*
Removed particle	5.50e-4	1.47e-4	1.49	3.48e-3	3.34e-3
Attached particle	2.72e-4	1.47e-4	1.49	2.11e-3	2.08e-3

Therefore, the effects of gravity and lift force for particle (lift-off) detachment are negligible. The magnitudes of drag force and hydrodynamic torque are comparable and, thus, they both play an important role in the particle rolling detachment process. (Note that here $d^+ = 2.67$ for 40 μm particles.)

Turbulent Near Wall Structure

Hinze [19] and Smith and Schwartz [65] summarized the streaky structures of turbulent near-wall flows. In the earlier works of Ounis *et al.* [39], Soltani and Ahmadi [36] and Zhang and Ahmadi [47], it was shown that the turbulence near-wall coherent eddies play a dominant role in particle deposition and resuspension processes. We performed several simulations with ensembles of 8192 particles of different sizes (from 30 μm to 60 μm) that are initially uniformly distributed on the lower surface of the duct. Figure 5 shows the locations of particles that remained attached on the lower wall at $t^+ = 40$. For 30 μm particles, Figure 5a shows that the particles are removed in certain bands. The attached particles also form roughly distinct bands. Similar trends are also observed in Figure 5b for 40 μm particles. Here, the structure of bands is more pronounced. In this figure, the distances between the nearby bands are about 100 – 150 wall units, which is consistent with spacing between high-speed and low-speed streaks as was also noted by Soltani and Ahmadi [42]. Similar band structures also exist for particles with $d = 50$ μm and $d = 60$ μm as shown in Figures 5c and 5d. The number of particles that remain attached to the wall, however, decreases rapidly as particle diameter

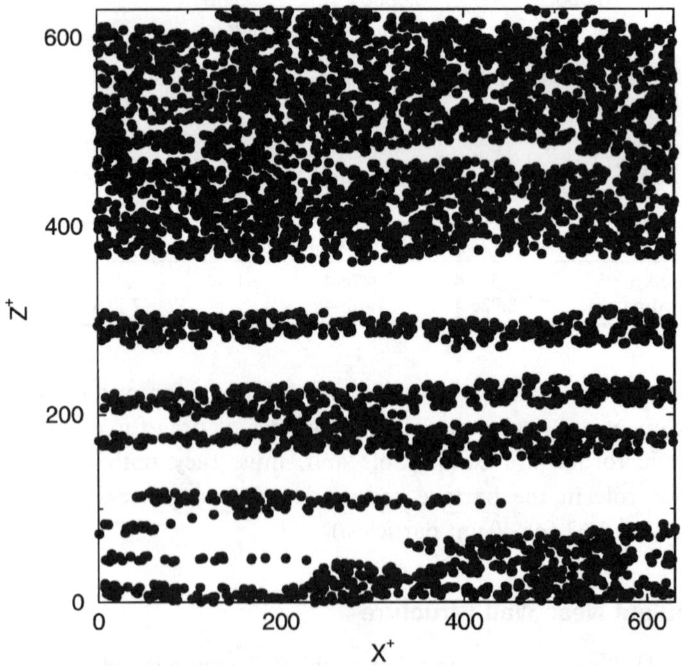

FIGURE 5a Distribution of the locations of 30 μm particles on the surface.

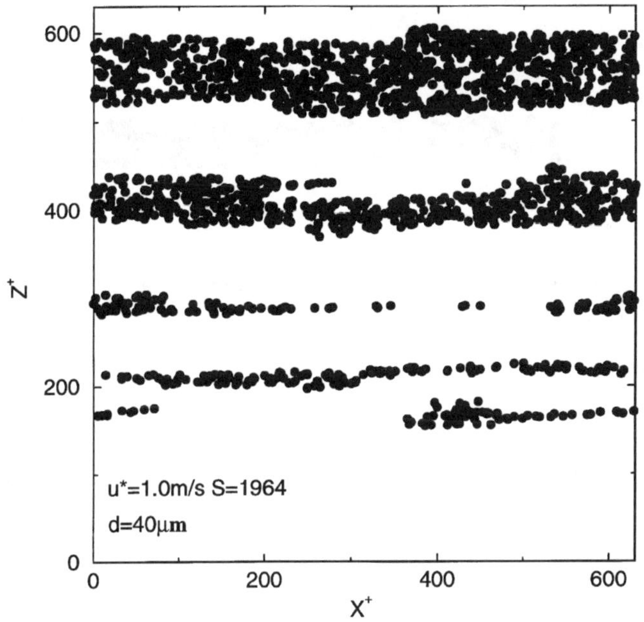

FIGURE 5b Distribution of the locations of 40 μm particles on the surface.

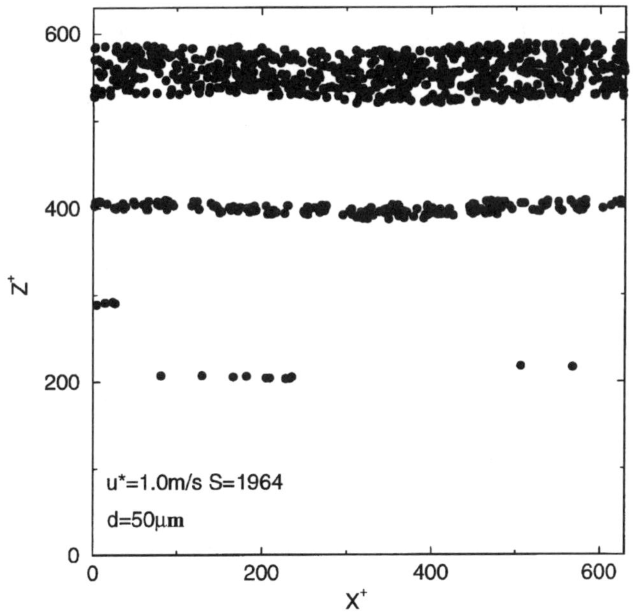

FIGURE 5c Distribution of the locations of 50 μm particles on the surface.

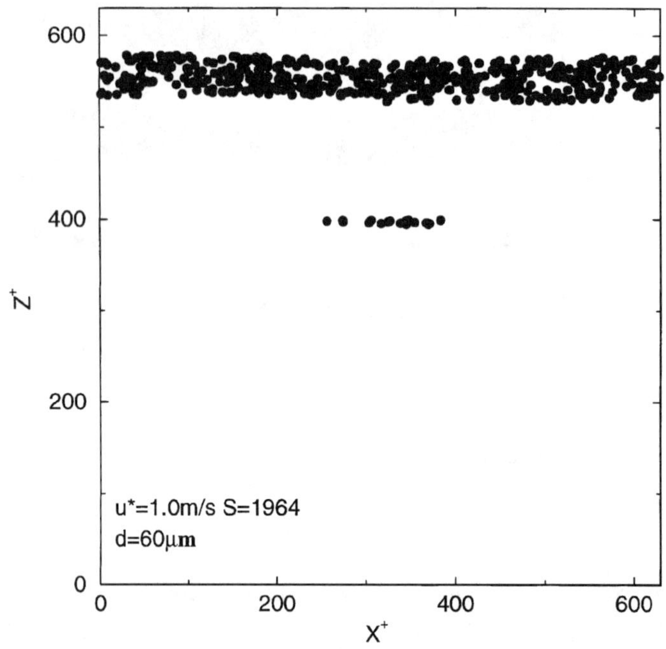

FIGURE 5d Distribution of the locations of 60 μm particles on the surface.

increases. This is because the hydrodynamic forces and torques acting on the particles increase faster than the adhesion force as d increases, while the effect of weight is negligible. Figure 5 further confirms the importance of near-flow structure (coherent vortices, high- and low-speed streaks) in particle removal and re-entrainment processes in turbulent flows.

Variations of mean velocity components averaged spatially over one periodic cycle in the streamwise direction at a distance of 1.33 wall units from the lower wall (corresponding to the centroid of a 40 μm particle) and temporally in a time period of (0, 40) wall units are shown in Figure 6a. It is observed that the streamwise velocity varies between 0.5 to 2.2 with a mean value of about 1.3. Both u^+ and w^+ components exhibit roughly periodical fluctuations in the spanwise direction. The amplitude of the streamwise velocity fluctuation is about 0.6 while that of w^+ varies between 0.1 to 0.3. This observation is consistent with the well-known streaky structure of turbulence in the near-wall region. Figure 6a shows that the peaks and valleys of u are

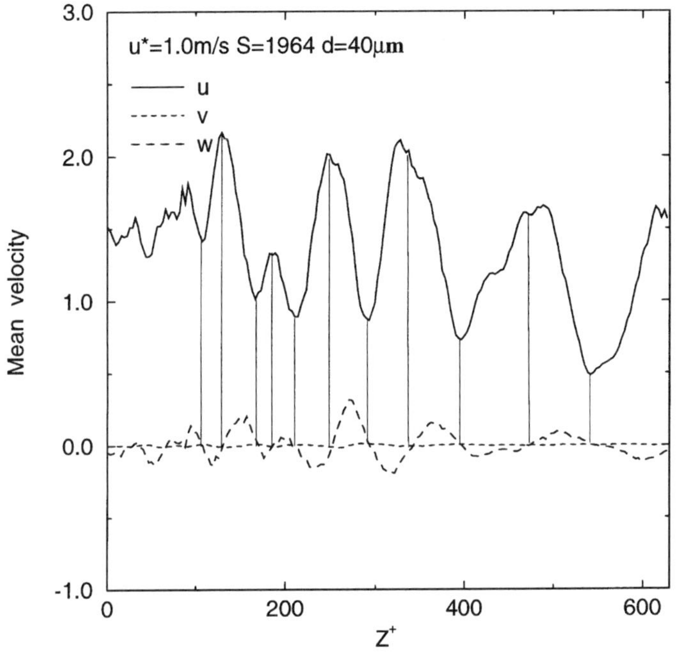

FIGURE 6a Spanwise variation of the averaged nondimensional velocities.

roughly at the points that w^+ becomes zero. The mean normal velocity is comparatively quite small. The periodic (positive and negative) variation of w^+ clearly indicates the presence of counter-rotating near-wall vorticies in the average sense. Figure 6b shows the close up of the v^+-velocity. The nearly-periodic fluctuation structure of v^+ is clearly observed from this figure. Comparing the locations of the peaks and valleys of the averaged velocity components in these figures shows that the high-speed streaks (peaks of u^+) correspond to the down-flow region (v^+ toward the wall). Similarly, the low-speed streaks correspond to upflow regions. This result is consistent with the early observation of periodic averaged vorticity reported by Soltani and Ahmadi [42].

Table VII showed that the hydrodynamic drag and torque acting on particles are the dominant factors for particle detachment. The magnitudes of hydrodynamic drag and torque are directly proportional to the particle-fluid slip velocity as indicated by Eqs. (41) and

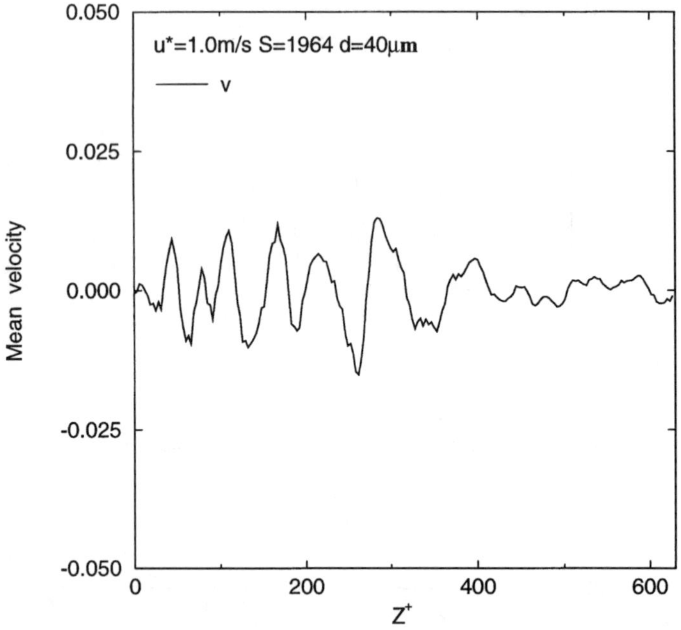

FIGURE 6b The close up of the spanwise variation of *v*-velocity.

(58). The magnitude of the streamwise velocity is much larger than the spanwise component. Thus, the streamwise velocity is the key parameter for particle removal from the wall. A careful examination of the result reveals that the bands in Figure 5 for which the particles are removed coincide with the regions in which u^+ is near its peaks in Figure 6. These observations show that the turbulence near-wall structures play an important role on the particle removal process.

Removal Rate

Reeks *et al.* [30] proposed a model for particle removal based on the influence of turbulence fluctuation energy transferred to particles attached to the wall. They found two distinctive regimes: an "initial" resuspension region in which about 90% of the particles are removed during a short initial burst lasting typically less than 10^{-2} s, which is followed by a "longer-term" gradual resuspension at a rate inversely proportional to time. The resuspension rate obtained in our simulations for 40 μm and 50 μm particles are shown in Figure 7. The

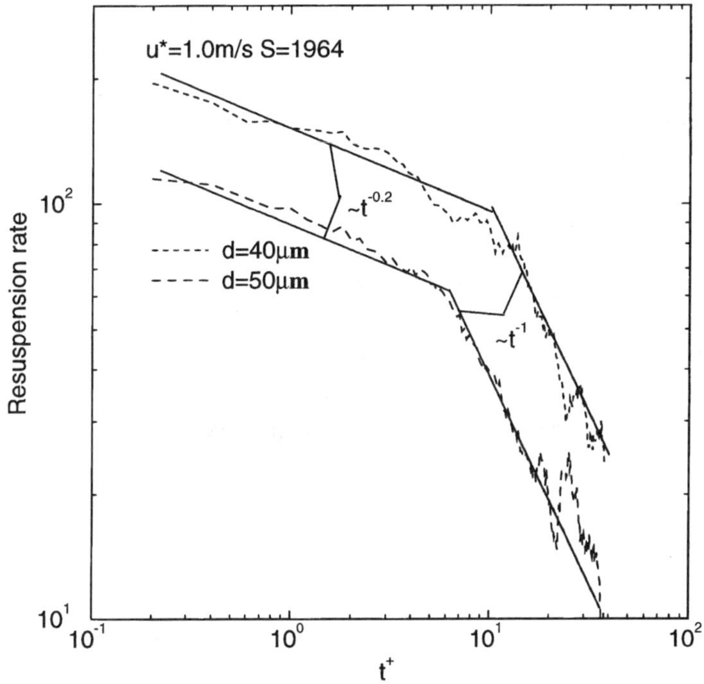

FIGURE 7 Variations of the resuspension rate *versus* time.

resuspension rate is defined as

$$\Lambda(t^+) = \frac{\Delta N}{\Delta t^+} \tag{64}$$

where ΔN is the number of particles resuspended in a time interval of Δt^+. Figure 7 shows a decreasing removal rate as the particle concentration on the wall is depleted. The trend in the variation of Λ, however, markedly changes. For 50 μm particles, when $t^+ < 5$, Λ varies as $t^{-0.2}$; while, for $t^+ > 7$, the removal rate varies roughly as t^{-1}. For 40 μm particles, similar trends in the variation are seen, but the change in slope occurs at $t^+ \approx 12$. The dimensional time duration used for evaluating the resuspension rate in these simulations is of the order of 10^{-3} s, which corresponds to the initial resuspension region suggested by Reeks *et al.* [30]. There are no experimental data available for such a short time duration, and the results can only be compared with the trend of Reeks' resuspension model which is in qualitative agreement with the field observation. Figure 7 shows that the variation of resuspension rate with time is comparable with the model of Reeks *et al.* [30]. The simulated "long-term" t^{-1} variations of resuspension rate, however, seem to initiate at a much short time when compared with the model.

RE-ENTRAINMENT TRAJECTORIES

In this section, ensembles of particle trajectories which are removed from the wall are computed and statistically analyzed. Here, a particle-to-fluid density ratio of $S = 1964$ and a shear velocity of $u^* = 1.0\,\mathrm{m/s}$ are assumed.

Re-entrainment Process

As noted before, an ensemble of 8192 particles with $d = 40\,\mu\mathrm{m}$ are initially randomly distributed on the lower wall. The particles are assumed to be resuspended when the detachment condition given by Eq. (15) is satisfied. The subsequent trajectories of detached particles are then evaluated using Eq. (17). In this case, $\tau_p = 0.012\,\mathrm{sec}$ and $\tau_p^+ = 780$. Figure 8 shows the instantaneous locations of particles in the $y-z$ plane at different times. For $t^+ = 250$, it is observed from Figure 8a that particles begin to move away from the wall. Their transport, however, is not a uniform diffusion process and shows definite structures in the spanwise direction. Figure 8b shows instantaneous particle locations at $t^+ = 500$. The periodic spanwise structure can clearly be seen from this figure. The instantaneous particle locations at $t^+ = 750$ and $t^+ = 1000$ are shown in Figures 8c and 8d, respectively. Figure 8c shows that particles seem to move away from the wall on certain distinct bands. While the bands are also noticeable in Figure 8d, the structure tends to smear out due to the movements of the near wall coherent eddies and turbulence dispersion in the core region.

Sample trajectories in different planes are shown in Figure 9. Trajectories in the $y-z$ plane in Figure 9a indicate that 40 μm particles first move away from the wall roughly straight up to about 7 to 10 wall

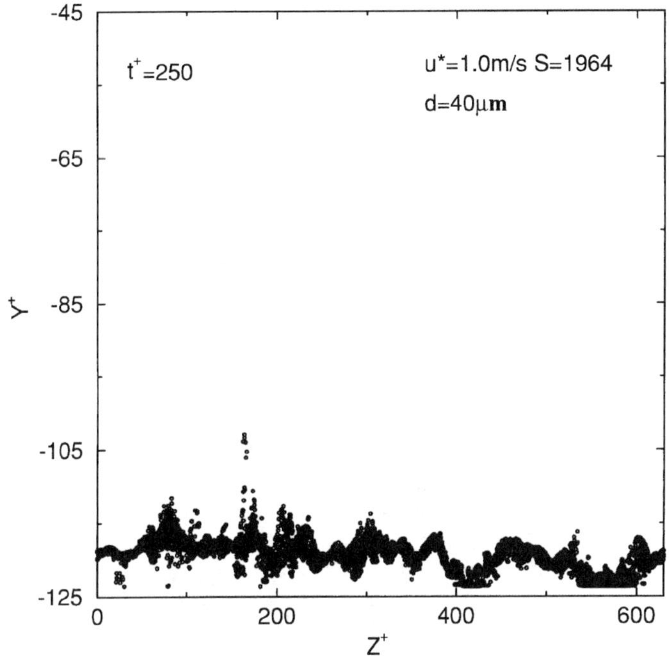

FIGURE 8a Distribution of 40 μm particles in the $y-z$ plane at $t^+ = 250$.

units from the wall and then begin to disperse. Figure 9b shows the close up of a few particle trajectories in the $y-z$ plane. Particle removal, deposition, and resuspension processes are seen from this figure. The roughly straight upward motion of detached particles is also clearly noticeable. Figure 9c shows the $x-y$ projection of 40 μm particle trajectories that are initially on the wall on a line at about $x^+ = 500$. It is observed that some particles are swept away by the streamwise fluid velocity and are entrained in the core flow. A number of removed particles deposit on the wall and are resuspended again. Figure 9d shows that 40 μm particles move in the streamwise direction and away from the wall roughly in their vertical planes with little dispersion in wall region, but disperse as they enter the core flow region. The simulation results presented in Figures 8 and 9 suggest that streaky axial flow structure together with the lift force play a key role in the detachment of particles larger than 40 μm and their reentrainment processes from the wall region.

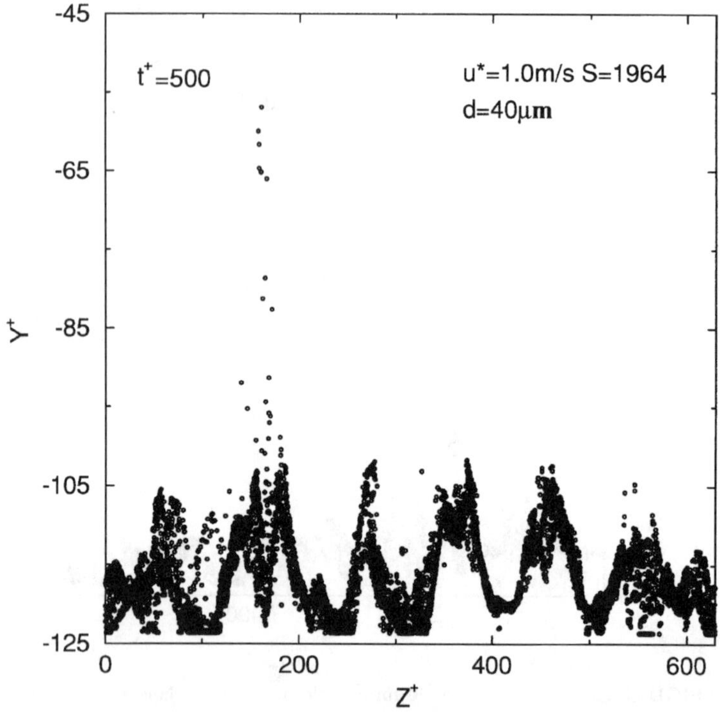

FIGURE 8b Distribution of 40 μm particles in the $y-z$ plane at $t^+ = 500$.

To analyze the re-entrainment process for sub-micrometer particles, the simulation is repeated for an ensemble of 8192 particles with $d = 0.14\,\mu\text{m}$. In this case, $\tau_p = 3.4 \times 10^{-7}$ sec and $\tau_p^+ = 0.023$. The particles are initially randomly distributed at a distance of one wall unit from the lower wall, and their subsequent trajectories are analyzed. Figure 10 shows particle positions in the $y-z$ plane at different times. Figure 10a indicates that the 0.14 μm particle concentration in the $y-z$ plane is nonuniform, and the particles tend to move away from the wall in certain bands. Similar structures in the distribution of particles are also noticeable at $t^+ = 500$, 750, 1000 shown in Figures 10b–10d. These small particles, however, exhibit more dispersion due to the Brownian motion effects.

Figures 11a–11c show sample particle trajectories for $d = 0.14\,\mu\text{m}$ in different planes in a time duration of (0, 1000) wall units. Contrary to the large 40 μm particles, these small particles do not move straight up,

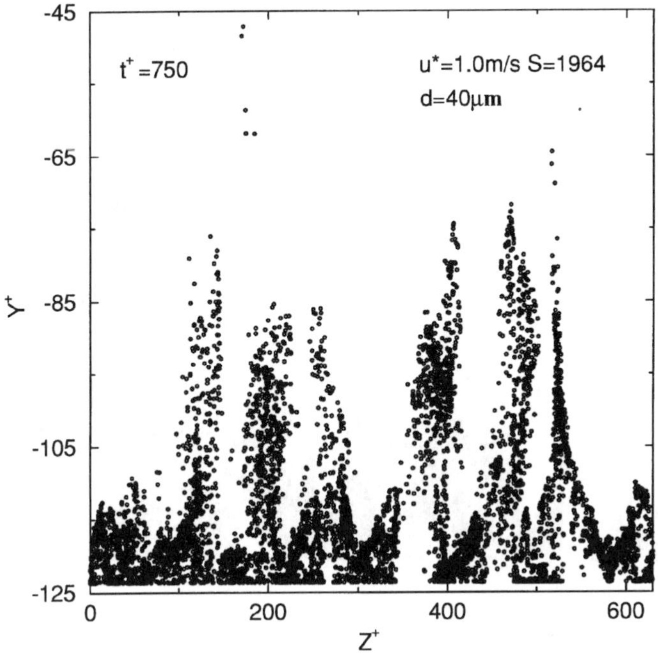

FIGURE 8c Distribution of 40 μm particles in the $y-z$ plane at $t^+ = 750$.

and their trajectories in the $y-z$ plane are curved and roughly follow the coherent near-wall vortices. Figure 11b shows that the resuspended 0.14 μm particles are, generally, dispersed in the flow with little re-deposition. Random trajectories of these particles (due to turbulence and Brownian motion) are clearly seen from Figures 11a–11c.

To clarify further the effect of near-wall turbulence structure on the particle re-entrainment process, the statistics of the simulated fluid velocity components near the wall are obtained. Averaging is performed spatially along the streamwise direction and temporally in a time period of (0, 100). Figure 12a shows the contours of the mean u^+-velocity in the $y-z$ plane. The periodic pairing of high-speed and low-speed streaks near the wall are clearly seen from this figure. The similar result for the v^+-velocity shown in Figure 12b indicates the periodic upward and downward flow structures. Comparing Figure 12a and 12b, it is found that the locations of the high-speed axial flow stream ($z^+ = 40, 140, 260, 350, 480, 600$) correspond to those of

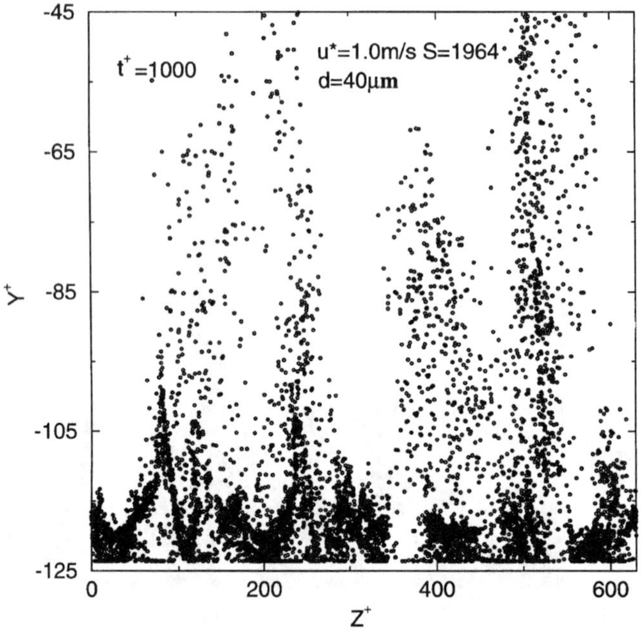

FIGURE 8d Distribution of 40 μm particles in the $y-z$ plane at $t^+ = 1000$.

downward flows (negative v^+); and the locations of the low-speed stream ($z^+ = 80, 200, 300, 400, 550$) correspond to those of upward flows.

The formation of U-shaped vortex loops, also known as horseshoe or hairpin vortices, is one important feature of turbulent boundary-layer flow in the models of Acarlar and Smith [66] and Robinson [67]. In this model, vortices play a major role in producing sweeps and ejections in the near-wall shear layers. A careful examination of the results shows that the locations of peaks of concentration of 40 μm particles in Figure 8b roughly correspond to those of high-speed axial streams in Figure 12a and downward flow regions in Figure 12b. In contrast, the peaks in the distribution of 0.14 μm particles shown in Figures 10a and 10b correspond to the upward flow regions in Figure 12b (and low u^+ in Fig. 12a). That is, the re-entrainment of large and small particles is controlled by different mechanisms. Large particles with $d = 40$ μm ($\tau_p^+ = 780$) move roughly straight from the wall up to about 10 wall units due to the shear-induced lift force, then begin to disperse. As a result, these particles move away faster from the wall in

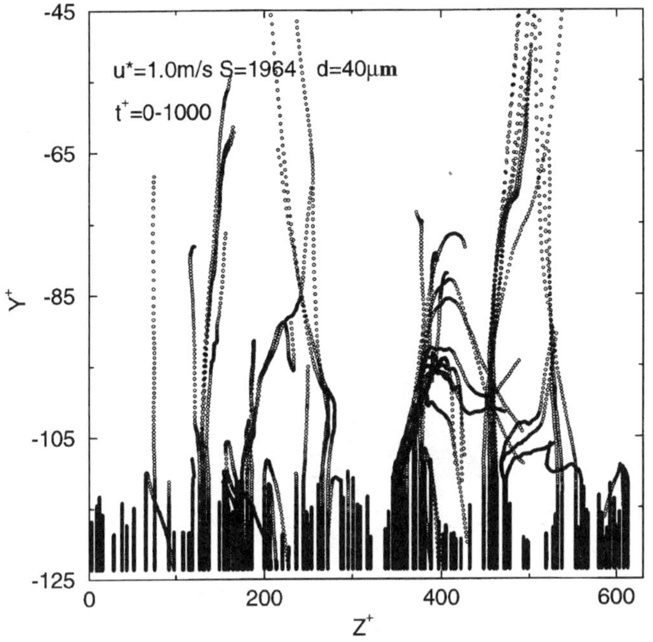

FIGURE 9a Sample 40 μm particle trajectories in the $y-z$ plane.

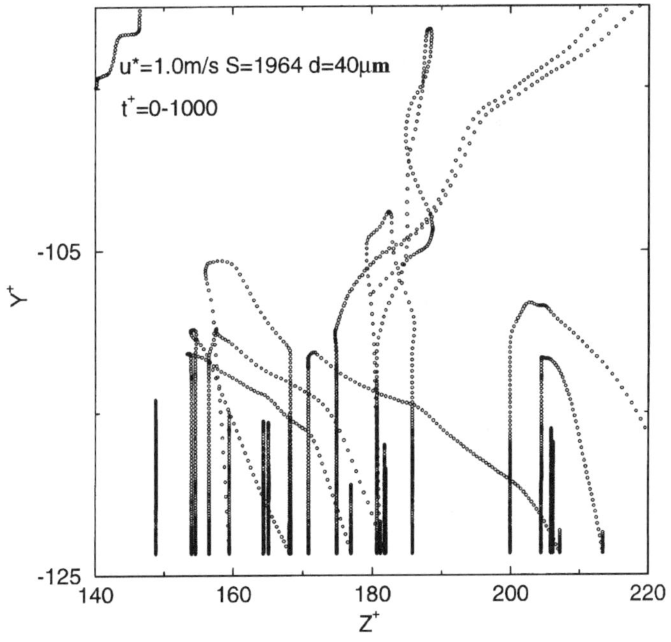

FIGURE 9b Close up of 40 μm particle trajectories in the $y-z$ plane.

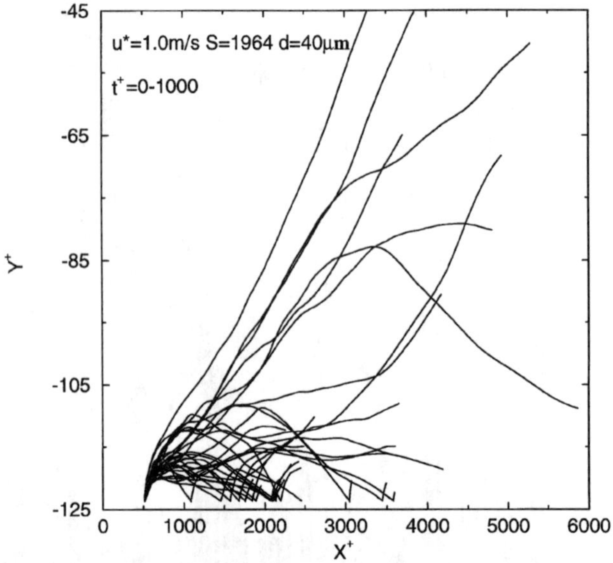

FIGURE 9c Sample 40 μm particle trajectories in the $x-y$ plane.

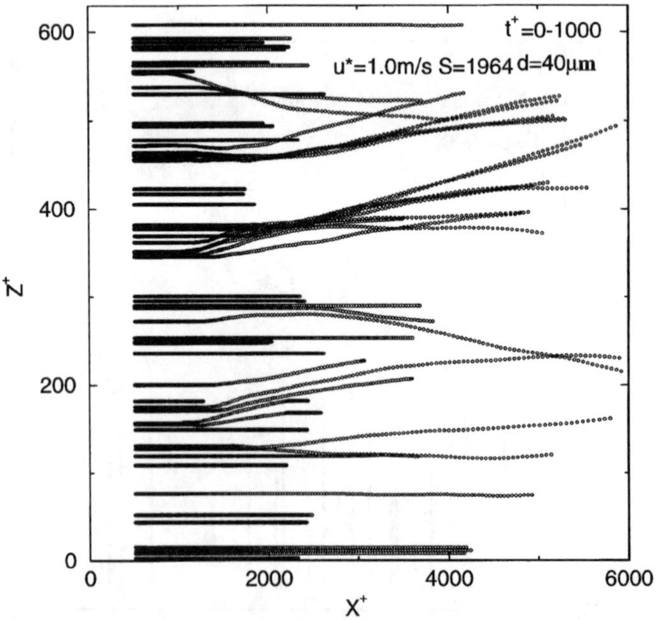

FIGURE 9d Sample 40 μm particle trajectories in the $x-z$ plane.

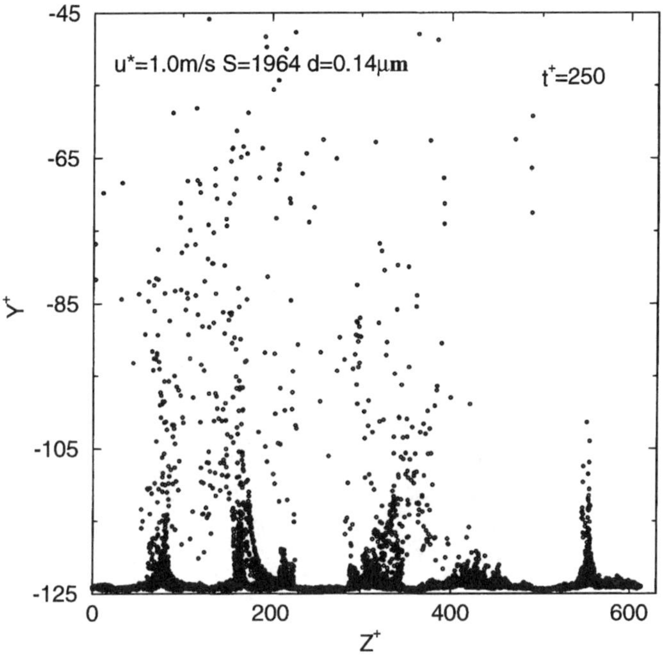

FIGURE 10a Distribution of 0.14 μm particles in the $y-z$ plane at $t^+ = 250$.

the high streamwise velocity bands, which correspond to downward flow regions (negative v^+). The small particles of the order of 0.1 μm ($\tau_p^+ = 0.023$), however, follow the near-wall upward flows formed by the near-wall coherent eddies. These regions correspond to the low streamwise velocities.

To the authors' knowledge there is no experimental data that could provide insight in the micro-mechanics of the particle removal process. The difficulty is the time scale of the dynamic rearrangement of the near-wall eddies, which is quite small compared with the macroscopic time, but sufficiently large compared with the time scale of particle removal from the wall region.

Plane Source

Additional simulations are performed to provide further understanding of the particle re-entrainment process. An ensemble of 8192

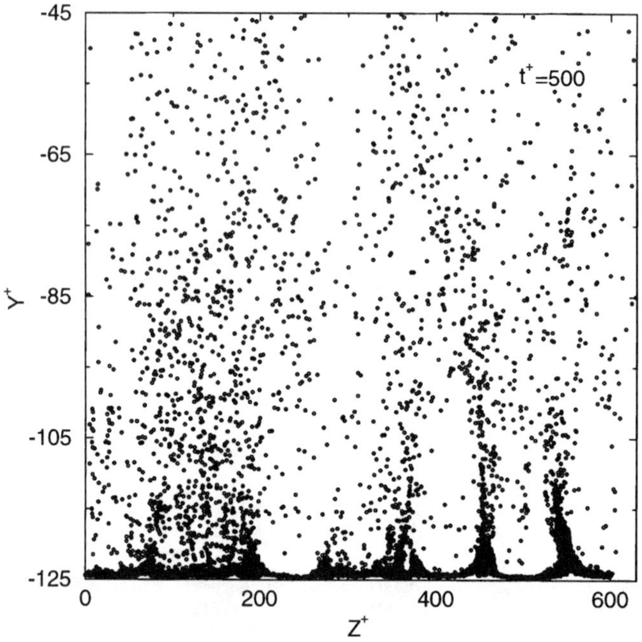

FIGURE 10b Distribution of 0.14 μm particles in the $y-z$ plane at $t^+ = 500$.

particles with $d = 0.14\,\mu\text{m}$ ($\tau_p^+ = 0.023$) and $d = 40\,\mu\text{m}$ ($\tau_p^+ = 780$) are initially released randomly in a plane at a distance of 1.5 wall units from the lower wall in a horizontal duct. At every time step, the particle positions are evaluated and statistically analyzed. The mean, the maximum, the minimum and the mean standard deviation (mean $\pm\,\sigma$) trajectories are computed; and the results are displayed in Figure 13. The mean paths for 0.14 μm and 40 μm particles are very close, while 0.14 μm particles disperse away from the wall much faster than 40 μm particles. Therefore, under the same flow conditions, small particles can more easily follow the streaming flows away from the wall and are re-entrained faster into the core flows when compared with larger particles. The gravitational sedimentation effect of large particles, however, reduces their rate of dispersion away from the wall.

Figure 13 also shows that the mean-σ for the 0.14 μm particles becomes smaller than the sample absolute minimum for $t^+ \geq 45$. This is because of dispersion of particles in the core region. Thus,

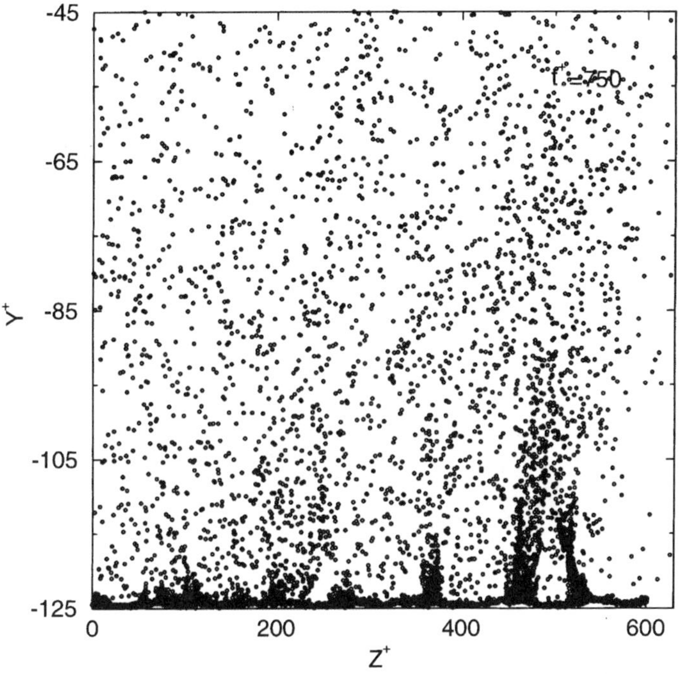

FIGURE 10c Distribution of 0.14 μm particles in the y–z plane at $t^+ = 750$.

σ becomes quite large and the mean-σ curve crosses the absolute minimum curve.

Velocity Statistics

In this section, simulation results for the mean and root-mean-square (RMS) velocities of 0.14 μm($\tau_p^+ = 0.023$), 3.21 μm($\tau_p^+ = 5$) and 40 μm($\tau_p^+ = 780$) particles during the re-entrainment process are reported. For each size, an ensemble of 8192 particles is initially uniformly distributed on a plane at a distance of 1.5 wall units from the lower wall. The particle mean and RMS velocities with the distance from the lower wall are obtained *via* averaging over a distance of 630 wall units in the streamwise direction for a time period of 0 to 100 wall units. Variations of the mean streamwise particle velocities with distance from the wall are shown in Figure 14a and are compared with the fluid velocity. It is observed that the mean streamwise particle

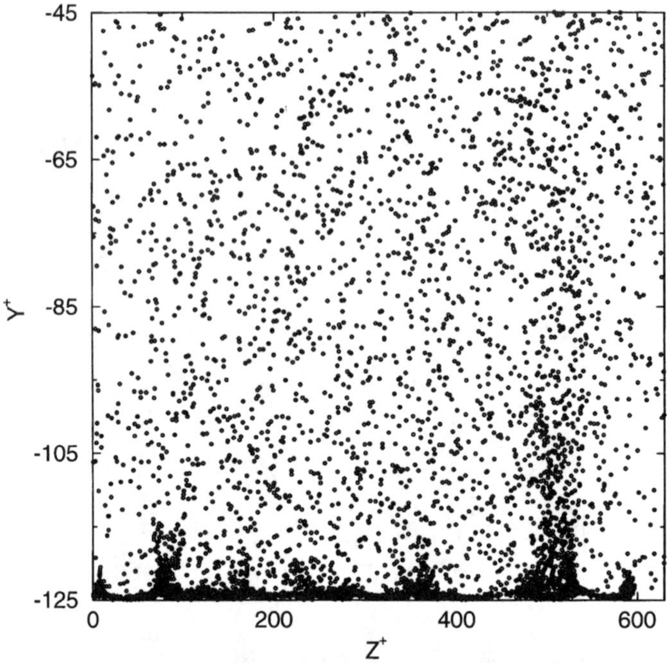

FIGURE 10d Distribution of 0.14 μm particles in the $y-z$ plane at $t^+ = 1000$.

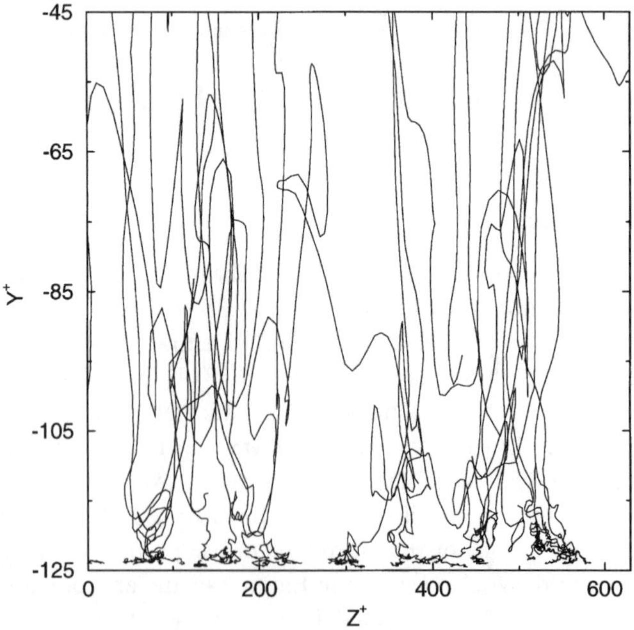

FIGURE 11a Sample 0.14 μm particle trajectories in the $y-z$ plane.

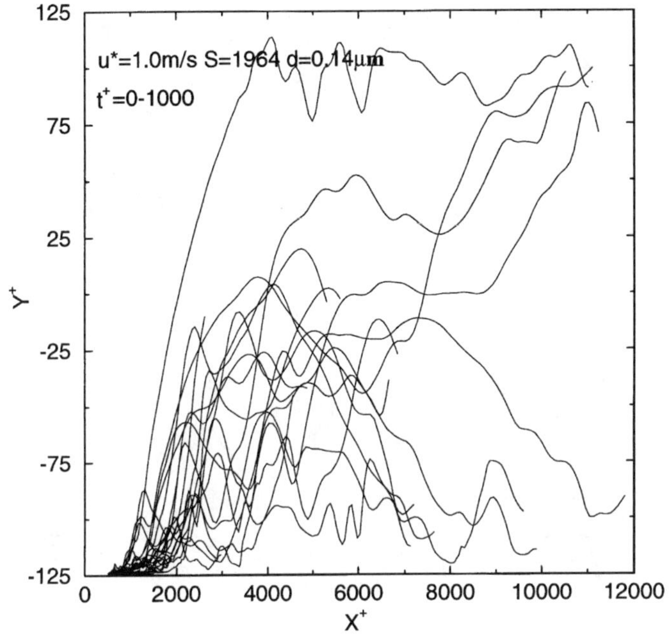

FIGURE 11b Sample 0.14 μm particle trajectories in the $x-y$ plane.

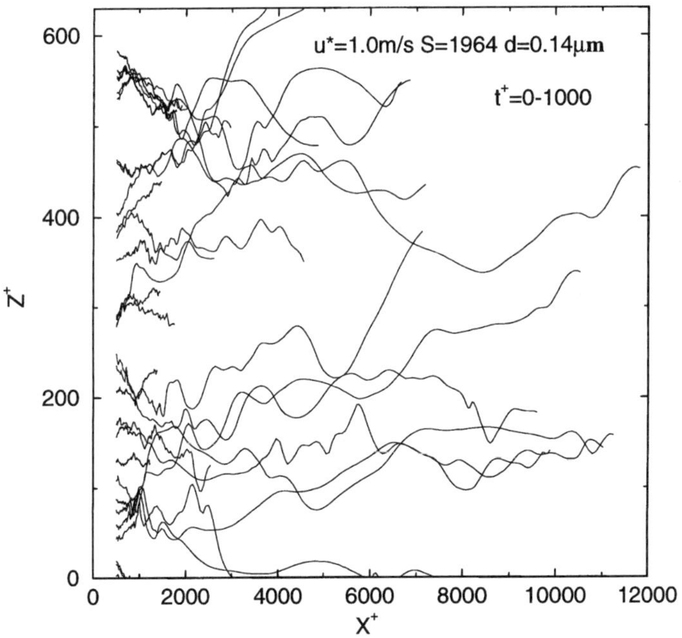

FIGURE 11c Sample 0.14 μm particle trajectories in the $x-z$ plane.

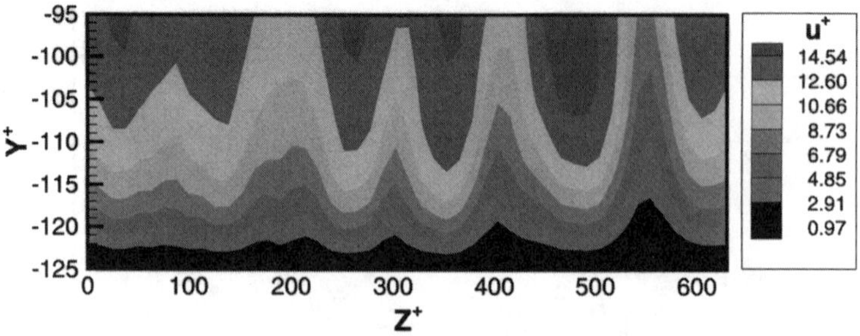

FIGURE 12a Contour plot of the mean *u*-velocity in the *y*–*z* plane. (See Color Plate VII).

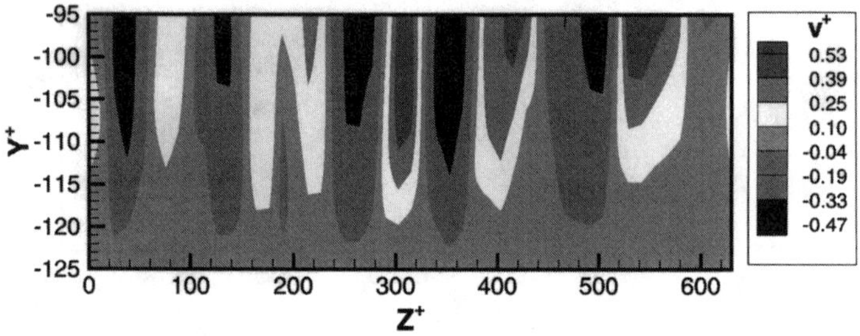

FIGURE 12b Contour plot of the mean *v*-velocity in the *y*–*z* plane. (See Color Plate VIII).

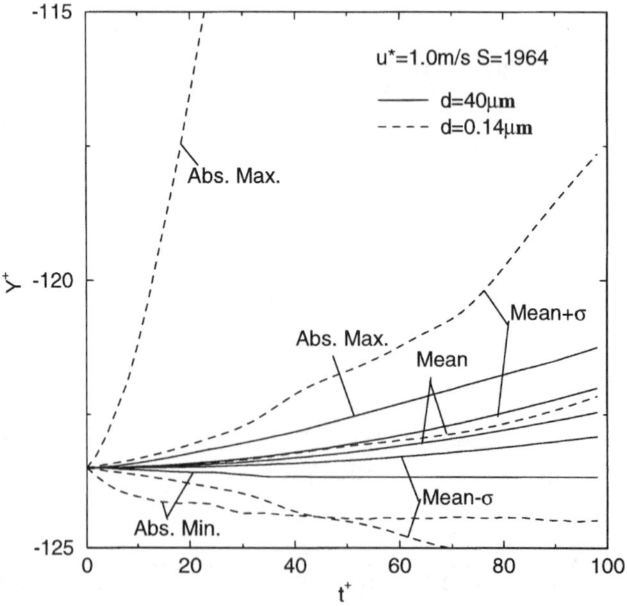

FIGURE 13 Trajectory statistics for 40 and 0.14 μm particles released from a plane source.

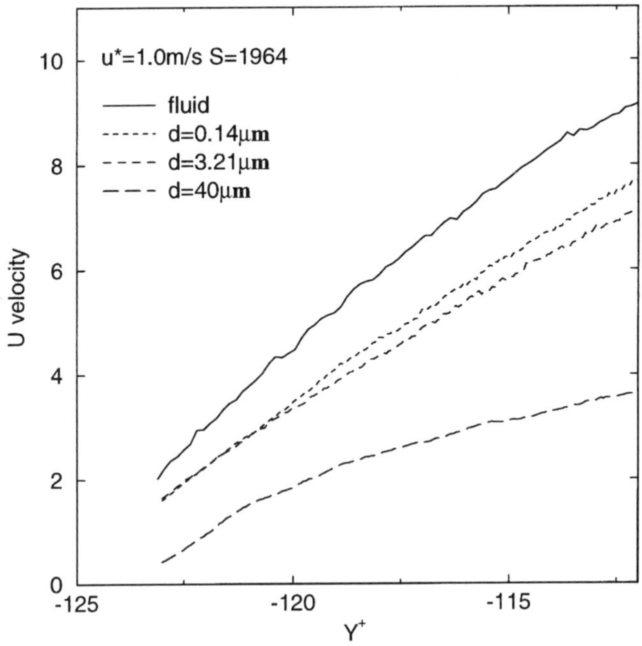

FIGURE 14a Variations of mean velocities of fluid and particles near the wall.

velocities are generally lower than that of the fluid, and smaller particles move faster than the larger ones in the streamwise direction. This is because particles are released very near the wall, and they move away from the wall by the upward motion of the near-wall eddies (for small particles) or by the lift force. (for large size particles). Thus, they carry lower streamwise velocities compared with that of the surrounding fluid.

RMS velocities for the fluid and different size particles are shown in Figure 14b. This figure shows that the RMS velocities of particles are lower than that of the fluid, and the smaller particles have larger RMS velocities compared with those of larger particles. The trend of variation of RMS particle velocities with distance from the wall is similar to that of the fluid. The exception is the streamwise RMS of 40 μm particles that increases up to about 3 wall units away from the wall and then decreases gradually. This may be due to the effect of the initial condition of these large particles with $\tau_p^+ = 780$ and the fact

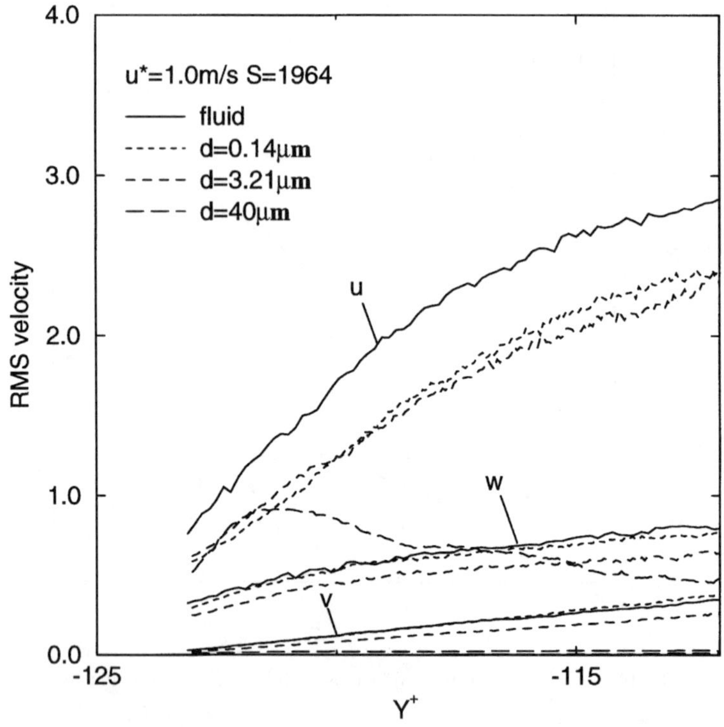

FIGURE 14b Variations of RMS velocities of fluid and particles near the wall.

that very few particles move away from the wall in the time duration of 100 wall units.

Mean Force

The statistics of various nondimensional forces acting on particles during the re-entrainment process are described in this section. Conditions of these numerical experiments are the same as those in the previous section. At every time step, ensemble averages of the y-component of drag, the lift forces, and the absolute value of Brownian and drag forces acting on particles that are moving in the region within 12 wall units from the lower wall are computed. (Positive sign denotes the direction is away from the lower wall.) The simulation results for the time duration of 50 to 100 wall units are shown in Figure 15.

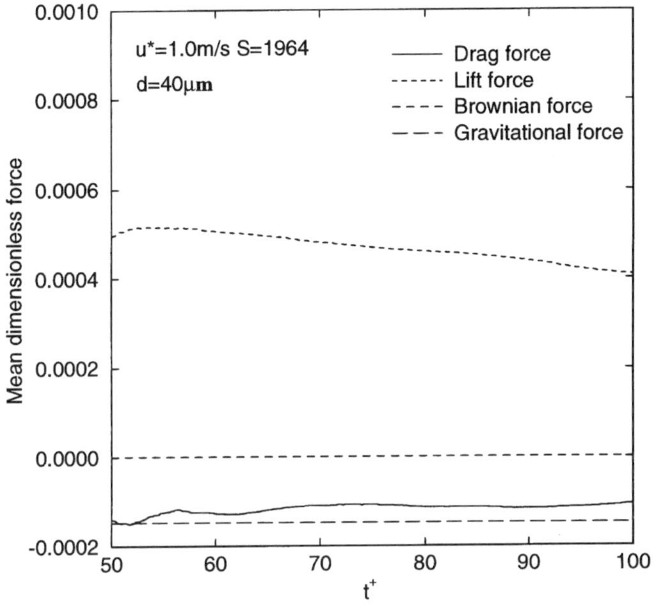

FIGURE 15a Time variations of averaged forces in the cross stream direction for 40 μm particles.

(The time after the startup to 50 wall units is omitted to eliminate the effect of initial conditions.)

Figure 15a shows the variation of various forces for 40 μm ($\tau_p^+ = 780$) in wall units. It is observed that the mean lift force is positive and relatively large indicating that the lift force makes particles move away from the wall. As noted before, this is the main mechanism for large size particle re-entrainment in turbulent flows. The mean drag force in the y-direction is negative, which also indicates that the particles are moving away from the wall. The magnitude of gravity is comparable with that of the mean drag force in this figure, and it also opposes the particle movement away from the wall. Figure 15a also shows that the Brownian force is negligible for 40 μm particles.

For 0.14 μm particles ($\tau_p^+ = 0.023$), Figure 15b shows the variation of averaged values of the Brownian, the drag, the lift and the gravitational forces. It is observed that the Brownian force plays an important role on the small-particle transport process. The lift force

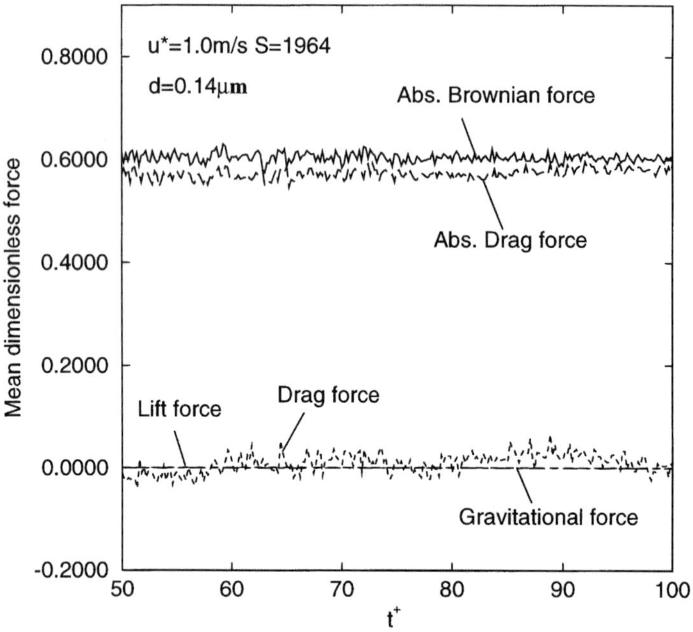

FIGURE 15b Time variations of averaged forces in the cross stream direction for 0.14 μm particles.

and the gravitational force for 0.14 μm particles are negligibly small. The mean absolute value of drag force is comparable with that of the Brownian force. The mean drag force, however, is oscillatory. These two figures also show that the magnitude of nondimensional drag force acting on 0.14 μm particles is about of an order of magnitude larger than that for 40 μm particles due to the significance of Brownian motion.

CONCLUSIONS

In this work, particle removal mechanisms from smooth surfaces in turbulent channel flows are studied. The theories of rolling and sliding detachments are used, and the particle removal process is studied. The effects of various forces, as well as near-wall turbulence flow structures are investigated. An ensemble of 8192 particles are used in these simulations for studying the re-entrainment process for each size and flow condition. Based on the present results, the following conclusions are drawn:

Removal and Adhesion

- The rolling detachment is the dominant mechanism for particle removal in turbulent flows.
- Drag and hydrodynamic torques are dominant, and the effect of lift and gravitational forces on particle detachment from the wall are negligible.
- The turbulence near-wall flow structure plays an important role in the particle detachment process.
- The simulated resuspension rates are in good agreement with the trend of model predictions of Reeks *et al.* [31]

Re-entrainment

- The instantaneous particle distribution in the $y-z$ plane during the re-entrainment process forms a periodic spanwise structure due to the turbulence near-wall coherent eddies.
- The present DNS of turbulent near-wall flows further shows that high-speed streamwise velocities correspond to those of downward flows (toward the wall), and the low-speed streamwise velocity regions correspond to upward flows (away from the wall).
- Turbulence near-wall flow structure plays an important role in both large and small particle re-entrainment processes but with different mechanisms.
- Large particles of the order of $d = 40\,\mu m$ ($\tau_p^+ = 780$) move roughly straight from the wall up to about 10 wall units due to the lift force, and then begin to disperse. These particles move away from the wall faster in the high-speed streamwise flow regions.
- Small particles follow the near-wall upward flows formed by the coherent near-wall vortices in the low-speed streamwise velocity regions during their re-entrainment process.
- Large particle dispersion perpendicular to the wall is slower than that for small particles due to the gravitational sedimentation effect and particle inertia.

- Particles move slower than the surrounding fluid in the streamwise direction and experience a lift force in the direction away from the wall.
- Small particles generally move faster than larger ones in the flow direction.
- For large particles, the Brownian force is negligible, and the Brownian motion is the dominating dispersion mechanism for small particles in the vicinity of the wall.

Acknowledgments

The authors would like to thank Professor John McLaughlin. This work was supported by the U.S. Department of Energy and New York State Science and Technology Foundation through the Center for Advanced Materials Processing (CAMP) of Clarkson University. The use of the NSF National Supercomputer Facility of the San Diego University is also gratefully acknowledged.

References

[1] Mittal, K. L., Ed., *Particles on Surfaces: Detection, Adhesion and Removal* **1–3**, (Plenum Press, New York, 1988, 1989, 1991).
[2] Corn, M., In: *Aerosol Sci.*, Davies, C. N. Ed. (Academic Press, New York, 1966), p. 359.
[3] Krupp, H., "Particle adhesion: Theory and Experiment", *Adv. Colloid Interface Sci.* **1**, 111–140 (1967).
[4] Visser, J.,"Adhesion of colloidal particles", In: *Surface and Colloid Sci.*, Matijevic, E. Ed., **8**, 3–84 (1976).
[5] Tabor, D., *Fluid Dynamics of Multiphase Systems* (Blaisdell Pub. Co., 1977).
[6] Bowling, R. A., "An analysis of particle adhesion on semiconductor surfaces", *J. Electrochem. Soc. Solid State Science Technol.* **132**, 2208–2219 (1985).
[7] Ranade, M. B., "Adhesion and removal of fine particles on surfaces", *J. Aerosol Sci. and Technol.* **7**, 161–176 (1987).
[8] Derjaguin, B. V., "*Untersuchungen ü ber die Reibung und adhäsion*", *IV, Koll. Z.* **69,** 155–164 (1934).
[9] Johnson, K. L., Kendall, K. and Roberts, A. D., "Surface energy and contact of elastic solids", *Proc. Royal. Soc. Lond.* **324**, 301–313 (1971).
[10] Derjaguin, B. V., Muller, V. M. and Toporov, Y. P. T., "Effect of contact deformation on the adhesion of particles", *J. Colloid Interface Sci.* **53**, 314–326 (1975).
[11] Muller, V. M., Yu, V. S. and Derjaguin, B. V., "On the influence of molecular forces on the deformation of an elastic sphere and its sticking to a rigid plane", *J. Colloid Interface Sci.* **77**, 91–101 (1980).
[12] Muller, V. M., Yu, V. S. and Derjaguin, B. V., "General theoretical consideration of the influence of structure forces on contact deformations and the reciprocal adhesion of elastic spherical particles", *J. Colloid Interface Sci.* **92**, 92–101 (1983).

[13] Maugis, D., "Adhesion of spheres: The JKR-DMT transition using a Dugale model", *J. Colloid Interface Sci.* **150**, 243–269 (1992).
[14] Maugis, D. and Pollock, H. M., *Acta Metall.* **32**, 1322 (1984).
[15] Tsai, C. J., Pui, D. Y. H., and Liu, B. Y. H., "Elastic flattening and particle adhesion", *J. Aerosol Sci. Technol.* **15**, 239–255 (1991).
[16] Rimai, D. S., DeMejo, L. P. and Verrland, W., "The effect of Young's modulus on the Surface–Force–Induced contact radius of spherical particles on polyurethane substrates", *J. Appl. Phys.* **71**, 2253–2258 (1992).
[17] Soltani, M. and Ahmadi, G., "Particle removal mechanism under base acceleration", *J. Adhesion* **44**, 161–175 (1994a).
[18] Soltani, M. and Ahmadi, G., "Particle detachment mechanisms from rough surfaces under substrate acceleration", *J. Adhesion Sci. Technol.* **9**(4), 453–373 (1995).
[19] Hinze, J. O., *Turbulence* (McGraw-Hill, New York, 1975).
[20] Ahmadi, G., "Overview of digital simulation procedures for aerosol transport in turbulent flows", In: *Particles in Gases and Liquids 3: Detection, Characterization, and Control*, Mittal, K. L. Ed. (Plenum, New York, 1993).
[21] Healy, J. W., "A review of resuspension models", In: *Transuranics in Natural Environments*, White, M. G. and Dunaway, P. B. Eds. (USERDA, Las Vegas, Nevada, 1977), pp. 211–222.
[22] Sehmel, G. A., "Particle resuspension: A review", *Environ. Int.* **4**, 107–127 (1980).
[23] Nicholson, K. W., "A review of particle resuspension", *Atmospheric Environment* **22**, 2639–2651 (1988).
[24] Smith, W. J., Whicher, F. W. and Meyer, H. R., "Review and Categorization of saltation, suspension and resuspension models", *Nuclear Safety* **23**, 685–699 (1982).
[25] Braaten, D. A., Paw, U. K. T. and Shaw, R. H., "Coherent turbulent structures and particle detachment in boundary layer flows", *J. Aerosol Sci.* **19**, 1183–1186 (1988).
[26] Yung, B. P. K., Merry, H. and Bott, T. R., "The role of turbulent bursts in particle re-entrainment in aqueous system", *Chem. Engng. Sci.* **44**, 873–882 (1989).
[27] Cleaver, J. W. and Yates, B., "Mechanism of detachment of colloid particles from a flat substrate in turbulent flow", *J. Colloid Interface Sci.* **44**, 464–474 (1973).
[28] Cleaver, J. W. and Yates, B., "A sublayer model for deposition of the particles from turbulent flow", *Chem. Eng. Sci.* **30**, 983–992 (1975).
[29] Cleaver, J. W. and Yates, B., "The effect of re-entrainment on particle deposition", *Chem. Eng. Sci.* **31**, 147–151 (1976).
[30] Reeks, M. W., Reed, J. and Hall, D., "On the resuspension of small particles by a turbulent flow", *J. Phys. D: Appl. Phys.* **21**, 574–589 (1988).
[31] Reeks, M. W. and Hall, D., "Deposition and resuspension of gas borne particles in recirculating turbulent flows", *J. Fluid Engng.* **110**, 165–171 (1988).
[32] Wen, H. Y. and Kasper, G., "On the kinetics of particle re-entrainment from surfaces", *J. Aerosol Sci.* **20**(4), 483–398 (1989).
[33] Wang, H. C., "Effect of inceptive motion on particle detachment from surfaces", *J. Aerosol Sci. Technol.* **13**, 386–396 (1990).
[34] Masironi, L. A. and Fish, B. R., "Direct observation of particle re-entrainment from surfaces", In: *Surface Contamination*, Fish, B. R. Ed., *Proc. of a Symp. at Gatlinburgh, Tennessee* (Pergamon Press, oxford, 1967), pp. 55–59.
[35] Soltani, M. and Ahmadi, G., "On particle adhesion and removal mechanisms turbulent flows", *J. Adhesion Sci. Tech.* **8**, 763–785 (1994).
[36] Soltani, M. and Ahmadi, G., "Particle detachment from rough surfaces in turbulent flows", *J. Adhesion* **51**, 105–123 (1995).
[37] McLaughlin, J. B., "Aerosol particle deposition in numerically simulated turbulent channel flow", *Phys. Fluids* **A1**, 1211–1224 (1989).

[38] Ounis, H., Ahmadi, G. and McLaughlin, J. B., "Dispersion and deposition of Brownian particles from point sources in a simulated turbulent channel flow", *J. Colloid Interface Sci.* **147**, 233–250 (1991).

[39] Ounis, H., Ahmadi, G. and McLaughlin, J. B., "Brownian particle deposition a directly simulated turbulent channel flow", *Phys. Fluids* **A5**, 1427–1432 (1993).

[40] Brooke, J. W., Kontomaris, K., Hanratty, T. J. and McLaughlin, J. B., "Turbulent deposition and trapping of aerosols at a wall", *Phys. Fluids* **A4**, 825–834 (1992).

[41] Pedinotti, S., Mariotti, G. and Banerjee, S., "Direct numerical simulation of particle behavior in the wall region of turbulent flows in horizontal channels", *Int. J. Multiphase Flow* **18**, 927–941 (1992).

[42] Soltani, M. and Ahmadi, G., "Direct numerical simulation of particle entrainment in turbulent channel flow", *Phys. Fluids* **7**, 647–657 (1995).

[43] Squires, K. D. and Eaton, J. K., "Measurements of particle dispersion obtained from direct numerical simulations of isotropic turbulence", *J. Fluid Mechanics* **226**, 1–35 (1991).

[44] Squires, K. D. and Eaton, J. K., "Preferential concentration of particles by turbulence", *Physics of Fluids* **A3**, 1169–1178 (1991b).

[45] Rashidi, M., Hetsroni, G. and Banerjee, S., "Particle-turbulent interaction in a boundary layer", *Int. J. Multiphase Flow* **16**(6), 935–949 (1990).

[46] Kim, J., Moin, P. and Moser, R., "Turbulent statistics in fully developed channel flow at low Reynolds number", *J. Fluid Mechanics* **177**, 133–166 (1987).

[47] Zhang, H. and Ahmadi, G., "Particle transport and deposition in vertical and horizontal turbulent duct flows", *J. Fluid Mechanics* **406**, 55–80 (2000).

[48] Goldman, A. J., Cox, R. G. and Brenner, H., "Slow viscous motion of a sphere parallel to a plane wall–II. Couette flow", *Chem. Engng. Sci.* **22**, 653–660 (1967b).

[49] Hinds, W. C., *Aerosol Technology, Properties, Behavior and Measurement of Airborne Particles* (John Wiley and Sons, New York, 1982).

[50] Clift, R., Grace, J. R. and Weber, M. E., *Drops and Particles* (Academic Press, New York, 1978).

[51] Goldman, A. J., Cox, R. G. and Brenner, H., "Slow viscous motion of a sphere parallel to a plane wall–I. Motion through a quiescent fluid", *Chem. Engng. Sci.* **22**, 637–651 (1967).

[52] Brenner, H., "The slow motion of a sphere through a viscous fluid towards a plane surface", *Chem. Eng. Sci.* **16**, 242–251 (1961).

[53] Li, Y., "Simulation of particle-laden channel flow with two-way coupling", *M. S. Dissertation*, Department of Chemical Engineering, Clarkson University (1998).

[54] Cox, R. G. and Brenner, H., "The slow motion of a sphere through a viscous fluid towards a plane surface–II. Small gap widths, including inertial effects", *Chem. Eng. Sci.* **22**, 1753–1777 (1967).

[55] O'Neill, M. E., "A sphere in contact with a plane wall in a slow linear shear flow", *Chem. Engng. Sci.* **23**, 1293–1298 (1968).

[56] Saffman, P. G., "The lift on a small sphere in a slow shear flow", *J. Fluid Mech.* **22**, 385–400 (1965).

[57] Saffman, P. G., Corrigendum to "The lift on a small sphere in a slow shear flow", *J. Fluid Mech.* **31**, 264 (1968).

[58] McLaughlin, J. B., "Inertial migration of a small sphere in linear shear flows", *J. Fluid Mech.* **224**, 261–274 (1991).

[59] Vasseur, P. and Cox, T. G., "The lateral migration of a spherical particle sedimenting in a stagnant bounded fluid", *J. Fluid Mech.* **80**, part 3, 561–591 (1977).

[60] Asmolov, E. S., "Transverse force acting on a spherical particle in a laminar boundary layer", *Izv. Akad. Nauk. SSSR Mekn. Zhidk. Gaza.* **5**, 66–71 (1989).

[61] McLaughlin, J. B., "The lift on a small sphere in wall-bounded linear shear flows", *J. Fluid Mech.* **246**, 249–265 (1993).

[62] Cox, R. G. and Hsu, S. K., "The lateral migration of solid particles in a laminar flow near a plane", *Int. J. Multiphase Flow* **3**, 201–222 (1977).
[63] Cherukat, P. and McLaughlin, J. B., "Wall induced lift on a sphere", *Int. J. Multiphase Flow* **16**, 899–907 (1994).
[64] Leighton, D. and Acrivos, A., "The lift on a small sphere touching a plane in the presence of a simple shear flow", *J. Applied Mathematics and Physics* (*ZAMP*), **36**, 174–178 (1985).
[65] Smith, C. R. and Schwartz, S. P., "Observation of streamwise rotation in the near-wall region of a turbulent boundary layer", *Phys. Fluid* **26**, 641–652 (1983).
[66] Acarlar, M. S. and Smith, C. R., "A study of hairpin vortices in a laminar boundary layer", Part II: Hairpin vortices generated by fluid injection, *J. Fluid. Mech.* **175**, 43–83 (1987).
[67] Robinson, S. K., "Kinematics of turbulent boundary layer structure", *Ph.D. Dissertation*, Stanford University (1990).

AUTHOR INDEX

SUBJECT INDEX

(b)

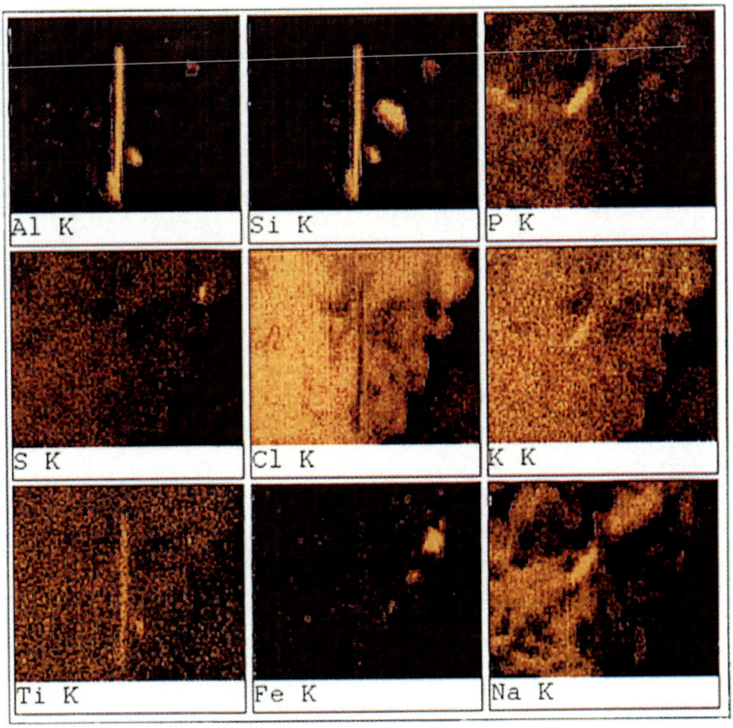

COLOR PLATE I. See R. Baier *et al.*, Figure 1(b).

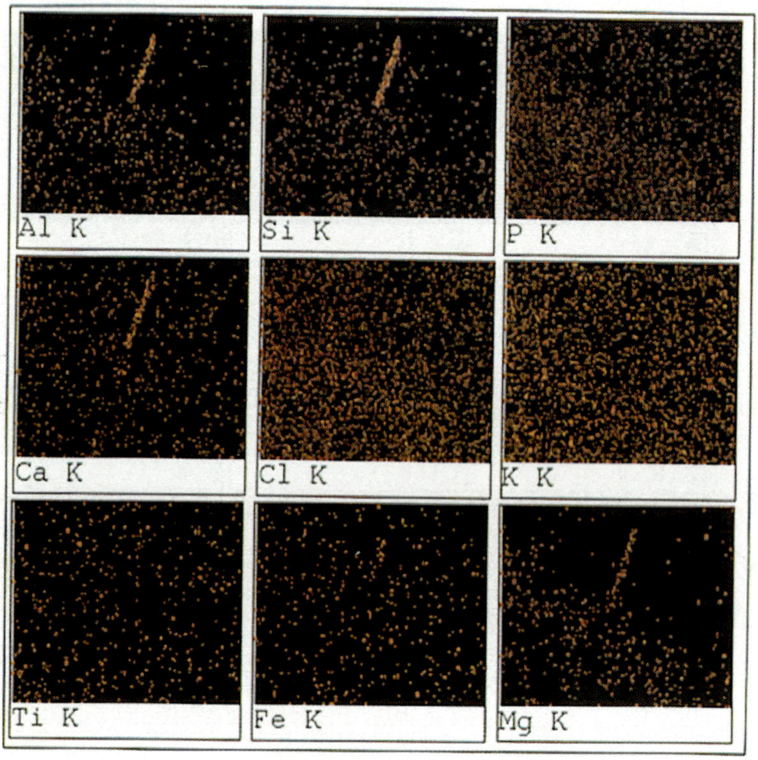

COLOR PLATE II. See R. Baier *et al.*, Figure 5.

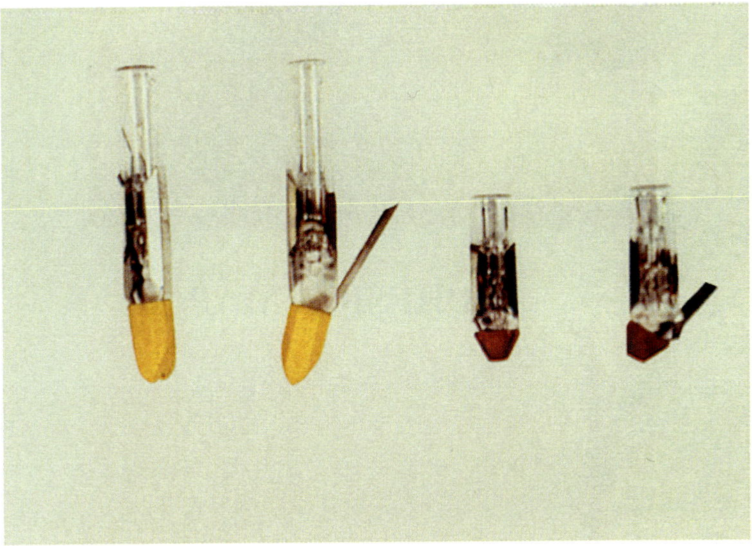

COLOR PLATE III. See D. W. Marshall, Figure 4.

TOPOGRAPHY

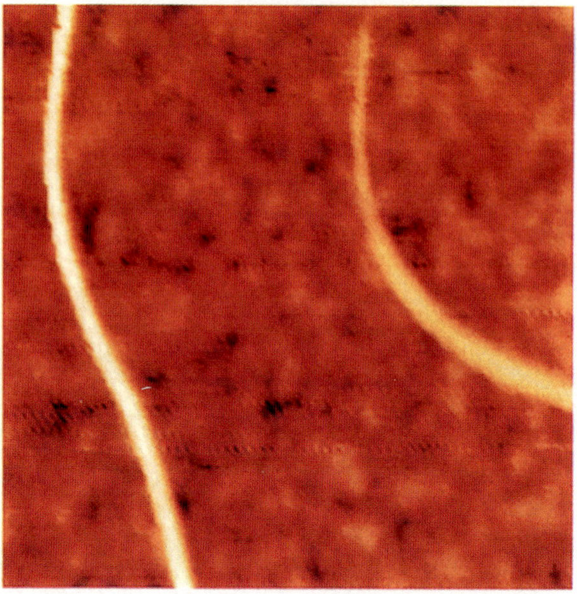

(a)

ADHESION MAP

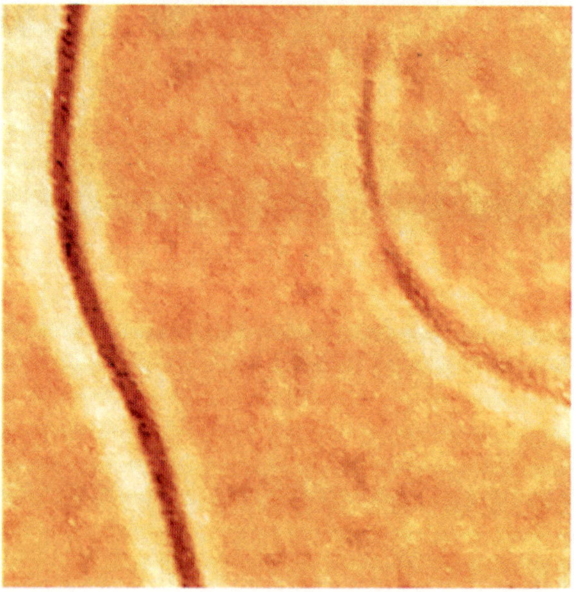

(b)

COLOR PLATE IV. See D. M. Schaefer and J. Gomez, Figure 5.

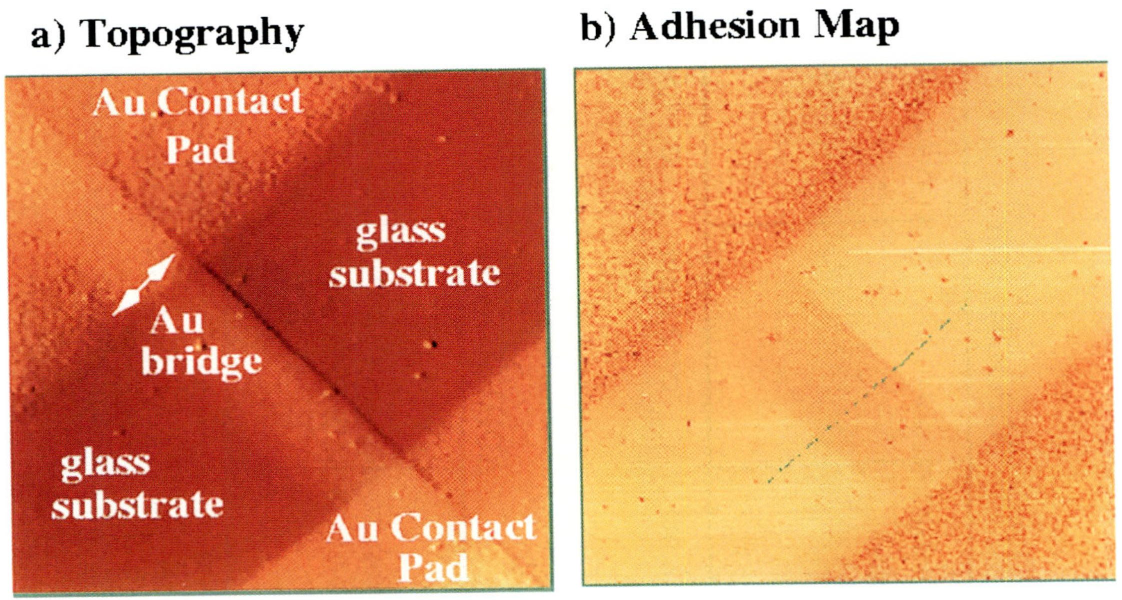

COLOR PLATE V. See D. M. Schaefer and J. Gomez, Figure 6.

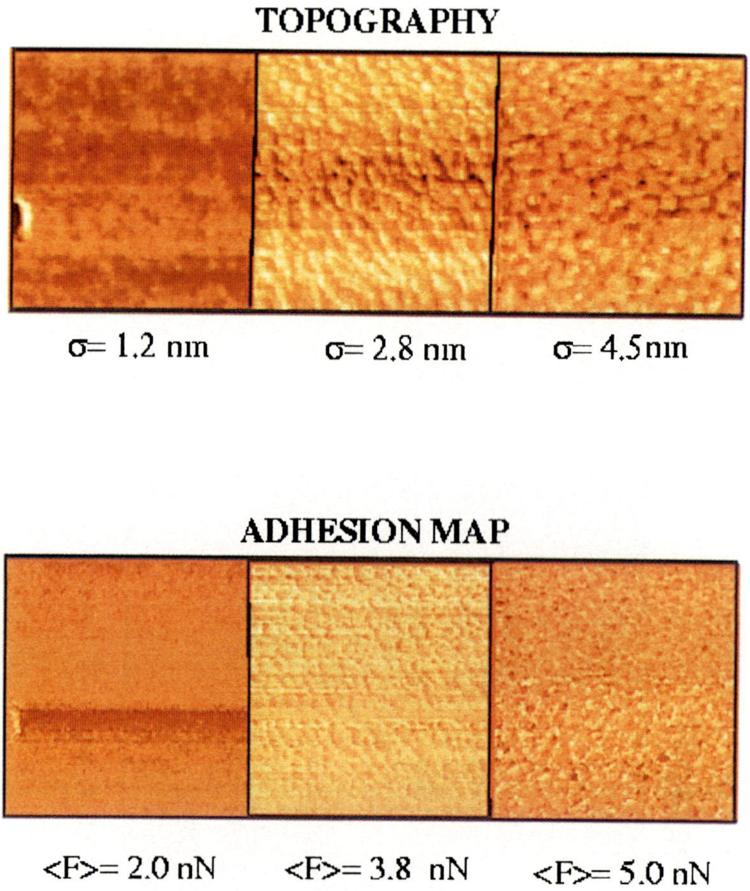

COLOR PLATE VI. See D. M. Schaefer and J. Gomez, Figure 8.

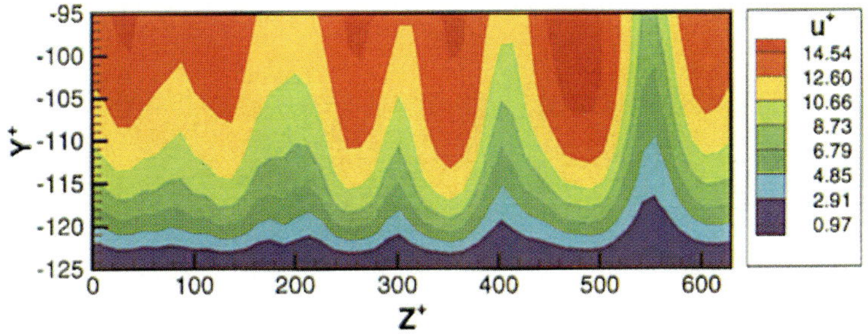

COLOR PLATE VII. See H. Zhang and G. Ahmadi, Figure 12a.

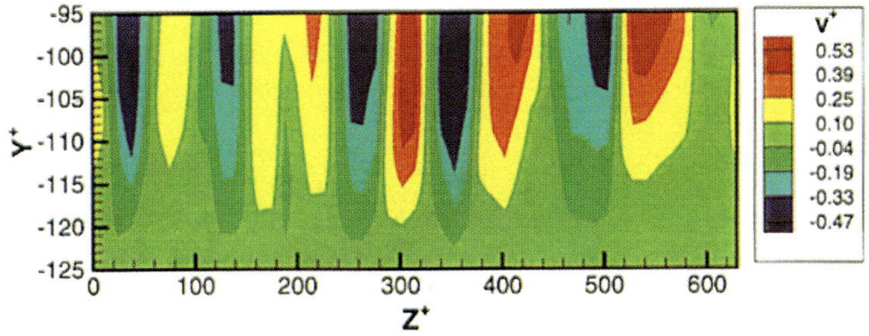

COLOR PLATE VIII. See H. Zhang and G. Ahmadi, Figure 12b.